Race, Class, and Gender

An Anthology

SEVENTH EDITION

MARGARET L. ANDERSEN
University of Delaware

PATRICIA HILL COLLINS
University of Maryland

WADSWORTH
CENGAGE Learning™

Australia • Brazil • Japan • Korea • Mexico • Singapore • Spain • United Kingdom • United States

WADSWORTH
CENGAGE Learning

Race, Class, and Gender: An Anthology, Seventh Edition
Margaret L. Andersen and
Patricia Hill Collins

Acquisition Editor: Chris Caldeira

Assistant Editor: Melanie Cregger

Editorial Assistant: Rachael Krapf

Technology Project Manager:
Lauren Keyes

Marketing Manager: Kim Russell

Marketing Assistant: Jillian Myers

Marketing Communications Manager:
Martha Pfeiffer

Project Manager, Editorial Production:
Cheri Palmer

Creative Director: Rob Hugel

Art Director: Caryl Gorska

Print Buyer: Karen Hunt

Permissions Editor: Roberta Broyer

Production Service: The Book Company

Copy Editor: Susan Gall

Cover Designer: Riezebos Holzbaur
Design Group

Cover Image: Helion, Jean (1904–1987)

Compositor: Macmillan Publishing
Solutions

© ARS, NY
Brilliant Figure, 1936
Location: Musee Departemental de l'Oise,
Beauvais, France
Photo Credit: Banque d'Images, ADAGP/
Art Resource, NY

© 2010, 2007 Wadsworth, Cengage Learning

ALL RIGHTS RESERVED. No part of this work covered by the copyright herein may be reproduced, transmitted, stored, or used in any form or by any means graphic, electronic, or mechanical, including but not limited to photocopying, recording, scanning, digitizing, taping, Web distribution, information networks, or information storage and retrieval systems, except as permitted under Section 107 or 108 of the 1976 United States Copyright Act, without the prior written permission of the publisher.

For product information and technology assistance, contact us at
Cengage Learning Customer & Sales Support, 1-800-354-9706.
For permission to use material from this text or product, submit all requests online at
www.cengage.com/permissions.
Further permissions questions can be e-mailed to
permissionrequest@cengage.com.

Library of Congress Control Number: 2008937985

Student Edition:

ISBN-13: 978-0-495-59882-4

ISBN-10: 0-495-59882-8

Wadsworth
10 Davis Drive
Belmont, CA 94002-3098
USA

Cengage Learning is a leading provider of customized learning solutions with office locations around the globe, including Singapore, the United Kingdom, Australia, Mexico, Brazil, and Japan. Locate your local office at **international.cengage.com/region**.

Cengage Learning products are represented in Canada by Nelson Education, Ltd.

For your course and learning solutions, visit **academic.cengage.com**.

Purchase any of our products at your local college store or at our preferred online store **www.ichapters.com.**

Printed in Canada
1 2 3 4 5 6 7 12 11 10 09

Contents

Preface

We write this preface at a time when the social dynamics of race—and its relationship to gender and class-is—changing. Barack Obama has become the first African American President of the United States. This followed a historic campaign in which women also had a major place[1]—Hillary Clinton in the Democratic primary and Sarah Palin as the Republican nominee for Vice President. People have asked, "Is this a major change in the nation's race relations?" "Has the glass ceiling for women finally been shattered?" or "Do the defeats of Clinton and Palin mean that women cannot be elected to high office?" And, "Where does social class fit?" Barack Obama was raised by a White, single mother. He had to work hard to gain the support of White, working-class men. Age played a role, too. Obama ran against an older, White man, John McCain, age seventy-two; younger voters were especially drawn to Obama. Ethnicity was aslo important in the election. Although originally more supportive of Clinton, Latinos/as supported Obama. But, no one of these factors alone—race, gender, class, age, or ethnicity—can explain the election outcome. Instead, understanding you have to understand how they worked together to understand how people voted as they did. In other words, to explain this historic election, you need to know how race, class, and gender—and their relationship to other social factors—are systematically interrelated.

That is the theme of this book: race, class, and gender simultaneously shape social issues and experiences in the United States. Central to the book is the idea that race, class, and gender are interconnected and that they must be understood as operating together if you want to understand the experiences of diverse groups and particular issues and events in society. We want this book to help students

---

[1]Geraldine Ferrari ran as the vice presidential nominee in 1984 with Walter Mondale as the presidential nominee. In 1972, Shirley Chisholm was the major African American candidate for president, although she did not receive the nomination; she had earlier been the first African American woman elected to Congress.

see how the lives of different groups develop in the context of their race, class, and gender location in society.

Since the publication of the first edition of this book, the study of race, class, and gender has become much more present in people's thinking and teaching. Yet, as our introductory essay will argue, not all of the work on race, class, and gender is centered in the intersectional framework that is central to this book. Often, people continue to treat one or the other of these social factors in isolation. Many studies of race, class, and gender also treat race, class, and gender as if they were equivalent experiences. Although we see them as interrelated—and sometimes similar in how they work—we also understand that each has its own dynamic, but a dynamic that can only be truly understood in relationship to the others. Moreover, with the growth of race, class, and gender studies, other social factors, such as sexuality, nationality, age, and disability, are also connected to the structural intersections of race, class, and gender. We hope that this book helps students understand how these structural phenomena—that is, the social forces of race, class, and gender and their connection with other social variables—are deeply embedded in the social structure of society.

We hope we achieve this by presenting an anthology that is more than a collection of readings. Our book is strongly centered in an analytical framework about the interconnections among race, class, and gender. In this edition we continue our efforts to further develop a framework of the *intersectionality* of race, class, and gender, or as Patricia Hill Collins has labeled it, the *matrix of domination*. The organization of the book features this framework. Our introductory essay distinguishes what we mean by this intersectional framework from other models of studying "difference." The four parts of the book are intended to help students see the importance of this intersectional framework, to engage critically the core concepts on which the framework is based, to analyze different social institutions and current social issues using this framework, including being able to apply it to understanding processes of social change.

The four major parts of the book reflect these goals. We open with a section and essay entitled "Why Race, Class, and Gender Still Matter." As in past editions, we include essays here that engage students in personal narratives, as a way of helping them step beyond their own social location and to see how race, class, gender, sexuality, and other social factors shape people's lives differently. We also intend for this section to show students the very different experiences that anchor the study of race, class, and gender. Being able to center oneself in the lived experience of others is, we think, important to grasping the important of race, class, gender, and other social factors, such as sexuality, ethnicity, religion, and so forth. Even when dominant groups may claim that race, class, and gender no longer matter in society, we know this is not true, We begin our book with essays that show their continuing, even if changing, significance.

The organization of our book also links ethnicity, nationality, and sexuality to the study of race, class, and gender. Although we caution readers against thinking that all forms of difference are the same, we know that race, class, and gender are interconnected with other systems of power and inequality. The title of Part II of this book ("Systems of Power and Inequality") reflects how we conceptualize race, class, gender, ethnicity, and sexuality as linked through systems of power and inequality.

Part III ("The Structure of Social Institutions") focuses on how race, class, and gender structure social institutions—now and historically. In this edition we have particularly strengthened the section on "Media and Popular Culture" to reflect the growing significance of the media and popular culture in the lives of our students. In fact, we also include more articles based on the media throughout this edition. In addition to the continued sections on work, families, media, and the state, we have created a new section on "Education and Health" to reflect the growing importance of these issues in national social policy.

Finally, we again revised Part IV of the book ("Pulling It All Together"). Many anthologies use their final section to show how students can make a difference in society, once they understand the importance of race, class, and gender. We think this is a tall order for students who may have had only a few weeks to even begin to understand how race, class, and gender matter—and matter together. Thus, rather than using our final section to make it seem that change can be easily accomplished, we use this section to show the different ways that one can *apply* understanding the intersectionality of race, class, and gender in different forms of social action and in different social sites.

Throughout, our book is grounded in a sociological perspective, although the articles come from different perspectives, disciplines, and experiences. Several articles provide a historical foundation for understanding how race, class, and gender have emerged. We also include materials that bring a global dimension to the study of race, class, and gender—not just by looking comparatively at other cultures but also by analyzing how the process of globalization is shaping life in the United States.

As in earlier editions, we have selected articles based primarily on two criteria: (1) accessibility to undergraduate readers and the general public, not just highly trained specialists, and (2) articles that are grounded in race *and* class *and* gender—in other words, intersectionality. Not all articles accomplish this as much as we would like, but we try not to select articles that focus exclusively on one issue while ignoring the others. In this regard, our book differs significantly from other anthologies on race, class, and gender that include a lot of articles on each, but do less to show how they are connected. We also distinguish our book from those that are centered in a multicultural perspective. Although multiculturalism is important, we do not think that the appreciation of cultural differences is the only thing that race, class, and gender are about. Rather, we see race, class, and gender as embedded in the structure of society and significantly influencing group cultures. But race, class, and gender are also structures of group opportunity, power, and privilege. We searched for articles that are conceptually and theoretically informed and at the same time accessible to undergraduate readers. Although it is important to think of race, class, and gender as analytical categories, we do not want to lose sight of how they affect human experiences and feelings; thus, we have included personal narratives that are reflective and analytical. We think that personal accounts generate empathy and also help us connect personal experiences to social structural conditions.

ORGANIZATION OF THE BOOK

As noted previously, we have organized the book to reflect how we approach the study of race, class, and gender. The first part establishes the framework with which we approach the study of race, class, and gender. We introduce each of the four parts with an essay by us that analyzes the issues raised by the reading selections. These essays are an important part of this book because they establish the framework that we use to think about race, class, and gender.

Part I, "Why Race, Class, and Gender Still Matter," contains several narrative pieces that reflect on how race, class, and gender shape individual and collective experiences. These different accounts are not meant to represent everyone who has been excluded, but to ground readers in the practice of thinking about the lives of people who rarely occupy the center of dominant group thinking. When excluded groups are placed in the forefront of thought, they are typically stereotyped and distorted. When you look at society through the different perspectives of persons reflecting on their location within systems of race, class, and gender oppression, you begin to see just how much race, class, and gender really matter—and continue to matter, despite claims to the contrary. Because the essays in this part are grounded in personal experiences, they capture student interest, but we also use them analytically to think beyond any one person's life.

Part II, "Systems of Power and Inequality," provides the conceptual foundation for understanding how race, class, and gender are linked together and how they link with other systems of power and inequality, especially ethnicity and sexuality. We treat each of these separately here, not because we think they stand alone, but to show students how each operates so they can better see their interlocking nature. The introductory essay provides working definitions for these major concepts and presents some of the contemporary data through which students can see how race, class, and gender stratify contemporary society.

In Part III, "The Structure of Social Institutions," we examine how intersecting systems of race, class, and gender shape the organization of social institutions and how, as a result, these institutions affect group experience. Social scientists routinely document how Latinos, African Americans, women, workers, and other distinctive groups are affected by institutional structures. We know this is true but want to go beyond these analyses to scrutinize how institutions are constructed through race, class, and gender relations. The articles in Part III show how we should not relegate the study of racial-ethnic groups, the working class, and women only to subjects marked explicitly as race, class, or gender studies. As categories of social experience, race, class, and gender shape all social institutions and systems of meaning; thus, it is important to think about people of color, different class experiences, and women in analyzing all social institutions and belief systems.

Finally, in Part IV, "Pulling It All together" we provide different examples of how using an intersectional perspective can help you understand the different social locations where social change and social action occur. Throughout the book, we have avoided a "social problems" approach, as if race, class, and gender are solely about victimization. People are not just victims; they are creative and visionary. The articles here show different ways that people given the realities of race, class, and gender. We think that, together, these articles illustrate for students how to apply a race, class, and gender framework adapt and resist understanding social change and how people negotiate race, class, and gender in their everyday lives.

We know that developing a complex understanding of the interrelationships between race, class, and gender is not easy and involves a long-term process engaging personal, intellectual, and political change. We do not claim to be models of perfection in this regard. We have been pleased by the strong response to the first six editions of this book, and we are fascinated by how race, class, and gender studies have developed since the publication of our first edition. We know further work is needed. Our own teaching and thinking has been transformed by developing this book. We imagine many changes still to come.

CHANGES TO THE SEVENTH EDITION

We have made several changes in the seventh edition that strengthen and refresh *Race, Class, and Gender*. These include the following:

- Twenty-six new readings.

- A stronger focus on social class throughout the book, as well as improved readings in the section on class.

- More focus on the media and popular culture, both in the section on this topic (in Part III) and in the selection of new articles.

- A stronger section on sexuality, including an essay specifically on the connections among racism, sexism, and homophobia, as well as new work on young people and heterosexuality, sex trafficking, and connections among race, ethnicity, and sexuality.

- A revised final part that helps students apply an understanding of race, class, and gender in different social locations and in the process of social change.

- Four revised introductions by us; this is one of the noted strengths of our book compared to others. These introductions articulate the analytical model we use to understand the intersections among race, class, and gender.

- New material on race, class, gender and important current issues, including military recruitment, health care policy, and the re-segregation of public schools.

PEDAGOGICAL FEATURES

We realize that the context in which you teach matters. If you teach in an institution where students are more likely to be working class, perhaps how the class system works will be more obvious to them than it is for students in a more privileged college environment. Many of those who use this book will be teaching in segregated environments, given the high degree of race and class—and even gender—segregation in education. Thus, how one teaches this book should reflect the different environments where faculty work. Ideally, the material in this book should be discussed in a multiracial, multicultural atmosphere, but we realize that is not always the case. We hope that the content of the book and the pedagogical features that enhance it will help bring a more inclusive analysis to educational settings than might be there to start with.

We see this book as more than just a collection of readings. The book has an analytical logic to its organization and content, and we think it can be used to format a course. Of course, some faculty will use the articles in an order different from how we present them, but we hope the four parts will help people develop the framework for their course. We also provide pedagogical tools to help people expand their teaching and learning beyond the pages of the book.

We have included features with this edition that provide faculty with additional teaching tools. They include the following:

- *Instructor's manual.* This edition includes an instructor's manual with suggestions for class exercises, discussion and examination questions, and course assignments.

- *Index.* The index will help students and faculty locate particular topics in the book quickly and easily.

- *Website.* Online support is available through the Wadsworth Sociology Resource Center. We encourage people to browse the *Race, Class, and Gender* page at the Wadsworth website or related resources. The address is www.cengage.com/sociology.

A NOTE ON LANGUAGE

Reconstructing existing ways of thinking to be more inclusive requires many transformations. One transformation needed involves the language we use when referring to different groups. Language reflects many assumptions about race, class, and gender; and for that reason, language changes and evolves as knowledge changes. The term *minority,* for example, marginalizes groups, making them seem somehow outside the mainstream or dominant culture. Even worse, the phrase *non-White,* routinely used by social scientists, defines groups in terms of what they are not and assumes that Whites have the universal experiences against which the experiences of all other groups are measured. We have consciously avoided using both of these terms throughout this book, although this is sometimes unavoidable.

We have capitalized Black in our writing because of the specific historical experience, varied as it is, of African Americans in the United States. We also capitalize White when referring to a particular group experience; however, we recognize that White American is no more a uniform experience than is African

American. We use *Hispanic* and *Latina/o* interchangeably, though we recognize that is not how groups necessarily define themselves. When citing data from other sources (typically government documents), we use *Hispanic* because that is usually how such data are reported.

Language becomes especially problematic when we want to talk about features of experience that different groups share. Using shortcut terms like Hispanic, Latina/o, Native American, and women of color homogenizes distinct historical experiences. Even the term *White* falsely unifies experiences across such factors as ethnicity, region, class, and gender, to name a few. At times, though, we want to talk of common experiences across different groups, so we have used labels such as Latina/o, Asian American, Native American, and women of color to do so. Unfortunately, describing groups in this way reinforces basic categories of oppression. We do not know how to resolve this problem but want readers to be aware of the limitations and significance of language as they try to think more inclusively about diverse group experiences.

ACKNOWLEDGMENTS

An anthology rests on the efforts of more people than the editors alone. This book has been inspired by our work with scholars and teachers from around the country who are working to make their teaching and writing more inclusive and sensitive to the experiences of all groups. Over the years of our own collaboration, we have each been enriched by the work of those trying to make higher education a more equitable and fair institution. In that time, our work has grown from many networks that have generated new race, class, and gender scholars. These associations continue to sustain us. Many people contributed to the development of this book.

We especially thank Michelle Beadle at the University of Maryland, College Park for her expert research assistance; she was a real partner in developing this book and we could not have completed it on time without her. We also thank Michelle Wilcox and Nena Craven at the University of Delaware for their help. And special thanks to Chang Won Lee at the University of Maryland for such careful work in proofreading. And, we appreciate the support given by our institutions, with special thanks to Ronet Bachman, Chair of the Department of Sociology at the University of Delaware and Suzanne Bianchi, Chair of the Sociology Department at the University of Maryland. Erin Parkins at Cengage helped with many of the details, and we sincerely thank Dusty Friedman and Gary Kliewer for their expert work in guiding us through the production process. As always, Chris Caldeira has proven to be an exceptional editor. We gratefully acknowledge her support and her willingness to work with us through our busy schedules. We also thank the reviewers who provided valuable commentary on the prior edition and thus helped enormously in the development of the seventh edition:

Julie A. Raulli, *Wilson College*

Susan L. Thomas, *Hollins University*

Beau Beaudoin, *Columbia College*

Peter Meiksins, *Cleveland State University*

Eleanor LaPointe, *Georgian Court University*

Lisa Handler, *Community College of Philadelphia*

Katja Guenther, *California State University, Fullerton*

Developing this book has been an experience based on friendship, hard work, travel, and fun. We especially thank Valerie, Roger, and Richard for giving us the support, love, and time we needed to do this work well. This book has deepened our friendship, as we have grown more committed to transformed ways of thinking and being. We are lucky to be working on a project that continues to enrich our work and our friendship.

About the Editors

Margaret L. Andersen (B.A. Georgia State University; M.A., Ph.D., University of Massachusetts, Amherst) is the Edward F. and Elizabeth Goodman Rosenberg Professor of Sociology at the University of Delaware where she also holds joint appointments in Black American Studies and Women's Studies. She has received two teaching awards at the University of Delaware. She has published numerous books and articles, including *Thinking about Women: Sociological Perspectives on Sex and Gender* (8th ed., Allyn and Bacon, 2009); *Race and Ethnicity in Society: The Changing Landscape* (edited with Elizabeth Higginbotham, 2nd ed., Cengage, 2009); and *Sociology: The Essentials* (co-authored with Howard F. Taylor, Cengage, 2008). She received the American Sociological Association's Jessie Bernard Award for expanding the horizons of sociology to include the study of women and the Eastern Sociological Society Robin Williams Lecturer Award. In 2008–09 she served as Vice President of the American Sociological Association.

Patricia Hill Collins (B.A., Brandeis University; M.A.T., Harvard University; Ph.D., Brandeis University) is Distinguished University Professor of Sociology at the University of Maryland, College Park, and Charles Phelps Taft Emeritus Professor of African American Studies and Sociology at the University of Cincinnati. She is the author of numerous articles and books including *From Black Power to Hip Hop: Racism, Nationalism and Feminism* (Temple University, 2006); *Black Sexual Politics: African Americans, Gender and the New Racism* (Routledge, 2004), which won the Distinguished Publication Award from the American Sociological Association; *Fighting Words* (University of Minnesota, 1998); and *Black Feminist Thought: Knowledge, Consciousness, and the Politics of Empowerment* (Routledge, 1990, 2000), which won the Jessie Bernard Award of the American Sociological Association and the C. Wright Mills Award of the Society for the Study of Social Problems. In 2008, she became the 100th President of the American Sociological Association.

About the Contributors

Joan Acker is Professor Emerita of Sociology at the University of Oregon. She founded and directed the University of Oregon's Center for the Study of Women in Society and is the recipient of the American Sociological Association's Career of Distinguished Scholarship Award as well as the Jessie Bernard Award for feminist scholarship. She is the author of *Class Questions, Feminist Answers* as well as many other works in the areas of gender, institutions, and class.

Teresa Amott is Provost and Dean of Faculty at Hobart and William Smith Colleges. She is the co-author of *Race, Gender, and Work: A Multicultural Economic History of Women in the United States* (with Julie Matthaei) and *Caught in Crisis: Women in the U.S. Economy Today*.

Rita Arditti is a Core Faculty member of the College of Graduate Studies of the Union Institute. She is also a founding member of the activist organizations Science for the People and the Women's Community Cancer Project. She is the author of *Searching for Life: The Grandmothers of the Plaza de Mayo and the Disappeared Children of Argentina*.

Maxine Baca Zinn is Professor of Sociology at Michigan State University and Senior Research Associate with the Julian Samora Research Institute. She is co-author of *Diversity in Families, In Conflict and Order*, and *Social Problems* (with D. Stanley Eitzen), and the co-editor of *Gender through the Prism of Difference: A Sex and Gender Reader* (with Pierrette Hondagneu-Sotelo and Michael A. Messner).

Fabricio E. Balcazar is Associate Professor in the Department of Psychology at the University of Illinois at Chicago (UIC). He is also the Director of the Center for Capacity Building on Minorities with Disabilities Research at UIC.

Pamela Block is Associate Professor at the State University of New York at Stony Brook. She is also an adjunct faculty member at the Center for Alcohol and Addiction Studies at Brown University.

Rebecca Brasfield is a Ph.D. graduate student in the Department of Sociology at the University of Illinois at Urbana-Champaign.

Denise Brennan is an Associate Professor of Anthropology in the Sociology and Anthropology Department at Georgetown University. She is the author of several books and articles about the global sex trade, human trafficking, and women's labor, primarily in Latin America and the Caribbean. She was recently awarded an American Association of University Women fellowship to conduct field research for her next book on the process of recovery and resettlement after trafficking.

Kenneth W. Brown is a contributor to *Essence* magazine. While unemployed, he completed an associate's degree; later, he worked in financial services while pursuing a bachelor's degree.

Erica Chito Childs is Assistant Professor of Sociology at Eastern Connecticut State University. She is a council member for the American Sociological Association section on Racial and Ethnic Minorities.

Ward Churchill (enrolled Keetoowah Band Cherokee) is a long-time native rights activist and Professor of Ethnic Studies and Coordinator of American Indian Studies at the University of Colorado. His books include *Agents of Repression, Fantasies of the Master Race, From a Native Son,* and *A Little Matter of Genocide: Holocaust and Denial in the Americas.*

Judith Ortiz Cofer is the Franklin Professor of English and Creative Writing at the University of Georgia. She is the author of numerous books of poetry and essays, and a novel, *The Line of the Sun.* She has won the PEN/Martha Albrand Special Citation in nonfiction for her book *Silent Dancing.*

Gary David Comstock is University Protestant Chaplain and visiting Associate Professor at Wesleyan University. He has published numerous books, including *Gay Theology Without Apology; Unrepentant, Self-Affirming, Practicing: Lesbian/ Bisexual/Gay People Within Organized Religion; Que(e)rying Religion: A Critical Anthology;* and *Violence against Lesbians and Gay Men.*

Mary Yu Danico is Vice Chair of the Psychology and Sociology Department at California Polytechnic State University, Pomona. Her research interests include immigration and migration, Korean American diaspora, Asian diaspora, and transnational families and communities. She is the co-editor of *Transforming the Academy: People of Color, Queers, and Women in Higher Education.*

Jessie Daniels teaches at the City University of New York, Hunter College. She has undertaken research on the intersections of race, class, gender, and sexuality in such settings as White supremacist movements, urban streets and jails, and the Internet. She is currently working on a book entitled *Cyber Lies* about White supremacy, globalization, and the Internet.

Bonnie Thornton Dill is Professor of Women's Studies and Director of the Research Consortium on Gender, Race, and Ethnicity at the University of Maryland, College Park. Her books include *Women of Color in U.S. Society,* co-edited with Maxine Baca Zinn, and *Across the Boundaries of Race and Class: Work and Family among Black Female Domestic Servants.*

Rachel E. Dubrofsky is an Assistant Professor in the Department of Communication at the University of South Florida. Her current research projects focus on studies of the media and women in popular culture, especially the phenomenon of reality television. She is the author of the article "Ally McBeal as Postfeminist Icon: The Aestheticizing and Fetishizing of the Independent Working Woman" in *Communication Review.*

Barbara Ehrenreich is the author of *The New York Times* best seller *Nickel and Dimed,* as well as a dozen other books on a variety of topics such as class, gender, war, and human rituals. She has also contributed to such publications as *Harper's, The Nation, The New York Times,* and *Time* magazine.

Kim Fellner is one of the founders of the National Organizers' Alliance. She was an anti-war activist during the Vietnam War era and became involved in community organization during graduate school.

Abby Ferber is Associate Professor and Director of Women's Studies at the University of Colorado at Colorado Springs. She is the author of *White Man Falling: Race, Gender and White Supremacy* and *Hate Crime in America: What Do We Know?* She is co-author of *Making a Difference: University Students of Color Speak Out* and co-editor of the anthology *Privilege: A Reader* with Michael Kimmel.

Nicholas Freudenberg is a Distinguished Professor of Urban Public Health at Hunter College, as well as Professor of Social Psychology at the Graduate Center of the City University of New York. He is the author of numerous publications concerning such public health topics as health promotion and disease prevention in low income communities, the health and social consequences of incarceration and jails, and how public health advocates can affect potentially damaging corporate practices. He is also the Chair of the Agenda for a Healthy New York project.

Marilyn Frye is Professor and Associate Chair of the Department of Philosophy at Michigan State University. She is the author of numerous books and essays, including *The Politics of Reality* and *Willful Virgin.* With her partner, Carolyn Shafer, she created and manages Bare Bones Studios for Women's Art—space and facilities for art making in a wide range of media.

Charles A. Gallagher is the Chair of the Department of Sociology at LaSalle University with research specialities in race and ethnic relations, urban sociology, and inequality. He has published several articles on subjects such as colorblind political narratives, racial categories within the context of interracial marriages, and perceptions of privilege based on ethnicity.

Herbert J. Gans has been a prolific and influential sociologist for more than fifty years. His published works on urban renewal and suburbanization are intertwined with his personal advocacy and participant observation, including a stint as consultant to the National Advisory Commission on Civil Disorder. He is the author of the classic *The Urban Villagers* as well as the more recent *Democracy and the News.*

Chong-suk Han teaches in the Sociology Department at Temple University and is currently a doctoral candidate in Social Welfare at the University of Washington. His research focuses on shared experiences of marginalization and the intersections of race and sexuality.

Adia M. Harvey is an Assistant Professor of Sociology at Georgia State University specializing in race, class, and gender, work and occupations, and social theory. She is the author of multiple articles in these areas, including "Personal Satisfaction and Economic Improvement: Examining Working-Class Black Women's Experiences with Entrepreneurship" in the *Journal of Black Studies.*

Daisy Hernandez is the managing editor of *ColorLines,* "The national newsmagazine on race and politics." She has contributed writings on race, gender, and sexuality to such diverse publications as *The New York Times, Ms. Magazine,* the *National Catholic Reporter, Curve,* and *In These Times.* She is the co-editor of *Colonize This! Young Women of Today's Feminism.*

Arlie Russell Hochschild is a Professor in the Department of Sociology at the University of California, Berkeley. She is the author of numerous books and articles, including, most recently, *The Commercialization of Intimate Life and Global Woman.* Her areas of interest include the family, market culture, global patterns of carework, and the relationship between culture and emotion.

Pierrette Hondagneu-Sotelo is Professor of Sociology, American Studies, and Ethnicity at the University of Southern California. She is author of *Gendered Transitions: Mexican Experiences of Immigration* and *Doméstica: Immigrant Workers Cleaning and Caring in the Shadow of Affluence,* which won the Society for Social Problems. C. Wright Mills Award.

Barbara Jensen is a community and counseling psychologist in private practice in Minneapolis–St. Paul. She also teaches sociology and psychology as an adjunct at Metropolitan State University and serves on the steering committee of the Minnesota Center for Labor and Working Class Studies.

Jonathan Ned Katz was the first tenured Professor of Lesbian and Gay Studies in the United States. (Department of Lesbian and Gay Studies, City College of San Francisco). He is the founder of the Queer Caucus of the College Art Association. He is also the co-founder of the activist group Queer Nation.

Robin D. G. Kelley is a Professor of History and Africana Studies at New York University. He is the author of numerous books, including *Freedom Dreams: The Black Radical Imagination; Three Strikes: The Fighting Spirit of Labor's Last Century; Race Rebels: Culture, Politics and the Black Working Class;* and *Yo' Mama's DisFunktional!*

Christopher B. Keys is a Department Chair and Professor of Community Psychology at DePaul University.

Nazli Kibria is Associate Professor of Sociology at Boston University. She is the author of *Becoming Asian American: Identities of Second Generation Chinese and Korean Americans* and *Family Tightrope: The Changing Lives of Vietnamese Americans.*

Celene Krauss is Professor of Sociology and Women's Studies at Kean University. She has written extensively on women's issues and toxic waste protests, with a focus on race, ethnicity, and class. Her work has appeared in *Sociological Forum, Qualitative Sociology,* and other journals and books.

Annette Lareau is Professor of Sociology at the University of Pennsylvan. She is the author of *Unequal Childhoods: Class, Race, and Family Life.*

Chungmei Lee is a Research Associate at the Civil Rights Project. She has served on several projects concerning the training and professional development of teachers, including Harvard's Program for Professional Education.

Pandora K. Leong is the Director of the Teen Health Initiative at the New York Civil Liberties Union.

David Leonhardt writes "Economic Scene," a regular column on economics and business in *The New York Times.* He is also a staff writer for the *New York Times Magazine,* and he has contributed to such publications as *Business Week* and *The Washington Post.*

Ariel Levy writes regularly for *New York Magazine* on topics such as gender, sexuality, celebrity, and politics. She is the author of *Female Chauvinist Pigs: Women and the Rise of Raunch Culture.*

Audre Lorde, who passed away in 1992, grew up in the West Indian community of Harlem in the 1930s, the daughter of immigrants from Grenada. She was Professor of English at Hunter College and a major figure in the lesbian and feminist movements. Among her works are *Sister Outsider; Zami: A New Spelling of My Name; Uses of the Erotic;* and numerous other books and essays.

Arturo Madrid is Murchison Distinguished Professor of the Humanities at Trinity University and the recipient of the Charles Frankel Prize in the Humanities, National Endowment for the Humanities, in 1996. From 1984 until 1993 he served as founding president of The Tomás Rivera Center, a national institute for policy studies on Latino issues.

Julianne Malveaux holds a Ph.D. in economics and is a syndicated columnist who writes regularly for *Essence* and also appears regularly on CNN, BET, Fox News, MSNBC, and other television news shows. She is the author of *Wall*

Street, Main Street, and the Side Street: A Mad Economist Takes a Stroll. She is president and CEO of her own multimedia company.

Gregory Mantsios is the Director of Worker Education at Queens College, the City University of New York.

Elizabeth Martinez is a Chicana scholar and activist. She is the co-founder of the Institute for MultiRacial Justice. Her latest book is *DeColores Means All of Us: Latina Views for a Multi-Colored Century.*

Julie Matthaei is a Professor of Economics at Wellesley College. Her books include *An Economic History of Women in America: Women's Work, The Sexual Division of Labor, and the Development of Capitalism,* and (with Teresa Amott) *Race, Gender, and Work: A Multicultural Economic History of Women in the United States.*

Peggy McIntosh is Associate Director of the Wellesley College Center for Research on Women. She is the founder and co-director of the National SEED Project on Inclusive Curriculum—a project that helps teachers make school climates fair and equitable. She is the co-founder of the Rocky Mountain Women's Institute.

Michael Messner is Professor and Chair of the Department of Sociology at the University of Southern California. His latest book is *Taking the Field: Women, Men, and Sports.* His other books include *Masculinties, Gender Relations, and Sport, Power at Play: Sports and the Problem of Masculinity,* and *The Politics of Masculinities: Men in Movements.*

Roslyn Arlin Mickelson is Professor of Sociology and Women's Studies at the University of North Carolina at Charlotte. She is the author of *Children on the Streets of the Americas: Globalization, Homelessness, and Education in the United States, Brazil, and Cuba.*

Robert B. Moore writes frequently on feminist issues and is the author of the classic "Racist Stereotyping in the English Language," which is included in this volume.

Philip Moss is a Professor in the Department of Regional Economic and Social Development at the Universiy of Massachusetts, Lowell. He studies the impacts of structural change on economic opportunity. With Chris Tilly he has published *Stories Employers Tell: Race, Skill, and Hiring in America.*

Katherine S. Newman is Malcolm Sterenson Forbes Professor of Sociology at Princeton University. She is the author of *Rampage: The Social Roots of School Shootings; No Shame in My Game: The Working Poor in the Inner City;* and *Falling from Grace: Downward Mobility in the Age of Affluence.*

Gary Orfield is a Professor of Education and the co-director of the Civil Rights Project at the UCLA Graduate School of Education & Information Studies. His research interests include civil rights, education policy, urban policy, and minority opportunity. He is the author of several books and articles in these areas, as well as the recipient of the 2007 "Social Justice in Education" Award by the American Educational Research Association.

Gina M. Pérez is an Associate Professor of Comparative American Studies at Oberlin College. Her research focuses on topics such as U.S. Latinas/os, gender, political economy, migration, transnationalism, urban anthropology, and poverty. She is the author of *The Near Northwest Side Story: Migration, Displacement, and Puerto Rican Families.*

Imani Perry is a Professor in the Rutgers School of Law. Her scholarship is informed by studies of how race becomes re-inscripted in law and culture after periods of social upheaval, as well as analyses of how Black cultural products relate to aesthetic and political images of Blackness. She is the author of *Prophets of the Hood: Politics and Poetics in Hip Hop.*

Barbara Ransby is an Associate Professor of History and African American studies at the University of Illinois at Chicago. She is the author of *Ella Baker and the Black Radical Tradition*, which has received numerous awards including the Lillian Smith Book Award and the Gustavas Myer Outstanding Book Award.

Lillian Rubin lives and works in San Francisco. She is an internationally known lecturer and writer. Some of her books include *The Man with the Beautiful Voice, Tangled Lives, the Transcendent Child,* and *Intimate Strangers.*

Almas Sayeed is the Child Welfare Policy Coordinator at the Kansas Coalition Against Sexual and Domestic Violence.

Amy J. Schulz is a Research Associate Professor in both the Health Behavior and Health Education Department and the Institute for Research on Women and Gender at the University of Michigan School of Public Health. In addition, she is the associate director of the Center for Research on Ethnicity, Culture and Health (CRECH). She is the author of numerous articles concerning the relationship between health and cultural factors such as race.

Janny Scott is a staff writer for *The New York Times.* She is the author of numerous articles dealing with the interactions of race, class, and politics.

Thomas M. Shapiro is Pokross Chair of Law and Social Policy at Brandeis University. His latest book is *The Hidden Cost of Being African American.* His earlier work with Melvin L. Oliver, *Black Wealth/White Wealth,* has won several awards including the American Sociological Association Distinguished Scholarly Award.

Jael Silliman is Associate Professor of Women's Studies at the University of Iowa. She is the co-editor of *Dangerous Intersections: Feminist Perspectives on Population, Environment, and Development* and the author of *Jewish Portraits, Indian Frames: Women's Narratives from a Diaspora of Hope.*

Stephen Samuel Smith is Professor of Political Science at Winthrop University. His interests include the politics of education, urban political economy, social movements, and program evaluation.

C. Matthew Snipp is Professor of Sociology at Stanford University. He is the author of *American Indians: The First of This Land* and *Public Policy Impacts on American Indian Economic Development*. His tribal heritage is Oklahoma Cherokee and Choctaw.

Ronald T. Takaki is Professor of Ethnic Studies at the University of California, Berkeley. He is the author of several books, including *Iron Cages: Race and Culture in 19th Century America; Strangers from a Different Shore: A History of Asian Americans;* and *A Different Mirror: A History of Multicultural America*.

Chris Tilly is a university professor in the Department of Regional Economic and Social Development at the University of Massachusetts, Lowell. He is the author of *Half a Job: Bad and Good Part-Time Jobs in a Changing Labor Market; Glass Ceilings and Bottomless Pits: Women's Work, Women's Poverty* (with Randy Albelda); and *Stories Employers Tell: Race, Skill, and Hiring in America* (with Philip Moss).

Jeremiah Torres is a graduate of Stanford University, where he studied symbolic systems. His article "Label Us Angry" appeared in the book *Asian American X*, a collection of essays about the experiences of contemporary Asian Americans.

Haunani-Kay Trask is a Hawaiian scholar and poet and has been an activist in the Native Hawaiian community for over 20 years. She is a professor and former director of the Center for Hawaiian Studies at the University of Hawaii at Manoa.

Eisa Nefertari Ulen is a teacher of English at Hunter College in New York City as well as an essayist and a journalist. She has contributed to many anthologies and other publications, including *The Washington Post, Ms.,* and *Health*. In addition, she is the recipient of a Frederick Douglass Creative Arts Center Fellowship for Young African American Fiction Writers and a Hunter College Presidential Award for Excellence for Teaching.

Linda Trinh Võ is Associate Professor and Chair of the Department of Asian American Studies School of Humanities at the University of California, Irvine. Her research interests include Asian American studies, race and ethnic relations, immigration theory, and social stratification and inequality. She also participates in community activism as a member of such organizations as the Orange County Asian and Pacific Islander Community Alliance (OCAPICA) and Project MotiVATe, a mentoring program for Vietnamese American teens.

Mary C. Waters is M.E. Zukerman Professor of Sociology and Harvard College Professor at Harvard University. She is the author of *Black Identities: West Indian Immigrant Dreams* and *American Realities; Ethnic Options: Choosing Identities in America;* and numerous articles on race, ethnicity, and immigration.

Kath Weston is a sociological anthropologist who has written several books, including *Families We Choose: Lesbians, Gays, and Kinship; Render Me, Gender Me; Long Slow Burn;* and *Gender in Real Times*.

Christine L. Williams is a Professor of Sociology at the University of Texas at Austin. Her research interests include gender and sexuality; work, occupations, and organizations; qualitative methodology; and sociological theory. She is the author of *Still a Man's World: Men Who Do "Women's Work"* as well as numerous articles and book chapters.

Patricia J. Williams is the James L. Dohn Professor of Law at Columbia University and the author of *The Alchemy of Race and Rights* and *The Diary of a Law Professor*. She is a regular contributor to *The Nation*.

Jennifer Wriggins is Associate Professor of Law at the University of Southern Maine. She specializes in the practice of civil litigation.

PART I

Why Race, Class, and Gender Still Matter

MARGARET L. ANDERSEN AND PATRICIA HILL COLLINS

The United States is a nation where people are supposed to be able to rise above their origins. Those who want to succeed, it is believed, can do so with hard enough work and good enough effort because the nation is founded on the principle of equality. Although equality has historically been denied to many, there is now a legal framework in place that guarantees protection from discrimination and equal treatment for all citizens. Historic social movements, such as the civil rights movement and the feminist movement, raised people's consciousness about the rights of minority groups and women. Moreover, these movements have generated new opportunities for multiple groups—African Americans, Latinos, White women, disabled people, lesbian, gay, bisexual, transgendered (LGBT) peoples, and older people, to name some of the groups who have been beneficiaries of civil rights action and legislation. And, if you ask Americans if they support non-discrimination policies, the overwhelming number will say yes. Why, then, do race, class, and gender still matter?

Race, class, and gender still matter because they continue to structure society in ways that value some lives more than others. Currently, some groups have more opportunities and resources, while other groups struggle. Race, class and gender matter because they remain the foundations for systems of power and inequality that, despite our nation's diversity, continue to be among the most significant social facts of people's lives. Thus, despite having removed the formal barriers to opportunity, the United States is still highly stratified along lines of race, class, and gender.

1

The persistent reality of race, class, and gender inequality was made apparent to the U.S. public when Hurricane Katrina struck New Orleans and the Gulf Coast in the fall of 2005. Many were surprised by the televised images of poor, overwhelmingly African American people—abandoned in the storm, completely without resources for several days, and without the basic necessities of food and water. It was shocking for some people to see U.S. citizens in such great need—images that people more commonly associate with poor nations, not the most affluent and powerful nation in the world. Even while many in the nation ignored and denied the realities of poverty, though, the national poverty rate had in fact steadily increased for the five years prior to Katrina. And, in many U.S. cities, the infant mortality rate—thought to be a good indicator of a nation's well-being—exceeds that in far-less-developed nations.

Still, Hurricane Katrina was said by many to be a "leveling experience"—a storm that displaced hundreds of thousands of people—a mix of women and men, adults and children, people of different races, rich and poor. But did it affect all equally? The widespread devastation did displace people of many different class backgrounds. Expensive waterfront homes, often second homes for families, were destroyed along with substandard housing in the poorest areas of New Orleans and rural, coastal Mississippi. Nature does not discriminate when it comes to where a storm hits or where a flood occurs. On the other hand, though, you might ask, why was the most low-lying section of the city of New Orleans a place where mostly poor, Black people lived? Why were African Americans most of those who could not evacuate and were forced to live—and, in some cases, die—in the horrid squalor of the Superdome and the convention center? Why were the poor left behind? Why was the government response so slow? And, why have low-income, largely African American communities been the slowest to recover?

Many people asked these questions following Hurricane Katrina, and the connection between race and class was once again brought to the forefront of national consciousness—even if only temporarily. Yet as Hurricane Katrina faded from the public eye, it was replaced by a new set of race, class, and gender dynamics in the 2008 presidential election. First were the groundbreaking campaigns run during the Democratic primary by a White woman (Hillary Clinton) and a Black man (Barack Obama). That was followed by the surprise inclusion of White woman, Sarah Palin, as Vice President on the Republican ticket. Many people believe that the success of these individuals means that the United States is quickly moving beyond the inequalities of race, class, and gender stripped bare by Hurricane Katrina. Barack Obama represents the idealized American story of hard work, stellar academic achievement, and upward social mobility; Hillary Clinton represents the final achievement of cracking the glass ceiling of women in positions of power. Sarah Palin suggested that her candidacy shattered the glass ceiling for women in politics that Hillary Clinton says she cracked. All three stories suggest that the social inequalities of race, class, and gender revealed by Hurricane Katrina may be a fading memory.

Despite this optimism, we think that race, class, and gender still matter and that both of these stories suggest the need for continued attention to race, class, and gender. In the absence of an inclusive perspective that took race *and* class *and* gender into account, during the presidential campaigns, the media increasingly cast Obama as the "race" candidate and Clinton and Palin the "gender" candidates. This framing of the campaign suggests that Obama has a race but no gender and that Clinton and Palin have gender but not race. This type of unidimensional thinking effectively reduced the complexity of all three candidates to a simple debate of pitting race against gender. This either/or thinking limited the terms of public debate. For example, Clinton and Palin could be accused of playing the gender card if they supported policies that took into account issues that largely affect women, and Obama could be considered to be playing the so-called race card if he talked about issues of concern to African Americans. Furthermore, all four candidates on the final tickets, including John McCain, persistently focused on their families, emphasizing—without ever saying so—that heterosexual marriage and family systems are an important part of a candidate's qualifications.

Both of these examples—Katrina and the presidential election—are stories of race, class, *and* gender. And, by the time this book is published, we will have seen the politics of race, class, and gender played out further as Senator Barack Obama ran against an older, White man, Senator John McCain.

In this book, we ask students to think about race, class, and gender as *systems of power.* We want to encourage readers to imagine ways to transform, rather than reproduce, existing social arrangements. This starts with shifting one's thinking so that groups who are often silenced or ignored become heard. All social groups are located in a system of power relationships wherein your social location can shape what you know—and what others know about you. As a result, dominant forms of knowledge have been constructed largely from the experiences of the most powerful—that is, those who have the most access to systems of education and communication. Thus, to acquire a more inclusive view—one that pays attention to group experiences that may differ from your own—requires that you take a new frame of vision.

You can think of this as if you were taking a photograph. For years, poor people, women, and people of color—and especially poor women of color—were totally outside the frame of vision of more powerful groups or distorted by their views. If you move your angle of sight to include those who have been overlooked, however, some accepted points of view may seem less revealing or just plain wrong. Completely new subjects can also appear. This is more than a matter of sharpening one's focus, although that is required for clarity. Instead, this new angle of vision means actually seeing things differently, perhaps even changing the lens you look through—thereby removing the filters (or stereotypes and misconceptions) that you bring to what you see and think.

DEVELOPING A RACE, CLASS, AND GENDER PERSPECTIVE

In this book we ask you to think about how race, class, and gender matter in shaping everyone's lived experiences. We focus on the United States, but increasingly the inclusive vision we present here matters on a global scale as well. Thinking from a perspective that engages race, class, and gender is not just about illuminating the experiences of oppressed groups. It also changes how we understand groups who are on both sides of power and privilege. For example, the development of women's studies has changed what we know and how we think about women; at the same time, it has changed what we know and how we think about men. This does not mean that women's studies is about "male-bashing." It means taking the experiences of women and men seriously and analyzing how race, class, and gender shape the experiences of both men and women—in different, but interrelated, ways. Likewise, the study of racial and ethnic groups begins by learning the diverse histories and experiences of these groups. In doing so, we also transform our understanding of white experiences. Rethinking class means seeing the vastly different experiences of both wealthy, middle-class, working class, and poor people in the United States and learning to think differently about privilege and opportunity. The exclusionary thinking that comes from past frames of vision simply does not reveal the intricate interconnections that exist among the different groups composing U.S. society.

It is important to stress that thinking about race, class, and gender is not just a matter of studying victims. Relying too heavily on the experiences of poor people, women, and people of color can erase our ability to see race, class, and gender as an integral part of everyone's experiences. We remind students that race, class, and gender have affected the experiences of all individuals and groups. As a result, we do not think we should talk only about women when talking about gender or only about poor people when talking about class. Because race, class, and gender affect the experiences of all, it is important to study Whites when analyzing race, the experiences of the affluent when analyzing class, and to study men when analyzing gender. Furthermore, we should not forget women when studying race or think only about Whites when studying gender.

So you might ask, how does reconstructing knowledge about excluded groups matter? To begin with, knowledge is not just some abstract thing—good to have, but not all that important. There are real consequences to having partial or distorted knowledge. First, learning about other groups helps you realize the partiality of your own perspective; furthermore, this is true for both dominant and subordinate groups. Knowing only the history of Puerto Rican women, for example, or seeing their history only in single-minded terms will not reveal the historical linkages between the oppression of Puerto Rican women and the exclusionary and exploitative treatment of African Americans, working-class whites, Asian American men, and similar groups. This is discussed by Ronald T. Takaki in his essay included here ("A Different Mirror") on the multicultural history of American society.

Second, having misleading and incorrect knowledge leads to the formation of bad social policy—policy that then reproduces, rather than solves, social problems. U.S. immigration policy has often taken a one-size-fits-all approach, failing to recognize that vast differences among groups coming to the United States privilege some and disadvantage others. Finally, knowledge is not just about content and information; it provides an orientation to the world. What you know frames how you behave and how you think about yourself and others. If what you know is wrong because it is based on exclusionary thought, you are likely to act in exclusionary ways, thereby reproducing the racism, anti-Semitism, sexism, class oppression, and homophobia of society. This may not be because you are intentionally racist, anti-Semitic, sexist, elitist, or homophobic; it may simply be because you do not know any better. Challenging oppressive race, class, and gender relations in society requires reconstructing what we know so that we have some basis from which to change these damaging and dehumanizing systems of oppression.

RACE, CLASS, AND GENDER AS A MATRIX OF DOMINATION

Race, class, and gender shape the experiences of all people in the United States. This fact has been widely documented in research and, to some extent, is commonly understood. Thus, for years, social scientists have studied the consequences of race, class, and gender inequality for different groups in society. The framework of race, class, and gender studies presented here, however, explores how race, class, and gender operate *together* in people's lives. Fundamentally, race, class, and gender are *intersecting* categories of experience that affect all aspects of human life; thus, they *simultaneously* structure the experiences of all people in this society. At any moment, race, class, or gender may feel more salient or meaningful in a given person's life, but they are overlapping and cumulative in their effects.

In this volume we focus on several core features of this intersectional framework for studying race, class, and gender. First, we emphasize *social structure* in our efforts to conceptualize intersections of race, class, and gender. We use the approach of a *matrix of domination* to analyze race, class, and gender. A matrix of domination sees social structure as having multiple, interlocking levels of domination that stem from the societal configuration of race, class, and gender relations. This structural pattern affects individual consciousness, group interaction, and group access to institutional power and privileges (Collins 2000). Within this structural framework, we focus less on comparing race, class, and gender as separate systems of power than on investigating the structural patterns that join them. Because of the simultaneity of race, class, and gender in people's lives, intersections of race, class, and gender can be seen in individual stories and personal experience. In fact, much exciting work on the intersections of race, class, and gender appears in autobiographies, fiction, and personal essays. We do recognize the significance of these individual narratives and include many here, but we also emphasize social structures that provide the context for individual experiences.

Second, studying interconnections among race, class, and gender within a context of social structures helps us understand how race, class, and gender are manifested differently, depending on their configuration with the others. Thus, one might say African American men are privileged *as men,* but this may not be true when their race and class are also taken into account. Otherwise, how can we possibly explain the particular disadvantages African American men experience in the criminal justice system, in education, and in the labor market? For that matter, how can we explain the experiences that Native American women undergo—disadvantaged by the unique experiences that they have based on race, class, *and* gender—none of which is isolated from the effects of the others? Studying the connections among race, class, and gender reveals that divisions by race and by class and by gender are not as clear-cut as they may seem. White women, for example, may be disadvantaged because of gender but privileged by race and perhaps (but not necessarily) by class. And increasing class differentiation within racial-ethnic groups reminds us that race is not a monolithic category, as can be seen in the fact that white poverty is increasing more than poverty among other groups, even while some whites hold the most power in society.

Third, the matrix of domination approach to race, class, and gender studies is historically grounded. We have chosen to emphasize the intersections of race, class, and gender as institutional systems that have had a special impact in the United States. Yet race, class, and gender intersect with other categories of experience, such as sexuality, ethnicity, age, ability, religion, and nationality. Historically, these intersections have taken varying forms from one society to the next; within any given society, the connections among them also shift. Thus, race is not inherently more important than gender, just as sexuality is not inherently more significant than class and ethnicity.

Given the complex and changing relationships among these categories of analysis, we ground our analysis in the historical, institutional context of the United States. Doing so means that race, class, and gender emerge as fundamental categories of analysis in the U.S. setting, so significant that in many ways they influence all of the other categories. Systems of race, class, and gender have been so consistently and deeply codified in U.S. laws that they have had intergenerational effects on economic, political, and social institutions. For example, the capitalist class relations that have characterized all phases of U.S. history have routinely privileged or penalized groups organized by gender and by race. U.S. social institutions have reproduced economic equalities for poor people, women, and people of color from one generation to the next. Thus, in the United States, race, class, and gender demonstrate visible, long-standing, material effects that in many ways foreshadow more recently visible categories of ethnicity, religion, age, ability, and/or sexuality.

DIFFERENCE, DIVERSITY, AND MULTICULTURALISM

How does the matrix of domination framework differ from other ways of conceptualizing race, class, and gender relationships? We think this can be best understood by contrasting the *matrix of domination framework* to what might be called a *difference framework* of race, class, and gender studies as well as related frameworks that emphasize diversity and multiculturalism. A difference framework, though viewing some of the common processes in race, class, and gender relations, tends to focus on unique group experiences. Books that use a framework of difference (or diversity or multiculturalism) will likely include writings by diverse groups of people, but on closer inspection, you will see that many of these writings treat race, class, and gender separately. Although we think such studies are valuable and add to the body of knowledge about race, class, and gender, we distinguish our work by looking at the *interrelationships* among race, class, and gender, not just their unique ways of being experienced.

You might think of the distinction between the two approaches as one of thinking comparatively, one core feature of a difference framework, versus thinking relationally, the hallmark of the matrix of domination approach. For example, in the difference framework individuals are encouraged to compare their experiences with those supposedly unlike them. When you think comparatively, you might look at how different groups have, for example, encountered prejudice and discrimination or you might compare laws prohibiting interracial marriage to current debates about same-sex marriage. These are important and interesting questions, but they are taken a step further when you think beyond comparison to the structural relationships between different group experiences. In contrast, when you think relationally, you see the social structures that *simultaneously* generate unique group histories and link them together in society. You then untangle the workings of social systems that shape the experiences of different people and groups, and you move beyond just comparing (for example) gender oppression with race oppression or the oppression of gays and lesbians with that of racial groups. Recognizing how intersecting systems of power shape different groups' experiences positions you to think about changing the system, not just documenting the effects of such systems on different people.

The language of difference encourages comparative thinking. People think comparatively when they learn about experiences other than their own and begin comparing and contrasting the experiences of different groups. This is a step beyond centering one's thinking in a single group (typically one's own), but it is nonetheless limited. For example, when students encounter studies of race, class, and gender for the first time, they often ask, "How is this group's experience like or not like my own?" This is an important question and a necessary first step, but it is not enough. For one thing, it tends to promote ranking the oppression of one group compared to another, as if the important thing were to determine who is most victimized. Furthermore, it frames one's understanding of different groups only within the context of other groups' experiences; thus, it can assume an artificial norm against which different groups are judged. Thinking comparatively tends to assume that race, class, and gender constitute separate and independent components of human experience that can be compared for their similarities and differences.

We should point out that comparative thinking can foster greater understanding and tolerance, but comparative thinking alone can also leave intact the power relations that create race, class, and gender relations. Because the concept of difference contains the unspoken question "different from what?" this framework can privilege those who are deemed to be "normal" and stigmatize people who are labeled as "different." And because it is based on comparison, the very concept of difference fosters dichotomous, either/or thinking. Some approaches to difference place people in either/or categories, as if one is either Black or White, oppressed or oppressor, powerful or powerless, normal or different when few of us fit neatly into any of these restrictive categories.

Some difference frameworks try to move beyond comparing systems of race, class, and gender by thinking in terms of an *additive* approach. The additive approach is reflected in terms such as *double* and *triple jeopardy*. Within this logic, poor African American women seemingly experience the triple oppression of race, gender, and class, whereas poor Latina lesbians encounter quadruple oppression, and so on. But social inequality cannot necessarily be quantified in this fashion. Adding together "differences" (thought to lie in one's difference from the norm) produces a hierarchy of difference that ironically reinstalls those who are additively privileged at the top while relegating those who are additively oppressed to the bottom. We do not think of race and gender oppression in the simple additive terms implied by phrases such as double and triple jeopardy. The effects of race, class, and gender do "add up," both over time and in intensity of impact, but seeing race, class, and gender only in additive terms misses the social structural connections among them and the particular ways in which different configurations of race, class, and gender affect group experiences.

Within difference frameworks, this additive thinking can foster another troubling outcome. One can begin with the concepts of race, class, and gender and continue to "add on" additional types of difference. Ethnicity, sexuality, religion, age, and ability all can be added on to race, class, and gender in ways that suggest that any of these forms of difference can substitute for others. This use of difference fosters a view of oppressions as equivalent and as being the same. Recognizing that difference encompasses more than race, class, and gender is a step in the right direction. But continuing to add on many distinctive forms of difference can be a never-ending process. After all, there are as many forms of difference as there are individuals. Ironically, this form of recognizing difference can erase the workings of power just as effectively as diversity initiatives.

When it comes to conceptualizing race, class, and gender relations, the matrix of domination approach also differs from another version of the focus on difference, namely, thinking about diversity. *Diversity* has become a catchword for trying to understand the complexities of race, class, and gender in the United States. What does *diversity* mean? Because the American public has become a more heterogeneous population, *diversity* has become a buzzword—popularly used, but loosely defined. People use *diversity* to mean cultural variety, numerical representation, changing social norms, and the inequalities that characterize the status of different groups. In thinking about diversity, people have recognized that race, class, gender, age, sexual orientation, and ethnicity matter; thus, groups who have previously been invisible, including people of color, gays, lesbians, and bisexuals, older people, and immigrants, are now in some ways more visible. At the same time that diversity is more commonly recognized, however, these same groups continue to be defined as "other"; that is, they are perceived through dominant group values, treated in exclusionary ways, and subjected to social injustice and economic inequality.

The movement to "understand diversity" has made many people more sensitive and aware of the intersections of race, class, and gender. Thinking about diversity has also encouraged students and social activists to see linkages to other categories of analysis, including sexuality, age, religion, physical disability, national identity, and ethnicity. But appreciating diversity is not the only point. The very term *diversity* implies that understanding race, class, and gender is simply a matter of recognizing the plurality of views and experiences in society—as if race, class, and gender were benign categories that foster diverse experiences instead of systems of power that produce social inequalities.

Diversity initiatives hold that the diversity created by race, class, and gender differences are pleasing and important, both to individuals and to society as a whole—so important, in fact, that diversity should be celebrated. Under diversity initiatives, ethnic foods, costumes, customs, and festivals are celebrated, and students and employees receive diversity training to heighten their multicultural awareness. Diversity initiatives also advance a notion that, despite their differences, "people are really the same." Under this view, the diversity created by race, class, and gender constitutes cosmetic differences of style, not structural opportunities.

Certainly, opening our awareness of distinct group experiences is important, but some approaches to diversity can erase the very real differences in power that race, class, and gender create. For example, diversity initiatives have asked people to challenge the silence that has surrounded many group experiences. In this framework, people think about diversity as "listening to the voices" of a multitude of previously silenced groups. This is an important part of coming to understand race, class, and gender, but it is not enough. One problem is that people may begin hearing the voices as if they were disembodied from particular historical and social conditions. This perspective can make experience seem to be just a matter of competing discourses, personifying "voice" as if the voice or discourse itself constituted lived experience. Second, the "voices" approach suggests that any analysis is incomplete unless every voice is heard. In a sense, of course, this is true, because inclusion of silenced people is one of the goals of race/class/gender work. But in a situation where it is impossible to hear every voice, how does one decide which voices are more important than others? One might ask, who are the privileged listeners within these voice metaphors?

We think that the matrix of domination model is more analytical than either the difference or diversity frameworks *because of its focus on structural systems of power and inequality*. This means that race, class, and gender involve more than either comparing and adding up oppressions or privileges or appreciating cultural diversity. The matrix of domination model requires analysis and criticism of existing systems of power and privilege; otherwise, understanding diversity becomes just one more privilege for those with the greatest access to education— something that has always been a mark of the elite class. Therefore, race, class, and gender studies mean more than just knowing the cultures of an array of human groups. Instead, studying race, class, and gender means recognizing and analyzing the hierarchies and systems of domination that permeate society and limit our ability to achieve true democracy and social justice.

Finally, the matrix of domination framework challenges the idea that race, class, and gender are important only at the level of culture—an implication of the catchword *multiculturalism*. *Culture* is traditionally defined as the "total way of life" of a group of people. It encompasses both material and symbolic components and is an important dimension of understanding human life. Analysis of culture per se, however, tends to look at the group itself rather than at the broader conditions within which the group lives. Of course, as anthropologists know, a sound analysis of culture situates group experience within these social structural conditions. Nonetheless, a narrow focus on culture tends to ignore social conditions of power, privilege, and prestige. The result is that multicultural studies often seem tangled up with notions of cultural pluralism—as if knowing a culture other than one's own is the only goal of a multicultural education.

Because we approach the study of race, class, and gender with an eye toward transforming thinking, we see our work as differing somewhat from the concepts implicit in the language of difference, diversity, or multiculturalism. Although we think it is important to see the diversity and plurality of different cultural forms, in our view this perspective, taken in and of itself, misses the broader point of understanding how racism, class relations, and sexism have shaped the experience of groups. Imagine, for example, looking for the causes of poverty solely within the culture of currently poor people, as if patterns of unemployment, unexpected health care costs, rising gas prices, and home mortgage foreclosures had no effect on people's opportunities and life decisions. Or imagine trying to study the oppression of LGBT people in terms of gay culture only. Obviously, doing this turns attention onto the group itself and away from the dominant society. Likewise, studying race only in terms of Latino culture or Asian American culture or African American culture, or studying women's oppression only by looking at women's culture, encourages thinking that blames the victims for their own oppression. For all of these reasons, the focus of this book is on the institutional, or structural, bases for race, class, and gender relations.

Thus, this book is not just about comparing differences, understanding diversity, or describing multicultural societies. Instead, we attempt to develop a structural perspective on the relationships among race, class, and gender as systems of power, the hallmark of the matrix of domination framework. We recognize that reaching these goals will require rejecting the kind of exclusionary thinking that virtually erased some groups' experiences and embracing an inclusive perspective that incorporates neglected groups and themes. Inclusive perspectives begin the recognition that the United States is a multicultural and diverse society. Population data and even casual observations reveal that obvious truth, but developing an inclusive perspective requires more than recognizing the plurality of experiences in this society. Understanding race, class, and gender means coming to see the systematic exclusion and exploitation of some groups as well as the intergenerational privileges of others. This is more than just adding in different group experiences to already established frameworks of thought. It means constructing new analyses that are focused on the centrality of race, class, and gender in the experiences of us all.

DEVELOPING AN INCLUSIVE PERSPECTIVE

We want readers to understand that race, class, and gender are linked experiences, no one of which is more important than the others; the three are interrelated and together configure the structure of U.S. society. You can begin to develop a more inclusive perspective by asking: How does the world look different if we put the experiences of those who have been excluded at the center of our thinking? At first, people might be tempted to simply assert the perspective and experience of their own group. Initially, this claiming of one's experience can be valuable and empowering, but ultimately centering exclusively in one's own experiences discourages inclusionary, relational thinking.

Developing an inclusive perspective calls for more than just seeing the world through the perspective of any one group whose views have been distorted or ignored. Remember that group membership cuts across race, class, and gender categories. For example, one may be an Asian American working-class woman or a Black middle-class man or a gay, White working-class woman. Inclusive perspectives see the interconnections between these experiences and do not reduce a given person's or group's life to a single factor. In addition, developing an inclusive perspective entails more than just summing up the experiences of individual groups, as in the additive model discussed previously. Race, class, and gender are social structural categories. This means that they are embedded in the institutional structure of society. Understanding them requires a social structural analysis—by which we mean revealing the race, class, and gender patterns and processes that form the very framework of society.

We believe that thinking about the experiences of those who have been excluded from knowledge changes how we think about society, history, and culture. No longer do different groups seem "different," "deviant," or "exotic." Rather, specific patterns of the intersections of race, class, and gender are revealed, as are the connections that exist among groups. We then learn how our different experiences are linked, both historically and now.

Once you understand that race, class, and gender are *simultaneous* and *intersecting* systems of relationship and meaning, you also can see the distinctive ways that other categories of experience intersect in society. Age, religion, sexual orientation, nationality, physical ability, region, and ethnicity also shape systems of privilege and inequality. We have tried to integrate these different experiences throughout the book, although we could not include as much as we would have liked.

In the opening readings in this section, we begin by providing personal accounts of what exclusion means and how it feels. We rely heavily on personal accounts that reflect the diverse experiences of race, class, gender, and/or sexual orientation. We intend for the personal nature of these accounts to build empathy among groups—an emotional stance that we think is critical to relational thinking and developing an inclusive perspective.

Arturo Madrid ("Missing People and Others"), for example, shows how his experience as a young Latino student was silenced throughout his educational curriculum, leaving him to feel like an "other" in a society where he seemingly had no place, no history, no culture. For him, this involves more than just acknowledging the diverse histories, cultures, and experiences of groups who have been defined as marginal in society—what we have come to think of as "valuing diversity." But there is something more important than just valuing the diverse histories and cultures of the different groups who constitute society and that is to recognize how groups whose experiences have been vital in the formation of society and culture have also been silenced in the construction of knowledge about this society. The result is that what we know—about the experiences of both these silenced groups and the dominant culture—is distorted and incomplete. Indeed, for that matter, ignoring such experiences also gives us a distorted view of how the nation itself has developed.

How much did you learn about the history of group oppression in your formal education? You probably touched briefly on topics such as the labor movement, slavery, women's suffrage, perhaps even the Holocaust, but most likely these were brief excursions from an otherwise dominant narrative that ignored working-class people, women, and people of color, along with others. For that matter, how much of what you study now is centered in the experiences of the most dominant groups in society? Think about the large number of social science studies that routinely make general conclusions about the population when they have been based on research done primarily on middle-class college students, or on men. Or, how much of the literature you read and artistic creations that you study are the work of new immigrant populations, or Asian Americans, Latinos/as, African Americans, Native Americans, gays, lesbians, or women?

By minimizing the experiences and creations of these different groups, we communicate that their work and creativity is less important and less central to the development of culture than is the history of White American men. What false or incomplete conclusions does this exclusionary thinking generate? When you learn, for example, that democracy and egalitarianism were central cultural beliefs in the early history of the United States, how do you explain the enslavement of millions of African Americans? the genocide of Native Americans? the absence of laws against child labor? the presence of laws forbidding intermarriage between Asian Americans and White Americans?

This book asks you to think more inclusively. Without doing so you are prone to understand society, your own life within it, and the experiences of others through stereotypes and the misleading information that is all around you. For example, Barbara Ransby ("Katrina, Black Women, and the Deadly Discourse on Black Poverty in America") shows how conservative thinkers employed dominant stereotypes about Black women and African American families to explain the crisis after Katrina. Instead of offering their sympathy and support to Black women and their families, media figures argued that had there been fathers present in the home, many Black families could have survived the crisis. Jeremiah Torres ("Label Us Angry") shows us how sometimes the pain of living with stereotypes is very personal and painful. Torres describes how his seemingly trouble-free childhood changed overnight when he became the victim of a hate crime in a community known for its acceptance of multiculturalism.

What new experiences, understandings, theories, histories, and analyses do these readings inspire? What does it take for a member of one group (say a Latino male) to be willing to learn from and value the experiences of another (for example, an Indian Muslim woman)? These essays show that, although we are caught in multiple systems, we can learn to see our connection to others.

This is not just an intellectual exercise. As Haunani-Kay Trask shows ("From a Native Daughter"), there can also be a gap between dominant cultural narratives and people's actual experiences. As she, a native Hawaiian, tells it, the official history she learned in schools was not what she was taught in her family and community. Dominant narratives can try to justify the oppression of different groups, but the unwritten, untold subordinated truth can be a source for knowledge in pursuit of social justice.

Engaging oneself at the personal level is critical to this process of thinking differently about race, class, and gender. Changing one's mind is not just a matter of assessing facts and data, though that is important; it also requires examining one's feelings. That is why we incorporate several personal narratives into this opening section of the book. Almas Sayeed's narrative ("Chappals and Gym Shorts"), for example, illustrates the challenges facing a young woman who tries to explain her growing feminist perspective to her father who loves her but who also wants her to get married. Unlike more conventional forms of sociological data (such as surveys, interviews, and even direct observations), personal accounts such as those by Sayeed, Trask, and Torres, are more likely to elicit emotional responses. Traditionally, social science has defined emotional engagement as an impediment to objectivity. Sociology, for example, has emphasized rational thought as the basis for social action and has often discouraged more personalized reflection, but the capacity to reflect on one's experience makes us distinctly human. Personal documents tap the private, reflective dimension of life, enabling us to see the inner lives of others and, in the process, revealing our own lives more completely.

The idea that objectivity is best reached only through rational thought is a specifically Western and masculine way of thinking—one that we challenge throughout this book. Including personal narratives is not meant to limit our level of understanding only to individuals. As sociologists, we study individuals in groups as a way of revealing the social structures shaping collective experiences. In doing so, we discover our common experiences and see the impact of the social structures of race, class, and gender on our experiences. Marilyn Frye's article, "Oppression," introduces the structural perspective that lies at the heart of this book. She distinguishes oppression from suffering, pointing out that many individuals suffer in society but that oppression is structured into the fabric of social institutions. Using the metaphor of a birdcage, Frye artfully explains the concept of social structure. Looking only at an individual wire in a cage does not reveal the network of wires that forms a cage; likewise, social structure refers to the patterns of behavior, belief, resource distribution, and social control that constitute society.

Analysis of the historical role of diverse groups is critical to understanding who we are as a society and a culture. As Ronald T. Takaki says in "A Different Mirror," this involves a debate over our national identity. Takaki makes a point of showing the common connections in the histories of African Americans, Chicanos, Irish Americans, Jews, and Native Americans. He argues that only when we understand this multidimensional history will we see ourselves in the full complexity of our humanity. Moreover, we hope that understanding the significance of race, class, and gender as advanced here by Takaki and others will also encourage readers to put the experiences of the United States itself into a broader context. Knowing how race, class, and gender operate within U.S. national borders should help you see beyond those borders. We hope that developing an awareness of how the increasingly global basis of society influences the configuration of race, class, and gender relationships in the United States will encourage readers to cast an increasingly inclusive perspective on the world itself.

REFERENCE

Collins, Patricia Hill. 2000. *Black Feminist Thought: Knowledge, Consciousness, and Empowerment.* New York: Routledge.

1

Missing People and Others
Joining Together to Expand the Circle

ARTURO MADRID

I am a citizen of the United States, as are my parents and as were their parents, grandparents, and great-grandparents. My ancestors' presence in what is now the United States antedates Plymouth Rock, even without taking into account any American Indian heritage I might have.

I do not, however, fit those mental sets that define America and Americans. My physical appearance, my speech patterns, my name, my profession (a professor of Spanish) create a text that confuses the reader. My normal experience is to be asked, "And where are *you* from?"

My response depends on my mood. Passive-aggressive, I answer, "From here." Aggressive-passive, I ask, "Do you mean where am I originally from?" But ultimately my answer to those follow-up questions that ask about origins will be that we have always been from here.

Overcoming my resentment I will try to educate, knowing that nine times out of ten my words fall on inattentive ears. I have spent most of my adult life explaining who I am not. I am exotic, but—as Richard Rodriguez of *Hunger of Memory* fame so painfully found out—not exotic enough ... not Peruvian, or Pakistani, or Persian, or whatever.

I am, however, very clearly *the other*, if only your everyday, garden-variety, domestic *other*. I've always known that I was *the other*, even before I knew the vocabulary or understood the significance of being *the other*.

I grew up in an isolated and historically marginal part of the United States, a small mountain village in the state of New Mexico, the eldest child of parents native to that region and whose ancestors had always lived there. In those vast and empty spaces, people who look like me, speak as I do, and have names like mine predominate. But the *americanos* lived among us: the descendants of those nineteenth-century immigrants who dispossessed us of our lands; missionaries who came to convert us and stayed to live among us; artists who became enchanted with our land and humanscape and went native; refugees from unhealthy climes, crowded spaces, unpleasant circumstances; and, of course, the inhabitants of Los Alamos, whose socio-cultural distance from us was moreover accentuated by the fact that they occupied a space removed from and proscribed

SOURCE: From Change 20 (May/June 1988): 55–59. Reprinted by permission.

to us. More importantly, however, they—*los americanos*—were omnipresent (and almost exclusively so) in newspapers, newsmagazines, books, on radio, in movies and, ultimately, on television.

Despite the operating myth of the day, school did not erase my otherness. It did try to deny it, and in doing so only accentuated it. To this day, schooling is more socialization than education, but when I was in elementary school—and given where I was—socialization was everything. School was where one became an American. Because there was a pervasive and systematic denial by the society that surrounded us that we were Americans. That denial was both explicit and implicit. My earliest memory of the former was that there were two kinds of churches: theirs and ours. The more usual was the implicit denial, our absence from the larger cultural, economic, political and social spaces—the one that reminded us constantly that we were *the other*. And school was where we felt it most acutely.

Quite beyond saluting the flag and pledging allegiance to it (a very intense and meaningful action, given that the United States was involved in a war and our brothers, cousins, uncles, and fathers were on the front lines) becoming American was learning English and its corollary—not speaking Spanish. Until very recently ours was a proscribed language—either *de jure* (by rule, by policy, by law) or *de facto* (by practice, implicitly if not explicitly; through social and political and economic pressure). I do not argue that learning English was not appropriate. On the contrary. Like it or not, and we had no basis to make any judgments on that matter, we were Americans by virtue of having been born Americans, and English was the common language of Americans. And there was a myth, a pervasive myth, that said that if we only learned to speak English well—and particularly without an accent—we would be welcomed into the American fellowship.

Senator Sam Hayakawa notwithstanding, the true text was not our speech, but rather our names and our appearance, for we would always have an accent, however perfect our pronunciation, however excellent our enunciation, however divine our diction. That accent would be heard in our pigmentation, our physiognomy, our names. We were, in short, *the other*.

Being *the other* means feeling different; is awareness of being distinct; is consciousness of being dissimilar. It means being outside the game, outside the circle, outside the set. It means being on the edges, on the margins, on the periphery. Otherness means feeling excluded, closed out, precluded, even disdained and scorned. It produces a sense of isolation, of apartness, of disconnectedness, of alienation.

Being *the other* involves a contradictory phenomenon. On the one hand being *the other* frequently means being invisible. Ralph Ellison wrote eloquently about that experience in his magisterial novel *The Invisible Man*. On the other hand, being *the other* sometimes involves sticking out like a sore thumb. What is she/he doing here?

If one is *the other*, one will inevitably be perceived unidimensionally; will be seen stereotypically; will be defined and delimited by mental sets that may not bear much relation to existing realities. There is a darker side to otherness as well. *The other* disturbs, disquiets, discomforts. It provokes distrust and suspicion. *The other* makes people feel anxious, nervous, apprehensive, even fearful. *The other* frightens, scares.

For some of us being *the other* is only annoying; for others it is debilitating; for still others it is damning. Many try to flee otherness by taking on protective colorations that provide invisibility, whether of dress or speech or manner or name. Only a fortunate few succeed. For the majority, otherness is permanently sealed by physical appearance. For the rest, otherness is betrayed by ways of being, speaking or of doing.

I spent the first half of my life downplaying the significance and consequences of otherness. The second half has seen me wrestling to understand its complex and deeply ingrained realities; striving to fathom why otherness denies us a voice or visibility or validity in American society and its institutions; struggling to make otherness familiar, reasonable, even normal to my fellow Americans.

I am also a missing person. Growing up in northern New Mexico I had only a slight sense of our being missing persons. *Hispanos*, as we called (and call) ourselves in New Mexico, were very much a part of the fabric of the society and there were Hispano professionals everywhere about me: doctors, lawyers, school teachers, and administrators. My people owned businesses, ran organizations and were both appointed and elected public officials.

To be sure, we did not own the larger businesses, nor at the time were we permitted to be part of the banking world. Other than that, however, people who looked like me, spoke like me, and had names like mine, predominated. There was, to be sure, Los Alamos, but as I have said, it was removed from our realities.

My awareness of our absence from the larger institutional life of society became sharper when I went off to college, but even then it was attenuated by the circumstances of history and geography. The demography of Albuquerque still strongly reflected its historical and cultural origins, despite the influx of Midwesterners and Easterners. Moreover, many of my classmates at the University of New Mexico in Albuquerque were Hispanos, and even some of my professors were.

I thought that would change at UCLA, where I began graduate studies in 1960. Los Angeles already had a very large Mexican population, and that population was visible even in and around Westwood and on the campus. Many of the groundskeepers and food-service personnel at UCLA were Mexican. But Mexican-American students were few and mostly invisible, and I do not recall seeing or knowing a single Mexican-American (or, for that matter, black, Asian, or American Indian) professional on the staff or faculty of that institution during the five years I was there.

Needless to say, persons like me were not present in any capacity at Dartmouth College—the site of my first teaching appointment—and, of course, were not even part of the institutional or individual mind-set. I knew then that we—a "we" that had come to encompass American Indians, Asian Americans, black Americans, Puerto Ricans, and women—were truly missing persons in American institutional life.

Over the past three decades, the *de jure* and *de facto* segregations that have historically characterized American institutions have been under assault. As a consequence, minorities and women have become part of American institutional life, and although there are still many areas where we are not to be found, the missing persons phenomenon is not as pervasive as it once was.

However, the presence of *the other*, particularly minorities, in institutions and in institutional life, is, as we say in Spanish, *a flor de tierra*; spare plants whose roots do not go deep, a surface phenomenon, vulnerable to inclemencies of an economic, political, or social nature.

Our entrance into and our status in institutional life is not unlike a scenario set forth by my grandmother's pastor when she informed him that she and her family were leaving their mountain village to relocate in the Rio Grande Valley. When he asked her to promise that she would remain true to the faith and continue to involve herself in the life of the church, she assured him that she would and asked him why he thought she would do otherwise.

"Doña Trinidad," he told her, "in the Valley there is no Spanish church. There is only an American church." "But," she protested, "I read and speak English and would be able to worship there." Her pastor's response was: "It is possible that they will not admit you, and even if they do, they might not accept you. And that is why I want you to promise me that you are going to go to church. Because if they don't let you in through the front door, I want you to go in through the back door. And if you can't get in through the back door, go in the side door. And if you are unable to enter through the side door I want you to go in through the window. What is important is that you enter and that you stay."

Some of us entered institutional life through the front door; others through the back door; and still others through side doors. Many, if not most of us, came in through windows and continue to come in through windows. Of those who entered through the front door, some never made it past the lobby; others were ushered into corners and niches. Those who entered through back and side doors inevitably have remained in back and side rooms. And those who entered through windows found enclosures built around them. For despite the lip service given to the goal of the integration of minorities into institutional life, what has occurred instead is ghettoization, marginalization, isolation.

Not only have the entry points been limited: in addition, the dynamics have been singularly conflictive. Gaining entry and its corollary—gaining space—have frequently come as a consequence of demands made on institutions and institutional officers. Rather than entering institutions more or less passively, minorities have, of necessity, entered them actively, even aggressively. Rather than taking, they have demanded. Institutional relations have thus been adversarial, infused with specific and generalized tensions.

The nature of the entrance and the nature of the space occupied have greatly influenced the view and attitudes of the majority population within those institutions. All of us are put into the same box; that is, no matter what the individual reality, the assessment of the individual is inevitably conditioned by a perception that is held of the class. Whatever our history, whatever our record, whatever our validations, whatever our accomplishments, by and large we are perceived unidimensionally and are dealt with accordingly.

My most recent experience in this regard is atypical only in its explicitness. A few years ago I allowed myself to be persuaded to seek the presidency of a large and prestigious state university. I was invited for an interview and presented myself before the selection committee, which included members of the board of trustees. The opening question of the brief but memorable interview was directed at me by a member of that august body. "Dr. Madrid," he asked, "why does a one-dimensional person like you think he can be the president of a multi-dimensional institution like ours?"

If, as I happen to believe, the well-being of a society is directly related to the degree and extent to which all of its citizens participate in its institutions, we have a challenge before us. One of the strengths of our society—perhaps its main strength—has been a tradition of struggle against clubbishness, exclusivity, and restriction.

Today, more than ever, given the extraordinary changes that are taking place in our society, we need to take up that struggle again—irritating, grating, troublesome, unfashionable, unpleasant as it is. As educated and educator members of this society, we have a special responsibility for leading the struggle against marginalization, exclusion, and alienation.

Let us work together to assure that all American institutions, not just its pre-collegiate educational and penal institutions, reflect the diversity of our society. Not to do so is to risk greater alienation on the part of a growing segment of our society. It is to risk increased social tension in an already conflictive world. And ultimately it is to risk the survival of a range of institutions that, for all their defects and deficiencies, permit us the space, the opportunity, and the freedom to improve our individual and collective lot; to guide the course of our government, and to redress whatever grievances we have. Let us join together to expand, not to close the circle.

2

Chappals and Gym Shorts
An Indian Muslim Woman in the Land of Oz

ALMAS SAYEED

It was finals week during the spring semester of my sophomore year at the University of Kansas, and I was buried under mounds of papers and exams. The stress was exacerbated by long nights, too much coffee and a chronic, building pain in my permanently splintered shins (left over from an old sports injury). Between attempting to understand the nuances of Kant's *Critique of Pure Reason* and applying the latest game-theory models to the 1979 Iranian revolution, I was regretting my decision to pursue majors in philosophy, women's studies *and* international studies.

My schedule was not exactly permitting much down time. With a full-time school schedule, a part-time job at Lawrence's domestic violence shelter and preparations to leave the country in three weeks, I was grasping to hold onto what little sanity I had left. Wasn't living in Kansas supposed to be more laid-back than this? After all, Kansas was the portal to the magical land of Oz, where wicked people melt when doused with mop water and bright red, sparkly shoes could substitute for the services of American Airlines, providing a quick getaway. Storybook tales aside, the physical reality of this period was that my deadlines were inescapable. Moreover, the most pressing of these deadlines was completely non-school related: my dad, on his way home to Wichita, was coming for a brief visit. This would be his first stay by himself, without Mom to accompany him or act as a buffer.

Dad visited me the night before my most difficult exam. Having just returned from spending time with his family—a group of people with whom he historically had an antagonistic relationship—Dad seemed particularly relaxed in his stocky six-foot-four frame. Wearing one of the more subtle of his nineteen cowboy hats, he arrived at my door, hungry, greeting me in Urdu, our mother tongue, and laden with gifts from Estée Lauder for his only daughter. Never mind that I rarely wore makeup and would have preferred to see the money spent on my electric bill or a stack of feminist theory books from my favorite used bookstore. If Dad's visit was going to include a conversation about how little I use beauty products, I was not going to be particularly receptive.

SOURCE: From Daisy Hernández and Bushra Rehman, eds., *Colonize This! Young Women of Color on Today's Feminism* (New York: Seal Press, 2002), pp. 203–14. Copyright © 2002 Daisy Hernández and Bushra Rehman. Reprinted by permission of the publisher, Seal Press.

"Almas," began my father from across the dinner table, speaking in his British-Indian accent infused with his love of Midwestern colloquialisms, "You know that you won't be a spring chicken forever. While I was in Philadelphia, I realized how important it is for you to begin thinking about our culture, religion and your future marriage plans. I think it is time we began a two-year marriage plan so you can find a husband and start a family. I think twenty-two will be a good age for you. You should be married by twenty-two."

I needed to begin thinking about the "importance of tradition" and be married by twenty-two? This, from the only Indian man I knew who had Alabama's first album on vinyl and loved to spend long weekends in his rickety, old camper near Cheney Lake, bass fishing and listening to traditional Islamic Quavali music? My father, in fact, was in his youth crowned "Mr. Madras," weightlifting champion of 1965, and had left India to practice medicine and be an American cowboy in his spare time. But he wanted *me* to aspire to be a "spring chicken," maintaining some unseen hearth and home to reflect my commitment to tradition and culture.

Dad continued, "I have met a boy that I like for you very much. Masoud's son, Mahmood. He is a good Muslim boy, tells great jokes in Urdu and is a promising engineer. We should be able to arrange something. I think you will be very happy with him!" Dad concluded with a satisfied grin.

Masoud, Dad's cousin? This would make me and Mahmood distant relatives of some sort. And Dad wants to "arrange something"? I had brief visions of being paraded around a room, serving tea to strangers in a sari or a shalwar kameez (a traditional South Asian outfit for women) wearing a long braid and chappals (flat Indian slippers), while Dad boasted of my domestic capabilities to increase my attractiveness to potential suitors. I quickly flipped through my mental Rolodex of rhetorical devices, acquired during years of women's studies classes and found the card blank. No doubt, even feminist scholar Catherine MacKinnon would have been rendered speechless sitting across the table in a Chinese restaurant speaking to my overzealous father.

It is not that I hadn't already dealt with the issue. In fact, we had been here before, ever since the marriage proposals began (the first one came when I was fourteen). Of course, when they first began, it was a family joke, as my parents understood that I was to continue my education. The jokes, however, were always at my expense: "You received a proposal from a nice boy living in our mosque. He is studying medicine," my father would come and tell me with a huge, playful grin. "I told him that you weren't interested because you are too busy with school. And anyway you can't cook or clean." My father found these jokes particularly funny, given my dislike of household chores. In this way, the eventuality of figuring out how to deal with these difficult issues was postponed with humor.

Dad's marriage propositions also resembled conversations that we had already had about my relationship to Islamic practices specific to women, some negotiated in my favor and others simply shelved for the time being. Just a year ago, Dad had come to me while I was home for the winter holidays, asking me to begin wearing *hijab,* the traditional headscarf worn by Muslim women. I categorically refused, maintaining respect for those women who chose to do so. I understood that for numerous women, as well as for Dad, hijab symbolized something much more than covering a woman's body or hair; it symbolized a way to adhere to religious and cultural traditions in order to prevent complete Western immersion. But even my sympathy for this concern didn't change my feeling that hijab constructed me as a woman first and a human being second. Veiling seemed to reinforce the fact that inequality between the sexes was a natural, inexplicable phenomenon that is impossible to overcome, and that women should cover themselves, accommodating an unequal hierarchy, for the purposes of modesty and self-protection. I couldn't reconcile these issues and refused my father's request to don the veil. Although there was tension—Dad claimed I had yet to have my religious awakening—he chose to respect my decision.

Negotiating certain issues had always been part of the dynamic between my parents and me. It wasn't that I disagreed with them about everything. In fact, I had internalized much of the Islamic perspective of the female body while simultaneously admitting to its problematic nature (to this day, I would rather wear a wool sweater than a bathing suit in public, no matter how sweltering the weather). Moreover, Islam became an important part of differentiating myself from other American kids who did not have to find a balance between two opposing cultures. Perhaps Mom and Dad recognized the need to concede certain aspects of traditional Islamic norms, because for all intents and purposes, I had been raised in the breadbasket of America.

By the time I hit adolescence, I had already established myself outside of the social norm of the women in my community. I was an athletic teenager, a competitive tennis player and a budding weightlifter. After a lot of reasoning with my parents, I was permitted to wear shorts to compete in tennis tournaments, but I was not allowed to show my legs or arms (no tank tops) outside of sports. It was a big deal for my parents to have agreed to allow me to wear shorts in the first place. The small community of South Asian Muslim girls my age, growing up in Wichita, became symbols of the future of our community in the United States. Our bodies became the sites to play out cultural and religious debates. Much in the same way that Lady Liberty had come to symbolize idealized stability in the *terra patria* of America, young South Asian girls in my community were expected to embody the values of a preexisting social structure. We were scrutinized for what we said, what we wore, being seen with boys in public and for lacking grace and piety. Needless to say, because of disproportionate muscle mass, crooked teeth, huge Lucy glasses, and a disposition to walk pigeon-toed, I was not among the favored.

To add insult to injury, Mom nicknamed me "Amazon Woman," lamenting the fact that she—a beautiful, petite lady—had produced such a graceless, unfeminine creature. She was horrified by how freely I got into physical fights with my younger brother and arm wrestled boys at school. She was particularly frustrated by the fact that I could not wear her beautiful Indian jewelry, especially her bangles and bracelets, because my wrists were too big. Special occasions, when I had to slather my wrists with tons of lotion in order to squeeze my hands into her tiny bangles, often bending the soft gold out of shape, caused us both infinite amounts of grief. I was the snot-nosed, younger sibling of the Bollywood (India's Hollywood) princess that my mother had in mind as a more appropriate representation of an Indian daughter. Rather, I loved sports, sports figures and books. I hated painful makeup rituals and tight jewelry.

It wasn't that I had a feminist awakening at an early age. I was just an obnoxious kid who did not understand the politics raging around my body. I did not possess the tools to analyze or understand my reaction to this process of social conditioning and normalization until many years later, well after I had left my parents' house and the Muslim community in Wichita. By positioning me as a subject of both humiliation and negotiation, Mom and Dad had inadvertently laid the foundations for me to understand and scrutinize the process of conditioning women to fulfill particular social obligations.

What was different about my dinner conversation with Dad that night was a sense of immediacy and detail. Somehow discussion about a "two-year marriage plan" seemed to encroach on my personal space much more than had previous jokes about my inability to complete my household chores or pressure to begin wearing hijab. I was meant to understand that when it came to marriage, I was up against an invisible clock (read: social norms) that would dictate how much time I had left: how much time I had left to remain desirable, attractive and marriageable. Dad was convinced that it was his duty to ensure my long-term security in a manner that reaffirmed traditional Muslim culture in the face of an often hostile foreign community. I recognized that the threat was not as extreme as being shipped off to India in order to marry someone I had never met. The challenge was far more subtle than this. I was being asked to choose my community; capitulation through arranged marriage would show my commitment to being Indian, to being a good Muslim woman and to my parents by proving that they had raised me with a sense of duty and the willingness to sacrifice for my culture, religion, and family.

There was no way to tell Dad about my complicated reality. Certain characteristics of my current life already indicated failure by such standards. I was involved in a long-term relationship with a white man, whose father was a prison guard on death row, an occupation that would have mortified my upper-middle-class, status-conscious parents. I was also struggling with an insurmountable crush on an *actress* in the Theater and Film Department. I was debating my sexuality in terms of cultural compatibility as well as gender. Moreover, there was no way to tell Dad that my social circle was supportive of these nontraditional romantic explorations. My friends in college had radically altered my perceptions of marriage and family. Many of my closest friends, including my roommates, were coming to terms with their own life-choices, having recently come out of the closet but unable to tell their families about their decisions. I felt inextricably linked to this group of women, who, like me, often had to lead double lives. The immediacy of fighting for issues such as queer rights, given the strength and beauty of my friends' romantic relationships, held far more appeal for me than the topics of marriage and security that my father broached over our Chinese dinner. There was no way to explain to my loving, charismatic, steadfastly religious father, who was inclined to the occasional violent outburst, that a traditional arranged marriage not only conflicted with the feminist ideology I had come to embrace, but it seemed almost petty in the face of larger, more pressing issues.

Although I had no tools to answer my father that night at dinner, feminist theory had provided me with the tools to understand *why* my father and I were engaged in the conversation in the first place. I understood that in his mind, Dad was fulfilling his social obligation as father and protector. He worried about my economic stability and, in a roundabout way, my happiness. Feminism and community activism had enabled me to understand these things as part of a proscribed role for women. At the same time, growing up in Kansas and coming to feminism here meant that I had to reconcile a number of different issues. I am a Muslim, first-generation Indian, feminist woman studying in a largely homogeneous white, Christian community in Midwestern America. What sacrifices are necessary for me to retain my familial relationships as well as a sense of personal autonomy informed by Western feminism?

The feminist agenda in my community is centered on ending violence against women, fighting for queer rights and maintaining women's reproductive choices. As such, the way that I initially became involved with this community was through community projects such as "Womyn Take Back the Night," attending pride rallies and working at the local domestic violence shelter. I am often the only woman of color in feminist organizations and at feminist events. Despite having grown up in the Bible belt, it is difficult for me to relate to stories told by my closest friends of being raised on cattle ranches and farms, growing up Christian by default and experiencing the strict social norms of small, religious communities in rural Kansas. Given the context of this community—a predominantly white, middle-class, college town—I have difficulty explaining that my feminism has to address issues like, "I should be able to wear *both* hijab *and* shorts if I choose to." The enormity of our agenda leaves little room to debate issues equally important but applicable only to me, such as the meaning of veiling, arranged marriages versus dating and how the north-south divide uniquely disadvantages women in the developing world.

It isn't that the women in my community ever turned to me and said, "Hey you, brown girl, stop diluting our priorities." To the contrary, the majority of active feminists in my community are eager to listen and understand my sometimes divergent perspective. We have all learned to share our experiences as women, students, mothers, partners and feminists. We easily relate to issues of male privilege, violence against women and figuring out how to better appreciate the sacrifices made by our mothers. From these commonalities we have learned to work together, creating informal social networks to complete community projects.

The difficulty arises when trying to put this theory and discussion into practice. Like last year, when our organization, the Womyn's Empowerment Action Coalition, began plans for the Womyn Take Back the Night march and rally, a number of organizers were eager to include the contribution of a petite, white belly dancer in the pre-march festivities. When I voiced my concern that historically belly dancing had been used as a way to objectify women's bodies in the Middle East, one of my fellow organizers (and a very good friend) laughed and called me a prude: "We're in Kansas, Almas," she said. "It doesn't mean the same thing in our culture. It is different here than over *there*." I understood what she meant, but having just returned from seven months in the West Bank, Palestine two months before, for me over there *was* over here. In the end, the dance was included while I wondered about our responsibility to women outside of the United States and our obligation to address the larger social, cultural issues of the dance itself.

To reconcile the differences between my own priorities and those of the women I work with, I am learning to bridge the gap between the Western white women (with the occasional African American or Chicana) feminist canon and my own experience as a first-generation Indian Muslim woman living in the Midwest. I struggle with issues like cultural differences, colonialism, Islam and feminism and how they relate to one another. The most difficult part has been to get past my myopic vision of simply laying feminist theory written by Indian, Muslim or postcolonial theorists on top of American-Western feminism. With the help of feminist theory and other feminists, I am learning to dissect Western models of feminism, trying to figure out what aspects of these models can be applied to certain contexts. To this end, I have had the privilege of participating in projects abroad, in pursuit of understanding feminism in other contexts.

For example, while living with my extended family in India, I worked for a micro-credit affiliate that advised women on how to get loans and start their own businesses. During this time I learned about the potential of micro-enterprise as a weapon against the feminization of poverty. Last year, I spent a semester in the West Bank, Palestine, studying the link between women and economics in transitional states and beginning to understand the importance of women's efforts during revolution. These experiences have been invaluable to me as a student of feminism and women's mobilization efforts. They have also shaped my personal development, helping me understand where the theoretical falls short of solving for the practical. In Lawrence, I maintain my participation in local feminist projects. Working in three different contexts has highlighted the amazing and unique ways in which feminism develops in various cultural settings yet still maintains certain commonalities.

There are few guidebooks for women like me who are trying to negotiate the paradigm of feminism in two different worlds. There is a delicate dance here that I must master—a dance of negotiating identity within interlinking cultural spheres. When faced with the movement's expectations of my commitment to local issues, it becomes important for me to emphasize that differences in culture and religion are also "local issues." This has forced me to change my frame of reference, developing from a rebellious tomboy who resisted parental imposition to a budding social critic, learning how to be a committed feminist and still keep my cultural, religious and community ties. As for family, we still negotiate despite the fact that Dad's two-year marriage plan has yet to come to fruition in this, my twenty-second year.

3

From a Native Daughter

HAUNANI-KAY TRASK

E noi'i wale mai nō ka haole, a,
'a'ole e pau nō hana a Hawai'i 'imi loa
Let the haole freely research us in detail
But the doings of deep delving Hawai'i
will not be exhausted.

—*Kepelino*
19th-century Hawaiian historian

When I was young, the story of my people was told twice: once by my parents, then again by my school teachers. From my *'ohana* (family), I learned about the life of the old ones: how they fished and planted by the moon; shared all the fruits of their labors, especially their children; danced in great numbers for long hours; and honored the unity of their world in intricate genealogical chants. My mother said Hawaiians had sailed over thousands of miles to make their home in these sacred islands. And they had flourished, until the coming of the *haole* (whites).

At school, I learned that the "pagan Hawaiians" did not read or write, were lustful cannibals, traded in slaves, and could not sing. Captain Cook had "discovered" Hawai'i and the ungrateful Hawaiians had killed him. In revenge, the Christian god had cursed the Hawaiians with disease and death.

I learned the first of these stories from speaking with my mother and father. I learned the second from books. By the time I left for college, the books had won out over my parents, especially since I spent four long years in a missionary boarding school for Hawaiian children.

When I went away I understood the world as a place and a feeling divided in two: one *haole* (white), and the other *kānaka* (Native). When I returned ten years later with a Ph.D., the division was sharper, the lack of connection more painful. There was the world that we lived in—my ancestors, my family, and my people—and then there was the world historians described. This world, they had written, was the truth. A primitive group, Hawaiians had been ruled by blood-thirsty priests and despotic kings who owned all the land and kept our people in feudal subjugation. The chiefs were cruel, the people poor.

SOURCE: From Haunani-Kay Trask, *From a Native Daughter* (Monroe, ME: Common Courage Press), 1993.

But this was not the story my mother told me. No one had owned the land before the *haole* came; everyone could fish and plant, except during sacred periods. And the chiefs were good and loved their people.

Was my mother confused? What did our *kūpuna* (elders) say? They replied: Did these historians (all *haole*) know the language? Did they understand the chants? How long had they lived among our people? Whose stories had they heard?

None of the historians had ever learned our mother tongue. They had all been content to read what Europeans and Americans had written. But why did scholars, presumably well-trained and thoughtful, neglect our language? Not merely a passageway to knowledge, language is a form of knowing by itself; a people's way of thinking and feeling is revealed through its music.

I sensed the answer without needing to answer. From years of living in a divided world, I knew the historian's judgment: *There is no value in things Hawaiian; all value comes from things haole.*

Historians, I realized, were very much like missionaries. They were a part of the colonizing horde. One group colonized the spirit; the other, the mind. Frantz Fanon had been right, but not just about Africans. He had been right about the bondage of my own people: "By a kind of perverted logic, [colonialism] turns to the past of the oppressed people, and distorts, disfigures, and destroys it" (1963:210). The first step in the colonizing process, Fanon had written, was the deculturation of a people. What better way to take our culture than to remake our image? A rich historical past became small and ignorant in the hands of Westerners. And we suffered a damaged sense of people and culture because of this distortion.

Burdened by a linear, progressive conception of history and by an assumption that Euro-American culture flourishes at the upper end of that progression, Westerners have told the history of Hawai'i as an inevitable if occasionally bittersweet triumph of Western ways over "primitive" Hawaiian ways. A few authors—the most sympathetic—have recorded with deep-felt sorrow the passing of our people. But in the end, we are repeatedly told, such an eclipse was for the best.

Obviously it was best for Westerners, not for our dying multitudes. This is why the historian's mission has been to justify our passing by celebrating Western dominance. Fanon would have called this missionizing, intellectual colonization. And it is clearest in the historian's insistence that *pre-haole* Hawaiian land tenure was "feudal"—a term that is now applied, without question, in every monograph, in every schoolbook, and in every tour guide description of my people's history.

From the earliest days of Western contact my people told their guests that *no one* owned the land. The land—like the air and the sea—was for all to use and share as their birthright. Our chiefs were *stewards* of the land; they could not own or privately possess the land any more than they could sell it.

But the *haole* insisted on characterizing our chiefs as feudal landlords and our people as serfs. Thus, a European term which described a European practice founded on the European concept of private property—feudalism—was imposed upon a people halfway around the world from Europe and vastly different from her in every conceivable way. More than betraying an ignorance of Hawaiian culture and history, however, this misrepresentation was malevolent in design.

By inventing feudalism in ancient Hawai'i, Western scholars quickly transformed a spiritually based, self-sufficient economic system of land use and occupancy into an oppressive, medieval European practice of divine right ownership, with the common people tied like serfs to the land. By claiming that a Pacific people lived under a European system—that the Hawaiians lived under feudalism—Westerners could then degrade a successful system of shared land use with a pejorative and inaccurate Western term. Land tenure changes instituted by Americans and in line with current Western notions of private property were then made to appear beneficial to the Hawaiians. But in practice, such changes benefited the *haole*, who alienated the people from the land, taking it for themselves.

The prelude to this land alienation was the great dying of the people. Barely half a century after contact with the West, our people had declined in number by eighty percent. Disease and death were rampant. The sandalwood forests had been stripped bare for international commerce between England and China. The missionaries had insinuated themselves everywhere. And a debt-ridden Hawaiian king (there had been no king before Western contact) succumbed to enormous pressure from the Americans and followed their schemes for dividing the land.

This is how private property land tenure entered Hawai'i. The common people, driven from their birthright, received less than one percent of the land. They starved while huge haole-owned sugar plantations thrived.

And what had the historians said? They had said that the Americans "liberated" the Hawaiians from an oppressive "feudal" system. By inventing a false feudal past, the historians justify—and become complicitous in—massive American theft.

Is there "evidence"—as historians call it—for traditional Hawaiian concepts of land use? The evidence is in the sayings of my people and in the words they wrote more than a century ago, much of which has been translated. However, historians have chosen to ignore any references here to shared land use. But there is incontrovertible evidence in the very structure of the Hawaiian language. If the historians had bothered to learn our language (as any American historian of France would learn French) they would have discovered that we show possession in two ways: through the use of an "a" possessive, which reveals acquired status, and through the use of an "o" possessive, which denotes inherent status. My body (*ko'u kino*) and my parents (*ko'u mākua*), for example, take the "o" form; most material objects, such as food (*ka'u mea'ai*) take the "a" form. But land, like one's body and one's parents, takes the "o" possessive (*ko'u 'āina*). Thus, in our way of speaking, land is inherent to the people; it is like our bodies and our parents. The people cannot exist without the land, and the land cannot exist without the people.

Every major historian of Hawai'i has been mistaken about Hawaiian land tenure. The chiefs did not own the land: they *could not* own the land. My mother was right and the *haole* historians were wrong. If they had studied our language they would have known that no one owned the land. But was their failing merely ignorance, or simple ethnocentric bias?

No, I did not believe them to be so benign. As I read on, a pattern emerged in their writing. Our ways were inferior to those of the West, to those of the historians' own culture. We were "less developed," or "immature," or "authoritarian." In some tellings we were much worse. Thus, Gavan Daws (1968), the most famed modern historian of Hawai'i, had continued a tradition established earlier by missionaries Hiram Bingham (1848; reprinted, 1981) and Sheldon Dibble (1909), by referring to the old ones as "thieves" and "savages" who regularly practiced infanticide and who, in contrast to "civilized" whites, preferred "lewd dancing" to work. Ralph Kuykendall (1938), long considered the most thorough if also the most boring of historians of Hawai'i, sustained another fiction—that my ancestors owned slaves, the outcast *kauwā*. This opinion, as well as the description of Hawaiian land tenure as feudal, had been supported by respected sociologist Andrew Lind.... Finally, nearly all historians had refused to accept our genealogical dating of A.D. 400 or earlier for our arrival from the South Pacific. They had, instead, claimed that our earliest appearance in Hawai'i could only be traced to A.D. 1100. Thus at least seven hundred years of our history were repudiated by "superior" Western scholarship. Only recently have archaeological data confirmed what Hawaiians had said these many centuries (Tuggle 1979).[1]

Suddenly the entire sweep of our written history was clear to me. I was reading the West's view of itself through the degradation of my own past. When historians wrote that the king owned the land and the common people were bound to it, they were saying that ownership was the only way human beings in their world could relate to the land, and in that relationship, some one person had to control both the land and the interaction between humans.

And when they said that our chiefs were despotic, they were telling of their own society, where hierarchy always results in domination. Thus any authority or elder is automatically suspected of tyranny.

And when they wrote that Hawaiians were lazy, they meant that work must be continuous and ever a burden.

And when they wrote that we were promiscuous, they meant that lovemaking in the Christian West is a sin.

And when they wrote that we were racist because we preferred our own ways to theirs, they meant that their culture needed to dominate other cultures.

And when they wrote that we were superstitious, believing in the *mana* of nature and people, they meant that the West has long since lost a deep spiritual and cultural relationship to the earth.

And when they wrote that Hawaiians were "primitive" in their grief over the passing of loved ones, they meant that the West grieves for the living who do not walk among their ancestors.

For so long, more than half my life, I had misunderstood this written record, thinking it described my own people. But my history was nowhere present. For we had not written. We had chanted and sailed and fished and built and prayed. And we had told stories through the great blood lines of memory: genealogy.

To know my history, I had to put away my books and return to the land. I had to plant *taro* in the earth before I could understand the inseparable bond between people and *'āina*. I had to feel again the spirits of nature and take gifts of plants and fish to the ancient altars. I had to begin to speak my language with our elders and leave long silences for wisdom to grow. But before anything else, I needed to learn the language like a lover so that I could rock within her and lie at night in her dreaming arms.

There was nothing in my schooling that had told me of this, or hinted that somewhere there was a longer, older story of origins, of the flowing of songs out to a great but distant sea. Only my parents' voices, over and over, spoke to me of a Hawaiian world. While the books spoke from a different world, a Western world.

And yet, Hawaiians are not of the West. We are of *Hawai'i Nei*, this world where I live, this place, this culture, this *'āina*.

What can I say, then, to Western historians of my place and people? Let me answer with a story.

A while ago I was asked to appear on a panel on the American overthrow of our government in 1893. The other panelists were all *haole*. But one was a *haole* historian from the American continent who had just published a book on what he called the American anti-imperialists. He and I met briefly in preparation for the panel. I asked him if he knew the language. He said no. I asked him if he knew the record of opposition to our annexation to America. He said there was no real evidence for it, just comments here and there. I told him that he didn't understand and that at the panel I would share the evidence. When we met in public and spoke, I said this:

There is a song much loved by our people. It was written after Hawai'i had been invaded and occupied by American marines. Addressed to our dethroned Queen, it was written in 1893, and tells of Hawaiian feelings for our land and against annexation to the United States. Listen to our lament:

Kaulana nā pua a'o Hawai'i	Famous are the children of Hawai'i
Kupa'a ma hope o ka 'āina	Who cling steadfastly to the land
Hiki mai ka 'elele o ka loko 'ino	Comes the evil-hearted with
Palapala 'ānunu me ka pākaha	A document greedy for plunder
Pane mai Hawai'i moku o Keawe	Hawai'i, island of Keawe, answers
Kokua nā hono a'o Pi'ilani	The bays of Pi'ilani [of Maui, Moloka'i, and Lana'i] help
Kako'o mai Kaua'i Mano	Kaua'i of Mano assists
Pau pu me ke one o Kakuhihewa	Firmly together with the sands of Kakuhihewa
'A'ole a'e kau i ka pūlima	Do not put the signature
Maluna o ka pepa o ka 'enemi	On the paper of the enemy
Ho'ohui 'āina kū'ai hewa	Annexation is wicked sale
I ka pono sivila a'o ke kānaka	Of the civil rights of the Hawaiian people
Mahope mākou o Lili'ūlani	We support Lili'uokalani
A loa'a 'e ka pono o ka 'āina	Who has earned the right to the land
Ha'ina 'ia mai ana ka puana	The story is told
'O ka po'e i aloha i ka 'āina	Of the people who love the land

This song, I said, continues to be sung with great dignity at Hawaiian political gatherings, for our people still share the feelings of anger and protest that it conveys.

But our guest, the *haole* historian, answered that this song, although beautiful, was not evidence of either opposition or of imperialism from the Hawaiian perspective.

Many Hawaiians in the audience were shocked at his remarks, but, in hindsight, I think they were predictable. They are the standard response of the historian who does not know the language and has no respect for its memory.

Finally, I proceeded to relate a personal story, thinking that surely such a tale could not want for authenticity since I myself was relating it. My *tūtū* (grandmother) had told my mother who had told me that at the time of the overthrow a great wailing went up throughout the islands, a wailing of weeks, a wailing of impenetrable grief, a wailing of death. But he remarked again, this too is not evidence.

And so, history goes on, written in long volumes by foreign people. Whole libraries begin to form, book upon book, shelf upon shelf. At the same time, the stories go on, generation to generation, family to family.

Which history do Western historians desire to know? Is it to be a tale of writings by their own countrymen, individuals convinced of their "unique" capacity for analysis, looking at us with Western eyes, thinking about us within Western philosophical contexts, categorizing us by Western indices, judging us by Judeo-Christian morals, exhorting us to capitalist achievements, and finally, leaving us an authoritative-because-Western record of their complete misunderstanding?

All this has been done already. Not merely a few times, but many times. And still, every year, there appear new and eager faces to take up the same telling, as if the West must continue, implacably, with the din of its own disbelief. But there is, as there has been always, another possibility. If it is truly our history Western historians desire to know, they must put down their books, and take up our practices. First, of course, the language. But later, the people, the *'āina*, the stories. Above all, in the end, the stories. Historians must listen, they must hear the generational connections, the reservoir of sounds and meanings.

They must come, as American Indians suggested long ago, to understand the land. Not in the Western way, but in the indigenous way, the way of living within and protecting the bond between people and *'āina*. This bond is cultural, and it can be understood only culturally. But because the West has lost any cultural understanding of the bond between people and land, it is not possible to know this connection through Western culture. This means that the history of indigenous people cannot be written from within Western culture. Such a story is merely the West's story of itself.

Our story remains unwritten. It rests within the culture, which is inseparable from the land. To know this is to know our history. To write this is to write of the land and the people who are born from her.

NOTES

1. See also Fornander (1878—85; reprinted, 1981). Lest one think these sources anti-
    quated, it should be noted that there exist only a handful of modern scholarly works
    on the history of Hawai'i. The most respected are those by Kuykendall (1938) and
    Daws (1968), and a social history of the 20th century by Lawrence Fuchs (1961). Of
    these, only Kuykendall and Daws claim any knowledge of pre-*haole* history, while
    concentrating on the 19th century. However, countless popular works have relied
    on these two studies which, in turn, are themselves based on primary sources written
    in English by extremely biased, anti-Hawaiian Westerners such as explorers, traders,
    missionaries (e.g., Bingham [1848; reprinted, 1981] and Dibble [1909]), and sugar
    planters. Indeed, a favorite technique of Daws's—whose *Shoal of Time* was once the
    most acclaimed and recent general history—is the lengthy quotation without com-
    ment of the most racist remarks by missionaries and planters. Thus, at one point, half
    a page is consumed with a "white man's burden" quotation from an 1886 *Planters
    Monthly* article ("It is better here that the white man should rule . . . ," etc., p. 213).
    Daws's only comment is, "The conclusion was inescapable." To get a sense of
    such characteristic contempt for Hawaiians, one has but to read the first few pages,
    where Daws refers several times to the Hawaiians as "savages" and "thieves" and
    where he approvingly has Captain Cook thinking, "It was a sensible primitive who
    bowed before a superior civilization" (p. 2). See also—among examples too
    numerous to cite—his glib description of sacred *hula* as a "frivolous diversion,"
    which, instead of work, the Hawaiians "would practice energetically in the hot
    sun for days on end . . . their bare brown flesh glistening with sweat" (pp. 65–66).
    Daws, who repeatedly displays an affection for descriptions of Hawaiian skin color,
    taught Hawaiian history for some years at the University of Hawai'i. He once held
    the Chair of Pacific History at the Australian National University's Institute of
    Advanced Studies.

    Postscript: Since this article was written, the first scholarly history by a
Native Hawaiian was published in English: *Native Land and Foreign Desires* by
Lilikalā Kame'eleihiwa (Honolulu: Bishop Museum Press, 1992).

REFERENCES

Bingham, Hiram. 1981. *A Residence of Twenty-one Years in the Sandwich Islands*. Tokyo:
    Charles E. Tuttle.

Daws, Gavan. 1968. *Shoal of Time: A History of the Hawaiian Islands*. Honolulu: University
    of Hawai'i Press.

Dibble, Sheldon. 1909. *A History of the Sandwich Islands*. Honolulu: Thrum Publishing.

Fanon, Frantz. 1963. *The Wretched of the Earth*. New York: Grove Press.

Fornander, Abraham. 1981. *An Account of the Polynesian Race, Its Origins, and Migrations
    and the Ancient History of the Hawaiian People to the Times of Kamehameha I*. Routledge,
    Vermont: Charles E. Tuttle.

Fuchs, Lawrence H. 1961. *Hawaii Pono: A Social History*. New York: Harcourt Brace & World.

Kuykendall, Ralph. 1938. *The Hawaiian Kingdom, 1778–1854: Foundation and Transformation*. Honolulu: University of Hawai'i Press.

Tuggle, H. David. 1979. "Hawai'i," in *The Prehistory of Polynesia*, ed. Jessie D. Jennings. Cambridge: Harvard University Press.

4

Katrina, Black Women, and the Deadly Discourse on Black Poverty in America

BARBARA RANSBY

INTRODUCTION

Most observers, even some of the most conservative and purportedly color-blind observers, have conceded the overwhelmingly racial character of the social disaster that followed in Hurricane Katrina's wake. Journalists covering the story could not help but acknowledge that those left behind to endure nature's wrath, and for whom little help and few resources were provided in the critical days following the hurricane, were disproportionately poor and Black. What does not often get added is that by most accounts those hardest hit and least able to rebound from it were also women: poor Black women who waded through chest-high water with sick and elderly parents, with young children on their hips and meager belongings in tow. This should not be surprising given the correlation between gender, race, and poverty. Black single mothers are more likely to be poor than any other demographic group, and New Orleans was no exception to the rule. In fact, a study by the Institute for Women's Policy Research points out that the percentage of women in poverty in New Orleans before the storm was considerably *higher* than in other parts of the country: more than half of the poor families of the city were headed by single mothers, and the median income for African American women workers in New Orleans before the storm was a paltry $19,951 a year (DeWeever 2005).

SOURCE: *Du Bois Review*, 3:1 (2006) 215–222. © 2006 W. E. B. Du Bois Institute for African and African American Research.

The effect of the hurricane on African American women was not merely a consequence of demographics; it was also fueled and framed by the rabid anti-poor discourse that has cast Black single mothers as unworthy of public aid or sympathy. In this paper, I will discuss several aspects of the gendered nature of the disaster: the effect of government inaction on Black women in New Orleans after the hurricane, the pre-Katrina discourse on Black female poverty that set the stage for that inaction, and how Black women activists have responded. Even though this was clearly a regional crisis, the various contradictions came together with particular vividness in New Orleans, so I will focus my observations on that city alone.

WOMEN IN KATRINA'S WAKE

When Katrina slapped New Orleans, she slapped everyone hard, but she slapped women especially hard. The impact is not simply measured in the number of injuries, deaths, and the amount of property loss, but in the kind of human currency that is difficult to measure. Women were more encumbered and less mobile. One gets a window into how women's lives were turned upside down by this crisis by looking at what women did and where women were situated in the community in ordinary times. As a number of commentators and experts have pointed out, there was a social crisis in New Orleans that had been fueled by the widespread prevalence of poverty and the absence of resources long before meteorologists sighted a category five hurricane bearing down on the Gulf Coast. There was already a 40% poverty rate among single mothers in the city. A state-by-state breakdown of poverty statistics ranked Louisiana number forty-seven out of fifty-one, and forty-third in terms of health-care insurance coverage. And 13% of Louisiana's children live in extreme poverty, which is defined as a family of four surviving on less than $10,000 a year.[1] A large percentage of New Orleans' poor single mothers also lived in the historic Ninth Ward, the low-lying area of the city most vulnerable to flooding. So, as in any crisis, those with few assets, little money, and even less maneuverability were hard-pressed to get out of the path of the storm and further compromised in their ability to recover after the blow.

ORAL HISTORIES

The impact of Katrina on women of African descent in New Orleans is best reflected in the stories and anecdotes that emerged from the storm. Our understanding of this tragedy and its aftermath is aided by the plethora of oral history projects that have emerged in response to the situation.[2] Some of them are an extension of preexisting archival or public history projects, and some are grass-roots interventions by students, artists, historians, and activists determined to document what actually happened and provide an outlet for those who want to tell their stories. The narratives, testimonies, and profiles of real flesh and blood people are the best rebuttal to one-dimensional stereotypes. One very powerful story, collected by *Alive in Truth: The New Orleans Disaster Oral History and Memory Project,* tells of the experience of a woman named Clarice B. (later identified as Clarice Butler), who describes her life before and after Katrina. "I worked all my life," she explains:

> I worked all my life for Metropolitan Homecare for 28 years: homecare, nurse's assistant. I took care of a lot of people in my life, a lot of people. I was good at my job, oh, yeah. It's not a clean job and it's not no dirty, dirty job. But no job is clean all the time, but it's a job. And I did good. I had to go to school: I went to school and wound up working in a nursing home.

Here is how she describes her ordeal after the levee broke and she found herself stranded on the interstate highway with thousands of others:

> And you want me to tell you the truth, my version of it? They tried to kill us. When you keep somebody on top of the Interstate for five days, with no food and water, that's killing people. And there ain't no ands, ifs, or buts about it, that was NOPD [New Orleans Police Department] killing people. Four people died around me. Four. Diabetes. I am a diabetic and I survived it, by the grace of God, but I survived it. . . . Look, I was on top of the Interstate. Five days, okay? Helicopters at night shining a light down on us. They know we was there. Policemen, the army, the whole nine yards, ambulance passing us up like we wasn't nothing. Drove by and by all day. At night when they got ready to pull out, they pulled out and left us in darkness. We was treated worse than an animal. People do leave a dog in a house, but they do leave him food and water. They didn't do that . . .

Clarice goes on to recall the trauma of leaving her home:

> And of course I had to leave my birds and my dog. Of course I didn't want to. But I didn't have no other choice. Didn't have a choice. So I brought my dogs and my bird to as far as I could bring them. And I left them there upstairs. And I'm hoping I can retrieve them. I'm hoping. I have to call the SPCA [Society for the Prevention of Cruelty to Animals] or somebody. I left them upstairs on the deck, and I think if they was captured I should get them back. I'm hoping, anyway. I had a little Chihuahua. He was 9 months old. I had five birds. Two parakeets and two cockatiels. And my cockatiels just had a baby bird which was five weeks old. So you know I'm heartbroken. But again, my life was more important at that moment.

Finally, she wonders aloud;

> Now why our Mayor and government did this I'll never understand it. I never would understand what happened to New Orleans. That is really a disaster. Nobody would never believe it until you get into that

situation. I go to bed one night with everything that I needed, and wake up the next morning with nothing.[3]

There was another woman's story that made a powerful impression on me, and which I could not get out of my mind for weeks after I saw it on CNN. The scene was of a middle-aged Black woman, dirty, desperate, and crying. She looked into the camera and said to the viewers, "We do not live like this."[4] She repeated it over and over again. Contradicting the image of slovenly, hapless, poor folk, this woman's face reflected not simply fatigue and hunger, but humiliation as well. Most poor people spend a lot of time and attention making sure their homes and their children are as neat and clean as possible so that they will not be straightjacketed into the stereotypes associated with poverty. And here this seemingly hard-working woman was left with nothing, not even her dignity.

Whatever circumstances led to poor Black women's lives being battered and devastated by this storm, as Clarice's story so painfully recounts, the real unforgivable disaster is the fact that they were abandoned by those whose job it was to intervene and help in such situations. The local government was paralyzed, and the federal government looked the other way. Despite the tens of millions of dollars spent on the various apparatuses of the Office of Homeland Security (OHS), no one seemed to have spent much time worrying about the widely predicted hurricane that terrorized the Gulf Coast region or the security of those who were its victims. There was no plan to help those who could not help themselves, and even after the failure of FEMA and the OHS to aid and coordinate relief efforts, the president's silence, and the federal government's inaction for days after the crisis occurred left tough veteran journalists dumbfounded, angry, and sometimes even in tears. Initially, President Bush seemed not to take the crisis and human suffering seriously. Perhaps his mother spoke for the family when she visited displaced families forced to flee to Houston. During that visit she made the following disturbing comment: "so many of the people in the area here, you know, were underprivileged anyway, so this, this is working very well for them."[5] In her mind, those poor families really didn't need to have real homes or familiar communities; instead, like animals, they just needed basic shelter and food, no matter under what conditions.

On one level, many of us could not help but be surprised by the level of disregard for the collective well-being of New Orleans' Black poor, White poor, its elderly and infirm. However, when we zero in on the plight of Black women, again, the stage was set long before the scandalous treatment they received after Katrina. The dismantling of welfare for the poor in 1996, which climaxed with President Clinton's Personal Responsibility Act, was surrounded by a public discourse that dehumanized and denigrated the Black poor, charging them as the main culprits in their own misfortune. Black women were implicitly deemed lazy, promiscuous, and irresponsible; hence the withdrawal of public aid was ostensibly designed to jolt them into the labor force and into more responsible sexual behavior. Never mind that there were shrinking jobs for applicants with few skills and little education, and never mind that the president himself was breaking the very same sexual moral code that he was so mightily imposing on single mothers. Still, the problem was defined as that of Black women having babies out of the confines of heterosexual marriages, rather than the low pay and lack of jobs and affordable housing that marked their condition and compromised the future of their children.

Post-Katrina pundits continued the "blame the Black poor" rhetoric even as the blame clearly lay elsewhere. Six weeks after Katrina wreaked havoc on the Gulf Coast region, Mona Charen, columnist and former staffer for President Ronald Reagan, wrote: "Still it is true as the aftermath of Katrina underlined that parts of the black community remain poor and dysfunctional" (Charen 2005). The word *dysfunctional* is usually offered to modify *family,* and the association Charen is making is clear. The rest of her article goes on to make the case that the biggest problem facing the Black poor before and after Katrina is that of single-mother families. But Charen was just one of many making this argument. Conservative pundit Rich Lowry of the *National Review* argued that "If people are stripped of the most basic social support—the two parent family . . . they will be more vulnerable in countless ways" in times of crisis. He went on to propose government programs that "include greater attention to out of wedlock births" (Lowry 2005). Liberals such as *New York Times* columnist Nicholas Kristof even jumped on the bandwagon, giving a positive nod to Lowry's proposal (Kristof 2005).

BLACK CONSERVATIVES WEIGH IN

The attack on Black mothers did not stop with journalists such as Lowry. Black conservatives weighed in with a vengeance.

Syndicated columnist George Will (2005) was one of the most outrageous in his slander of the Black women of New Orleans. He first contended that there was too much obsession about race. In his words: "America's always fast-flowing river of race-obsessing has overflowed its banks," in discussions about Katrina. Those who are poor are poor because they don't follow the rules, Will insists, and those rules mean conforming to his code of sexual and social behavior. He offered "three not-at-all recondite rules for avoiding poverty: Graduate from high school, don't have a baby until you are married, don't marry while you are a teenager. Among people who obey those rules, poverty is minimal." If only things were so simple. Will drives the ill-conceived argument home, however, by making an even more explicit point:

> . . . it is a safe surmise that more than 80 percent of African American births in inner-city New Orleans—as in some other inner cities—were to women without husbands. That translates into a large and constantly renewed cohort of lightly parented adolescent males, and that translates into chaos in neighborhoods and schools, come rain or come shine.
> (Will 2005)

So, in Will's view, the chaos of Katrina was an extension of the self-inflicted chaos created by homes without strong father figures. These are the distorted realities that conservatives have to craft for themselves in order to sleep at night, I suppose. The reality on the ground is of course quite different, as the stories of Clarice B. and others illustrate so compellingly.

A powerful hurricane ravaged the lives of poor Black women and their families and neighbors, not because the women did not have wedding bands on their fingers, nor because their sons lacked strong paternal figures in the home to enforce curfew. To suggest as much is another way of devaluing the suffering and strivings of these families. Putting the issue in an international context, writer and activist Ritu Sharma, who works with the Washington-based advocacy group *Woman's Edge* (the Coalition for Women's Economic Development and Global Equity), writes that "women are the vast majority of the world's poor and money is the great protector" (Sharma 2006). Those who have little or none are more vulnerable than others to hurricanes, tsunamis, and all other forms of natural disasters that quickly escalate into human and social disasters.

WOMEN TAKE ACTION AFTER THE STORM

While they may have little else, poor Black women are creative and resilient. They have to be in order to survive in such difficult and challenging times. So, if one part of this story is what happened to African American women after Katrina hit, the other half of the story is how they responded. And if part one is depressing and disturbing, part two is uplifting and encouraging. African American women have responded to the crisis as individuals and in groups. One individual response was that of long-time New Orleans resident and organizer Diane "Momma D." Frenchcoat, an older resident who became a self-appointed relief worker after the storm. In the weeks following the hurricane, each day she would collect food, pile it in her cart, and navigate the flooded and filthy streets to deliver meals to hungry and isolated neighbors. She eventually recruited others to help in her efforts, dubbing them the "Soul Patrol." When asked by a newspaper reporter why she did not evacuate the city for safer ground, she replied: "Why would I leave now? Why would I leave my people when so many of them are still here and suffering?" The reporter described her in this way: "Graying dreadlocks flowed down the nape of her neck, spilling over her sturdy, sloping shoulders as she spoke of a city she hopes will be reborn" (Lee 2005).

Another inspiring story of determination against the odds is that of Beverly Wright, initiator of the volunteer-driven *A Safe Way Back Home* project. A professor of Environmental Studies at Dillard University and a lifelong New Orleans resident, Dr. Wright's project has educated New Orleans residents about the toxins still prevalent in the soil and in their homes. She was particularly concerned about the lawns of contaminated homes. With advice from the

Environmental Protection Agency (EPA) and donations from several founda-tions, *A Safe Way Back Home* has provided equipment, information, and protec-tive gear for dozens of residents to skim off toxic topsoil and replace it with healthy sod. The coalitions that Wright was able to forge were an interesting and important aspect of the project. Based on her past research and consulting for national labor unions about environmental dangers in the workplace, she was able to enlist the United Steelworkers Union to help train volunteers and to provide tools and equipment for the project. College students were recruited, and the National Black Environmental Justice Network, to which Wright belongs, lent its support and resources as well. A creative team effort of some unlikely allies is making a difference in the lives of dozens of families.

On the surface, a very masculine and muscular image of relief workers dom-inates popular images. Men are pictured lifting boxes, operating heavy equip-ment, and toting guns, ostensibly to keep the peace. However, women are working tirelessly and courageously in the trenches, as has so often been the case. Even within the larger coalitions and community-based organizations such as the People's Hurricane Relief Fund and Oversight Coalition (PHRFOC), women are important actors, leaders, and contributors. The PHRFOC even has a women's caucus to highlight and make visible the work of women, providing a forum where women can support one another within the larger effort. The work of Diane Frenchcoat, Beverly Wright, and the women of the PHRFOC are but three examples of African American women taking initiative, being imaginative, and acting boldly. These real stories stand in stark contrast to what the George Wills and Rich Lowrys of the world would have us believe.

A sober read of the situation in New Orleans . . . is still worrisome. Some biased and shortsighted city builders are trying to push Black women and chil-dren out of the picture altogether, to reconfigure a city without what must be perceived as the burden of the Black poor. PHRFOC and others have fought this scheme by demanding the right to return and by insisting upon voting rights for displaced citizens. A number of scholars and activists have referred to this period of rebuilding and remapping of this southern subregion as another "Reconstruction." At stake today, as they were in the years following the Civil War, are land rights, voting rights, control of the military, accountability, jobs, and the reconstitution of families and communities. Wherever New Orleans is headed in the future, hardworking Black women with big hearts and steel-willed determination will be a part of the picture. They have needs and problems to be sure, but their presence adds to rather than detracts from the strength and vitality of a remarkable American city.[6]

NOTES

1.   Statistics taken from DeWeever, Avis Jones (2005). *The Women of New Orleans and the Gulf Coast: Multiple Disadvantages and Key Assets for Recovery, Part I.* Published by the Institute for Women's Policy Research Washington, DC, October 11, 2005. www.iwpr. org/pdf/D464.pdf (accessed May 25, 2006).

2. A few of the projects that are attempting to document the stories of hurricane survivors include: the 1–10 Witness Project (www.i10witness.com), which emerged out of a group of artists, teachers, and activists; the Story Corps radio project; projects hosted by the Center for Cultural Resources in Baton Rouge (www.hurricanestories.org); and projects initiated by the University of Southwestern Mississippi and the Mississippi Humanities Council.

3. Oral history collected by AliveinTruth.org, and posted at www.Alternet.org on October 29, 2005.

4. Claudette Paul was also quoted in the *New York Times* as saying that "We need help. We don't live like this in America" (Appleborne et al., 2005).

5. Barbara Bush interviewed on Marketplace, National Public Radio (NPR), September 5, 2005.

6. Thanks to Joseph Lipari for his assistance with the research for this essay.

REFERENCES

Alive in Truth: The New Orleans Disaster Oral History and Memory Project.
www.aliveintruth.org/index.html (accessed May 25, 2006).

Appleborne, Peter, Christopher Drew, Jere Longman, and Andrew Revkin (2005). A Delicate Balance Is Undone in a Flash. *New York Times*, September 4, A25.

Bonavoglia, Angela (2005). Hurricane Pundits Blow Hot Air on Single Mothers. Women' Enews, September 14. (http://www.womensenews.org/article.cfm/dyn/aid/2449) (accessed May 25, 2006).

Charen, Mona (2005), No More Marches, Townhall.com, October 14. www.townhall.com (accessed May 25, 2006).

DeWeever, Avis Jones (2005). The Women of New Orleans and the Gulf Coast: Multiple Disadvantages and Key Assets for Recovery, Part I, October 11. Washington, DC: Institute for Women's Policy Research. www.iwpr.org/pdf/D464.pdf (accessed May 25, 2006).

Lee, Tymaine D. (2005). Momma's Mission. *Times-Picayune*, September 18, Metro Section, available at www.nola.com (accessed May 25, 2006).

Lowry, Rich (2005). The Coming Battle Over New Orleans. National Review Online, September 2. www.nationalreview.com (accessed September 4, 2005).

Peterson, Jesse Lee (2005a). Moral poverty costs Blacks in New Orleans. Worldnetdaily.com, September 21. (http://www.worldnetdaily.com) (accessed May 25, 2006).

Peterson, Jesse Lee (2005b). Truth: solution to Black America's moral poverty. Worldnetdaily.com, October 7. (http://www.worldnetdaily.com) (accessed May 25, 2006).

Rosen, Ruth (2005). Get Hitched, Young Woman. Tompaine.com, September 26. (http://www.tompaine.com) (accessed May 25, 2006).

Sharma, Ritu (2006). Disasters Dramatize How Women's Poverty Is Lethal. Women's eNews, January 5. www.womensenews.org (accessed May 25, 2006).

Will, George (2005). Poverty of Thought. *Washington Post*, September 14, A27.

5

Oppression

MARILYN FRYE

It is a fundamental claim of feminism that women are oppressed. The word "oppression" is a strong word. It repels and attracts. It is dangerous and dangerously fashionable and endangered. It is much misused, and sometimes not innocently.

The statement that women are oppressed is frequently met with the claim that men are oppressed too. We hear that oppressing is oppressive to those who oppress as well as to those they oppress. Some men cite as evidence of their oppression their much-advertised inability to cry. It is tough, we are told, to be masculine. When the stresses and frustrations of being a man are cited as evidence that oppressors are oppressed by their oppressing, the word "oppression" is being stretched to meaninglessness; it is treated as though its scope includes any and all human experience of limitation or suffering, no matter the cause, degree, or consequence. Once such usage has been put over on us, then if ever we deny that any person or group is oppressed, we seem to imply that we think they never suffer and have no feelings. We are accused of insensitivity; even of bigotry. For women, such accusation is particularly intimidating, since sensitivity is one of the few virtues that has been assigned to us. If we are found insensitive, we may fear we have no redeeming traits at all and perhaps are not real women. Thus are we silenced before we begin: the name of our situation drained of meaning and our guilt mechanisms tripped.

But this is nonsense. Human beings can be miserable without being oppressed, and it is perfectly consistent to deny that a person or group is oppressed without denying that they have feelings or that they suffer. . . .

The root of the word "oppression" is the element "press." *The press of the crowd; pressed into military service; to press a pair of pants; printing press; press the button.* Presses are used to mold things or flatten them or reduce them in bulk, sometimes to reduce them by squeezing out the gases or liquids in them. Something pressed is something caught between or among forces and barriers which are so related to each other that jointly they restrain, restrict or prevent the thing's motion or mobility. Mold. Immobilize. Reduce.

The mundane experience of the oppressed provides another clue. One of the most characteristic and ubiquitous features of the world as experienced by oppressed people is the double bind—situations in which options are reduced

SOURCE: From Marilyn Frye, *The Politics of Reality* (Trumansburg, NY: Crossing Press, 1983), pp. 1–16. Reprinted by permission.

to a very few and all of them expose one to penalty, censure, or deprivation. For example, it is often a requirement upon oppressed people that we smile and be cheerful. If we comply, we signal our docility and our acquiescence in our situation. We need not, then, be taken note of. We acquiesce in being made invisible, in our occupying no space. We participate in our own erasure. On the other hand, anything but the sunniest countenance exposes us to being perceived as mean, bitter, angry, or dangerous. This means, at the least, that we may be found "difficult" or unpleasant to work with, which is enough to cost one one's livelihood; at worst, being seen as mean, bitter, angry, or dangerous has been known to result in rape, arrest, beating, and murder. One can only choose to risk one's preferred form and rate of annihilation.

Another example: It is common in the United States that women, especially younger women, are in a bind where neither sexual activity nor sexual inactivity is all right. If she is heterosexually active, a woman is open to censure and punishment for being loose, unprincipled, or a whore. The "punishment" comes in the form of criticism, snide and embarrassing remarks, being treated as an easy lay by men, scorn from her more restrained female friends. She may have to lie and hide her behavior from her parents. She must juggle the risks of unwanted pregnancy and dangerous contraceptives. On the other hand, if she refrains from heterosexual activity, she is fairly constantly harassed by men who try to persuade her into it and pressure her to "relax" and "let her hair down"; she is threatened with labels like "frigid," "uptight," "man-hater," "bitch," and "cocktease." The same parents who would be disapproving of her sexual activity may be worried by her inactivity because it suggests she is not or will not be popular, or is not sexually normal. She may be charged with lesbianism. If a woman is raped, then if she has been heterosexually active she is subject to the presumption that she liked it (since her activity is presumed to show that she likes sex), and if she has not been heterosexually active, she is subject to the presumption that she liked it (since she is supposedly "repressed and frustrated"). Both heterosexual activity and heterosexual nonactivity are likely to be taken as proof that you wanted to be raped, and hence, of course, weren't *really* raped at all. You can't win. You are caught in a bind, caught between systematically related pressures.

Women are caught like this, too, by networks of forces and barriers that expose one to penalty, loss, or contempt whether one works outside the home or not, is on welfare or not, bears children or not, raises children or not, marries or not, stays married or not, is heterosexual, lesbian, both, or neither. Economic necessity; confinement to racial and/or sexual job ghettos; sexual harassment; sex discrimination; pressures of competing expectations and judgments about *women, wives,* and *mothers* (in the society at large, in racial and ethnic subcultures, and in one's own mind); dependence (full or partial) on husbands, parents, or the state; commitment to political ideas; loyalties to racial or ethnic or other "minority" groups; the demands of self-respect and responsibilities to others. Each of these factors exists in complex tension with every other, penalizing or prohibiting all of the apparently available options. And nipping at one's heels, always, is the endless pack of little things. If one dresses one way, one is subject to the assumption that one is advertising one's sexual availability; if one dresses another way, one appears to "not care about oneself" or to be "unfeminine." If one uses "strong language," one invites categorization as a whore or slut; if one does not, one invites categorization as a "lady"—one too delicately constituted to cope with robust speech or the realities to which it presumably refers.

The experience of oppressed people is that the living of one's life is confined and shaped by forces and barriers which are not accidental or occasional and hence avoidable, but are systematically related to each other in such a way as to catch one between and among them and restrict or penalize motion in any direction. It is the experience of being caged in: all avenues, in every direction, are blocked or booby trapped.

Cages. Consider a birdcage. If you look very closely at just one wire in the cage, you cannot see the other wires. If your conception of what is before you is determined by this myopic focus, you could look at that one wire, up and down the length of it, and be unable to see why a bird would not just fly around the wire any time it wanted to go somewhere. Furthermore, even if, one day at a time, you myopically inspected each wire, you still could not see why a bird would have trouble going past the wires to get anywhere. There is no physical property of any one wire, *nothing* that the closest scrutiny could discover, that will reveal how a bird could be inhibited or harmed by it except in the most accidental way. It is only when you step back, stop looking at the wires one by one, microscopically, and take a macroscopic view of the whole cage, that you can see why the bird does not go anywhere; and then you will see it in a moment. It will require no great subtlety of mental powers. It is perfectly *obvious* that the bird is surrounded by a network of systematically related barriers, no one of which would be the least hindrance to its flight, but which, by their relations to each other, are as confining as the solid walls of a dungeon.

It is now possible to grasp one of the reasons why oppression can be hard to see and recognize: one can study the elements of an oppressive structure with great care and some good will without seeing the structure as a whole, and hence without seeing or being able to understand that one is looking at a cage and that there are people there who are caged, whose motion and mobility are restricted, whose lives are shaped and reduced. . . .

As the cageness of the birdcage is a macroscopic phenomenon, the oppressiveness of the situations in which women live their various and different lives is a macroscopic phenomenon. Neither can be *seen* from a microscopic perspective. But when you look macroscopically you can see it—a network of forces and barriers which are systematically related and which conspire to the immobilization, reduction, and molding of women and the lives we live. . . .

6

Label Us Angry

JEREMIAH TORRES

It hurts to know that the most painful and shocking event of my life happened in part because of my race—something I can never change. On October 23, 1998, my friend and I experienced what would forever change our perceptions of our hometown and society in general.

We both attended elementary, middle, and high school in the quiet, prosperous, seemingly sophisticated college town of Palo Alto. In the third grade, we happily sang "It's a Small World," holding hands with the children of professors, graduate students, and professionals of the area, oblivious to our diversity in race, culture, or experience. Our small world grew larger as we progressed through the school system, each year learning more about what made us different from each other. But on that October evening, the world grew too large for us to handle.

Carlos and I were ready for a night out with the boys. It was his seventeenth birthday, and we were about to celebrate at the pool hall. I pulled out of the Safeway driveway as a speeding driver delivered a jolting honk. I followed him out, speeding to catch up with him, my immediate anger getting the better of me.

We lined up at the stoplight, and the passenger, a young white man dressed for the evening, rolled down his window; I followed. He looked irritated.

"He wasn't honking at you, you stupid fuck!"

His words slapped me across the face. I opened my stunned mouth, only to deliver an empty breath, so I gave him my middle finger until I could return some angry words. He grimaced and reached under his seat to pull out a bottle of mace, spraying it directly in my face, barely missing Carlos, who witnessed the bizarre scene in shock. It burned.

"Take that you fucking lowlifes! Stupid chinks!"

Carlos instinctively bolted out the door at those words. He started pounding the white guy without a second thought, with a new anger he had never known or felt before. Psssht! The white guy hit Carlos point blank in the face with the mace. He screamed; tires squealed; "fuck you's" were exchanged.

SOURCE: From Arar Hay and John Hsu, eds. *Asian American X: An Intersection of 21st Century Asian American Voices* (Ann Arbor, MI: University of Mechigan Press, 2004).

We spent the next ten minutes half-blind, clutching our eyes in the burning pain, cursing in raging anger that made us forget for moments the intense, throbbing fire on our faces. I crawled out of my car to follow Carlos's screams and curses, opening my eyes to the still, spectating traffic surrounding us. I stumbled to the sidewalk, where Carlos pounded the ground and recalled the words of the white guy. We needed water.

I stumbled further to a nearby house that had lights in the living room. I doorbelled frantically, but nobody answered. I appealed to the traffic for help. They just watched, forming a new route around my car to continue about their evening. The mucous membranes in our sinuses cut loose, and we spit every few seconds to sustain our gasping breaths. After nearly five minutes of appeals, a kind woman stopped to call the cops and give us water to quench the burning.

The cops came within minutes with advice for dealing with the mace. We tried to identify the car and the white guy who had sprayed us, and they sent out the obligatory all points bulletin. They questioned us soon after, asking if we were in a gang. I returned a blank stare with a silent "no." Apparently, two Filipino teenagers finding trouble on a Friday evening raised suspicions of a new Filipino gang in Palo Alto—yeah, all five of us.

I often ask myself if it would have been different had I been driving a BMW and dressed in an ironed polo shirt and slacks, like a typical Palo Alto kid. Maybe then the white guy would not have been afraid and called us lowlifes and chinks. I don't think so. He wasn't afraid of us; he initiated the curses and maced us from a safe distance. He reached out to hurt us because he was having a bad day and we looked different.

That night was our first encounter with overt racism that stems from a hatred of difference. We hadn't seen it through the smiles and happy songs of elementary school or the isolated cliques of middle and high school, but now we knew it was there. We hadn't seen it through the clean-cut, sophisticated facade of the Palo Alto white guy, but now we knew it was there. The "lowlife," "chink," and "gangster" labels made us different, marginalizing us from the town we called home.

Those labels made us angry, but we hesitated to project that anger. At first, we didn't tell anyone except our closest friends, afraid our parents would find out and react irrationally by locking us in our rooms to keep us away from trouble. But then we realized that the trouble had found us, and we decided to voice our anger.

We wrote an anonymous article in the school newspaper narrating the incident and the underlying racism that had come to surface. We noted that the incident wasn't purely racial, or a hate crime, but proof that racist tendencies still exist, even in open-minded suburban towns like Palo Alto. Parents, students, and teachers were shocked, maybe because they knew the truth in what we were saying. Many asked if it was Carlos and me who had been maced, but I responded, "Does it matter? What matters is that some people in this town still can't accept diversity. It's sad." We confronted the community with an issue previously reserved for hypothetical classroom discussions and brought it into the open. It was the least we could do to release our anger and expose its roots, hoping for a change in those who chose to label us.

After the article, Carlos and I took different routes. I continued with my studies, complying with my regimen of high school classes and activities as my anger subsided. I tried to lay the incident aside, having exposed it and promoted self-inspection and possible change in others through writing. Carlos remained angry. Why not? He got a face full of mace and racist labels for his seventeenth birthday. He alienated himself from the white majority and returned the mean gestures of the white guy to the yuppie congregation of Palo Alto. He became an outsider. Whenever someone would look at him funny, he would stare back, sometimes too harshly.

On the day after finals, he was making his way through the front parking lot of school when a parent looked at him funny. He stared back. The parent called him a punk. Carlos exploded. He cursed and gestured all he could at the father, and when he sped away in his Suburban, Carlos followed. Carlos couldn't keep up with the Suburban, so he took a quarter from his pocket and threw it at the back window, shattering it to pieces. Carlos ran away when the cops came to school.

Within two days, students had identified Carlos as the perpetrator, and he was suspended from school as the father called his lawyer, indicting Carlos of "assault with the intent to hurt." Weeks passed until a court hearing, and Carlos attended anger management counseling, but he was still angry—angry that he was being tried over throwing a quarter and that once again "the white guys were winning." His mother scraped up the little money she had to spare to afford him a lawyer for the trial, but there was no contesting the father's accusations. Carlos was sentenced to a night in juvenile hall and two hundred hours of community service over some angry words and throwing a quarter. He became a convicted felon.

He had learned once again that he couldn't win against the labels thrown at him, the labels that hurt him more than the mace or the night in juvy, and so he became more of an outsider. In both cases, the labels distanced us from the "normal" Palo Altans: white, clean-cut, wealthy. That division didn't always exist, however; it was created by the generalizations "normal" Palo Altans made through labels. To them, we looked like lowlifes, chinks, gangsters, and punks. In truth, we were two Filipino Americans headed toward Stanford and Berkeley, living in a town that swiftly disowned us with four reckless labels after raising us for ten years. Label us angry.

7

A Different Mirror

RONALD T. TAKAKI

I had flown from San Francisco to Norfolk and was riding in a taxi to my hotel to attend a conference on multiculturalism. Hundreds of educators from across the country were meeting to discuss the need for greater cultural diversity in the curriculum. My driver and I chatted about the weather and the tourists. The sky was cloudy, and Virginia Beach was twenty minutes away. The rearview mirror reflected a white man in his forties. "How long have you been in this country?" he asked. "All my life," I replied, wincing. "I was born in the United States." With a strong southern drawl, he remarked: "I was wondering because your English is excellent!" Then, as I had many times before, I explained: "My grandfather came here from Japan in the 1880s. My family has been here, in America, for over a hundred years." He glanced at me in the mirror. Somehow I did not look "American" to him; my eyes and complexion looked foreign.

Suddenly, we both became uncomfortably conscious of a racial divide separating us. An awkward silence turned my gaze from the mirror to the passing landscape, the shore where the English and the Powhatan Indians first encountered each other. Our highway was on land that Sir Walter Raleigh had renamed "Virginia" in honor of Elizabeth I, the Virgin Queen. In the English cultural appropriation of America, the indigenous peoples themselves would become outsiders in their native land. Here, at the eastern edge of the continent, I mused, was the site of the beginning of multicultural America. Jamestown, the English settlement founded in 1607, was nearby: the first twenty Africans were brought here a year before the Pilgrims arrived at Plymouth Rock. Several hundred miles offshore was Bermuda, the "Bermoothes" where William Shakespeare's Prospero had landed and met the native Caliban in *The Tempest*. Earlier, another voyager had made an Atlantic crossing and unexpectedly bumped into some islands to the south. Thinking he had reached Asia, Christopher Columbus mistakenly identified one of the islands as "Cipango" (Japan). In the wake of the admiral, many peoples would come to America from different shores, not only from Europe but also Africa and Asia. One of them would be my grandfather. My mental wandering across terrain and time ended abruptly as we arrived at my destination. I said good-bye to my driver and went into the hotel, carrying a vivid reminder of why I was attending this conference.

SOURCE: From Ronald T. Takaki, *A Different Mirror: A History of Multicultural America* (Boston: Little, Brown, 1993), pp. 1–17. Copyright © 1993 by Ronald T. Takaki. Reprinted by permission of the publisher.

Questions like the one my taxi driver asked me are always jarring, but I can understand why he could not see me as American. He had a narrow but widely shared sense of the past—a history that has viewed American as European in ancestry. "Race," Toni Morrison explained, has functioned as a "metaphor" necessary to the "construction of Americanness": in the creation of our national identity, "American" has been defined as "white."[1]

But America has been racially diverse since our very beginning on the Virginia shore, and this reality is increasingly becoming visible and ubiquitous. Currently, one-third of the American people do not trace their origins to Europe; in California, minorities are fast becoming a majority. They already predominate in major cities across the country—New York, Chicago, Atlanta, Detroit, Philadelphia, San Francisco, and Los Angeles.

This emerging demographic diversity has raised fundamental questions about America's identity and culture. In 1990, *Time* published a cover story on "America's Changing Colors." "Someday soon," the magazine announced, "white Americans will become a minority group." How soon? By 2056, most Americans will trace their descent to "Africa, Asia, the Hispanic world, the Pacific Islands, Arabia—almost anywhere but white Europe." This dramatic change in our nation's ethnic composition is altering the way we think about ourselves. "The deeper significance of America's becoming a majority nonwhite society is what it means to the national psyche, to individuals' sense of themselves and their nation—their idea of what it is to be American.". . .[2]

What is fueling the debate over our national identity and the content of our curriculum is America's intensifying racial crisis. The alarming signs and symptoms seem to be everywhere—the killing of Vincent Chin in Detroit, the black boycott of a Korean grocery store in Flatbush, the hysteria in Boston over the Carol Stuart murder, the battle between white sportsmen and Indians over tribal fishing rights in Wisconsin, the Jewish-black clashes in Brooklyn's Crown Heights, the black-Hispanic competition for jobs and educational resources in Dallas, which *Newsweek* described as "a conflict of the have-nots," and the Willie Horton campaign commercials, which widened the divide between the suburbs and the inner cities.[3]

This reality of racial tension rudely woke America like a fire bell in the night on April 29, 1992. Immediately after four Los Angeles police officers were found not guilty of brutality against Rodney King, rage exploded in Los Angeles. Race relations reached a new nadir. During the nightmarish rampage, scores of people were killed, over two thousand injured, twelve thousand arrested, and almost a billion dollars' worth of property destroyed. The live televised images mesmerized America. The rioting and the murderous melee on the streets resembled the fighting in Beirut and the West Bank. The thousands of fires burning out of control and the dark smoke filling the skies brought back images of the burning oil fields of Kuwait during Desert Storm. Entire sections of Los Angeles looked like a bombed city. "Is this America?" many shocked viewers asked. "Please, can we get along here," pleaded Rodney King, calling for calm. "We all can get along. I mean, we're all stuck here for a while. Let's try to work it out."[4]

But how should "we" be defined? Who are the people "stuck here" in America? One of the lessons of the Los Angeles explosion is the recognition of the fact that we are a multiracial society and that race can no longer be defined in the binary terms of white and black. "We" will have to include Hispanics and Asians. While blacks currently constitute 13 percent of the Los Angeles population, Hispanics represent 40 percent. The 1990 census revealed that South Central Los Angeles, which was predominantly black in 1965 when the Watts rebellion occurred, is now 45 percent Hispanic. A majority of the first 5,438 people arrested were Hispanic, while 37 percent were black. Of the fifty-eight people who died in the riot, more than a third were Hispanic, and about 40 percent of the businesses destroyed were Hispanic-owned. Most of the other shops and stores were Korean-owned. The dreams of many Korean immigrants went up in smoke during the riot: two thousand Korean-owned businesses were damaged or demolished, totaling about $400 million in losses. There is evidence indicating they were targeted. "After all," explained a black gang member, "we didn't burn our community, just *their* stores."[5]

"I don't feel like I'm in America anymore," said Denisse Bustamente as she watched the police protecting the firefighters. "I feel like I am far away." Indeed, Americans have been witnessing ethnic strife erupting around the world—the rise of neo-Nazism and the murder of Turks in Germany, the ugly "ethnic cleansing" in Bosnia, the terrible and bloody clashes between Muslims and Hindus in India. Is the situation here different, we have been nervously wondering, or do ethnic conflicts elsewhere represent a prologue for America? What is the nature of malevolence? Is there a deep, perhaps primordial, need for group identity rooted in hatred for the other? Is ethnic pluralism possible for America? But answers have been limited. Television reports have been little more than thirty-second sound bites. Newspaper articles have been mostly superficial descriptions of racial antagonisms and the current urban malaise. What is lacking is historical context; consequently, we are left feeling bewildered.[6]

How did we get to this point, Americans everywhere are anxiously asking. What does our diversity mean, and where is it leading us? *How* do we work it out in the post–Rodney King era?

Certainly one crucial way is for our society's various ethnic groups to develop a greater understanding of each other. For example, how can African Americans and Korean Americans work it out unless they learn about each other's cultures, histories, and also economic situations? This need to share knowledge about our ethnic diversity has acquired new importance and has given new urgency to the pursuit for a more accurate history. . . .

While all of America's many groups cannot be covered [here], the English immigrants and their descendants require attention, for they possessed inordinate power to define American culture and make public policy. What men like John Winthrop, Thomas Jefferson, and Andrew Jackson thought as well as did mattered greatly to all of us and was consequential for everyone. A broad range of groups [is important]: African Americans, Asian Americans, Chicanos, Irish, Jews, and Indians. While together they help to explain general patterns in our society, each has contributed to the making of the United States.

African Americans have been the central minority throughout our country's history. They were initially brought here on a slave ship in 1619. Actually, these first twenty Africans might not have been slaves; rather, like most of the white laborers, they were probably indentured servants. The transformation of Africans into slaves is the story of the "hidden" origins of slavery. How and when was it decided to institute a system of bonded black labor? What happened, while freighted with racial significance, was actually conditioned by class conflicts within white society. Once established, the "peculiar institution" would have consequences for centuries to come. During the nineteenth century, the political storm over slavery almost destroyed the nation. Since the Civil War and emancipation, race has continued to be largely defined in relation to African Americans—segregation, civil rights, the underclass, and affirmative action. Constituting the largest minority group in our society, they have been at the cutting edge of the Civil Rights Movement. Indeed, their struggle has been a constant reminder of America's moral vision as a country committed to the principle of liberty. Martin Luther King clearly understood this truth when he wrote from a jail cell: "We will reach the goal of freedom in Birmingham and all over the nation, because the goal of America is freedom. Abused and scorned though we may be, our destiny is tied up with America's destiny."[7]

Asian Americans have been here for over one hundred and fifty years, before many European immigrant groups. But as "strangers" coming from a "different shore," they have been stereotyped as "heathen," exotic, and unassimilable. Seeking "Gold Mountain," the Chinese arrived first, and what happened to them influenced the reception of the Japanese, Koreans, Filipinos, and Asian Indians as well as the Southeast Asian refugees like the Vietnamese and the Hmong. The 1882 Chinese Exclusion Act was the first law that prohibited the entry of immigrants on the basis of nationality. The Chinese condemned this restriction as racist and tyrannical. "They call us 'Chink,'" complained a Chinese immigrant, cursing the "white demons." "They think we no good! America cuts us off. No more come now, too bad!" This precedent later provided a basis for the restriction of European immigrant groups such as Italians, Russians, Poles, and Greeks. The Japanese painfully discovered that their accomplishments in America did not lead to acceptance, for during World War II, unlike Italian Americans and German Americans, they were placed in internment camps. Two-thirds of them were citizens by birth. "How could I as a 6-month-old child born in this country," asked Congressman Robert Matsui years later, "be declared by my own Government to be an enemy alien?" Today, Asian Americans represent the fastest-growing ethnic group. They have also become the focus of much mass media attention as "the Model Minority" not only for blacks and Chicanos, but also for whites on welfare and even middle-class whites experiencing economic difficulties.[8]

Chicanos represent the largest group among the Hispanic population, which is projected to outnumber African Americans. They have been in the United States for a long time, initially incorporated by the war against Mexico. The treaty had moved the border between the two countries, and the people of "occupied" Mexico suddenly found themselves "foreigners" in their "native land." As historian Albert Camarillo pointed out, the Chicano past is an integral part of America's westward expansion, also known as "manifest destiny." But while the early Chicanos were a colonized people, most of them today have immigrant roots. Many began the trek to El Norte in the early twentieth century. "As I had heard a lot about the United States," Jesus Garza recalled, "it was my dream to come here." "We came to know families from Chihuahua, Sonora, Jalisco, and Durango," stated Ernesto Galarza. "Like ourselves, our Mexican neighbors had come this far moving step by step, working and waiting, as if they were feeling their way up a ladder." Nevertheless, the Chicano experience has been unique, for most of them have lived close to their homeland—a proximity that has helped reinforce their language, identity, and culture. This migration to El Norte has continued to the present. Los Angeles has more people of Mexican origin than any other city in the world, except Mexico City. A mostly mestizo people of Indian as well as African and Spanish ancestries, Chicanos currently represent the largest minority group in the Southwest, where they have been visibly transforming culture and society.[9]

The Irish came here in greater numbers than most immigrant groups. Their history has been tied to America's past from the very beginning. Ireland represented the earliest English frontier: the conquest of Ireland occurred before the colonization of America, and the Irish were the first group that the English called "savages." In this context, the Irish past foreshadowed the Indian future. During the nineteenth century, the Irish, like the Chinese, were victims of British colonialism. While the Chinese fled from the ravages of the Opium Wars, the Irish were pushed from their homeland by "English tyranny." Here they became construction workers and factory operatives as well as the "maids" of America. Representing a Catholic group seeking to settle in a fiercely Protestant society, the Irish immigrants were targets of American nativist hostility. They were also what historian Lawrence J. McCaffrey called "the pioneers of the American urban ghetto," "previewing" experiences that would later be shared by the Italians, Poles, and other groups from southern and eastern Europe. Furthermore, they offer contrast to the immigrants from Asia. The Irish came about the same time as the Chinese, but they had a distinct advantage: the Naturalization Law of 1790 had reserved citizenship for "whites" only. Their compatible complexion allowed them to assimilate by blending into American society. In making their journey successfully into the mainstream, however, these immigrants from Erin pursued an Irish "ethnic" strategy: they promoted "Irish" solidarity in order to gain political power and also to dominate the skilled blue-collar occupations, often at the expense of the Chinese and blacks.[10]

Fleeing pogroms and religious persecution in Russia, the Jews were driven from what John Cuddihy described as the "Middle Ages into the Anglo-American world of the *goyim* 'beyond the pale.'" To them, America represented the Promised Land. This vision led Jews to struggle not only for themselves but also for other oppressed groups, especially blacks. After the 1917 East St. Louis race riot, the Yiddish *Forward* of New York compared this antiblack violence to a 1903 pogrom in Russia: "Kishinev and St. Louis—the same soil, the same people." Jews cheered when Jackie Robinson broke into the Brooklyn

Dodgers in 1947. "He was adopted as the surrogate hero by many of us growing up at the time," recalled Jack Greenberg of the NAACP Legal Defense Fund. "He was the way we saw ourselves triumphing against the forces of bigotry and ignorance." Jews stood shoulder to shoulder with blacks in the Civil Rights Movement: two-thirds of the white volunteers who went south during the 1964 Freedom Summer were Jewish. Today Jews are considered a highly successful "ethnic" group. How did they make such great socioeconomic strides? This question is often reframed by neoconservative intellectuals like Irving Kristol and Nathan Glazer to read: if Jewish immigrants were able to lift themselves from poverty into the mainstream through self-help and education without welfare and affirmative action, why can't blacks? But what this thinking overlooks is the unique history of Jewish immigrants, especially the initial advantages of many of them as literate and skilled. Moreover, it minimizes the virulence of racial prejudice rooted in American slavery.[11]

Indians represent a critical contrast, for theirs was not an immigrant experience. The Wampanoags were on the shore as the first English strangers arrived in what would be called "New England." The encounters between Indians and whites not only shaped the course of race relations, but also influenced the very culture and identity of the general society. The architect of Indian removal, President Andrew Jackson told Congress: "Our conduct toward these people is deeply interesting to the national character." Frederick Jackson Turner understood the meaning of this observation when he identified the frontier as our transforming crucible. At first, the European newcomers had to wear Indian moccasins and shout the war cry. "Little by little," as they subdued the wilderness, the pioneers became "a new product" that was "American." But Indians have had a different view of this entire process. "The white man," Luther Standing Bear of the Sioux explained, "does not understand the Indian for the reason that he does not understand America." Continuing to be "troubled with primitive fears," he has "in his consciousness the perils of this frontier continent. . . . The man from Europe is still a foreigner and an alien. And he still hates the man who questioned his path across the continent." Indians questioned what Jackson and Turner trumpeted as "progress." For them, the frontier had a different "significance": their history was how the West was lost. But their story has also been one of resistance. As Vine Deloria declared, "Custer died for your sins."[12]

By looking at these groups from a multicultural perspective, we can comparatively analyze their experiences in order to develop an understanding of their differences and similarities. Race, we will see, has been a social construction that has historically set apart racial minorities from European immigrant groups. Contrary to the notions of scholars like Nathan Glazer and Thomas Sowell, race in America has not been the same as ethnicity. A broad comparative focus also allows us to see how the varied experiences of different racial and ethnic groups occurred within shared contexts.

During the nineteenth century, for example, the Market Revolution employed Irish immigrant laborers in New England factories as it expanded cotton fields worked by enslaved blacks across Indian lands toward Mexico. Like blacks, the Irish newcomers were stereotyped as "savages," ruled by passions rather than "civilized" virtues such as self-control and hard work. The Irish saw themselves as the "slaves" of British oppressors, and during a visit to Ireland in the 1840s, Frederick Douglass found that the "wailing notes" of the Irish ballads reminded him of the "wild notes" of slave songs. The United States annexation of California, while incorporating Mexicans, led to trade with Asia and the migration of "strangers" from Pacific shores. In 1870, Chinese immigrant laborers were transported to Massachusetts as scabs to break an Irish immigrant strike; in response, the Irish recognized the need for interethnic working–class solidarity and tried to organize a Chinese lodge of the Knights of St. Crispin. After the Civil War, Mississippi planters recruited Chinese immigrants to discipline the newly freed blacks. During the debate over an immigration exclusion bill in 1882, a senator asked: If Indians could be located on reservations, why not the Chinese?[13]

Other instances of our connectedness abound. In 1903, Mexican and Japanese farm laborers went on strike together in California: their union officers had names like Yamaguchi and Lizarras, and strike meetings were conducted in Japanese and Spanish. The Mexican strikers declared that they were standing in solidarity with their "Japanese brothers" because the two groups had toiled together in the fields and were now fighting together for a fair wage. Speaking in impassioned Yiddish during the 1909 "uprising of twenty thousand" strikers in New York, the charismatic Clara Lemlich compared the abuse of Jewish female garment workers to the experience of blacks: "[The bosses] yell at the girls and 'call them down' even worse than I imagine the Negro slaves were in the South." During the 1920s, elite universities like Harvard worried about the increasing numbers of Jewish students, and new admissions criteria were instituted to curb their enrollment. Jewish students were scorned for their studiousness and criticized for their "clannishness." Recently, Asian American students have been the targets of similar complaints: they have been called "nerds" and told there are "too many" of them on campus.[14]

Indians were already here, while blacks were forcibly transported to America, and Mexicans were initially enclosed by America's expanding border. The other groups came here as immigrants: for them, America represented liminality—a new world where they could pursue extravagant urges and do things they had thought beyond their capabilities. Like the land itself, they found themselves "betwixt and between all fixed points of classification." No longer fastened as fiercely to their old countries, they felt a stirring to become new people in a society still being defined and formed.[15]

These immigrants made bold and dangerous crossings, pushed by political events and economic hardships in their homelands and pulled by America's demand for labor as well as by their own dreams for a better life. "By all means let me go to America," a young man in Japan begged his parents. He had calculated that in one year as a laborer here he could save almost a thousand yen—an amount equal to the income of a governor in Japan. "My dear Father," wrote an immigrant Irish girl living in New York, "Any man or woman without a family are fools that would not venture and come to this plentyful Country where no man or woman ever hungered." In the shtetls of Russia, the cry "To America!"

roared like "wildfire." "America was in everybody's mouth," a Jewish immigrant recalled. "Businessmen talked [about] it over their accounts; the market women made up their quarrels that they might discuss it from stall to stall; people who had relatives in the famous land went around reading their letters." Similarly, for Mexican immigrants crossing the border in the early twentieth century, El Norte became the stuff of overblown hopes. "If only you could see how nice the United States is," they said, "that is why the Mexicans are crazy about it."[16]

The signs of America's ethnic diversity can be discerned across the continent— Ellis Island, Angel Island, Chinatown, Harlem, South Boston, the Lower East Side, places with Spanish names like Los Angeles and San Antonio or Indian names like Massachusetts and Iowa. Much of what is familiar in America's cultural landscape actually has ethnic origins. The Bing cherry was developed by an early Chinese immigrant named Ah Bing. American Indians were cultivating corn, tomatoes, and tobacco long before the arrival of Columbus. The term *okay* was derived from the Choctaw word *oke*, meaning "it is so." There is evidence indicating that the name *Yankee* came from Indian terms for the English—from *eankke* in Cherokee and *Yankwis* in Delaware. Jazz and blues as well as rock and roll have African American origins. The "Forty-Niners" of the Gold Rush learned mining techniques from the Mexicans; American cowboys acquired herding skills from Mexican *vaqueros* and adopted their range terms—such as *lariat* from *la reata*, *lasso* from *lazo*, and *stampede* from *estampida*. Songs like "God Bless America," "Easter Parade," and "White Christmas" were written by a Russian-Jewish immigrant named Israel Baline, more popularly known as Irving Berlin.[17]

Furthermore, many diverse ethnic groups have contributed to the building of the American economy, forming what Walt Whitman saluted as "a vast, surging, hopeful army of workers." They worked in the South's cotton fields, New England's textile mills, Hawaii's canefields, New York's garment factories, California's orchards, Washington's salmon canneries, and Arizona's copper mines. They built the railroad, the great symbol of America's industrial triumph. . . .

Moreover, our diversity was tied to America's most serious crisis: the Civil War was fought over a racial issue—slavery. . . .

. . . The people in our study have been actors in history, not merely victims of discrimination and exploitation. They are entitled to be viewed as subjects—as men and women with minds, wills, and voices.

> *In the telling and retelling*
>   *of their stories,*
> *They create communities*
>   *of memory.*

They also re-vision history. "It is very natural that the history written by the victim," said a Mexican in 1874, "does not altogether chime with the story of the victor." Sometimes they are hesitant to speak, thinking they are only "little people." "I don't know why anybody wants to hear my history," an Irish maid said apologetically in 1900. "Nothing ever happened to me worth the tellin'."[18]

But their stories are worthy. Through their stories, the people who have lived America's history can help all of us, including my taxi driver, understand that Americans originated from many shores, and that all of us are entitled to dignity. "I hope this survey do a lot of good for Chinese people," an immigrant told an interviewer from Stanford University in the 1920s. "Make American people realize that Chinese people are humans. I think very few American people really know anything about Chinese." But the remembering is also for the sake of the children. "This story is dedicated to the descendants of Lazar and Goldie Glauberman," Jewish immigrant Minnie Miller wrote in her autobiography. "My history is bound up in their history and the generations that follow should know where they came from to know better who they are." Similarly, Tomo Shoji, an elderly Nisei woman, urged Asian Americans to learn more about their roots: "We got such good, fantastic stories to tell. All our stories are different." Seeking to know how they fit into America, many young people have become listeners; they are eager to learn about the hardships and humiliations experienced by their parents and grandparents. They want to hear their stories, unwilling to remain ignorant or ashamed of their identity and past.[19]

The telling of stories liberates. By writing about the people on Mango Street, Sandra Cisneros explained, "the ghost does not ache so much." The place no longer holds her with "both arms. She sets me free." Indeed, stories may not be as innocent or simple as they seem to be. Native-American novelist Leslie Marmon Silko cautioned:

> I will tell you something about stories . . .
> They aren't just entertainment.
>   Don't be fooled.

Indeed, the accounts given by the people in this study vibrantly re-create moments, capturing the complexities of human emotions and thoughts. They also provide the authenticity of experience. After she escaped from slavery, Harriet Jacobs wrote in her autobiography: "[My purpose] is not to tell you what I have heard but what I have seen—and what I have suffered." In their sharing of memory, the people in this study offer us an opportunity to see ourselves reflected in a mirror called history.[20]

In his recent study of Spain and the New World, *The Buried Mirror,* Carlos Fuentes points out that mirrors have been found in the tombs of ancient Mexico, placed there to guide the dead through the underworld. He also tells us about the legend of Quetzalcoatl, the Plumed Serpent: when this god was given a mirror by the Toltec deity Tezcatlipoca, he saw a man's face in the mirror and realized his own humanity. For us, the "mirror" of history can guide the living and also help us recognize who we have been and hence are. In *A Distant Mirror,* Barbara W. Tuchman finds "phenomenal parallels" between the "calamitous fourteenth century" of European society and our own era. We can, she observes, have "greater fellow-feeling for a distraught age" as we painfully recognize the "similar disarray," "collapsing assumptions," and "unusual discomfort."[21]

But what is needed in our own perplexing times is not so much a "distant" mirror, as one that is "different." While the study of the past can provide collective self-knowledge, it often reflects the scholar's particular perspective or view of the world. What happens when historians leave out many of America's peoples? What happens, to borrow the words of Adrienne Rich, "when someone with the authority of a teacher" describes our society, and "you are not in it"? Such an experience can be disorienting—"a moment of psychic disequilibrium, as if you looked into a mirror and saw nothing."[22]

Through their narratives about their lives and circumstances, the people of America's diverse groups are able to see themselves and each other in our common past. They celebrate what Ishmael Reed has described as a society "unique" in the world because "the world is here"—a place "where the cultures of the world crisscross." Much of America's past, they point out, has been riddled with racism. At the same time, these people offer hope, affirming the struggle for equality as a central theme in our country's history. At its conception, our nation was dedicated to the proposition of equality. What has given concreteness to this powerful national principle has been our coming together in the creation of a new society. "Stuck here" together, workers of different backgrounds have attempted to get along with each other.

> People harvesting
> Work together unaware
> Of racial problems,

wrote a Japanese immigrant describing a lesson learned by Mexican and Asian farm laborers in California.[23]

Finally, how do we see our prospects for "working out" America's racial crisis? Do we see it as through a glass darkly? Do the televised images of racial hatred and violence that riveted us in 1992 during the days of rage in Los Angeles frame a future of divisive race relations—what Arthur Schlesinger Jr. has fearfully denounced as the "disuniting of America"? Or will Americans of diverse races and ethnicities be able to connect themselves to a larger narrative? Whatever happens, we can be certain that much of our society's future will be influenced by which "mirror" we choose to see ourselves. America does not belong to one race or one group. . . . Americans have been constantly redefining their national identity from the moment of first contact on the Virginia shore. By sharing their stories, they invite us to see ourselves in a different mirror.[24]

NOTES

1. Toni Morrison, *Playing in the Dark: Whiteness in the Literary Imagination* (Cambridge, Mass., 1992), p. 47.

2. William A. Henry III, "Beyond the Melting Pot," in "America's Changing Colors," *Time,* vol. 135, no. 15 (April 9, 1990), pp. 28–31.

3. "A Conflict of the Have-Nots," *Newsweek,* December 12, 1988, pp. 28–29.

4. Rodney King's statement to the press, *New York Times,* May 2, 1992, p. 6.

5. Tim Rutten, "A New Kind of Riot," *New York Times Review of Books,* June 11, 1992, pp. 52–53; Maria Newman "Riots Bring Attention to Growing Hispanic Presence in South-Central Area," *New York Times,* May 11, 1992, p. A10; Mike Davis, "In L.A. Burning All Illusions," *The Nation,* June 1, 1992, pp. 744–745; Jack Viets and Peter Fimrite, "S.F. Mayor Visits Riot-Torn Area to Buoy Businesses," *San Francisco Chronicle,* May 6, 1992, p. A6.

6. Rick DelVecchio, Suzanne Espinosa, and Carl Nolte, "Bradley Ready to Lift Curfew," *San Francisco Chronicle,* May 4, 1992, p. A1.

7. Abraham Lincoln, "The Gettysburg Address," in *The Annals of America,* vol. 9, *1863–1865: The Crisis of the Union* (Chicago, 1968), pp. 462–463; Martin Luther King, *Why We Can't Wait* (New York, 1964), pp. 92–93.

8. Interview with old laundryman, in "Interviews with Two Chinese," circa 1924, Box 326, folder 325, Survey of Race Relations, Stanford University, Hoover Institution Archives; Congressman Robert Matsui, speech in the House of Representatives on the 442 bill for redress and reparations, September 17, 1987, *Congressional Record* (Washington, D.C., 1987), p. 7584.

9. Albert Camarillo, *Chicanos in a Changing Society: From Mexican Pueblos to American Barrios in Santa Barbara and Southern California, 1848–1930* (Cambridge, Mass., 1979), p. 2; Juan Nepornuceno Seguín, in David J. Weber (ed.), *Foreigners in Their Native Land: Historical Roots of the Mexican Americans* (Albuquerque, N. Mex., 1973), p. vi; Jesus Garza, in Manuel Garnio, *The Mexican Immigrant: His Life Story* (Chicago, 1931), p. 15; Ernesto Galarza, *Barrio Boy: The Story of a Boy's Acculturation* (Notre Dame, Ind., 1986), p. 200.

10. Lawrence J. McCaffrey, *The Irish Diaspora in America* (Washington, D.C., 1984), pp.6, 62.

11. John Murray Cuddihy, *The Ordeal of Civility: Freud, Marx, Levi Strauss, and the Jewish Struggle with Modernity* (Boston, 1987), p. 165; Jonathan Kaufman, *Broken Alliance: The Turbulent Times between Blacks and Jews in America* (New York, 1989), pp. 28, 82, 83–84, 91, 93, 106.

12. Andrew Jackson, First Annual Message to Congress, December 8, 1829, in James D. Richardson (ed.), *A Compilation of the Messages and Papers of the Presidents, 1789–1897* (Washington, D.C., 1897), vol. 2, p. 457; Frederick Jackson Turner, "The Significance of the Frontier in American History," in *The Early Writings of Frederick Jackson Turner* (Madison, Wis., 1938), pp. 185ff.; Luther Standing Bear, "What the Indian Means to America," in Wayne Moquin (ed.), *Great Documents in American Indian History* (New York, 1973), p. 307; Vine Deloria, Jr., *Custer Died for Your Sins: An Indian Manifesto* (New York, 1969).

13. Nathan Glazer, *Affirmative Discrimination: Ethnic Inequality and Public Policy* (New York, 1978); Thomas Sowell, *Ethnic America: A History* (New York, 1981); David R. Roediger, *The Wages of Whiteness: Race and the Making of the American Working Class* (London, 1991), pp. 134–136; Dan Caldwell, "The Negroization of the Chinese Stereotype in California," *Southern California Quarterly,* vol. 33 (June 1971), pp. 123–131.

14. Thomas Almaguer, "Racial Domination and Class Conflict in Capitalist Agriculture: The Oxnard Sugar Beet Workers' Strike of 1903," *Labor History,* vol. 25, no. 3 (summer 1984), p. 347; Howard M. Sachar, *A History of the Jews in America* (New York, 1992), p. 183.

15. For the concept of liminality, see Victor Turner, *Dramas, Fields, and Metaphors: Symbolic Action in Human Society* (Ithaca, N.Y., 1974), pp. 232, 237; and Arnold Van Gennep, *The Rites of Passage* (Chicago, 1960). What I try to do is to apply liminality to the land called America.

16. Kazuo Ito, *Issei: A History of Japanese Immigrants in North America* (Seattle, 1973), p. 33; Arnold Schrier, *Ireland and the American Emigration, 1850–1900* (New York, 1970), p. 24; Abraham Cahan, *The Rise of David Levinsky* (New York, 1960; originally published in 1917), pp. 59–61; Mary Antin, quoted in Howe, *World of Our Fathers* (New York, 1983), p. 27; Lawrence A. Cardoso, *Mexican Emigration to the United States, 1897–1931* (Tucson, Ariz., 1981), p. 80.

17. Ronald Takaki, *Strangers from a Different Shore: A History of Asian Americans* (Boston, 1989), pp. 88–89; Jack Weatherford, *Native Roots: How the Indians Enriched America* (New York, 1991), pp. 210, 212; Carey McWilliams, *North from Mexico: The Spanish-Speaking People of the United States* (New York, 1968), p. 154; Stephan Themstrom (ed.), *Harvard Encyclopedia of American Ethnic Groups* (Cambridge, Mass., 1980), p. 22; Sachar, *A History of the Jews in America,* p. 367.

18. Weber (ed.), *Foreigners in Their Native Land,* p. vi; Hamilton Holt (ed.), *The Life Stories of Undistinguished Americans as Told by Themselves* (New York, 1906), p. 143.

19. "Social Document of Pany Lowe, interviewed by C. H. Burnett, Seattle, July 5, 1924," p. 6, Survey of Race Relations, Stanford University, Hoover Institution Archives; Minnie Miller, "Autobiography," private manuscript, copy from Richard Balkin; Tomo Shoji, presentation, Obana Cultural Center, Oakland, California, March 4, 1988.

20. Sandra Cisneros, *The House on Mango Street* (New York, 1991), pp. 109–110; Leslie MarmonSilko, *Ceremony* (New York, 1978), p. 2; Harriet A. Jacobs, *Incidents in the Life of a Slave Girl, written by herself* (Cambridge, Mass., 1987; originally published in 1857), p. xiii.

21. Carlos Fuentes, *The Buried Mirror: Reflections on Spain and the New World* (Boston, 1992), pp. 10, 11, 109; Barbara W. Tuchman, *A Distant Mirror: The Calamitous 14th Century* (New York, 1978), pp. xiii, xiv.

22. Adrienne Rich, *Blood, Bread, and Poetry: Selected Prose, 1979–1985* (New York, 1986), p. 199.

23. Ishmael Reed, "America: The Multinational Society," in Rick Simonson and Scott Walker (eds.), *Multi-cultural Literacy* (St. Paul, 1988), p. 160; Ito, *Issei,* p. 497.

24. Arthur M. Schlesinger, Jr., *The Disuniting of America: Reflections on a Multicultural Society* (Knoxville, Tenn., 1991); Carlos Bulosan, *America Is in the Heart: A Personal History* (Seattle, 1981), pp. 188–189.

PART II

Systems of Power and Inequality

MARGARET L. ANDERSEN AND PATRICIA HILL COLLINS

One of the most important things to learn about race, class, and gender is that they are *systemic forms of inequality*. Although most people tend to think of them as individual characteristics (or identities), they are built into the very structure of society—and it is this social fact that drives our analysis of race, class, and gender as *intersectional systems of inequality*. This does not make them irrelevant as individual or group characteristics but points you to the analysis of *social structure* to think about how race, class, and gender operate, what they mean, and how they influence people's lives.

Locating racial oppression in the structure of social institutions provides a different frame of analysis from that which would be obtained by analyzing only individuals. Using a social structural analysis of race, class, and gender turns your attention to how they work as *systems of power*—systems that differentially advantage and disadvantage groups depending on their social location. Moreover, this means that no one of these social facts singularly predetermines where you will be situated within this system of power and social relationships. Thus, not all men are equally powerful and not all women are equally oppressed. When you focus on the intermingling structural relationships of race, class, *and* gender, you see a more complex, ever-changing, and multidimensional social order.

To repeat, neither race nor class nor gender operate alone. They do so within a system of simultaneous, interrelated social relationships—what we have earlier called the *matrix of domination*. This means that they also engage other social facts—ethnicity, sexuality, age, disability, even the region where you live, and so forth. In this section we examine race, class, and gender from an institutional or structural perspective. And, later in the section, we examine how race, class, and gender also intersect with ethnicity, nationality, and sexuality. You will see that each of these systems of power and inequality intertwine—and they do so in different ways at different points in time. Thus, now sexuality has become more visible as a system of social subordination, but beliefs about sexuality have long served to buttress the beliefs that support racial and gender subordination. Likewise, a system of racial subordination has historically been one of the ways that class structure was created: some accumulated property through the appropriation of other people's labor, even while groups who provided the labor that produced property for others were denied basic rights of citizenship, such as the right to vote, the right to marry, the right to own property, and the right to be considered a U.S. citizen.

Throughout Part II, you should keep the concept of *social structure* in mind. Remembering Marilyn Frye's analogy of the birdcage in Part I, be aware that race, class, and gender form a structure of social relations. This structure is supported by ideological beliefs that make things appear "normal" and "acceptable," thus clouding our awareness of how the structure operates. Thus, one of the prevailing beliefs about racism is that it is largely a thing of the past, now that formal barriers to racial discrimination have been removed. But, as Charles Gallagher ("Color-Blind Privilege: The Social and Political Functions of Erasing the Color Line in Post Race America") points out in his formulation of *color-blind racism*, racism persists, even when it takes on new forms.

Understanding the intersections among race, class, and gender—and how they interrelate with ethnicity and sexuality—requires knowing how to conceptualize each. Although we would rather not treat them separately, we do so initially here to learn what each means and how each is manifested in different group experiences. At the same time, the readings in this section examine the connections among race, class, gender, and other important social categories—namely, ethnicity and sexuality. As we review each in turn, you will also notice several common themes.

First, *each is a socially constructed category.* That is, their significance stems not from some "natural" state, but from what they have become as the result of social and historical processes. Second, notice how *each tends to construct groups in binary (or polar opposite) terms:* man/woman, Black/White, rich/poor, gay/straight, or citizen/alien. These binary constructions create the "otherness" that we examined in the first part of this book. Third, *each is a category of individual and group identity, but note—and this is important—they are also social structures.* That is, they are not just about identity but are about group location in a system of power and inequality. Thus, as we move into Part II, you will see how they are part of the institutional fabric of society. More than being about individual and group identity, race, class, and gender—and the other systems that they intersect with—shape patterns in the labor market, families, state institutions (such as the government and the law), mass media, and so forth. This is a key difference, as we have seen, in a model that focuses solely on difference and one that focuses on the matrix of domination. That is, the matrix of domination forces you to look at social structures, while models of difference often dissolve into individual or group identities. Finally, *neither race, class, nor gender is a fixed category.* Because they are social constructions, their form—and their interrelationship—changes over time. This also means that social change is possible.

As you learn about race, class, and gender, you should keep the intersectional model in mind. One way to think about their interrelationship in a social system is to imagine a typical college basketball game. This will probably seem familiar: the players on the court, the cheerleaders moving about on the side, the band playing, fans cheering, boosters watching from the best seats, and—if the team is ranked—perhaps a television crew. Everybody seems to have a place in the game. Everybody seems to be following the rules. But what explains the patterns that we see and don't see?

Race clearly matters. The predominance of young African American men on many college basketball teams is noticeable. Why do so many young Black men play basketball? Some people argue that African Americans are simply better in areas requiring physical skills such as sports, but there is another reality: for many young Black men, sports may seem the only hope for a good job, so sports, like the military, can seem like an attractive mobility route. Perceived promises of high salaries, endorsements, and merchandise can make young people believe sports are a path to success, thus, many young African American boys believe they can earn a good living playing professional sports. The odds of actually doing so are extremely slim. In truth, of the forty thousand African American boys playing high school basketball, only thirty-five will make it to the NBA (National Basketball Association) and only seven of those will be starters. This makes the odds of success 0.000175 (Eitzen 1999)! In the face of other systematic disadvantages posed by race, however, some will still think this is their best chance for success.

But does a racial analysis fully explain the "rules" of college basketball? Not really. Who benefits from college basketball? Yes, players (both Black and White, and, increasingly, immigrants) get scholarships and a chance to earn college degrees. Players reap the rewards, but who really benefits? College athletics is big business, and the players make far less from it than many people believe. As amateur athletes, they are forbidden to take any payment for their skills. They are offered the hope of an NBA contract when they turn pro, or at least a college degree if they graduate. But few actually turn pro. Indeed, many never graduate from college. Among Division I Black basketball players, 43 percent graduate from college, compared with 52 percent of White male basketball players. Hispanic and American Indian players (men) number so few that graduation rates are impossible to interpret. Interestingly, however, graduation rates among Black male college athletes (including all sports) are slightly higher than among nonathletes, most likely because of the scholarship support they receive. The reverse is true for Native American and Asian American students. Among women, student-athletes have much higher graduation rates than nonathletes as well as higher graduation rates than male student-athletes (National Collegiate Athletic Association 2007).

Winning teams also benefit educational institutions. Winning teams garner increased admissions applications, more alumni giving, higher levels of corporate support, and television revenues. Simply put, athletics is big business. Corporate sponsors want their names and products identified with winning teams and athletes; advertisers want their products promoted by winning players. Even though college athletes are forbidden to promote products, corporations create and market products in conjunction with prevailing excitement about basketball, sustained by the players' achievements. Athletic shoes, workout clothing, cars, and beer all target the consumer dollars of those who enjoy watching basketball. And how many jobs are supported by the revenues generated from the enterprise of college basketball? Of course, there are the sports reporters, team physicians, trainers, and coaches—most of them defined as professionals. But there are also the service workers—preparing and selling food, cleaning the toilets and stands, and maintaining the stadium. A class system defines one's place in this system of inequality, so much so that many of the service workers are invisible to the fans. So, while institutions profit the most—whether it be schools, product manufacturers, or advertisers—there are differential benefits depending on your "rank" within the system of basketball. And the stakes get even bigger when you move beyond college into the world of professional sports.

So, do race and class fully explain the "rules" of basketball? What about gender? Only in a few schools does women's basketball draw as large an audience as men's. And certainly in the media, men's basketball is generally the public's focal point, even though women's sports are increasingly popular. In college basketball, like the pros, most of the coaches and support personnel are men, as are the camera crew and announcers. Where are the women? A few are coaches, rarely paid what the men receive—even on the most winning teams. Those closest to the action on the court may be cheerleaders—tumbling, dancing, and being thrown into the air in support of the exploits of the athletes. Others may be in the band. Some women are in the stands, cheering the team—many of them accompanied by their husbands, partners, boyfriends, parents, and children. Many work in the concession stands, fulfilling women's roles of serving others. Still others are even more invisible, left to clean the restrooms, locker rooms, and stands after the crowd goes home.

Men's behavior reveals a gendered dimension to basketball, as well. Where else are men able to put their arms around one another, slap one another's buttocks, hug one another, or cry in public without having their "masculinity" questioned? Sportscasters, too, bring gender into the play of sports, such as when they talk about men's athletic achievements as heroic but talk about women athletes' looks or their connection to children. For that matter, look at the prominence given to men's teams in sports pages of the daily newspaper compared with sports news about women, who are typically relegated to the back pages—if their athletic accomplishments are reported at all. Sometimes what we don't see can be just as revealing as what we do see. Gender, as a feature of the game on the court, is so familiar that it may go unquestioned.

This discussion of college basketball demonstrates how race, class, and gender each provide an important, yet partial, perspective on social action. Using an intersectional model, we not only see each of them in turn but also the connections among them. In fact, race, class, and gender are so inextricably intertwined that gaining a comprehensive understanding of a basketball game requires thinking about all of them and how they work together. New questions then emerge: Why are most of those serving the food in concession stands likely to be women and men of color? How are norms of masculinity played out through sport? What class and racial ideologies are promoted through assuming that sports are a mobility route for those who try hard enough? If race, class, and gender relations are embedded in something as familiar and widespread as college basketball, to what extent are other social practices, institutions, relations, and social issues similarly structured?

Race, gender, and class divisions are deeply embedded in the structure of social institutions such as work, family, education, and the state. They shape human relationships, identities, social institutions, and the social issues that emerge from within institutions. Evelyn Nakano Glenn postulates that you can see the intersections of race, class, and gender in three realms of society: the representational realm, the social interaction realm, and the social structural realm. The representational realm includes the symbols, language, and images that convey racial meanings in society; the social interaction realm refers to the norms and behaviors observable in human relationships; and the social structural realm involves the institutional sites where power and resources are distributed in society (Glenn 2002: 12).

This means that race, class, and gender affect all levels of our experience—our consciousness and ideas, our interaction with others, and the social institutions we live within. And, because they are interconnected, no one can be subsumed under the other. In this section of the book, although we focus on each one to provide conceptual grounding, keep in mind that race, class, and gender are connected and overlapping—in all three realms of society: the realm of ideas, interaction, and institutions.

You might begin by considering a few facts:

- The United States is in the midst of a sizable redistribution of wealth, with a greater concentration of wealth and income in the hands of a few than at most previous periods of time. At the same time, a declining share of income is going to the middle class—a class that finds its position slipping, relative to years past (DeNavas-Walt, Proctor, and Smith 2007; Sullivan, Warren, and Westbrook 2001).

- Within class groups, racial group experiences are widely divergent. Thus, although there has been substantial growth of an African American and Latino middle class, they have a more tenuous hold on this class status than groups with more stable footing in the middle class. Furthermore, there is significant class differentiation within different racial groups (Charles 2006; Bean 2001; Pattillo 2007; Lacy 2007).

- Women in the top 25 percent of income groups have seen the highest wage growth of any group over the last twenty years; the lowest earning groups of women, like men, have seen wages fall, while the middle has remained flat (Mishel, Bernstein, and Allegretto 2005). Class differences within gender are hidden by thinking of women as a monolithic group.

- Women of color, including Latinas, African American women, Native American women, and Asian American women, are concentrated in the bottom rungs of the labor market along with recent immigrant women (www.bls.gov).

- Poverty has been steadily increasing in the United States since 2000; it is particularly severe among women, especially among women of color and their children, but in recent years, has risen most among White people (DeNavas-Walt et al. 2007).

- The shrinkage of social support is not only affecting the very poor, however. Job benefits in the form of health insurance, pensions, and so forth for all workers have declined. Following job loss, less than half of U.S. workers are currently eligible for unemployment insurance (Emsellem et al. 2002).

- At both ends of the economic spectrum, there is a growth of gated communities: well-guarded, locked neighborhoods for the rich and prisons for the poor—particularly Latinos and African American men. At the same time, growth in the rate of imprisonment is highest among women (Harrison and Beck 2006).

None of these facts can be explained through an analysis that focuses only on race or class or gender. Clearly, race matters. Class matters. Gender matters. And they matter together. We turn now to defining some of the basic concepts involving the systems of power and inequality that we examine in this section.

RACE AND RACISM

For many years, race and racism in the United States have been thought of in terms of Black/White relations. But changes in the racial landscape of the United States make such a framework now inadequate for thinking about all group experiences, a point made by Elizabeth Martinez (in "Seeing More than Black and White"). Martinez shows that Black/White relations have defined racism in the United States for centuries but that a rapidly changing population that includes diverse Latino groups is forcing Americans to reconsider the nature of racism. Color, she argues, has been the marker of race, but she challenges the dualistic thinking that has promoted this racist thinking. Moreover, other groups—Asian Americans, Native Americans, American Muslims, and others—are demonstrating the ethnic and racial diversity in the United States. This reality has resulted in shifting understandings of race and ethnic relations. Still, it is important to locate the study of race, racism, and ethnicity in the shifting relations of power that also characterize the treatment of different groups. This makes race and racism fundamentally about the character of U.S. social institutions.

Still, most people think about race and racism in the context of prejudice. How does prejudice differ from racism? *Prejudice* is a hostile attitude toward a person who is presumed to have negative characteristics associated with a group to which he or she belongs. Racism is more systematic than this and is not the same thing as prejudice. *Prejudice* refers to people's attitudes. *Racism* is a system of power and privilege; it can be manifested in people's attitudes but is rooted in society's structure. It is reflected in the different group advantages and disadvantages, based on their location in this societal system. Racism is structured into society, not just in people's minds. As such, it is built into the very fabric of dominant institutions in the United States and has been since the founding of the nation. Joe Feagin refers to this as *systemic racism,* meaning the "complex array of anti-Black practices, the unjustly gained political-economic power of Whites, the continuing economic and other resource inequalities along racial lines, and the White racial ideologies and attitudes created to maintain and rationalize White privilege and power" (2000: 6).

Thus, racism is part of society's structure, not just present in the minds of individual bigots, but racism also shapes everyday social relations. Patricia J. Williams, a noted African American legal scholar, illustrates this in her discussion of persistent discrimination in housing ("Of Race and Risk"). Despite her middle-class status, systemic racism confronts her—and other African Americans— in daily encounters. Practices of everyday racism are part of the edifice of institutional racism; yet, we often misread their meaning.

Understanding racism as systemic also means that people may not be individually racist but can still benefit from a system that is organized to benefit some at the expense of others, as Charles Gallagher points out in his essay on color-blind racism ("Color-Blind Privilege"). *Color-blind racism* is a new form of racism in which dominant groups assume that race no longer matters—even when society is highly racially segregated and when individual and group well-being is still strongly determined by race. Many people also believe that being nonracist means being color-blind—that is, refusing to recognize or treat as significant a person's racial background and identity. But to ignore the significance of race in a society where racial groups have distinct historical and contemporary experiences is to deny the reality of their group experience. Being color-blind in a society structured on racial privilege means assuming that everybody is "White," which is why people of color might be offended by friends who say, for example, "But I never think of you as Black." This blindness to the persistent realities of race also leads to an idea that there is nothing we should be doing about it—either individually or collectively—thus, racism is perpetuated. In "White Privilege: Unpacking the Invisible Knapsack," Peggy McIntosh describes how the system of racial privilege becomes invisible to those who benefit from it, even though it structures the everyday life of both White people and people of color.

Discrimination is one of the driving forces of racism. Though it is perhaps not as overt as, for example, during Jim Crow segregation in the South, discrimination can still be seen in various patterns and practices. Indeed, segregation, though not mandated by law as it was during Jim Crow, is as stark as ever. Segregation in schools has actually increased over recent years. And, in research known as *audit studies,* researchers, one White and one Black, are matched in credentials and appearance; they pose—in person—as job or housing applicants. Audit studies find significant levels of discrimination. White job applicants in such studies are offered the job almost half of the time; Black applicants, only 11 percent of the time. White applicants are often told things such as "You are just what we are looking for." Black applicants, on the other hand, are told they do not have the right attributes for the job. Studies of employers have also found that employers hold considerable stereotypes about Black workers that prevent them from hiring them (Moss and Tilly 2001).

Another dimension of racism is that its forms change over time. Racial discrimination is no longer legal, but racism nonetheless continues to structure relations among groups and to differentiate the power that different groups have. The changing character of racism is also evident in the fact that specific racial group histories differ, but different racial groups share common experiences of racial oppression. Thus, Chinese Americans were never enslaved, but they experienced forced residential segregation and economic exploitation based on their presumed racial characteristics. Mexican Americans were never placed on federal reservations, as Native Americans have been, but in some regards both groups share the experience of colonization by White settlers. Both have experienced having their lands appropriated by White people; Native Americans were removed from their lands and forced into reservations, if not killed. Chicanos originally held land in what is now the American Southwest, but it was taken following the Mexican American War; in 1848, Mexico ceded huge parts of what are now California, New Mexico, Nevada, Colorado, Arizona, and Utah to the United States for $15 million. Mexicans living there were one day Mexicans, the next living in the United States, though without all the rights of citizens.

Most people assume that race is biologically fixed, an assumption that is fueled by arguments about the presumed biological basis for different forms of inequality. The concept of race is more social than biological, however. Scientists working on the human genome project have even found that there is no "race" gene. But you should not conclude from this that race is not "real." It is just that its reality stems from its social significance. That is, the meaning and significance of *race* stems from specific social, historical, and political contexts. It is these contexts that make race meaningful, not just whatever physical differences may exist among groups.

To understand this, think about how racial categories are created, by whom, and for what purposes. Racial classification systems reflect prevailing views of race, thereby establishing groups that are presumed to be "natural." These constructed racial categories then serve as the basis for allocating resources; furthermore, once defined, the categories frame political issues and conflicts (Omi and Winant 1994). Omi and Winant define *racial formation* as "the sociohistorical process by which racial categories are created, inhabited, transformed, and destroyed" (1994: 55). In Nazi Germany, Jews were considered to be a race—a social construction that became the basis for the Holocaust. Abby L. Ferber's essay ("What White Supremacists Taught a Jewish Scholar about Identity") shows the complexities that evolve in the social construction of race. As someone who studies White supremacist groups, she sees how White racism defines her as Jewish, even while she lives in society as White. Her reflections reveal, too, the interconnections between racism and anti-Semitism (the hatred of Jewish people), reminding us of the interplay between different systems of oppression.

In understanding racial formation, we see that societies construct rules and practices that define groups in racial terms. Moreover, racial meanings constantly change as institutions evolve and as different groups contest prevailing racial definitions. Some groups are "racialized"; others are not. Where, for example, did the term *Caucasian* come from? Although many take it to be "real" and don't think about its racist connotations, the term has racist origins. It was developed in the late eighteenth century by a German anthropologist, Johann Blumenbach. He developed a racial classification scheme that put people from the Russian Caucases at the top of the racial hierarchy because he thought Caucasians were the most beautiful and sophisticated people; darker people were put on the bottom of the list: Asians, Africans, Polynesians, and Native Americans (Hannaford 1996). It is amazing when you think about it that this term remains with us, with few questioning its racist origin and connotations.

Consider also the changing definitions of *race* in the U.S. census. Given the large number of multiracial groups and the increasing diversity brought about by immigration, we can no longer think of race in mutually exclusive terms. In 1860, only three "races" were presumed to exist—Whites, Blacks, and mulattoes. By 1890, however, these original three races had been joined by five others—quadroon, octoroon, Chinese, Japanese, and Indian. Ten short years later, this list shrank to five races—White, Black, Chinese, Japanese, and Indian—a situation reflecting the growth of strict segregation in the South (Rodriquez 2000). Now people of mixed racial heritage present a challenge to census classifications. In the 2000 census, the U.S. government for the first time allowed people to check multiple boxes to identify themselves as more than one race. In addition, you could check "Hispanic" as a separate category. This change in the census reflects the growing number of multiracial people in the United States. The census categories are not just a matter of accurate statistics; they have significant consequences for the apportionment of societal resources. Thus, while some might argue that we should not "count" race at all, doing so is important because data on racial groups are used to enforce voting rights, to regulate equal employment opportunities, and to determine various governmental supports, among other things.

The overarching structure of racial power relations means that placement in this structure leads to differences in how people see racism—or don't—and what they are willing to do about it. Herbert J. Gans explores this idea in his essay, "Race as Class." He discusses how different groups are perceived within a racial hierarchy that stems from how social resources (both material and symbolic) are distributed in society. Some groups become less "racially marked" with upward class mobility, although, as Gans shows, this does not usually occur for African Americans—indicative of the strong hold that racial beliefs have on the general American public. As we will see, even with changes in attitudes and some optimism about racial progress, marked differences by race are still evident in employment, political representation, schooling, and other basic measures of group well-being.

CLASS

As Janny Scott and David Leonhardt show ("Shadowy Lines That Still Divide"), along with race, class is a major force in American society. Even with the belief that class mobility is possible and with such a strong consumer-based society, class continues to shape the life chances of Americans in very different ways. Moreover, class is not just about money; like race, it involves attitudes and behaviors. This is poignantly shown in the essay here by Barbara Jensen ("Across the Great Divide: Crossing Classes and Clashing Cultures") where she describes the experience of a working-class woman struggling to complete her education and, yet, feeling like an outsider in the college classroom and trying to live with the contradictions of her educational aspirations and working-class community expectations. Thus, class is both about material and symbolic dimensions of social life, but, like race, the social class system is grounded in social institutions and practices.

Rather than thinking of social class as a rank held by an individual, think of social class as a series of relations that pervade the entire society and shape our social institutions and relationships with one another. Although class shapes identity and individual well-being, class is a system that differentially structures group access to economic, political, cultural, and social resources. Within the United States, the class system evolves from patterns of capitalist development, and those patterns intersect with race and gender. To begin with, the class system in the United States is marked by striking differences in income. *Income* is the amount of money brought into a household in one year. Measures of income in the United States are based on annually reported census data drawn from a sample of the population. These data show quite dramatic differences in class standing when taking gender and race into account. *Median income* is the income level above and below which half of the population lies. It is the best measure of group income standing. Thus, in 2007, median income for non–Hispanic White households was $54,920 (meaning half of such households earned more than this and half below); this is the "middle." Black households had a median income of $33,916; Hispanic households, $38,679; Asian Americans, $66,103 (DeNavas-Walt et al. 2008; see also Figure 1).

But this tells only part of the story. Household income is the income of a total household. What about individual earners? This is where you can see the confounding influence of gender. Among full-time, year-round workers, men earned (in 2007) $46,224 and women, $36,167. When you also consider race, though, you see that income does not fall simply along lines of either race or gender (see Figure 2). White and Asian American women actually earn more than Black and, especially, Hispanic men, although there is a persistent gap *within* each racial-ethnic group between women and men (U.S. Census Bureau 2008). These data should

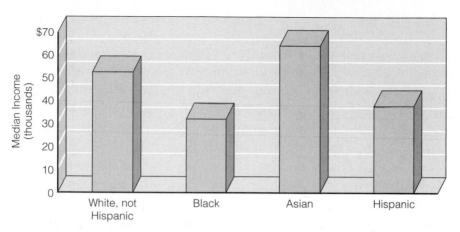

FIGURE 1 Median Household Income (2006)

SOURCE: DeNavas-Walt, Bernadette D. Proctor, and Jessica C. Smith. 2008. *Income, Poverty, and Health Insurance: Coverage in the United States: 2007*, Washington, DC: U.S. Census Bureau

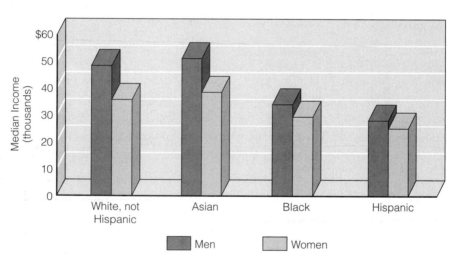

FIGURE 2 Median Income, Year-round, Full-time Workers

SOURCE: U.S. Census Bureau, 2008. *Detailed Historical Tables*. www.census.gov

lead you to be cautious about making general claims that women earn less than men—it matters *which women* and *which men* you are talking about.

Something to keep in mind is that because household income results from the income of individual workers, many households need more than one earner just to reach median levels of income. Some workers also have to hold multiple jobs to make ends meet. People who have more than one job typically do so to meet regular household expenses or to pay off debt. Although Whites (both men and women) are the group most likely to hold multiple jobs, Black and Hispanic jobholders are more likely to do so to meet basic household expenses (*Monthly Labor Review,* October 2000; www.bls.gov). In addition, workers at every class level have increased the number of hours they work (since 1980), although it is women and those at the lower ends of the class scale who have increased their working hours the most (Mishel et al. 2005).

Differences in patterns of wealth are even more revealing than differences in income when regarding how inequality is perpetuated in society. *Wealth* is determined by adding all of one's financial assets and subtracting all of one's debt. Income and wealth are related but are not the same thing. As important as income can be in determining one's class status, wealth is even more significant.

Consider this: Imagine two recent college graduates. They graduate in the same year, from the same college, with the same major and the identical grade point average. Both get jobs with the same salary in the same company. One student's parents paid all college expenses and gave her a car upon graduation. The other student worked while in school and has graduated with substantial debt from student loans. This student's family has no money to help support the new worker. Who is better off? Same salary, same credentials, but one person has a clear advantage—one that will be played out many times over as the young worker buys a home, finances her own children's education, and possibly inherits additional assets. This shows you the significance of wealth—not just income—in structuring social class.

Thus, income data indicate quite dramatic differences in race, class, and gender standing. Furthermore, wealth differences are also startling. The wealthiest 1 percent of the population control 33 percent of all wealth—the bottom 80 percent, only 16 percent (Mishel et al. 2005). For most Americans, debt, not wealth, is more common. Furthermore, 25 percent of White households, 61 percent of Black households, and 54 percent of Hispanic households have no financial assets at all (Oliver and Shapiro 2006), indicative of the vast differences in wealth holdings among different racial groups. In fact, the median net worth of White households is more than ten times that of African American and Latino households.

Wealth is especially significant because it provides a *cumulative* advantage to those who have it. Wealth helps pay for college costs for children and down payments on houses; it can cushion the impact of emergencies, such as unexpected unemployment or sudden health problems. Even small amounts of wealth can provide the cushion that averts economic disaster for families. Buying a home, investing, being free of debt, sending one's children to college, and transferring economic assets to the next generation are all instances of class advantage that add up over time and produce advantage, even beyond one's current income level. Sociologists Melvin Oliver and Thomas Shapiro (2006) have found, for example, that even Black and White Americans at the same income level, with the same educational and occupational assets, have a substantial difference in their financial assets—an average difference of $43,143 per year! This means that, even when earning the same income, the two groups are in quite different class situations—although both may be considered middle class. Furthermore, wealth produces more wealth because inheritance allows people to transmit economic status from one generation to the next. As Thomas M. Shapiro shows ("The Hidden Cost of Being African American"), these differences in wealth contribute to the perpetuation of racial inequality across generations.

When you study class in relationship to race, you will see that overall there are wide differences in the class status of Whites and people of color, but you should be careful not to see all Whites and Asian Americans as well off and all African Americans, Native Americans, and Latinos as poor. Consider the range of social class experiences just among Whites. Although on average White households possess higher accumulated wealth and have higher incomes than Black, Hispanic, and Native American households, large numbers of White households do not. White people also account for almost half (46 percent) of the nation's poor; and, as poverty has risen since the year 2000, it has risen most among White Americans (DeNavas-Walt et al. 2008). Of course, class experiences even *within* racial groups can vary widely, as we have seen an expansion of the Black and Latino middle class in recent years.

These facts should caution us about conclusions based on *aggregate data* (that is, data that represent whole groups). Such data give you a broad picture of group differences, but they are not attentive to the more nuanced picture you see when taking into account race, class, and gender (along with other factors, such as age, level of education, occupation, and so forth). Aggregate data on Asian Americans, for example, show them as a group to be relatively well off. This portrayal, like the stereotypes of a model minority, however, obscures significant differences both when making comparisons between Asian Americans and other groups and between Asian American groups. So, for example, although Asian American median income is, in the aggregate, higher than for White Americans, this does not mean all Asian American families are better off than White families. If you look at poverty rates, you get a different picture. Ten percent of Asian Americans are poor, compared with 8 percent of White, non-Hispanic families (DeNavas-Walt et al. 2008). Recent Asian immigrants have the highest rates of poverty, including the Hmong, Laotians, and Cambodians, whose rates of poverty match that of African Americans and Native Americans, about a 25 percent poverty rate (Le 2008).

As the U.S. public witnessed following the devastation of Hurricane Katrina, poverty is more widespread than many believe it to be, even though in recent years it has been increasing. Hurricane Katrina brought renewed attention to poverty and its location in the intersection of race and class, but poverty is also a matter of gender—evidenced by the fact that women and their children are especially hard hit by poverty. Thirty-nine percent of Black families and of Hispanic families headed by women are poor, as are 17 percent of Asian American and 21 percent of White families headed by women. Poverty rates among children (those under eighteen years of age) are especially disturbing: in 2006, 18 percent of all children in the United States lived below the *poverty line* ($24,366 for a family of four). When adding race, the figures are even more disturbing: 35 percent of African American children, 29 percent of Hispanic children, 13 percent of Asian/Pacific Islander children, and 19 percent of White (non-Hispanic) children are poor—astonishing figures for one of the most affluent nations in the world (DeNavas-Walt et al. 2008; see Figure 3).

Simplistic solutions suggested by current welfare policy imply that women would not be poor if they would just get married or get a job. But the root causes of poverty lie in the distribution of wealth and capital, coupled with low wages and high unemployment among certain groups—groups whose social location is the result of race, class, and gender stratification. Keep in mind that 16 percent of workers still end up being poor; even among those working year-round and full time, 8 percent are poor. And these figures count only those whose earnings fell below the official poverty line. If you calculate the income

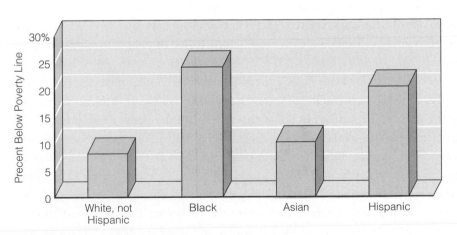

FIGURE 3 Percent Living in Poverty (2006)

SOURCE: DeNavas-Walt, Bernadette D. Proctor, and Jessica C. Smith. 2008. *Income, Poverty, and Health Insurance: Coverage in the United States: 2007*. Washington, DC: U.S. Census Bureau.

received by someone working full time (forty hours a week) and year-round (fifty-two weeks, no vacation) at the federal minimum wage ($6.55 per hour), you will see that the dollars earned ($13,624) do not even come close to the federal poverty line for a family of four ($24,366 in 2007).

In the United States, the social class system is also marked by differences in power. Social class is not just a matter of material difference; it is a pattern of domination in which some groups have more power than others. *Power* is the ability to influence and dominate others. This refers not just to interpersonal power but also to the structural power that some groups have because of their position in a racialized and gendered class system that is fundamental to the system of capitalism. This is well argued by Joan Acker in her article, "Is Capitalism Gendered and Racialized?" Her answer is yes. Under capitalism, groups with vast amounts of wealth also influence systems such as the media and the political process in ways that less-powerful groups cannot. Privilege in social class thus encompasses both a position of material advantage and the ability to control and influence others.

The class system is currently undergoing some profound changes, changes that are linked to patterns of economic transformation in the political economy—changes that are both global and domestic. Jobs are being exported overseas as vast multinational corporations seek to enhance their profits by promoting new markets and cutting the cost of labor. Within the United States, there is a shift from a manufacturing-based economy to a service economy, with corresponding changes in the types of jobs available and the wages attached to these jobs. Fewer skilled, decent-paying manufacturing jobs exist today than in the past. Fewer workers are covered by job benefits and unemployment insurance; only 40 percent of workers are now eligible for unemployment benefits following job loss (www.epi.org). Wages are flat for most workers, except those at the very top. Millions of people are left with jobs that do not pay enough, in part-time or temporary work, or without any work at all.

All told, class divisions in the United States are becoming more marked. There is a growing gap between the haves and have-nots. Income growth has been greatest for those at the top end of the population—both the upper 20 percent and the upper 5 percent of all income groups, regardless of race. For everyone else, income growth has remained flat. Although in every racial group, the top earners have seen the most growth in income, Black and Hispanic high-earners still earn less overall than Whites. At the same time, in the nation's cities and towns, homelessness has become increasingly apparent even to casual observers. Even before Hurricane Katrina displaced hundreds of thousands of families, the number of homeless in a given year was estimated to be about 2 million. Families are the largest segment of the homeless population. Half of the homeless are African American; about 39 percent are children (National Coalition for the Homeless 2008).

If social class is so important in shaping life chances, why don't more people realize its significance? The answer lies in how dominant groups use ideology to explain the class system and other systems of inequality. Ideology created by dominant groups refers to a system of beliefs that simultaneously distort reality and justify the status quo. The class system in the United States has been supported through the myth that we live in a classless society. This myth serves the dominant class, making class privilege seem like something that one earns, not something that is deeply embedded in the institutions of society. The system of privilege and inequality (by race, class, and gender) is least visible to those who are most privileged and who, in turn, control the resources to define the dominant cultural belief systems. Perhaps this is why the privileged, not the poor, are more likely to believe that one gets ahead through hard work. It also helps explain why men more than women deny that patriarchy exists and why Whites more than Blacks believe racism is disappearing.

Also, people do challenge the inequities they face. Sometimes this is through organized social action, as Robin D. G. Kelley shows in the essay "How the New Working Class Can Transform America." Kelley details some of the social action being taken by new coalitions of Latino, Black, Asian American, and some White workers. Kelley describes two organizations, the national Justice for Janitors unionization movement and the Bus Riders Union in Los Angeles. Kelley's article points to the multiple strategies used by these grassroots movements that draw upon traditional organizing principles but that also respond to the new challenges of urbanization.

GENDER

Gender, like race, is a social construction, not a biological imperative. *Gender* is rooted in social institutions and results in patterns within society that structure the relationships between women and men and that give them differing positions of advantage and disadvantage within institutions. As an identity, gender is learned; that is, through gender socialization, people construct definitions of themselves and others that are marked by gender. Like race, however, gender cannot be understood at the individual level alone. Gender is structured in social institutions, including work, families, mass media, and education.

You can see this if you think about the concept of a gendered institution. *Gendered institution* is now used to define the total patterns of gender relations that are "present in the processes, practices, images, and ideologies, and distribution of power in the various sectors of social life" (Acker 1992: 567). This term brings a much more structural analysis of gender to the forefront. Rather than seeing gender only as a matter of interpersonal relationships and learned identities, this framework focuses the analysis of gender on relations of power—just as thinking about institutional racism focuses on power relations and economic and political subordination—not just interpersonal relations. Changing gender relations is not just a matter of changing individuals. As with race and class, change requires transformation of institutional structures.

Gender, however, is not a monolithic category. Maxine Baca Zinn, Pierrette Hondagneu-Sotelo, and Michael Messner argue ("Gender Through the Prism of Difference") that, although gender is grounded in specific power relations, it is important to understand that gender is constructed differently depending on the specific social locations of diverse groups. Thus race, class, nationality, sexual orientation, and other factors produce varying social and economic consequences that cannot be understood by looking at gender differences alone. These authors ask us to move beyond studying differences and instead to use multiple "prisms" to see and comprehend the complexities of multiple systems of domination—each of which shapes and is shaped by gender. Seen in this way, men appear as a less monolithic and uni-dimensional group as well.

Race, class, gender, and sexuality *together* construct stereotypes. Each gains meaning in relationship to the others (Glenn 2002). Judith Ortiz Cofer ("The Myth of the Latin Woman: I Just Met a Girl Named María") shows how the combination of her gender identity with her status as a Latina results in quite hurtful stereotypes that she continuously encounters in everyday life. You cannot understand her experience as a woman without also locating her in the ethnic, racial, national, and migration experiences that are also part of her daily life. Thus, gender oppression is maintained through multiple systems—systems that are reflected in group stereotypes. These stereotypes also sexualize groups in different, but particular, ways.

One place where the stereotypes associated with gender are constantly reproduced is in the popular media. *Controlling images* of women (and men) in popular culture are both racialized and sexualized, especially (but not exclusively) for women of color. In the media, when they appear, Jewish American women, for example, are stereotyped as "JAPs" (Jewish American princesses), as if all are rich and privileged. This stereotype simultaneously promotes anti-Semitism and misogyny (defined as the hatred of women) at the same time that it uses an anti-Asian stereotype ("Jap") to denigrate others (Espiritu 2007). For other groups, race, gender, and sexual stereotypes intertwine in different ways. Thus, African American men are stereotyped as hypermasculine and oversexed, and African American women as promiscuous, bad mothers, and nurturing "mammies" who care for everyone else, but not their own children. Latinos are stereotyped as "macho" and, like African American men, sexually passionate, but out of control. Latinas are stereotyped as either "hot" or virgin-like. Similarly, White women are sexually stereotyped in dichotomous terms as "madonnas" or "whores." Working-class women are more likely to be seen as "sluts" and upper-class women as frigid and cold. We can see that controlling images of sexuality are part of the architecture of race, class, and gender oppression (Collins 2000, 2007), revealing the interlocking systems of race, class, gender, and sexuality.

Often, images in the media are unquestioned, unless examined with a critical eye. Rachel E. Dubrofsky ("*The Bachelor:* Whiteness in the Harem") does this well in her analysis of the popular television show *The Bachelor*. She equates the structure of this show with the harem: one man surrounded by highly sexualized, exotic women. Her essay shows not only the gender and sexual stereotypes that this show embeds, but also how racialized these images are, involving stereotypes of Middle Easterners, as well as implicitly denigrating African American and Latina women who systematically get eliminated from this sexualized competition. Perhaps having read her analysis, you will not see so-called "reality TV" in the same way again.

Gender ideologies abound in other ways, too. A common one is that women now have it made. The facts tell us otherwise. True, the gap between women's and men's income has closed, although most analysts agree that the narrowing of the gap reflects a drop in men's wages more than an increase in women's wages. And, women are more present in professional jobs—women who have become defined as the stereotypical "working woman." Despite this new image, though, most women remain concentrated in gender-segregated occupations with low wages, little opportunity for mobility, and stressful conditions. This is particularly true for women of color, who are more likely to be in occupations that are both race- and gender-segregated. And, as we have seen, among women heading their own households, poverty persists at alarmingly high rates. Income and occupational data, however, do not tell the full story for women. High rates of violence against women—whether in the home, on campus, in the workplace, or on the streets—indicate the continuing devaluation of and danger for women in this society.

Gender is also formed in the context of social institutions—family, workplaces, schools, religion, and so forth. Michael Messner shows (in "Masculinities and Athletic Careers") that sports are an important arena where gender identities and relations are formed. His study examines how boys from different racial and social class backgrounds learn about masculinity through their involvement in sports. Messner demonstrates that there is no one set of beliefs about masculinity to which all men subscribe; instead, sports shape understandings of masculinity and men's experiences in sports in race- and class-specific ways. Likewise, Julianne Malveaux ("Gladiators, Gazelles, and Groupies: Basketball Love and Loathing") captures how sports is a major vehicle not only for ideas about gender, but also about race. She discusses her own ambivalance about loving basketball even while knowing of the intersections of race, class, and gender in producing profit and entertainment for some and exploitation for others.

ETHNICITY AND MIGRATION

Sociologists traditionally define *ethnicity* as groups who share a common culture; however, like race, ethnicity develops within the context of systems of power. Thus, the meaning and significance of ethnicity can shift over time and in different social and political contexts. For example, groups may develop a sense of heightened ethnicity in the context of specific historical events; likewise, their feeling of sharing a common group identity can result from being labeled as "outsiders" by dominant groups. Think, for example, of the heightened sense of group identity that has developed for Arab Americans in the aftermath of 9/11.

Social and historical context also shapes who becomes defined as "ethnic." Asian Americans, as an example, have developed a *panethnic* identity—that is, an identity of common experiences, even across diverse ethnic groups. But this panethnic identity has arisen from social movements that defined different Asian American groups (Chinese, Japanese, Korean, and so forth) as having common social and political experiences within the United States (Espiritu 1992). Thus, even groups given a single group identity may have diverse ethnic and cultural experiences among them, as with African Americans who may be West Indian, Dominican, U.S. born, or from other cultural origins.

Patterns in the shifting meaning of ethnicity are being influenced by the changing contours of the U.S. population. Racial-ethnic groups comprise a larger proportion of the U.S. population, and their numbers are expected to rise over the years ahead. One-quarter of the U.S. population is now Black, Hispanic, Asian American, Pacific Islander, or Native American; by 2050, non-Hispanic Whites are predicted to make up only slightly more than half the total population. At the same time that the nation is becoming more diverse, changes are occurring regarding which groups predominate. Latinos have recently exceeded African Americans as the largest minority group in the population. Hispanic, Native American, Asian American, and African American populations are also growing more rapidly than White American populations, with the greatest growth among Hispanics and Asian Americans.

Immigration into the United States is only one facet of the increasing importance that ethnicity and migration play in contemporary social life. Changes in ethnicity and migration are being enacted on a world stage—a global context that increasingly links nations together in a global social order where diversity is more typical than not. You need only consult an older world atlas to get a sense of the widespread changes that characterize the current period. *World cities* have been formed that are linked through international systems of commerce; within these cities, migrants play an important role in providing the labor that is needed to keep pace with the global economy. Migrant (or "guest") workers may provide the labor for multinational corporations, or they may provide the service work that increasingly characterizes postindustrial society. Some workers may provide agricultural labor for multinational producers of food; others do domestic labor for middle- and upper-class families whose own lives are being transformed by changes in the world economic system. Wherever you look, the world is being changed by the increasingly global basis of modern life.

Such changes indicate that globalization is not just about what is happening elsewhere in the world—as if global studies were just a matter of comparing societies. Globalization is everywhere. No nation, including the United States, can really be understood without seeing how globalization is affecting life even at home. Patterns of life in any one society are now increasingly shaped by the connections between societies, and this is evident by looking at how ethnicity and migration are changing life in the United States. *Transnational families* are more common—that is, families whose members live and reside in different nations (often at a great distance from one another). Cultural features associated with different cultural traditions are increasingly evident, even in communities in the United States that have been thought of as "all-American" towns. Youth cultures embrace common features whether in Russia, Mexico, Japan, or the United States because of the penetration of world capitalist markets. Thus, hip-hop plays on the streets of Mexico City and salsa is heard on radio stations in the United States.

What do such changes mean for understanding ethnicity and migration in a transnational, global context? Barbara Ehrenreich and Arlie Russell Hochschild ("Global Woman") show how globalization results in increased migration—among many groups, although their focus is on women. Globalization also links the experiences of more privileged women in the West who are moving into professional jobs while utilizing the work of migrant women to do the domestic work still considered part of women's work. Patterns of family and work are thus disrupted around the world as globalization produces new forms of gendered migration and gendered family labor.

Within the borders of the United States, changes associated with migration are also raising new questions about national identity. How will the United States define itself as a nation-state in the changing global context? How will the increased visibility of people of color in the United States, in Latin America, in Africa, in Asia, and within the borders of former European colonial powers shape the future? How are cultural representations and institutions changed in a world context marked by such an ethnic mix—both within and between countries? In this context, who gets defined as a race, and how does that map onto concepts of citizenship and the division of labor in different nations? What will it mean to be American in a nation where racial-ethnic diversity is so much a part of the national fabric?

The events of September 11, 2001, give these questions additional importance. In response to terrorism, most in the United States were catapulted into a strong sense of national identity. Just as racial profiling was becoming increasingly criticized by the public, though, fears of terrorism have made it suddenly permissible to many. In the aftermath of 9/11, more restrictive immigration practices and policies have been employed; international and "ethnic" students are increasingly under surveillance; civil liberties have been restricted in the interests of national security. The events of 9/11 and the international tensions that have followed bring increasing urgency to our grappling with questions of how we can maintain a society of such great diversity within a framework of social justice.

Thus, several themes emerge in rethinking ethnicity and migration through the lens of race, class, and gender. Despite the ideology of the "melting pot," national identity in the United States has been closely linked to a history of racial privilege. As Lillian Rubin points out in "Is This a White Country, or What?" the term *American* is usually assumed to mean White. Other types of Americans, such as African Americans and Asian Americans, become distinguished from the "real" Americans by virtue of their race. More importantly, certain benefits are reserved for those deemed to be "deserving Americans."

Many Americans never "melted" into the melting pot. Many people think that race is like ethnicity and that the failure of people of color to assimilate into the mainstream as White ethnic groups purportedly have done represents an unwillingness to shed their culture. But as Mary C. Waters points out in "Optional Ethnicities: For Whites Only?" this view seriously misreads the meaning of both race and ethnicity in shaping American national identity. Waters suggests that White Americans of European ancestry have *symbolic ethnicity* in that their ethnic identity does not influence their lives unless they want it to. Waters contrasts this symbolic ethnic identity among many Whites with the socially enforced and imposed racial identity among African Americans. Because race operates as a physical marker in the United States, intersections of race and ethnicity operate differently for Whites and for people of color. White ethnics can thus have "ethnicity" without cost, but people of color pay the price for their ethnic identity.

Despite these structural forces, however, people are not just passive victims of these social forces. Nazli Kibria's research ("The Contested Meaning of 'Asian American': Racial Dilemmas in the Contemporary U.S.") shows how the very definition of what it means to be Asian American changes as people negoiate new identities in the context of racial, ethnic, and community boundaires. In the background are also the changes in immigration laws that stem from actions of the state. Kibria shows how people create and contest definitions of group identity—a process that can be full of contradictions in the complex context of immigration, racism, and class difference.

Within this context, various groups try to carve out spaces where they can safely express social identities and form relationships with peers. Such is the case for some Asian American youth, examined by Mary Yu Danico and Linda Trinh Võ ("No Lattés Here: Asian American Youth and the Cyber Café Obsession"). Like other groups who find themselves with limited social and community resources, for the Asian American young people in Danico and Võ's study, cybercafes become a place where they not only escape the "digital divide," but where they can form community with other immigrant youth and create a social and cultural outlet for themselves. You might ask yourself where you would find other such spaces among different groups.

SEXUALITY

The linkage between race, class, and gender is revealed within studies of sexuality, just as sexuality is a dimension of each. For example, constructing images about Black sexuality is central to maintaining institutional racism. Similarly, beliefs about women's sexuality structure gender oppression. Thus, sexuality operates as a system of power and inequality comparable to and intersecting with the systems of race, class, and gender.

The connection between race, class, gender, and sexuality is explored in the opening essay here by Patricia Hill Collins ("Prisons for Our Bodies, Closets for Our Minds: Racism, Heterosexism, and Black Sexuality"). Collins shows how historically racism has been buttressed by beliefs about Black sexuality. Sexuality has been used as the vehicle to support racial fears and racial subordination. Racial subordination was built on the exploitation of Black bodies—both as labor and as sexual objects. Strictures against certain interracial, sexual relationships (but not those between White men and African American women) are a way to maintain White patriarchy. Collins also shows how stigmatizing the sexuality of African Americans can in some ways be likened to the experiences of lesbian, gay, bisexual, and transgendered persons. This is not to say that racial and sexual oppression are the same, but it is to say that these systems of oppression operate together producing social beliefs and social actions that oppress and exploit African American men and women in particular ways, while also oppressing sexual minorites.

Underlying the analysis of sexuality here are the concepts of *homophobia* (the fear and hatred of homosexuality) and *heterosexism* (the institutionalized power and privilege accorded to heterosexual behavior and identification). If only heterosexual forms of gender identity are labeled "normal," then gays, lesbians, and bisexuals become ostracized, oppressed, and defined as "socially deviant." Homophobia affects heterosexuals as well because it is part of the gender ideology used to distinguish "normal" men and women from those deemed deviant. Thus, young boys learn a rigid view of masculinity—one often associated with violence, bullying, and degrading others—to avoid being perceived as a "fag." The oppression of lesbians and gay men is then linked to the structure of gender in everyone's lives.

The institutionalized structures and beliefs that define and enforce heterosexual behavior as the only natural and permissible form of sexual expression are what is meant by *heterosexism*. As Jonathan Ned Katz points out ("The Invention of Heterosexuality"), heterosexism is a specific historic construction; its meaning and presumed significance have evolved and changed at distinct historical times.

Understanding this rests on understanding sexuality, like race, class, gender, and ethnicity, as a social construction. Sexuality has a biological context because bodies make sex pleasurable, but the real significance of sexuality lies in its social dimensions. Thus, ideas about gender in society influence how we experience sexuality and how sexuality is controlled. Homophobia is used as a mechanism to enforce the construction of masculinity. The hatred directed toward lesbians, gays, and bisexuals is thus part of the system by which gender is created and maintained. In this regard, sexuality and gender are deeply linked.

Ariel Levy ("Get a Life, Girls") examines the link between gender and sexuality by observing contemporary sexual practices, such as in very public, highly sexualized displays by women. In her book, *Female Chauvinist Pigs,* Levy argues that during the 1960s and 1970s, the feminist and sexual liberation movements were allied, but that since then, sexual liberation has continued while the feminist revolution has stalled. To her, the result is that young women, in particular, take sexual freedom as a sign of their independence and freedom, even in the absence of a feminist critique of power and the sexual objectification of women. She is asking whether women can actually be sexually free in the presence of persistent inequities of race, class, and gender.

When you think about sexuality within an intersectional framework of race, class, and gender, many new questions and new perspectives come to light. One of these is examined by Chung-suk Han ("Darker Shades of Queer: Race and Sexuality at the Margins"). Han discusses the situation for gay men of color in a context where gay men are generally depicted as White (and stereotypically rich). His essay reveals the importance of remembering how the workings of race, class, gender, and sexuality can play out even within different more progressive communities.

Denise Brennan ("Selling Sex for Visas: Sex Tourism as a Stepping-stone to International Migration") reveals another dimension to the discussion of sexuality—sex as work. Sex workers sell their bodies or images of their bodies for money. Brennan emphasizes how sex workers are part of an international system of sex tourism, thus bringing a global perspective to the discussion of sexuality. Sex work on a global scale is also linked to world politics about race and class, given the specific position of women of color in international sex work. Women's sexuality is used to promote tourism and where images of the "exotic other" are used to attract more affluent classes to various regions of the world, thus linking sexuality to processes of migration and more global analyses of race, class, and gender.

REFERENCES

Acker, Joan. 1992. "Gendered Institutions: From Sex Roles to Gendered Institutions." *Contemporary Sociology* 21 (September): 565–569.

Bean, Frank D. 2001. *The Latino Middle Class: Myth, Reality, and Potential.* Los Angeles, CA: Tomas Rivera Policy Institute.

Charles, Camille Z. 2006. *Won't You Be My Neighbor?: Race, Class, and Residence in Los Angeles.* New York: Russell Sage Foundation.

Collins, Patricia Hill. 2007. *Black Sexual Politics: African Americans, Gender, and the New Racism.* New York: Routledge.

Collins, Patricia Hill. 2000. *Black Feminist Thought: Knowledge, Consciousness, and the Politics of Empowerment.* New York: Routledge.

DeNavas-Walt, Carmen, Bernadette D. Proctor, and Jessica C. Smith. 2008. *Income, Poverty, and Health Insurance: Coverage in the United States: 2007, P60-235.* Washington, DC: U.S. Census Bureau.

Eitzen, D. Stanley. 1999. *Fair and Foul: Beyond the Myths and Paradoxes of Sport.* Lanham, MD: Rowman and Littlefield.

Emsellem, Murray, Jessica Goldberg, Rick McHugh, Wendell Primus, Rebecca Smith, and Jeffrey Wenger. 2002. *Failing the Unemployed.* Washington, DC: Economic Policy Institute. www.epinet.org.

Espiritu, Yen Le. 2007. *Asian American Women and Men*, 2nd ed. Lanham, MD: Rowman and Littlefield.

Espiritu, Yen Le. 1992. *Asian American Panethnicity: Bridging Institutions and Identities.* Philadelphia, PA: Temple University Press.

Feagin, Joe R. 2000. *Racist America: Roots, Current Realities, and Future Reparations.* New York: Routledge.

Glenn, Evelyn Nakano. 2002. *Unequal Freedom: How Race and Gender Shaped American Citizenship and Labor.* Cambridge, MA: Harvard University Press.

Hannaford, Ivan. 1996. *Race: The History of an Idea in the West.* Baltimore: Johns Hopkins University Press.

Harrison, Paige M., and Allen J. Beck. 2006. *Prisoners in 2005.* Washington, DC: Bureau of Justice Statistics.

Lacy, Karyn R. 2007. *Blue-Chip Black: Race, Class, and Status in the New Black Middle Class.* Berkeley, CA: University of California Press.

Le, C.N. 2008. "Socioeconomic Statistics & Demographics." *Asian-Nation: The Landscape of Asian America.* www.asian-nation.org/demographics.html (April 11).

Mishel, Lawrence, Jared Bernstein, and Sylvia Allegretto. 2005. *The State of Working America 2004/2005.* Washington, DC: Economic Policy Institute.

Monthly Labor Review. October 2000. "At Issue: Reasons for Working Multiple Jobs." Vol. 123, No. 10.

Moss, Philip I., and Chris Tilly. 2001. *Stories Employers Tell: Race, Skill, and Hiring in America (Multi City Study of Urban Inequality).* New York: Russell Sage Foundation.

National Coalition for the Homeless. 2008. *Facts about Homelessness.* www.national homeless.org.

National Collegiate Athletic Association. 2007. *Graduation Rates Report for NCAA— Division I Schools.* Indianapolis, IN: NCAA. www.ncaa.org.

Oliver, Melvin L., and Thomas M. Shapiro. 2006. *Black Wealth/White Wealth: A New Perspective on Racial Inequality*, 2nd ed. New York: Routledge.

Omi, Michael, and Howard Winant. 1994. *Racial Formation in the United States: From the 1960s to the 1990s*, 2nd ed. New York: Routledge.

Pattillo, Mary. 2007. *Black on the Block: The Politics of Race and Class in the City.* Chicago, IL: University of Chicago Press.

Rodriguez, Clara. 2000. *Changing Race: Latinos, the Census, and the History of Ethnicity.* New York: New York University Press.

Sullivan, Teresa A., Elizabeth Warren, and Jay Westbrook. 2001. *The Fragile Middle Class: Americans in Debt.* New Haven, CT: Yale University Press.

U.S. Census Bureau. 2007. *Detailed Income Statistics.* Washington, DC: Department of Commerce.

8

Seeing More Than Black and White

ELIZABETH MARTINEZ

The racial and ethnic landscape has changed too much in recent years to view it with the same eyes as before. We are looking at a multi-dimensional reality in which race, ethnicity, nationality, culture and immigrant status come together with breathtakingly new results. We are also seeing global changes that have a massive impact on our domestic situation, especially the economy and labor force. For a group of Korean restaurant entrepreneurs to hire Mexican cooks to prepare Chinese dishes for mainly African-American customers, as happened in Houston, Texas, has ceased to be unusual.

The ever-changing demographic landscape compels those struggling against racism and for a transformed, non-capitalist society to resolve several strategic questions. Among them: doesn't the exclusively Black-white framework discourage the perception of common interests among people of color and thus sustain White Supremacy? Doesn't the view that only African Americans face serious institutionalized racism isolate them from potential allies? Doesn't the Black-white model encourage people of color to spend too much energy understanding our lives in relation to whiteness, obsessing about what white society will think and do?

That tendency is inevitable in some ways: the locus of power over our lives has long been white (although big shifts have recently taken place in the color of capital, as we see in Japan, Singapore and elsewhere). The oppressed have always survived by becoming experts on the oppressor's ways. But that can become a prison of sorts, a trap of compulsive vigilance. Let us liberate ourselves, then, from the tunnel vision of whiteness and behold the many colors around us! Let us summon the courage to reject outdated ideas and stretch our imaginations into the next century.

For a Latina to urge recognizing a variety of racist models is not, and should not be, yet another round in the Oppression Olympics. We don't need more competition among different social groups for the gold medal of "Most Oppressed." We don't need more comparisons of suffering between women and Blacks, the disabled and the gay, Latino teenagers and white seniors, or whatever. Pursuing some hierarchy of oppression leads us down dead-end streets where we will never find the linkage between different oppressions and how to

SOURCE: From Elizabeth Martinez, *De Colores Means All of Us: Latina Views for a Multi-Colored Century* (Cambridge, MA: South End Press, 1998). Reprinted by permission of the South End Press.

overcome them. To criticize the exclusively Black-white framework, then, is not some resentful demand by other people of color for equal sympathy, equal funding, equal clout, equal patronage or other questionable crumbs. Above all, it is not a devious way of minimizing the centrality of the African-American experience in any analysis of racism.

The goal in re-examining the Black-white framework is to find an effective strategy for vanquishing an evil that has expanded rather than diminished. Racism has expanded partly as a result of the worldwide economic recession that followed the end of the post-war boom in the early 1970s, with the resulting capitalist restructuring and changes in the international division of labor. Those developments generated feelings of insecurity and a search for scapegoats. In the United States racism has also escalated as whites increasingly fear becoming a weakened, minority population in the next century. The stage is set for decades of ever more vicious divide-and-conquer tactics.

What has been the response from people of color to this ugly White Supremacist agenda? Instead of uniting, based on common experience and needs, we have often closed our doors in a defensive, isolationist mode, each community on its own. A fire of fear and distrust begins to crackle, threatening to consume us all. Building solidarity among people of color is more necessary than ever—but the exclusively Black-white definition of racism makes such solidarity more difficult than ever.

We urgently need twenty-first century thinking that will move us beyond the Black-white framework without negating its historical role in the construction of U.S. racism. We need a better understanding of how racism developed both similarly and differently for various peoples, according to whether they experienced genocide, enslavement, colonization or some other structure of oppression. At stake is the building of a united anti-racist force strong enough to resist White Supremacist strategies of divide-and-conquer and move forward toward social justice for all. . . .

. . . African Americans have reason to be uneasy about where they, as a people, will find themselves politically, economically and socially with the rapid numerical growth of other folk of color. The issue is not just possible job loss, a real question that does need to be faced honestly. There is also a feeling that after centuries of fighting for simple recognition as human beings, Blacks will be shoved to the back of history again (like the back of the bus). Whether these fears are real or not, uneasiness exists and can lead to resentment when there's talk about a new model of race relations. So let me repeat: in speaking here of the need to move beyond the bipolar concept, the goal is to clear the way for stronger unity against White Supremacy. The goal is to identify our commonalities of experience and needs so we can build alliances.

The commonalities begin with history, which reveals that again and again peoples of color have had one experience in common: European colonization and/or neo-colonialism with its accompanying exploitation. This is true for all indigenous peoples, including Hawaiians. It is true for all Latino peoples, who were invaded and ruled by Spain or Portugal. It is true for people in Africa, Asia and the Pacific Islands, where European powers became the colonizers. People of color were victimized by colonialism not only externally but also through internalized racism—the "colonized mentality."

Flowing from this shared history are our contemporary commonalities. On the poverty scale, African Americans and Native Americans have always been at the bottom, with Latinos nearby. In 1995 the U.S. Census found that Latinos have the highest poverty rate, 24 percent. Segregation may have been legally abolished in the 1960s, but now the United States is rapidly moving toward resegregation as a result of whites moving to the suburbs. This leaves people of color—especially Blacks and Latinos—with inner cities that lack an adequate tax base and thus have inadequate schools. Not surprisingly, Blacks and Latinos finish college at a far lower rate than whites. In other words, the victims of U.S. social ills come in more than one color. Doesn't that indicate the need for new, inclusive models for fighting racism? Doesn't that speak to the absolutely urgent need for alliances among peoples of color?

With greater solidarity, justice for people of color could be won. And an even bigger prize would be possible: a U.S. society that advances beyond "equality," beyond granting people of color a respect equal to that given to Euro-Americans. Too often "equality" leaves whites still at the center, still embodying the Americanness by which others are judged, still defining the national character. . . .

. . . Innumerable statistics, reports and daily incidents should make it impossible to exclude Latinos and other non-Black populations of color when racism is discussed, but they don't. Police killings, hate crimes by racist individuals and murders with impunity by border officials should make it impossible, but they don't. With chilling regularity, ranch owners compel migrant workers, usually Mexican, to repay the cost of smuggling them into the United States by laboring the rest of their lives for free. The 45 Latino and Thai garment workers locked up in an El Monte, California, factory, working 18 hours a day seven days a week for $299 a month, can also be considered slaves (and one must ask why it took three years for the Immigration and Naturalization Service to act on its own reports about this horror) (*San Francisco Examiner,* August 8, 1995). Abusive treatment of migrant workers can be found all over the United States. In Jackson Hole, Wyoming, for example, police and federal agents rounded up 150 Latino workers in 1997, inked numbers on their arms and hauled them off to jail in patrol cars and a horse trailer full of manure (*Los Angeles Times,* September 6, 1997).

These experiences cannot be attributed to xenophobia, cultural prejudice or some other, less repellent term than racism. Take the case of two small Latino children in San Francisco who were found in 1997 covered from head to toe with flour. They explained they had hoped to make their skin white enough for school. There is no way to understand their action except as the result of fear in the racist climate that accompanied passage of Proposition 187, which denies schooling to the children of undocumented immigrants. Another example: Mexican and Chicana women working at a Nabisco plant in Oxnard, California, were not allowed to take bathroom breaks from the assembly line and were told to wear diapers instead. Can we really imagine white workers being treated that way? (The Nabisco women did file a suit and won, in 1997.)

No "model minority" myth protects Asians and Asian Americans from hate crimes, police brutality, immigrant-bashing, stereotyping and everyday racist prejudice. Scapegoating can even take their lives, as happened with the murder of Vincent Chin in Detroit some years ago. . . .

WHY THE BLACK-WHITE MODEL?

A bipolar model of racism has never been really accurate for the United States. Early in this nation's history, Benjamin Franklin perceived a tri-racial society based on skin color—"the lovely white" (Franklin's words), the Black, and the "tawny," as Ron Takaki tells us in *Iron Cages*. But this concept changed as capital's need for labor intensified in the new nation and came to focus on African slave labor. The "tawny" were decimated or forcibly exiled to distant areas; Mexicans were not yet available to be the main labor force. As enslaved Africans became the crucial labor force for the primitive accumulation of capital, they also served as the foundation for the very idea of whiteness—based on the concept of blackness as inferior.

Three other reasons for the Black-white framework seem obvious: numbers, geography and history. African Americans have long been the largest population of color in the United States; only recently has this begun to change. Also, African Americans have long been found in sizable numbers in most parts of the United States, including major cities, which has not been true of Latinos until recent times. Historically, the Black-white relationship has been entrenched in the nation's collective memory for some 300 years—whereas it is only 150 years since the United States seized half of Mexico and incorporated those lands and their peoples. Slavery and the struggle to end it formed a central theme in this country's only civil war—a prolonged, momentous conflict. Above all, enslaved Africans in the United States and African Americans have created an unmatched heritage of massive, persistent, dramatic and infinitely courageous resistance, with individual leaders of worldwide note.

We also find sociological and psychological explanations of the Black-white model's persistence. From the days of Jefferson onward, Native Americans, Mexicans and later the Asian/Pacific Islanders did not seem as much a threat to racial purity or as capable of arousing white sexual anxieties as did Blacks. A major reason for this must have been Anglo ambiguity about who could be called white. Most of the Mexican *ranchero* elite in California had welcomed the U.S. takeover, and Mexicans were partly European—therefore "semi-civilized"; this allowed Anglos to see them as white, unlike lower-class Mexicans. For years

Mexicans were legally white, and even today we hear the ambiguous U.S. Census term "Non–Hispanic Whites."

Like Latinos, Asian Americans have also been officially counted as white in some historical periods. They have been defined as "colored" in others, with "Chinese" being yet another category. Like Mexicans, they were often seen as not really white but not quite Black either. Such ambiguity tended to put Asian Americans along with Latinos outside the prevailing framework of racism.

Blacks, on the other hand, were not defined as white, could rarely become upper-class and maintained an almost constant rebelliousness. Contemporary Black rebellion has been urban: right in the Man's face, scary. Mexicans, by contrast, have lived primarily in rural areas until a few decades ago and "have no Mau-Mau image," as one Black friend said, even when protesting injustice energetically. Only the nineteenth-century resistance heroes labeled "bandits" stirred white fear, and that was along the border, a limited area. Latino stereotypes are mostly silly: snoozing next to a cactus, eating greasy food, always being late and disorganized, rolling big Carmen Miranda eyes, shrugging with self-deprecation "me no speek good eengleesh." In other words, *not serious*. This view may be altered today by stereotypes of the gangbanger, criminal or dirty immigrant, but the prevailing image of Latinos remains that of a debased white, at best. . . .

Among other important reasons for the exclusively Black–white model, sheer ignorance leaps to mind. The oppression and exploitation of Latinos (like Asians) have historical roots unknown to most Americans. People who learn at least a little about Black slavery remain totally ignorant about how the United States seized half of Mexico or how it has colonized Puerto Rico. . . .

One other important reason for the bipolar model of racism is the stubborn self-centeredness of U.S. political culture. It has meant that the nation lacks any global vision other than relations of domination. In particular, the United States refuses to see itself as one among some 20 countries in a hemisphere whose dominant languages are Spanish and Portuguese, not English. It has only a big yawn of contempt or at best indifference for the people, languages and issues of Latin America. It arrogantly took for itself alone the name of half the western hemisphere, America, as was its "Manifest Destiny," of course.

So Mexico may be nice for a vacation and lots of Yankees like tacos, but the political image of Latin America combines incompetence with absurdity, fat corrupt dictators with endless siestas. Similar attitudes extend to Latinos within the United States. My parents, both Spanish teachers, endured decades of being told that students were better off learning French or German. The mass media complain that "people can't relate to Hispanics (or Asians)." It takes mysterious masked rebels, a beautiful young murdered singer or salsa outselling ketchup for the Anglo world to take notice of Latinos. If there weren't a mushrooming, billion-dollar "Hispanic" market to be wooed, the Anglo world might still not know we exist. No wonder that racial paradigm sees only two poles.

The exclusively Black–white framework is also sustained by the "model minority" myth, because it distances Asian Americans from other victims of racism. Portraying Asian Americans as people who work hard, study hard, obey the established order and therefore prosper, the myth in effect admonishes Blacks and Latinos: "See, anyone can make it in this society if you try hard enough. The poverty and prejudice you face are all *your* fault."

The "model" label has been a wedge separating Asian Americans from others of color by denying their commonalities. It creates a sort of racial bourgeoisie, which White Supremacy uses to keep Asian Americans from joining forces with the poor, the homeless and criminalized youth. People then see Asian Americans as a special class of yuppie: young, single, college-educated, on the white-collar track—and they like to shop for fun. Here is a dandy minority group, ready to be used against others.

The stereotype of Asian Americans as whiz kids is also enraging because it hides so many harsh truths about the impoverishment, oppression and racist treatment they experience. Some do come from middle- or upper-class families in Asia, some do attain middle-class or higher status in the U.S., and their community must deal with the reality of class privilege where it exists. But the hidden truths include the poverty of many Asian/Pacific Islander groups, especially women, who often work under intolerable conditions, as in the sweatshops. . . .

THE DEVILS OF DUALISM

Yet another cause of the persistent Black-white conception of racism is dualism, the philosophy that sees all life as consisting of two irreducible elements. Those elements are usually oppositional, like good and evil, mind and body, civilized and savage. Dualism allowed the invaders, colonizers and enslavers of today's United States to rationalize their actions by stratifying supposed opposites along race, color or gender lines. So mind is European, male and rational; body is colored, female and emotional. Dozens of other such pairs can be found, with their clear implications of superior-inferior. In the arena of race, this society's dualism has long maintained that if a person is not totally white (whatever that can mean biologically), he or she must be considered Black. . . .

Racism evolves; our models must also evolve. Today's challenge is to move beyond the Black-white dualism that has served as the foundation of White Supremacy. In taking up this challenge, we have to proceed with both boldness and infinite care. Talking race in these United States is an intellectual minefield; for every observation, one can find three contradictions and four necessary qualifications from five different racial groups. Making your way through that complexity, you have to think: keep your eyes on the prize.

9

Of Race and Risk

PATRICIA J. WILLIAMS

Several years ago, at a moment when I was particularly tired of the unstable lifestyle that academic careers sometimes require, I surprised myself and bought a real house. Because the house was in a state other than the one where I was living at the time, I obtained my mortgage by telephone. I am a prudent little squirrel when it comes to things financial, always tucking away stores of nuts for the winter, and so I meet the criteria of a quite good credit risk. My loan was approved almost immediately.

A little while later, the contract came in the mail. Among the papers the bank forwarded were forms documenting compliance with the Fair Housing Act, which outlaws racial discrimination in the housing market. The act monitors lending practices to prevent banks from redlining—redlining being the phenomenon whereby banks circle certain neighborhoods on the map and refuse to lend in those areas. It is a practice for which the bank with which I was dealing, unbeknownst to me, had been cited previously—as well as since. In any event, the act tracks the race of all banking customers to prevent discrimination. Unfortunately, and with the creative variability of all illegality, some banks also use the racial information disclosed on the fair housing forms to engage in precisely the discrimination the law seeks to prevent.

I should repeat that to this point my entire mortgage transaction had been conducted by telephone. I should also note that I speak a Received Standard English, regionally marked as Northeastern perhaps, but not easily identifiable as black. With my credit history, my job as a law professor and, no doubt, with my accent, I am not only middle class but apparently match the cultural stereotype of a good white person. It is this, perhaps, that the loan officer of the bank, whom I had never met, had checked off the box on the fair housing form indicating that I *was* white.

Race shouldn't matter, I suppose, but it seemed to in this case, so I took a deep breath, crossed out "white" and sent the contract back. That will teach them to presume too much, I thought. A done deal, I assumed. But suddenly the transaction came to a screeching halt. The bank wanted more money, more points, a higher rate of interest. Suddenly I found myself facing great resistance and much more debt. To make a long story short, I threatened to sue under the act in question, the bank quickly backed down and I procured the loan on the original terms.

SOURCE: From *The Nation*, December 29, 1997. Reprinted with permission.

93

What was interesting about all this was that the reason the bank gave for its new-found recalcitrance was not race, heaven forbid. No, it was all about economics and increased risk: The reason they gave was that property values in that neighborhood were suddenly falling. They wanted more money to buffer themselves against the snappy winds of projected misfortune.

Initially, I was surprised, confused. The house was in a neighborhood that was extremely stable. I am an extremely careful shopper; I had uncovered absolutely nothing to indicate that prices were falling. It took my realtor to make me see the light. "Don't you get it," he sighed. "This is what always happens." And even though I suppose it was a little thick of me, I really hadn't gotten it: For of course, I was the reason the prices were in peril.

The bank's response was driven by demographic data that show that any time black people move into a neighborhood, whites are overwhelmingly likely to move out. In droves. In panic. In concert. Pulling every imaginable resource with them, from school funding to garbage collection to social workers who don't want to work in black neighborhoods. The imagery is awfully catchy, you had to admit: the neighborhood just tipping on over like a terrible accident, whoops! Like a pitcher, I suppose. All that nice fresh wholesome milk spilling out, running away...leaving the dark, echoing, upended urn of the inner city.

In retrospect, what has remained so fascinating to me about this experience was the way it so exemplified the problems of the new rhetoric of racism. For starts, the new rhetoric of race never mentions race. It wasn't race but risk with which the bank was so concerned.

Second, since financial risk is all about economics, my exclusion got reclassified as just a consideration of class. There's no law against class discrimination, goes the argument, because that would represent a restraint on that basic American freedom, the ability to contract or not. If schools, trains, buses, swimming pools and neighborhoods remain segregated, it's no longer a racial problem if someone who just happens to be white keeps hiking up the price for someone who accidentally and purely by the way happens to be black. Black people end up paying higher prices for the attempt to integrate, even as the integration of oneself threatens to lower the value of one's investment.

By this measure of mortgage-worthiness, the ingredient of blackness is cast not just as a social toll but as an actual tax. A fee, an extra contribution at the door, an admission charge for the high costs of handling my dangerous propensities, my inherently unsavory properties. I was not judged based on my independent attributes or financial worth; not even was I judged by statistical profiles of what my group actually does. (For in fact, anxiety-stricken, middle-class people make grovelingly good cake-baking neighbors when not made to feel defensive by the unfortunate historical strategies of bombs, burnings or abandonment.) Rather, I was being evaluated based on what an abstraction of White Society writ large thinks we—or I—do, and that imagined "doing" was treated and thus established as a self-fulfilling prophecy. It is a dispiriting message: that some in society apparently not only devalue black people but devalue *themselves* and their homes just for having us as part of their landscape.

"I bet you'll keep your mouth shut the next time they plug you into the computer as white," laughed a friend when he heard my story. It took me aback, this postmodern pressure to "pass," even as it highlighted the intolerable logic of it all. For by these "rational" economic measures, an investment in my property suggests the selling of myself.

10

Color-Blind Privilege

The Social and Political Functions of Erasing the Color Line in Post Race America

CHARLES A. GALLAGHER

The young white male sporting a FUBU (African-American owned apparel company "For Us By Us") shirt and his white friend with the tightly set, perfectly braided cornrows blended seamlessly into the festivities at an all white bar mitzvah celebration. A black model dressed in yachting attire peddles a New England, yuppie boating look in Nautica advertisements. It is quite unremarkable to observe white, Asian or African Americans with dyed purple, blond or red hair. White, black and Asian students decorate their bodies with tattoos of Chinese characters and symbols. In cities and suburbs young adults across the color line wear hip-hop clothing and listen to white rapper Eminem and black rapper 50-cent. It went almost unnoticed when a north Georgia branch of the NAACP installed a white biology professor as its president. Subversive musical talents like Jimi Hendrix, Bob Marley and The Who are now used to sell Apple Computers, designer shoes and SUVs. Du-Rag kits, complete with bandana headscarf and elastic headband, are on sale for $2.95 at hip-hop clothing stores and family centered theme parks like Six Flags. Salsa has replaced ketchup as the best selling condiment in the United States. Companies as diverse as Polo, McDonalds, Tommy Hilfiger, Walt Disney World, Master Card, Skechers sneakers, IBM, Giorgio Armani and Neosporin antibiotic ointment have each crafted advertisements that show an integrated, multiracial cast of characters interacting and consuming their products in [a] post-race, color-blind world.

SOURCE: From *Race, Gender & Class* 10/2003: 22–37. Reprinted by permission.

Americans are constantly bombarded by depictions of race relations in the media, which suggests that discriminatory racial barriers have been dismantled. Social and cultural indicators suggest that America is on the verge, or has already become, a truly color-blind nation. National polling data indicate that a majority of whites now believe discrimination against racial minorities no longer exists. A majority of whites believe that blacks have "as good a chance as whites" in procuring housing and employment or achieving middle class status while a 1995 survey of white adults found that a majority of whites (58%) believed that African Americans were "better off" finding jobs than whites (Gallup, 1997; Shipler, 1998). Much of white America now see[s] a level playing field, while a majority of black Americans sees a field [that] is still quite uneven. . . . The color-blind or race neutral perspective holds that in an environment where institutional racism and discrimination have been replaced by equal opportunity, one's qualifications, not one's color or ethnicity, should be the mechanism by which upward mobility is achieved. Color as a cultural style may be expressed and consumed through music, dress or vernacular but race as a system which confers privileges and shapes life chances is viewed as an atavistic and inaccurate accounting of U.S. race relations.

Not surprisingly, this view of society blind to color is not equally shared. Whites and blacks differ significantly, however, on their support for affirmative action, the perceived fairness of the criminal justice system, the ability to acquire the "American Dream," and the extent to which whites have benefited from past discrimination (Moore, 1995; Moore & Saad, 1995; Kaiser, 1995). This article examines the social and political functions colorblindness serves for whites in the United States. Drawing on interviews and focus groups with whites from around the country I argue that color-blind depictions of U.S. race relations serves to maintain white privilege by negating racial inequality. Embracing a color-blind perspective reinforces whites' belief that being white or black or brown has no bearing on an individual's or a group's relative place in the socio-economic hierarchy.

DATA AND METHOD

I use data from seventeen focus groups and thirty-individual interviews with whites from around the country. Thirteen of the seventeen focus groups were conducted in a college or university setting, five in a liberal arts college in the Rocky Mountains and the remaining eight at a large urban university in the Northeast. Respondents in these focus groups were selected randomly from the student population. Each focus group averaged six respondents. . . equally divided between males and females. An overwhelming majority of these respondents were between the ages of eighteen and twenty-two years of age. The remaining four focus groups took place in two rural counties in Georgia and were obtained through contacts from educational and social service providers in each county. One county was almost entirely white (99.54%) and in the other county whites constituted a racial minority. These four focus groups allowed me to tap rural attitudes about race relations in environments where whites had little or consistent contact with racial minorities. . . .

COLORBLINDNESS AS NORMATIVE IDEOLOGY

The perception among a majority of white Americans that the socio-economic playing field is now level, along with whites' belief that they have purged themselves of overt racist attitudes and behaviors, has made colorblindness the dominant lens through which whites understand contemporary race relations. Colorblindness allows whites to believe that segregation and discrimination are no longer [an] issue because it is now illegal for individuals to be denied access to housing, public accommodations or jobs because of their race. Indeed, lawsuits alleging institutional racism against companies like Texaco, Denny's, Coke, and Cracker Barrel validate what many whites know at a visceral level is true: firms which deviate from the color-blind norms embedded in classic liberalism will be punished. As a political ideology, the commodification and mass marketing of products that signify color but are intended for consumption across the color line further legitimate colorblindness. Almost every household in the United States has a television that, according to the U.S. Census, is on for seven hours every day (Nielsen 1997). Individuals from any racial background can wear hip-hop clothing, listen to rap music (both purchased at Wal-Mart) and root for their favorite, majority black, professional sports team. Within the context of racial symbols that are bought and sold in the market, colorblindness means that one's race has no bearing on who can purchase a Jaguar, live in an exclusive neighborhood, attend private schools or own a Rolex.

The passive interaction whites have with people of color through the media creates the impression that little, if any, socio-economic difference exists between the races. . . .

Highly visible and successful racial minorities like [former] Secretary of State Colin Powell and . . . [Secretary of State] Condelleeza Rice are further proof to white America that the state's efforts to enforce and promote racial equality have been accomplished.

The new color-blind ideology does not, however, ignore race; it acknowledges race while disregarding racial hierarchy by taking racially coded styles and products and reducing these symbols to commodities or experiences that whites and racial minorities can purchase and share. It is through such acts of shared consumption that race becomes nothing more than an innocuous cultural signifier. Large corporations have made American culture more homogenous through the ubiquitousness of fast food, television, and shopping malls but this trend has also created the illusion that we are all the same through consumption. Most adults eat at national fast food chains like McDonalds, shop at mall anchor stores like Sears and J.C. Penney's and watch major league sports, situation comedies or television drama. Defining race only as cultural symbols that are for sale allows whites to experience and view race as nothing more than a benign cultural marker that has been stripped of all forms of institutional, discriminatory or coercive power. The post-race, color-blind perspective allows whites to imagine that depictions of racial minorities working in high status jobs and consuming the same products, or at least appearing in commercials for products whites desire or consume, is the same as living in a society where color is no longer used to allocate resources or shape group outcomes. By constructing a picture of society where racial harmony is the norm, the color-blind perspective functions to make white privilege invisible while removing from public discussion the need to maintain any social programs that are race-based.

How then, is colorblindness linked to privilege? Starting with the deeply held belief that America is now a meritocracy, whites are able to imagine that the socio-economic success they enjoy relative to racial minorities is a function of individual hard work, determination, thrift and investments in education. The color-blind perspective removes from personal thought and public discussion any taint or suggestion of white supremacy or white guilt while legitimating the existing social, political and economic arrangements which privilege whites. This perspective insinuates that class and culture, and not institutional racism, are responsible for social inequality. Colorblindness allows whites to define themselves as politically and racially tolerant as they proclaim their adherence to a belief system that does not see or judge individuals by the "color of their skin." This perspective ignores, as Ruth Frankenberg puts it, how whiteness is a "location of structural advantage societies structured in racial dominance" (2001 p. 76).... Colorblindness hides white privilege behind a mask of assumed meritocracy while rendering invisible the institutional arrangements that perpetuate racial inequality. The veneer of equality implied in colorblindness allows whites to present their place in the racialized social structure as one that was earned.

OPPORTUNITY HAS NO COLOR

Given this norm of colorblindness it was not surprising that respondents in this study believed that using race to promote group interests was a form of (reverse) racism. . . .

Believing and acting as if America is now color-blind allows whites to imagine a society where institutional racism no longer exists and racial barriers to upward mobility have been removed. The use of group identity to challenge the existing racial order by making demands for the amelioration of racial inequities is viewed as racist because such claims violate the belief that we are a nation that recognizes the rights of individuals not rights demanded by groups. . . .

The logic inherent in the color-blind approach is circular; since race no longer shapes life chances in a color-blind world there is no need to take race into account when discussing differences in outcomes between racial groups. This approach erases America's racial hierarchy by implying that social, economic and political power and mobility is equally shared among all racial groups. Ignoring the extent or ways in which race shapes life chances validates whites' social location in the existing racial hierarchy while legitimating the political and economic arrangements that perpetuate and reproduce racial inequality and privilege. ⌐

REFERENCES

Frankenberg, R. (2001). The mirage of an unmarked whiteness. In B. B. Rasmussen, E. Klineberg, I. J. Nexica & M. Wray (eds.) *The making and unmaking of whiteness.* Durham: Duke University Press.

Gallup Organization. (1997). Black/white relations in the U.S. June 10, pp. 1–5.

Kaiser Foundation. (1995). *The four Americas: Government and social policy through the eyes of America's multi-racial and multi-ethnic society.* Menlo Park, CA: Kaiser Family Foundation.

Moore, D. (1995). "Americans' most important sources of information: Local news." *The Gallup Poll Monthly*, September, pp. 2–5.

Moore, D. & Saad, L. (1995). No immediate signs that Simpson trial intensified racial animosity. *The Gallup Poll Monthly*, October, pp. 2–5.

Nielsen, A. C. (1997). *Information please almanac.* Boston: Houghton Mifflin.

Shipler, D. (1998). *A country of strangers: Blacks and whites in America.* New York: Vintage Books.

11

White Privilege

Unpacking the Invisible Knapsack

PEGGY McINTOSH

Through work to bring materials from Women's Studies into the rest of the curriculum, I have often noticed men's unwillingness to grant that they are overprivileged, even though they may grant that women are disadvantaged. They may say they will work to improve women's status, in the society, the university, or the curriculum, but they can't or won't support the idea of lessening men's. Denials which amount to taboos surround the subject of advantages

SOURCE: Copyright © 1988 by Peggy McIntosh. Permission to duplicate must be obtained from the author. Excerpting is not authorized. A longer analysis and list of privileges, including heterosexual privilege, is available from Peggy McIntosh, Wellesley College, Center for Research on Women, Wellesley, MA, 02481-8203.Tel. (617) 283-2520; Fax (617) 283-2504.

which men gain from women's disadvantages. These denials protect male privilege from being fully acknowledged, lessened, or ended.

Thinking through unacknowledged male privilege as a phenomenon, I realized that since hierarchies in our society are interlocking, there was most likely a phenomenon of white privilege which was similarly denied and protected. As a white person, I realized I had been taught about racism as something which puts others at a disadvantage, but had been taught not to see one of its corollary aspects, white privilege, which puts me at an advantage.

I think whites are carefully taught not to recognize white privilege, as males are taught not to recognize male privilege. So I have begun in an untutored way to ask what it is like to have white privilege. I have come to see white privilege as an invisible package of unearned assets which I can count on cashing in each day, but about which I was "meant" to remain oblivious. White privilege is like an invisible weightless knapsack of special provisions, maps, passports, codebooks, visas, clothes, tools, and blank checks.

Describing white privilege makes one newly accountable. As we in Women's Studies work to reveal male privilege and ask men to give up some of their power, so one who writes about having white privilege must ask, "Having described it, what will I do to lessen or end it?"

After I realized the extent to which men work from a base of unacknowledged privilege, I understood that much of their oppressiveness was unconscious. Then I remembered the frequent charges from women of color that white women whom they encounter are oppressive. I began to understand why we are justly seen as oppressive, even when we don't see ourselves that way. I began to count the ways in which I enjoy unearned skin privilege and have been conditioned into oblivion about its existence.

My schooling gave me no training in seeing myself as an oppressor, as an unfairly advantaged person, or as a participant in a damaged culture. I was taught to see myself as an individual whose moral state depended on her individual moral will. My schooling followed the pattern my colleague Elizabeth Minnich has pointed out: whites are taught to think of their lives as morally neutral, normative, and average, and also ideal, so that when we work to benefit others, this is seen as work which will allow "them" to be more like "us."

I decided to try to work on myself at least by identifying some of the daily effects of white privilege in my life. I have chosen those conditions which I think in my case *attach somewhat more to skin-color privilege* than to class, religion, ethnic status, or geographical location, though of course all these other factors are intricately intertwined. As far as I can see, my African American co-workers, friends and acquaintances with whom I come into daily or frequent contact in this particular time, place, and line of work cannot count on most of these conditions.

1. I can if I wish arrange to be in the company of people of my race most of the time.

2. If I should need to move, I can be pretty sure of renting or purchasing housing in an area which I can afford and in which I would want to live.

3. I can be pretty sure that my neighbors in such a location will be neutral or pleasant to me.

4. I can go shopping alone most of the time, pretty well assured that I will not be followed or harassed.

5. I can turn on the television or open to the front page of the paper and see people of my race widely represented.

6. When I am told about our national heritage or about "civilization," I am shown that people of my color made it what it is.

7. I can be sure that my children will be given curricular materials that testify to the existence of their race.

8. If I want to, I can be pretty sure of finding a publisher for this piece on white privilege.

9. I can go into a music shop and count on finding the music of my race represented, into a supermarket and find the staple foods which fit with my cultural traditions, into a hairdresser's shop and find someone who can cut my hair.

10. Whether I use checks, credit cards, or cash, I can count on my skin color not to work against the appearance of financial reliability.

11. I can arrange to protect my children most of the time from people who might not like them.

12. I can swear, or dress in secondhand clothes, or not answer letters, without having people attribute these choices to the bad morals, the poverty, or the illiteracy of my race.

13. I can speak in public to a powerful male group without putting my race on trial.

14. I can do well in a challenging situation without being called a credit to my race.

15. I am never asked to speak for all the people of my racial group.

16. I can remain oblivious of the language and customs of persons of color who constitute the world's majority without feeling in my culture any penalty for such oblivion.

17. I can criticize our government and talk about how much I fear its policies and behavior without being seen as a cultural outsider.

18. I can be pretty sure that if I ask to talk to "the person in charge," I will be facing a person of my race.

19. If a traffic cop pulls me over or if the IRS audits my tax return, I can be sure I haven't been singled out because of my race.

20. I can easily buy posters, postcards, picture books, greeting cards, dolls, toys, and children's magazines featuring people of my race.

21. I can go home from most meetings of organizations I belong to feeling somewhat tied in, rather than isolated, out-of-place, outnumbered, unheard, held at a distance, or feared.

22. I can take a job with an affirmative action employer without having co-workers on the job suspect that I got it because of my race.

23. I can choose public accommodation without fearing that people of my race cannot get in or will be mistreated in the places I have chosen.

24. I can be sure that if I need legal or medical help, my race will not work against me.

25. If my day, week, or year is going badly, I need not ask of each negative episode or situation whether it has racial overtones.

26. I can choose blemish cover or bandages in "flesh" color and have them more or less match my skin.

I repeatedly forgot each of the realizations on this list until I wrote it down. For me white privilege has turned out to be an elusive and fugitive subject. The pressure to avoid it is great, for in facing it I must give up the myth of meritocracy. If these things are true, this is not such a free country; one's life is not what one makes it; many doors open for certain people through no virtues of their own.

In unpacking this invisible knapsack of white privilege, I have listed conditions of daily experience which I once took for granted. Nor did I think of any of these perquisites as bad for the holder. I now think that we need a more finely differentiated taxonomy of privilege, for some of these varieties are only what one would want for everyone in a just society, and others give license to be ignorant, oblivious, arrogant and destructive.

I see a pattern running through the matrix of white privilege, a pattern of assumptions which were passed on to me as a white person. There was one main piece of cultural turf; it was my own turf, and I was among those who could control the turf. *My skin color was an asset for any move I was educated to want to make.* I could think of myself as belonging in major ways, and of making social systems work for me. I could freely disparage, fear, neglect, or be oblivious to anything outside of the dominant cultural forms. Being of the main culture, I could also criticize it fairly freely.

In proportion as my racial group was being made confident, comfortable, and oblivious, other groups were likely being made inconfident, uncomfortable, and alienated. Whiteness protected me from many kinds of hostility, distress, and violence, which I was being subtly trained to visit in turn upon people of color.

For this reason, the word "privilege" now seems to me misleading. We usually think of privilege as being a favored state, whether earned or conferred by birth or luck. Yet some of the conditions I have described here work to systematically overempower certain groups. Such privilege simply *confers dominance* because of one's race or sex.

I want, then, to distinguish between earned strength and unearned power conferred systemically. Power from unearned privilege can look like strength when it is in fact permission to escape or to dominate. But not all of the privileges on my list are inevitably damaging. Some, like the expectation that neighbors will be decent to you, or that your race will not count against you in court, should be the norm in a just society. Others, like the privilege to ignore less powerful people, distort the humanity of the holders as well as the ignored groups.

We might at least start by distinguishing between positive advantages which we can work to spread, and negative types of advantages which unless rejected will always reinforce our present hierarchies. For example, the feeling that one belongs within the human circle, as Native Americans say, should not be seen as privilege for a few. Ideally it is an *unearned entitlement*. At present, since only a few have it, it is an *unearned advantage* for them. This paper results from a process of coming to see that some of the power which I originally saw as attendant on being a human being in the U.S. consisted in *unearned advantage* and *conferred dominance*.

I have met very few men who are truly distressed about systemic, unearned male advantage and conferred dominance. And so one question for me and others like me is whether we will be like them, or whether we will get truly distressed, even outraged, about unearned race advantage and conferred dominance and if so, what we will do to lessen them. In any case, we need to do more work in identifying how they actually affect our daily lives. Many, perhaps most, of our white students in the U.S. think that racism doesn't affect them because they are not people of color; they do not see "whiteness" as a racial identity. In addition, since race and sex are not the only advantaging systems at work, we need similarly to examine the daily experience of having age advantage, or ethnic advantage, or physical ability, or advantage related to nationality, religion, or sexual orientation.

Difficulties and dangers surrounding the task of finding parallels are many. Since racism, sexism, and heterosexism are not the same, the advantaging associated with them should not be seen as the same. In addition, it is hard to disentangle aspects of unearned advantage which rest more on social class, economic class, race, religion, sex and ethnic identity than on other factors. Still, all of the oppressions are interlocking, as the Combahee River Collective Statement of 1977 continues to remind us eloquently.

One factor seems clear about all of the interlocking oppressions. They take both active forms which we can see and embedded forms which as a member of the dominant group one is taught not to see. In my class and place, I did not see myself as a racist because I was taught to recognize racism only in individual acts of meanness by members of my group, never in invisible systems conferring unsought racial dominance on my group from birth.

Disapproving of the systems won't be enough to change them. I was taught to think that racism could end if white individuals changed their attitudes. [But] a "white" skin in the United States opens many doors for whites whether or not we approve of the way dominance has been conferred on us. Individual acts can palliate, but cannot end, these problems.

To redesign social systems we need first to acknowledge their colossal unseen dimensions. The silences and denials surrounding privilege are the key political tool here. They keep the thinking about equality or equity incomplete, protecting unearned advantage and conferred dominance by making these taboo subjects. Most talk by whites about equal opportunity seems to me now to be about equal opportunity to try to get into a position of dominance while denying that *systems* of dominance exist.

It seems to me that obliviousness about white advantage, like obliviousness about male advantage, is kept strongly inculturated in the United States so as to maintain the myth of meritocracy, the myth that democratic choice is equally available to all. Keeping most people unaware that freedom of confident action is there for just a small number of people props up those in power, and serves to keep power in the hands of the same groups that have most of it already.

Though systemic change takes many decades, there are pressing questions for me and I imagine for some others like me if we raise our daily consciousness on the perquisites of being light-skinned. What will we do with such knowledge? As we know from watching men, it is an open question whether we will choose to use unearned advantage to weaken hidden systems of advantage, and whether we will use any of our arbitrarily-awarded power to try to reconstruct power systems on a broader base.

12

What White Supremacists Taught a Jewish Scholar About Identity

ABBY L. FERBER

A few years ago, my work on white supremacy led me to the neo-Nazi tract *The New Order,* which proclaims: "The single serious enemy facing the white man is the Jew." I must have read that statement a dozen times. Until then, I hadn't thought of myself as the enemy.

When I began my research for a book on race, gender, and white supremacy, I could not understand why white supremacists so feared and hated Jews. But after being immersed in newsletters and periodicals for months, I learned that white supremacists imagine Jews as the masterminds behind a great plot to mix races and, thereby, to wipe the white race out of existence.

The identity of white supremacists, and the white racial purity they espouse, requires the maintenance of secure boundaries. For that reason, the literature I read described interracial sex as "the ultimate abomination." White supremacists

SOURCE: From *The Chronicle of Higher Education,* May 7, 1999, pp. B6–B7. Reprinted with permission of the author.

see Jews as threats to racial purity, the villains responsible for desegregation, integration, the civil-rights movement, the women's movement, and affirmative
action—each depicted as eventually leading white women into the beds of
black men. Jews are believed to be in control everywhere, staging a multi-
pronged attack against the white race. For *WAR,* the newsletter of White
Aryan Resistance, the Jew "promotes a thousand social ills . . . [f]or which you'll
have to foot the bills."

Reading white-supremacist literature is a profoundly disturbing experience,
and even more difficult if you are one of those targeted for elimination. Yet, as a
Jewish woman, I found my research to be unsettling in unexpected ways. I had
not imagined that it would involve so much self-reflection. I knew white supremacists were vehemently anti-Semitic, but I was ambivalent about my Jewish
identity and did not see it as essential to who I was. Having grown up in a
large Jewish community, and then having attended a college with a large
Jewish enrollment, my Jewishness was invisible to me—something I mostly
ignored. As I soon learned, to white supremacists, that is irrelevant.

Contemporary white supremacists define Jews as non-white: "not a religion,
they are an Asiatic *race,* locked in a mortal conflict with Aryan man," according
to *The New Order.* In fact, throughout white-supremacist tracts, Jews are
described not merely as a separate race, but as an impure race, the product of
mongrelization. Jews, who pose the ultimate threat to racial boundaries, are
themselves imagined as the product of mixed-race unions.

Although self-examination was not my goal when I began, my research
pushed me to explore the contradictions in my own racial identity.
Intellectually, I knew that the meaning of race was not rooted in biology or
genetics, but it was only through researching the white-supremacist movement
that I gained a more personal understanding of the social construction of race.
Reading white-supremacist literature, I moved between two worlds: one
where I was white, another where I was the non-white seed of Satan; one
where I was privileged, another where I was despised; one where I was safe
and secure, the other where I was feared and thus marked for death.

According to white-supremacist ideology, I am so dangerous that I must be
eliminated. Yet, when I put down the racist, anti-Semitic newsletters, leave my
office, and walk outdoors, I am white.

Growing up white has meant growing up privileged. Sure, I learned about
the historical persecutions of Jews, overheard the hushed references to distant
relatives lost in the Holocaust. I knew of my grandmother's experiences with
anti-Semitism as a child of the only Jewish family in a Catholic neighborhood.
But those were just stories to me. Reading white supremacists finally made the
history real.

While conducting my research, I was reminded of the first time I felt like an
"other." Arriving in the late 1980s for the first day of graduate school in the
Pacific Northwest, I was greeted by a senior graduate student with the welcome:
"Oh, you're the Jewish one." It was a jarring remark, for it immediately set me

apart. This must have been how my mother felt, I thought, when, a generation earlier, a college classmate had asked to see her horns. Having lived in predominantly Jewish communities, I had never experienced my Jewishness as "otherness." In fact, I did not even *feel* Jewish. Since moving out of my parents' home, I had not celebrated a Jewish holiday or set foot in a synagogue. So it felt particularly odd to be identified by this stranger as a Jew. At the time, I did not feel that the designation described who I was in any meaningful sense.

But whether or not I define myself as Jewish, I am constantly defined by others that way. Jewishness is not simply a religious designation that one may choose, as I once naïvely assumed. Whether or not I see myself as Jewish does not matter to white supremacists.

I've come to realize that my own experience with race reflects the larger historical picture for Jews. As whites, Jews today are certainly a privileged group in the United States. Yet the history of the Jewish experience demonstrates precisely what scholars mean when they say that race is a social construction.

At certain points in time, Jews have been defined as a non-white minority. Around the turn of the last century, they were considered a separate, inferior race, with a distinguishable biological identity justifying discrimination and even genocide. Today, Jews are generally considered white, and Jewishness is largely considered merely a religious or ethnic designation. Jews, along with other European ethnic groups, were welcomed into the category of "white" as beneficiaries of one of the largest affirmative-action programs in history—the 1944 GI Bill of Rights. Yet, when I read white-supremacist discourse, I am reminded that my ancestors were expelled from the dominant race, persecuted, and even killed.

Since conducting my research, having heard dozens of descriptions of the murders and mutilations of "race traitors" by white supremacists, I now carry with me the knowledge that there are many people out there who would still wish to see me dead. For a brief moment, I think that I can imagine what it must feel like to be a person of color in our society ... but then I realize that, as a white person, I cannot begin to imagine that.

Jewishness has become both clearer and more ambiguous for me. And the questions I have encountered in thinking about Jewish identity highlight the central issues involved in studying race today. I teach a class on race and ethnicity, and usually, about midway through the course, students complain of confusion. They enter my course seeking answers to the most troubling and divisive questions of our time, and are disappointed when they discover only more questions. If race is not biological or genetic, what is it? Why, in some states, does it take just one black ancestor out of 32 to make a person legally black, yet those 31 white ancestors are not enough to make that person white? And, always, are Jews a race?

I have no simple answers. As Jewish history demonstrates, what is and is not a racial designation, and who is included within it, is unstable and changes over time—and that designation is always tied to power. We do not have to look far to find other examples: The Irish were also once considered non-white in the United States, and U.S. racial categories change with almost every census.

My prolonged encounter with the white-supremacist movement forced me to question not only my assumptions about Jewish identity, but also my assumptions about whiteness. Growing up "white," I felt raceless. As it is for most white people, my race was invisible to me. Reflecting the assumption of most research on race at the time, I saw race as something that shaped the lives of people of color—the victims of racism. We are not used to thinking about whiteness when we think about race. Consequently, white people like myself have failed to recognize the ways in which our own lives are shaped by race. It was not until others began identifying me as the Jew, the "other," that I began to explore race in my own life.

Ironically, that is the same phenomenon shaping the consciousness of white supremacists: They embrace their racial identity at the precise moment when they feel their privilege and power under attack. Whiteness historically has equaled power, and when that equation is threatened, their own whiteness becomes visible to many white people for the first time. Hence, white supremacists seek to make racial identity, racial hierarchies, and white power part of the natural order again. The notion that race is a social construct threatens that order. While it has become an academic commonplace to assert that race is socially constructed, the revelation is profoundly unsettling to many, especially those who benefit most from the constructs.

My research on hate groups not only opened the way for me to explore my own racial identity, but also provided insight into the question with which I began this essay: Why do white supremacists express such hatred and fear of Jews? This ambiguity in Jewish racial identity is precisely what white supremacists find so threatening. Jewish history reveals race as a social designation, rather than a God-given or genetic endowment. Jews blur the boundaries between whites and people of color, failing to fall securely on either side of the divide. And it is ambiguity that white supremacists fear most of all.

I find it especially ironic that, today, some strict Orthodox Jewish leaders also find that ambiguity threatening. Speaking out against the high rates of intermarriage among Jews and non-Jews, they issue dire warnings. Like white supremacists, they fear assaults on the integrity of the community and fight to secure its racial boundaries, defining Jewishness as biological and restricting it only to those with Jewish mothers. For both white supremacists and such Orthodox Jews, intermarriage is tantamount to genocide.

For me, the task is no longer to resolve the ambiguity, but to embrace it. My exploration of white-supremacist ideology has revealed just how subversive doing so can be: Reading white-supremacist discourse through the lens of Jewish experience has helped me toward new interpretations. White supremacy is not a movement just about hatred, but even more about fear: fear of the vulnerability and instability of white identity and privilege. For white supremacists, the central goal is to naturalize racial identity and hierarchy, to establish boundaries.

Both my own experience and Jewish history reveal that to be an impossible task. Embracing Jewish identity and history, with all their contradictions, has given me an empowering alternative to white-supremacist conceptions of race. I have found that eliminating ambivalence does not require eliminating ambiguity.

13

Race as Class

HERBERT J. GANS

Humans of all colors and shapes can make babies with each other. Consequently most biologists, who define races as subspecies that cannot interbreed, argue that scientifically there can be no human races. Nonetheless, lay people still see and distinguish between races. Thus, it is worth asking again why the lay notion of race continues to exist and to exert so much influence in human affairs.

Lay persons are not biologists, nor are they sociologists, who argue these days that race is a social construction arbitrary enough to be eliminated if "society" chose to do so. The laity operates with a very different definition of race. They see that humans vary, notably in skin color, the shape of the head, nose, and lips, and quality of hair, and they choose to define the variations as individual races.

More important, the lay public uses this definition of race to decide whether strangers (the so-called "other") are to be treated as superior, inferior, or equal. Race is even more useful for deciding quickly whether strangers might be threatening and thus should be excluded. Whites often consider dark-skinned strangers threatening until they prove otherwise, and none more than African Americans.

Scholars believe the color differences in human skins can be traced to climatic adaptation. They argue that the high levels of melanin in dark skin originally protected people living outside in hot, sunny climates, notably in Africa and South Asia, from skin cancer. Conversely, in cold climates, the low amount of melanin in light skins enabled the early humans to soak up vitamin D from a sun often hidden behind clouds. These color differences were reinforced by millennia of inbreeding when humans lived in small groups that were geographically and socially isolated. This inbreeding also produced variations in head and nose

SOURCE: *Contexts*, Vol. 4, Issue 4, pp. 17–21. © 2005 by the American Sociological Association. All rights reserved.

shapes and other facial features so that Northern Europeans look different from people from the Mediterranean area, such as Italians and, long ago, Jews. Likewise, East African faces differ from West African ones, and Chinese faces from Japanese ones. (Presumably the inbreeding and isolation also produced the DNA patterns that geneticists refer to in the latest scientific revival and redefinition of race.)

Geographic and social isolation ended long ago, however, and human population movements, intermarriage, and other occasions for mixing are eroding physical differences in bodily features. Skin color stopped being adaptive too after people found ways to protect themselves from the sun and could get their vitamin D from the grocery or vitamin store. Even so, enough color variety persists to justify America's perception of white, yellow, red, brown, and black races.

Never mind for the moment that the skin of "whites," as well as many East Asians and Latinos is actually pink; that Native Americans are not red; that most African Americans come in various shades of brown; and that really black skin is rare. Never mind either that color differences within each of these populations are as great as the differences between them, and that, as DNA testing makes quite clear, most people are of racially mixed origins, even if they do not know it. But remember that this color palette was invented by whites. Nonwhite people would probably divide the range of skin colors quite differently.

Advocates of racial equality use these contradictions to fight against racism. However, the general public also has other priorities. As long as people can roughly agree about who looks "white," "yellow," or "black" and find that their notion of race works for their purposes, they ignore its inaccuracies, inconsistencies, and other deficiencies.

Note, however, that only some facial and bodily features are selected for the lay definition of race. Some, like the color of women's nipples or the shape of toes (and male navels), cannot serve because they are kept covered. Most other visible ones, like height, weight, hairlines, ear lobes, finger or hand sizes—and even skin texture—vary too randomly and frequently to be useful for categorizing and ranking people or judging strangers. After all, your own child is apt to have the same stubby fingers as a child of another skin color or, what is equally important, a child from a very different income level.

RACE, CLASS, AND STATUS

In fact, the skin colors and facial features commonly used to define race are selected precisely because, when arranged hierarchically, they resemble the country's class-and-status hierarchy. Thus, whites are on top of the socioeconomic pecking order as they are on top of the racial one, while variously shaded non-whites are below them in socioeconomic position (class) and prestige (status).

The darkest people are for the most part at the bottom of the class-status hierarchy. This is no accident, and Americans have therefore always used race as a marker or indicator of both class and status. Sometimes they also use it to enforce class position, to keep some people "in their place." Indeed, these uses are a major reason for its persistence.

Of course, race functions as more than a class marker, and the correlation between race and the socioeconomic pecking order is far from statistically perfect: All races can be found at every level of that order. Still, the race-class correlation is strong enough to utilize race for the general ranking of others. It also becomes more useful for ranking dark-skinned people as white poverty declines so much that whiteness becomes equivalent to being middle or upper class.

The relation between race and class is unmistakable. For example, the 1998– 2000 median household income of non-Hispanic whites was $45,500; of Hispanics (currently seen by many as a race) as well as Native Americans, $32,000; and of African Americans, $29,000. The poverty rates for these same groups were 7.8 percent among whites, 23.1 among Hispanics, 23.9 among blacks, and 25.9 among Native Americans. (Asians' median income was $52,600—which does much to explain why we see them as a model minority.)

True, race is not the only indicator used as a clue to socioeconomic status. Others exist and are useful because they can also be applied to ranking co-racials. They include language (itself a rough indicator of education), dress, and various kinds of taste, from given names to cultural preferences, among others.

American English has no widely known working-class dialect like the English Cockney, although "Brooklynese" is a rough equivalent, as is "black vernacular." Most blue-collar people dress differently at work from white-collar, professional, and managerial workers. Although contemporary American leisure-time dress no longer signifies the wearer's class, middle-income Americans do not usually wear Armani suits or French haute couture, and the people who do can spot the knockoffs bought by the less affluent.

Actually, the cultural differences in language, dress, and so forth that were socially most noticeable are declining. Consequently, race could become yet more useful as a status marker, since it is so easily noticed and so hard to hide or change. And in a society that likes to see itself as classless, race comes in very handy as a substitute.

THE HISTORICAL BACKGROUND

Race became a marker of class and status almost with the first settling of the United States. The country's initial holders of cultural and political power were mostly WASPs (with a smattering of Dutch and Spanish in some parts of what later became the United States). They thus automatically assumed that their kind of whiteness marked the top of the class hierarchy. The bottom was assigned to the most powerless, who at first were Native Americans and slaves. However, even before the former had been virtually eradicated or pushed to the country's edges, the skin color and related facial features of the majority of colonial America's slaves had become the markers for the lowest class in the colonies.

Although dislike and fear of the dark are as old as the hills and found all over the world, the distinction between black and white skin became important in America only with slavery and was actually established only some decades after the first importation of black slaves. Originally, slave owners justified their enslavement of black Africans by their being heathens, not by their skin color.

In fact, early Southern plantation owners could have relied on white indentured servants to pick tobacco and cotton or purchased the white slaves that were available then, including the Slavs from whom the term slave is derived. They also had access to enslaved Native Americans. Blacks, however, were cheaper, more plentiful, more easily controlled, and physically more able to survive the intense heat and brutal working conditions of Southern plantations.

After slavery ended, blacks became farm laborers and sharecroppers, de facto indentured servants, really, and thus they remained at the bottom of the class hierarchy. When the pace of industrialization quickened, the country needed new sources of cheap labor. Northern industrialists, unable and unwilling to recruit southern African Americans, brought in very poor European immigrants, mostly peasants. Because these people were near the bottom of the class hierarchy, they were considered nonwhite and classified into races. Irish and Italian newcomers were sometimes even described as black (Italians as "guineas"), and the eastern and southern European immigrants were deemed "swarthy."

However, because skin color is socially constructed, it can also be reconstructed. Thus, when the descendants of the European immigrants began to move up economically and socially, their skins apparently began to look lighter to the whites who had come to America before them. When enough of these descendants became visibly middle class, their skin was seen as fully white. The biological skin color of the second and third generations had not changed, but it was socially blanched or whitened. The process probably began in earnest just before the Great Depression and resumed after World War II. As the cultural and other differences of the original European immigrants disappeared, their descendants became known as white ethnics.

This pattern is now repeating itself among the peoples of the post–1965 immigration. Many of the new immigrants came with money and higher education, and descriptions of their skin color have been shaped by their class position. Unlike the poor Chinese who were imported in the 19th century to build the West and who were hated and feared by whites as a "yellow horde," today's affluent Asian newcomers do not seem to look yellow. In fact, they are already sometimes thought of as honorary whites, and later in the 21st century they may well turn into a new set of white ethnics. Poor East and Southeast Asians may not be so privileged, however, although they are too few to be called a "yellow horde."

Hispanics are today's equivalent of a "swarthy" race. However, the children and grandchildren of immigrants among them will probably undergo "whitening" as they become middle class. Poor Mexicans, particularly in the Southwest, are less likely to be whitened, however. (Recently a WASP Harvard professor came close to describing these Mexican immigrants as a brown horde.)

Meanwhile, black Hispanics from Puerto Rico, the Dominican Republic, and other Caribbean countries may continue to be perceived, treated, and mistreated as if they were African American. One result of that mistreatment is their low median household income of $35,000, which was just $1,000 more than that of non-Hispanic blacks but $4,000 below that of so-called white Hispanics.

Perhaps South Asians provide the best example of how race correlates with class and how it is affected by class position. Although the highly educated Indians and Sri Lankans who started coming to America after 1965 were often darker than African Americans, whites only noticed their economic success. They have rarely been seen as nonwhites, and are also often praised as a model minority.

Of course, even favorable color perceptions have not ended racial discrimination against newcomers, including model minorities and other affluent ones. When they become competitors for valued resources such as highly paid jobs, top schools, housing, and the like, they also become a threat to whites. California's Japanese-Americans still suffer from discrimination and prejudice four generations after their ancestors arrived here.

AFRICAN-AMERICAN EXCEPTIONALISM

The only population whose racial features are not automatically perceived differently with upward mobility are African Americans: Those who are affluent and well educated remain as visibly black to whites as before. Although a significant number of African Americans have become middle class since the civil rights legislation of the 1960s, they still suffer from far harsher and more pervasive discrimination and segregation than nonwhite immigrants of equivalent class position. This not only keeps whites and blacks apart but prevents blacks from moving toward equality with whites. In their case, race is used both as a marker of class and, by keeping blacks "in their place," an enforcer of class position and a brake on upward mobility.

In the white South of the past, African Americans were lynched for being "uppity." Today, the enforcement of class position is less deadly but, for example, the glass ceiling for professional and managerial African Americans is set lower than for Asian Americans, and on-the-job harassment remains routine.

Why African-American upward economic mobility is either blocked or, if allowed, not followed by public blanching of skin color remains a mystery. Many explanations have been proposed for the white exceptionalism with which African Americans are treated. The most common is "racism," an almost innate prejudice against people of different skin color that takes both personal and institutional forms. But this does not tell us why such prejudice toward African Americans remains stronger than that toward other nonwhites.

A second explanation is the previously mentioned white antipathy to blackness, with an allegedly primeval fear of darkness extrapolated into a primordial fear of dark-skinned people. But according to this explanation, dark-skinned immigrants such as South Asians should be treated much like African Americans.

A better explanation might focus on "Negroid" features. African as well as Caribbean immigrants with such features—for example, West Indians and Haitians—seem to be treated somewhat better than African Americans. But this remains true only for new immigrants; their children are generally treated like African Americans.

Two additional explanations are class-related. For generations, a majority or plurality of all African Americans were poor, and about a quarter still remain so. In addition, African Americans continue to commit a proportionally greater share of the street crime, especially street drug sales—often because legitimate job opportunities are scarce. African Americans are apparently also more often arrested without cause. As one result, poor African Americans are more often considered undeserving than are other poor people, although in some parts of America, poor Hispanics, especially those who are black, are similarly stigmatized.

The second class-based explanation proposes that white exceptionalist treatment of African Americans is a continuing effect of slavery: They are still perceived as ex-slaves. Many hateful stereotypes with which today's African Americans are demonized have changed little from those used to dehumanize the slaves. (Black Hispanics seem to be equally demonized, but then they were also slaves, if not on the North American continent.) Although slavery ended officially in 1864, ever since the end of Reconstruction subtle efforts to discourage African-American upward mobility have not abated, although these efforts are today much less pervasive or effective than earlier.

Some African Americans are now millionaires, but the gap in wealth between average African Americans and whites is much greater than the gap between incomes. The African-American middle class continues to grow, but many of its members barely have a toehold in it, and some are only a few paychecks away from a return to poverty. And the African-American poor still face the most formidable obstacles to upward mobility. Close to a majority of working-age African-American men are jobless or out of the labor force. Many women, including single mothers, now work in the low-wage economy, but they must do without most of the support systems that help middle-class working mothers. Both federal and state governments have been punitive, even in recent Democratic administrations, and the Republicans have cut back nearly every antipoverty program they cannot abolish.

Daily life in a white-dominated society reminds many African Americans that they are perceived as inferiors, and these reminders are louder and more relentless for the poor, especially young men. Regularly suspected of being criminals, they must constantly prove that they are worthy of equal access to the American Dream. For generations, African Americans have watched immigrants pass them in the class hierarchy, and those who are poor must continue to compete with current immigrants for the lowest-paying jobs. If unskilled African

Americans reject such jobs or fail to act as deferentially as immigrants, they justify the white belief that they are less deserving than immigrants. Blacks' resentment of such treatment gives whites additional evidence of their unworthiness, thereby justifying another cycle of efforts to keep them from moving up in class and status.

Such practices raise the suspicion that the white political economy and white Americans may, with the help of nonwhites who are not black, use African Americans to anchor the American class structure with a permanently lower-class population. In effect, America, or those making decisions in its name, could be seeking, not necessarily consciously, to establish an undercaste that cannot move out and up. Such undercastes exist in other societies: the gypsies of Eastern Europe, India's untouchables, "indigenous people" and "aborigines" in yet other places. But these are far poorer countries than the United States.

SOME IMPLICATIONS

The conventional wisdom and its accompanying morality treat racial prejudice, discrimination, and segregation as irrational social and individual evils that public policy can reduce but only changes in white behavior and values can eliminate. In fact, over the years, white prejudice as measured by attitude surveys has dramatically declined, far more dramatically than behavioral and institutional discrimination.

But what if discrimination and segregation are more than just a social evil? If they are used to keep African Americans down, then they also serve to eliminate or restrain competitors for valued or scarce resources, material and symbolic. Keeping African Americans from decent jobs and incomes as well as quality schools and housing makes more of these available to all the rest of the population. In that case, discrimination and segregation may decline significantly only if the rules of the competition change or if scarce resources, such as decent jobs, become plentiful enough to relax the competition, so that the African-American population can become as predominantly middle class as the white population. Then the stigmas, the stereotypes inherited from slavery, and the social and other arrangements that maintain segregation and discrimination could begin to lose their credibility. Perhaps "black" skin would eventually become as invisible as "yellow" skin is becoming.

THE MULTIRACIAL FUTURE

One trend that encourages upward mobility is the rapid increase in interracial marriage that began about a quarter century ago. As the children born to parents of different races also intermarry, more and more Americans will be multiracial, so that at some point far in the future the current quintet of skin colors will be irrelevant. About 40 percent of young Hispanics and two-thirds of young Asians now "marry out," but only about 10 percent of blacks now marry nonblacks—yet another instance of the exceptionalism that differentiates blacks.

Moreover, if race remains a class marker, new variations in skin color and in other visible bodily features will be taken to indicate class position. Thus, multiracials with "Negroid" characteristics could still find themselves disproportionately at the bottom of the class hierarchy. But what if at some point in the future everyone's skin color varied by only a few shades of brown? At that point, the dominant American classes might have to invent some new class markers.

If in some utopian future the class hierarchy disappears, people will probably stop judging differences in skin color and other features. Then lay Americans would probably agree with biologists that race does not exist. They might even insist that race does not need to exist.

14

Shadowy Lines That Still Divide

JANNY SCOTT AND DAVID LEONHARDT

There was a time when Americans thought they understood class. The upper crust vacationed in Europe and worshiped an Episcopal God. The middle class drove Ford Fairlanes, settled the San Fernando Valley, and enlisted as company men. The working class belonged to the AFL–CIO, voted Democratic, and did not take cruises to the Caribbean.

Today, the country has gone a long way toward an appearance of classlessness. Americans of all sorts are awash in luxuries that would have dazzled their grandparents. Social diversity has erased many of the old markers. It has become harder to read people's status in the clothes they wear, the cars they drive, the votes they cast, the god they worship, the color of their skin. The contours of class have blurred; some say they have disappeared.

But class is still a powerful force in American life. Over the past three decades it has come to play a greater, not lesser, role in important ways. At a time when education matters more than ever, success in school remains linked tightly to class. At a time when the country is increasingly integrated racially, the rich are isolating themselves more and more. At a time of extraordinary advances in medicine, class differences in health and life span are wide and appear to be widening.

SOURCE: The New York Times, 2005; *Class Matters,* New York: Henry Holt & Co.

And new research on mobility, the movement of families up and down the economic ladder, shows there is far less of it than economists once thought and less than most people believe. In fact, mobility, which once buoyed the working lives of Americans as it rose in the decades after World War II, has lately flattened out or possibly even declined, many researchers say.

Mobility is the promise that lies at the heart of the American dream. It is supposed to take the sting out of the widening gulf between the have-mores and the have-nots. There are poor and rich in the United States, of course, the argument goes; but as long as one can become the other, as long as there is something close to equality of opportunity, the differences between them do not add up to class barriers. . . .

The trends are broad and seemingly contradictory: the blurring of the landscape of class and the simultaneous hardening of certain class lines; the rise in standards of living while most people remain moored in their relative places.

Even as mobility seems to have stagnated, the ranks of the elite are opening. Today, anyone may have a shot at becoming a United States Supreme Court justice or a CEO, and there are more and more self-made billionaires. Only thirty-seven members of last year's Forbes 400, a list of the richest Americans, inherited their wealth, down from almost two hundred in the mid-1980s.

So it appears that while it is easier for a few high achievers to scale the summits of wealth, for many others it has become harder to move up from one economic class to another. Americans are arguably more likely than they were thirty years ago to end up in the class into which they were born.

A paradox lies at the heart of this new American meritocracy. Merit has replaced the old system of inherited privilege, in which parents to the manner born handed down the manor to their children. But merit, it turns out, is at least partly class-based. Parents with money, education, and connections cultivate in their children the habits that the meritocracy rewards. When their children then succeed, their success is seen as earned.

The scramble to scoop up a house in the best school district, channel a child into the right preschool program, or land the best medical specialist are all part of a quiet contest among social groups that the affluent and educated are winning in a rout.

"The old system of hereditary barriers and clubby barriers has pretty much vanished," said Eric Wanner, president of the Russell Sage Foundation, a social science research group in New York City that has published a series of studies on the social effects of economic inequality.

In place of the old system, Wanner said, have arisen "new ways of transmitting advantage that are beginning to assert themselves."

FAITH IN THE SYSTEM

Most Americans remain upbeat about their prospects for getting ahead. A recent *New York Times* poll on class found that 40 percent of Americans believed that the chance of moving up from one class to another had risen over the last thirty years, a period in which the new research shows that it has not. Thirty-five percent said it had not changed, and only 23 percent said it had dropped.

More Americans than twenty years ago believe it possible to start out poor, work hard, and become rich. They say hard work and a good education are more important to getting ahead than connections or a wealthy background.

"I think the system is as fair as you can make it," Ernie Frazier, a sixty-five-year-old real estate investor in Houston, said in an interview after participating in the poll. "I don't think life is necessarily fair. But if you persevere, you can overcome adversity. It has to do with a person's willingness to work hard, and I think it's always been that way."

Most say their standard of living is better than their parents' and imagine that their children will do better still. Even families making less than $30,000 a year subscribe to the American dream; more than half say they have achieved it or will do so.

But most do not see a level playing field. They say the very rich have too much power, and they favor the idea of class-based affirmative action to help those at the bottom. Even so, most say they oppose the government's taxing the assets a person leaves at death. . . .

THE ATTRIBUTES OF CLASS

One difficulty in talking about class is that the word means different things to different people. Class is rank, it is tribe, it is culture and taste. It is attitudes and assumptions, a source of identity, a system of exclusion. To some, it is just money. It is an accident of birth that can influence the outcome of a life. Some Americans barely notice it; others feel its weight in powerful ways.

At its most basic, class is one way societies sort themselves out. Even societies built on the idea of eliminating class have had stark differences in rank. Classes are groups of people of similar economic and social position; people who, for that reason, may share political attitudes, lifestyles, consumption patterns, cultural interests, and opportunities to get ahead. Put ten people in a room and a pecking order soon emerges.

When societies were simpler, the class landscape was easier to read. Marx divided nineteenth-century societies into just two classes; Max Weber added a few more. As societies grew increasingly complex, the old classes became more heterogeneous. As some sociologists and marketing consultants see it, the commonly accepted big three—the upper, middle, and working classes—have broken down into dozens of microclasses, defined by occupations or lifestyles. . . .

One way to think of a person's position in society is to imagine a hand of cards. Everyone is dealt four cards, one from each suit: education, income, occupation, and wealth, the four commonly used criteria for gauging class. Face cards in a few categories may land a player in the upper middle class. At first, a person's class is his parents' class. Later, he may pick up a new hand of his own; it is likely to resemble that of his parents, but not always.

Bill Clinton traded in a hand of low cards with the help of a college education and a Rhodes scholarship and emerged decades later with four face cards.
Bill Gates, who started off squarely in the upper middle class, made a fortune
without finishing college, drawing three aces.

Many Americans say that they too have moved up the nation's class ladder.
In the *Times* poll, 45 percent of respondents said they were in a higher class than
when they grew up, while just 16 percent said they were in a lower one.
Overall, 1 percent described themselves as upper class, 15 percent as upper middle class, 42 percent as middle, 35 percent as working, and 7 percent as lower.

"I grew up very poor and so did my husband," said Wanda Brown, the fifty-
eight-year-old wife of a retired planner for the Puget Sound Naval Shipyard who
lives in Puyallup, Washington, near Tacoma. "We're not rich but we are comfortable and we are middle class and our son is better off than we are."

THE AMERICAN IDEAL

The original exemplar of American social mobility was almost certainly Benjamin
Franklin, one of seventeen children of a candle maker. About twenty years ago,
when researchers first began to study mobility in a rigorous way, Franklin seemed
representative of a truly fluid society, in which the rags-to-riches trajectory was the
readily achievable ideal, just as the nation's self-image promised.

In a 1987 speech, Gary S. Becker, a University of Chicago economist who
would later win a Nobel Prize, summed up the research by saying that mobility
in the United States was so high that very little advantage was passed down from
one generation to the next. In fact, researchers seemed to agree that the grandchildren of privilege and of poverty would be on nearly equal footing.

If that had been the case, the rise in income inequality beginning in the
mid-1970s should not have been all that worrisome. The wealthy might have
looked as if they were pulling way ahead, but if families were moving in and
out of poverty and prosperity all the time, how much did the gap between the
top and bottom matter?

But the initial mobility studies were flawed, economists now say. Some
studies relied on children's fuzzy recollections of their parents' income. Others
compared single years of income, which fluctuate considerably. Still others misread the normal progress people make as they advance in their careers, like from
young lawyer to senior partner, as social mobility.

The new studies of mobility, which methodically track people's earnings
over decades, have found far less movement. The economic advantage once
believed to last only two or three generations is now believed to last closer to
five. Mobility happens, just not as rapidly as was once thought. . . .

One study, by the Federal Reserve Bank of Boston, found that fewer families moved from one quintile, or fifth, of the income ladder to another during
the 1980s than during the 1970s and that still fewer moved in the 1990s than in
the 1980s. A study by the Bureau of Labor Statistics also found that mobility
declined from the 1980s to the 1990s.

The incomes of brothers born around 1960 have followed a more similar path than the incomes of brothers born in the late 1940s, researchers at the Chicago Federal Reserve and the University of California, Berkeley, have found. Whatever children inherit from their parents—habits, skills, genes, contacts, money—seems to matter more today.

Studies on mobility over generations are notoriously difficult, because they require researchers to match the earnings records of parents with those of their children. Some economists consider the findings of the new studies murky; it cannot be definitively shown that mobility has fallen during the last generation, they say, only that it has not risen. The data will probably not be conclusive for years.

Nor do people agree on the implications. Liberals say the findings are evidence of the need for better early-education and antipoverty programs to try to redress an imbalance in opportunities. Conservatives tend to assert that mobility remains quite high, even if it has tailed off a little.

But there is broad consensus about what an optimal range of mobility is. It should be high enough for fluid movement between economic levels but not so high that success is barely tied to achievement and seemingly random, economists on both the right and left say. . . .

One surprising finding about mobility is that it is not higher in the United States than in Britain or France. It is lower here than in Canada and some Scandinavian countries but not as low as in developing countries like Brazil, where escape from poverty is so difficult that the lower class is all but frozen in place.

Those comparisons may seem hard to believe. Britain and France had hereditary nobilities; Britain still has a queen. The founding document of the United States proclaims all men to be created equal. The American economy has also grown more quickly than Europe's in recent decades, leaving an impression of boundless opportunity.

But the United States differs from Europe in ways that can gum up the mobility machine. Because income inequality is greater here, there is a wider disparity between what rich and poor parents can invest in their children. Perhaps as a result, a child's economic background is a better predictor of school performance in the United States than in Denmark, the Netherlands, or France, one study found.

"Being born in the elite in the U.S. gives you a constellation of privileges that very few people in the world have ever experienced," Levine said. "Being born poor in the U.S. gives you disadvantages unlike anything in Western Europe and Japan and Canada."

BLURRING THE LANDSCAPE

Why does it appear that class is fading as a force in American life?

For one thing, it is harder to read position in possessions. Factories in China and elsewhere churn out picture-taking cellphones and other luxuries that are now affordable to almost everyone. Federal deregulation has done the same for plane tickets and long-distance phone calls. Banks, more confident about measuring risk, now extend credit to low-income families, so that owning a home or driving a new car is no longer evidence that someone is middle class.

The economic changes making material goods cheaper have forced businesses to seek out new opportunities so that they now market to groups they once ignored. Cruise ships, years ago a symbol of the high life, have become the oceangoing equivalent of the Jersey Shore. BMW produces a cheaper model with the same insignia. Martha Stewart sells chenille jacquard drapery and scallop-embossed ceramic dinnerware at Kmart.

"The level of material comfort in this country is numbing," said Paul Bellew, executive director for market and industry analysis at General Motors. "You can make a case that the upper half lives as well as the upper 5 percent did fifty years ago."

Like consumption patterns, class alignments in politics have become jumbled. In the 1950s, professionals were reliably Republican; today they lean Democratic. Meanwhile, skilled labor has gone from being heavily Democratic to almost evenly split.

People in both parties have attributed the shift to the rise of social issues, like gun control and same-sex marriage, which have tilted many working-class voters rightward and upper-income voters toward the left. But increasing affluence plays an important role, too. When there is not only a chicken, but an organic, free-range chicken, in every pot, the traditional economic appeal to the working class can sound off-key.

Religious affiliation, too, is no longer the reliable class marker it once was. The growing economic power of the South has helped lift evangelical Christians into the middle and upper middle classes, just as earlier generations of Roman Catholics moved up in the mid-twentieth century. It is no longer necessary to switch one's church membership to Episcopal or Presbyterian as proof that one has arrived. . . .

The once tight connection between race and class has weakened, too, as many African-Americans have moved into the middle and upper middle classes. Diversity of all sorts—racial, ethnic, and gender—has complicated the class picture. And high rates of immigration and immigrant success stories seem to hammer home the point: The rules of advancement have changed.

The American elite, too, is more diverse than it was. The number of corporate chief executives who went to Ivy League colleges has dropped over the past fifteen years. There are many more Catholics, Jews, and Mormons in the Senate than there were a generation or two ago. Because of the economic earthquakes of the last few decades, a small but growing number of people have shot to the top. . . .

These success stories reinforce perceptions of mobility, as does cultural myth-making in the form of television programs like *American Idol* and *The Apprentice.*

But beneath all that murkiness and flux, some of the same forces have deepened the hidden divisions of class. Globalization and technological change have shuttered factories, killing jobs that were once stepping-stones to the middle class. Now that manual labor can be done in developing countries for two dollars a day, skills and education have become more essential than ever.

This has helped produce the extraordinary jump in income inequality. The after-tax income of the top 1 percent of American households jumped 139 percent, to more than $700,000, from 1979 to 2001, according to the Congressional Budget Office, which adjusted its numbers to account for inflation. The income of the middle fifth rose by just 17 percent, to $43,700, and the income of the poorest fifth rose only 9 percent.

For most workers, the only time in the last three decades when the rise in hourly pay beat inflation was during the speculative bubble of the 1990s. Reduced pensions have made retirement less secure.

Clearly, a degree from a four-year college makes even more difference than it once did. More people are getting those degrees than did a generation ago, but class still plays a big role in determining who does or does not. At 250 of the most selective colleges in the country, the proportion of students from upper-income families has grown, not shrunk. . . .

Class differences in health, too, are widening, recent research shows. Life expectancy has increased overall; but upper-middle-class Americans live longer and in better health than middle-class Americans, who live longer and in better health than those at the bottom.

Class plays an increased role, too, in determining where and with whom affluent Americans live. More than in the past, they tend to live apart from everyone else, cocooned in their exurban châteaus. Researchers who have studied census data from 1980, 1990, and 2000 say the isolation of the affluent has increased.

Family structure, too, differs increasingly along class lines. The educated and affluent are more likely than others to have their children while married. They have fewer children and have them later, when their earning power is high. On average, according to one study, college-educated women have their first child at age thirty, up from twenty-five in the early 1970s. The average age among women who have never gone to college has stayed at about twenty-two.

Those widening differences have left the educated and affluent in a superior position when it comes to investing in their children. . . .

The benefits of the new meritocracy do come at a price. It once seemed that people worked hard and got rich in order to relax, but a new class marker in upper-income families is having at least one parent who works extremely long hours (and often boasts about it). In 1973, one study found, the highest-paid tenth of the country worked fewer hours than the bottom tenth. Today, those at the top work more.

In downtown Manhattan, black cars line up outside Goldman Sachs's headquarters every weeknight around nine. Employees who work that late get a free ride home, and there are plenty of them. Until 1976, a limousine waited at 4:30 p.m. to ferry partners to Grand Central Terminal. But a new management team eliminated the late-afternoon limo to send a message: 4:30 is the middle of the workday, not the end.

A RAGS-TO-RICHES FAITH

Will the trends that have reinforced class lines while papering over the distinctions persist?

The economic forces that caused jobs to migrate to low-wage countries are still active. The gaps in pay, education, and health have not become a major political issue. The slicing of society's pie is more unequal than it used to be, but most Americans have a bigger piece than they or their parents once did. They appear to accept the trade-offs.

Faith in mobility, after all, has been consciously woven into the national self-image. Horatio Alger's books have made his name synonymous with rags-to-riches success, but that was not his personal story. He was a second-generation Harvard man, who became a writer only after losing his Unitarian ministry because of allegations of sexual misconduct. Ben Franklin's autobiography was punched up after his death to underscore his rise from obscurity.

The idea of fixed class positions, on the other hand, rubs many the wrong way. Americans have never been comfortable with the notion of a pecking order based on anything other than talent and hard work. Class contradicts their assumptions about the American dream, equal opportunity, and the reasons for their own successes and even failures. Americans, constitutionally optimistic, are disinclined to see themselves as stuck.

Blind optimism has its pitfalls. If opportunity is taken for granted, as something that will be there no matter what, then the country is less likely to do the hard work to make it happen. But defiant optimism has its strengths. Without confidence in the possibility of moving up, there would almost certainly be fewer success stories.

15

Across the Great Divide

Crossing Classes and Clashing Cultures

BARBARA JENSEN

The blonde curls of Shelly's home permanent stuck to the tears on her face as she dashed from the classroom. "Oh God, I'm so sorry," she cried out. Just twenty minutes earlier she had been in the midst of an animated class discussion in a college course she liked, the psychology of women. Shelly had never thought about being a woman much before; she found it exciting and comforting to do so.

The class was having a discussion about relationships between women and men. The subject was intimacy, and the students were discussing some of the

SOURCE: Michael Zwerg (ed), 2004. *What's Class Got to Do with It? American Society in the Twenty-first Century.* Ithaca, NY: Cornell University Press, pp. 168–183.

different ways men and women understand and express it. Shelly felt she was starting to understand some of the problems in her marriage. Maybe she could make things better. She was eager and animated in the discussion.

But something went wrong. Shelly was talking about the declining intimacy in her marriage and how college "made things really weird" between her and her husband. It wasn't just his complaining about the time she was gone; he was starting to make fun of her studying, saying she was turning into a "geek" and an "egghead." She told the class, "He even picked up a textbook and threw it against the wall, smashing the spine of a $65 book!" Then he hollered, "This shit means more to you than me and the kids!" and stomped out of the house. She said that later, when they "talked it out," he said *she* wasn't any fun anymore, *she* wasn't interested in anything. The class laughed out loud, because in class she was interested in everything. Encouraged, she exclaimed, "I couldn't believe it! That's just what I think about him! He's the one ... I'm interested in things now that I never even *thought* of before, you know what I mean? I asked him, 'What am I not interested in?' and he said, 'Bowling with Georgie and Bill and watching TV'! Like I have time for that now! Like *he* has shown any interest at all in all the things I've been studying."

Shelly's eyes blurred with tears and she fell silent; her pale skin was flushed. A couple of older women in the class started to talk, gently and with warmth, about how they had had to leave their husbands because they needed to "find themselves" and "get a new start." A forty-something woman offered that her spouse really wanted her to go to school, and that Shelly deserved to have that support. A man, who was on the board of a battered women's shelter, emphasized that she had a *right* to expect that support, that men have to learn to give women the things they have always had. He went on to warn her about "offender psychology" and how "they can't stand for their women to be independent, that's how they keep control." The other women from blue-collar backgrounds were uncharacteristically quiet.

"But that's not it!" Shelly insisted in frustration. "You don't understand ... ," she trailed off, struggling for words and understanding. "He, he's a good husband, you know.... He was my *only* support at first ... when my family was lecturing me about my duty to him and the kids. He was great, he—"

A woman who had identified herself as a former battered woman and the man who worked for the shelter exchanged glances with each other. Shelly saw this and scrambled to undo the impression she had given: "That's not it! He really doesn't mind me going to school. I know how it must sound ... he doesn't normally yell and he's never hit me or even thrown anything like that before, you know? My girlfriends always envy me because he's so sweet and he's great with the kids and he's *so* handsome, I mean ... he always knows what to say to people, I mean, not *college* people, but ... you know, regular people. And it never really got stale, I mean, I was still crazy about him until ... until ... I don't know..."

Shelly stumbled to a halt and fell silent. Just when someone else was about to speak she blurted out, "I love him! When I think of losing him ..." Her eyes teared up and she started shaking her head. "It's like the whole world is turned upside down!" Tears steamed down her red-hot face and she ran from the class-room to the lavatory down the hall.

Shelly is a college student at a small, urban university that mostly serves "returning" older and first-generation college students. She is close to her extended Swedish American and German American farm family and is the first one to go to college. Her husband and friends are all working class. She never really thought about going to college before her boss said she might lose her job if she didn't and that the company would pay for it. To her surprise, she loves it. She eagerly reads the class materials; she finds it surprisingly easy to talk in class, and other people often seem to appreciate what she has to say. She suspects she talks way too much, but she "just get[s] so excited." She wonders how could she not know before that she "loves ideas," as another woman in class put it. She was thirty-two years old and had had two children. "Where was I all those years?" she asked once in class.

After the class had left, Shelly came back in to apologize to me. She assured me that in more than two years of college, she had never behaved so "unprofessional before." She apologized a few more times. Her shoulders sank, deflated. She bowed her head and stared at her shoes. "Maybe he's right, maybe I don't belong here." She was embarrassed, and afraid.

Shelly is experiencing a confusing, exciting, and debilitating situation both in her outer life and within her. She is by turns excited, lost, elated, angry, bewildered, shameful, grateful, and "numb." All of a sudden, her past won't cohere with her present, her future has become uncertain; nothing quite "fits right" anymore. Shelly knows that no one she knew before seems to understand what she is going through, and some even resent it. That night she realized that her new friends don't understand either. She is in the midst of a working class "crossover" experience, something she never expected when she went back to school to "get my piece of paper" so she could keep her job as a legal secretary. She had no idea what she was getting into, she had no idea she would fall in love with a new world. She certainly didn't know that she might actually begin to *become someone else*. Though she is delighted with all the new things she has learned, nothing she has learned in this new world helps explain her situation to her. With no language or concepts to bridge or even explain this experience, she is falling prey to the contradictions within it.

There is suffering in this private passage, unvoiced and unseen, a particularly confused suffering in the midst of outward success. This struggle to figure out "who I am anymore," as Shelly once put it, the crossover's collection of contradictory experiences, emotions, and values are the subject of this [article]. I have come to believe Shelly's struggle constitutes a particular inner and outer (psychological and sociological) constellation that many working class people who enter the middle class experience. The psychological similarities among "upwardly mobile" working class people are striking to me. So is the invisible and "privatized" nature of this potentially painful experience. I am a (counseling and community) psychologist, a teacher of first-generation college students, and a person from the working class who has spent my adult life jostling back and forth between different worlds.

Like me, [many] people ... have bumped uneasily between professional middle and working class cultures. We engage (or struggle to avoid engaging) with these often opposing worldviews. This often creates a state of *cognitive dissonance,* or an inner clashing of values and experiences that create emotional and mental confusion. Common emotional reactions to this are anger, shame, sorrow (loss), "impostor syndrome," and substance abuse. These are often so muffled as to be invisible to crossovers themselves. Common behavioral reactions I have seen are distancing, resisting, and creating/bridging.

I believe that central to the "crossover experience" is an existential dilemma. By "existential" I mean a problem of existence: of living one's life, of how best to live, and of the human need to make meaning in and of our lives. And central to this dilemma, though not its only feature, is the presence of cultural differences between the professional middle class and working class people. There are stories, sacrifices, and secret shame that have no ear and precious little voice. The hearing and seeing of these cultural differences—the ability to see *outside* the cultural biases of the professional middle class—is crucial to any meaningful understanding of working class life. Without this, all the well-meaning "solidarity" one may feel for the working class is ineffective. "I feel like they're always talking down to me," said one of my working class students, who is active in the political Left, "but maybe I *am* stupid, because, honestly, half the time I don't know what the hell they're talking about."

People in or from the professional/managerial class will likely be the vehicle of change for the "upwardly mobile" working class person in higher education, job promotions, marriage, psychotherapy, and other crossover experiences. They can show Shelly how to write and speak in Standard English, how to put her napkin in her lap instead of on the table, and how to negotiate with difficult clients. But they can't tell her where she's been and how it has made her who she is, or where it is she might be going.

In this article, I address the less obvious ways that class hurts working class people in higher education (and other avenues of upward mobility). I point toward unfair, unjust advantage and disadvantage that cuts across lines of gender, ethnicity, and "race." In higher education (as elsewhere), this unacknowledged crossover challenge serves to exclude working class people from certain opportunities and privileges, even from their own inner lives. Their counterparts, from the professional middle class, find in higher education the cultural rules, values, language, and community mores that are familiar to them. Working class people must do psychosocial back flips through a maze of new rules, new values, and new language. My concern is twofold: I am worried about the Shellys, and I am worried about the society we all live in that creates, mystifies, and personalizes unequal opportunity and the cultural (as well as economic) domination of one class of people over another. The painful distance between the ideal and the real is felt by those who fall between the cracks; working class crossovers bear it as personal "stupidity," lack of "ambition," "failure," and even psychopathology (depression, substance abuse, and more). These constitute significant invisible costs that working class crossovers are forced to pay. Visibility and voice are the first practical antidotes to this invisible identity crisis.

"SURVIVOR GUILT" AND CULTURAL COLLISION

Working class crossovers are likely to be completely invisible to people from the professional middle class, because middle class people have learned to assume their inner and outer lives are "normal." If you have learned to walk and talk middle class well enough to "fake it," middle class people will assume you have always been one of them, at least if you have white skin. Successful crossovers can't necessarily help you either. As likely as not, they have already been "made safe," ... via the cultural processes and decisions they have gone through to get where they are.

The *invisibility* and the *unconsciousness* of the crossover experience, in my view, can make it painful, debilitating, even devastating. The dilemma manifests in a multitude of so-called (and genuine) personal problems. If crossovers are not conscious of the cross-class experience, the problems it creates can hide behind many personal perspectives. For Shelly, it is a marriage problem. For someone else it is a problem with her "unenlightened" parents. For yet another it is a "chemical imbalance." For many it is a compulsion to ditch class or get "loaded," or to suddenly "blow off" an important exam. Maybe it is simply having "the blues" all the time. For marriage problems, depression, chemical abuse, fear of success, and family-of-origin problems there is, at least, a certain amount of collective wisdom about coping, changing, treating, managing. In my experience, the process of moving from the working class to the professional middle class is a highly personalized and tangled mess of psychological, sociological, and cultural confusion. As philosopher Ludwig Wittgenstein said, that which one has no language for is often not even perceived. . . .

For Shelly, who was previously happy with her husband and her working class life, there is a blazing new star on her horizon, a life of the mind. It complicates things because she wants both to keep her working class roots and to develop her intellectual abilities. She loves her husband, and she can barely stand the strain of not "doing it together." The ambivalence in her relationship is a mirror of her own gathering ambivalence, her own feeling of being "torn"—*torn not just between success and failure in college but between two different notions of what it means to succeed in life.*

I, and others, have tried to articulate what seem to be some of the valuable and central features of working class cultures and how these contrast with middle class culture. . . . Here's my snapshot: working class people are raised with a more here-and-now sensibility, in activities and worldview; individuality (but not necessarily self) is downplayed in favor of a powerful sense of community and loyalty, and an internal sense of "belonging." Working class cultures also tend to be more embedded in ethnic (non-Anglo) traditions, and so are more diverse by nature. Conversely, middle class culture is more homogeneous (and Anglo) and tends to put a premium on individual accomplishments, on the achievement of planned (and publicly recognized) goals in general, and on earning self-definition by way of these achievements—what I called "becoming." These are fundamental differences in outlooks and approaches to life. . . .

As rules start to change for the crossover, family problems abound. Some parents push their children toward "good, clean jobs" as a way to show love and to make their own difficult work lives feel more meaningful. But their reward for this sacrifice is sometimes poignant. If their children succeed, more likely than not, they have adopted the culture, style, and *classism* of the professional/managerial class. Many parents shrink back in shame and confusion while the children they worked so hard to send to school become cultural strangers to them. They fear what is too often true; their children have become embarrassed by their "low class," "backward" family. . . .

DOMINATION AND CULTURAL "CAPITAL"

My mother resisted my evangelical efforts to improve her perfectly good and colorful English. She was a fighter, and the struggle I'm describing is a matter not simply of *different* cultures but of one dominating the other. Is it any wonder that working class families do not easily surrender their children to the people who they know help make their own lives difficult? In the world of work the professional/managerial class is employed by the very wealthy to inflict appalling abuse and neglect on "lower class" workers. Working class people do not have equality of economic opportunity. . . .

This domination also happens, in more genteel settings, by way of what Bourdieu calls "cultural capital." Professional middle class social style, language, and knowledge constitute a kind of social currency. People who have learned these things can use it for entrance into, and access to some amount of power in, the academy (as in business and government). Cultural barriers may be as effective in shutting out working class people as are the (significant) economic ones, perhaps more so. I have said elsewhere that most working class people's native tongue is more metaphoric than literal, more personal and particular than abstract and universal. It is more implicit than explicit, more for members of a defined social group, also more pithy, colorful, and narrative. It reflects cultural differences from the middle class. It is the opposite of how students are expected to write and speak to get good grades in school. This makes trying to "make it" in school considerably more difficult. Indeed, successful working class students are not necessarily "making it" in the sense that their parents, partners, and former peers understand that term. Cultural difference and prejudice against working class culture combine to frustrate the "upwardly" mobile student.

To succeed in higher education (and, often, in a middle class marriage) working class people must learn to adopt and represent middle class culture as their own. This culture does not grant dual citizenship. You must "leave behind" your "low class" ways, your "bad" English, your values of humility and inclusion (don't show off and be a "big shot," because it says you think you're better than others), and much more—not least the people you love! In early adulthood there are developmental tasks of differentiation at play that I suspect help fuel the leap the young crossover student is trying to make. But it is a cruel and unsuspected consequence to have that process set up a chasm that may never be bridged again.

CONCLUDING WITH CREATION

... Somewhere at the center of all these arguments and abstractions sits Shelly, hiding and crying in a bathroom stall at a midwestern state university. It would be wrong-headed to try to tell her what her own decision and solutions might be. What we *can* do is clarify what those decision points are by seeing her dilemma more clearly. We can start by illuminating and validating both her past and her present. If she can see her and her husband's dilemma as a clash of cultures rather than a battle of good and bad, better and worse, normal and abnormal, she may even be able to avoid choosing decisively *between* the cultures. If she has someone to talk with about how she might reconcile them in her own life, she and her family might move forward in a way right for her and them.

Working class cultures have many humane, healthy, and life-giving qualities for which people from the middle class pine and search, at no small consequence to their bank accounts. Like most counseling psychologists, I have spent a part of my career helping both "failures" and "successes" from the professional middle class improve their mental health. We help them, often, to embrace a kind of humanity that values warmth over brilliance, "connectedness" over competition, and that helps them to find a self that exists in spite of personal achievements or failures. I do not intend to romanticize what can be, in many ways, a difficult working class life with limited options, but it is also easy for me to remember, and enjoy to this day, many positive aspects of working class life. . . .

What college *should* give Shelly (and what it gives her middle class peers) is additions to herself, as Phillips points out—not subtractions from herself, or a "transformation." What she needs is an *integration* of new abilities and aware-nesses, not a compartmentalized half-self plagued with doubt and addicted to success. This work is "creating" or "bridging," a third response to the clashing of cultures and the resulting confusion. If this is difficult, if it seems contradictory to people (like me) who long for a classless society, we still need to apply our-selves to the task.

16

The Hidden Cost of Being African American

THOMAS M. SHAPIRO

I met Frank and Suzanne Conway during the late-afternoon rush hour at a restaurant in Los Angeles. Recently laid off from a communications marketing
firm and now taking courses to become certified to teach elementary school,
Frank arrived after picking up their daughter, Logan, from day care. Suzanne
arrived from her job as an operations supervisor for a money management company. The Conways loved their home in the diverse urban neighborhood of
Jefferson Park, near the University of Southern California, but were gravely concerned about sending Logan to weak public schools. They talked to me at length
over coffee about this community-school dilemma, their high educational hopes,
and their future plans. The Conways' story and their solution to their dilemma
turned out to be more common than anticipated. Because they receive generous
help from their families, they are considering moving to a suburban community
with highly regarded schools. Home prices there start at four times those where
they live now, and Logan would grow up and go to school in a far more
homogenous community—family wealth makes these decisions logical and desirable for some families.

Of course, as with the nearly one in three American families without financial assets, many of the family interviews did not brim over with optimistic
choices and options but rather turned on how lack of family wealth severely
restricts community, housing, and schooling opportunities. Like the Conways,
Alice and Bob Bryant work at professional jobs and earn a middle-class income,
but they do not have access to family wealth—they are asset-poor. Living in the
working-class Dorchester section of Boston, they are frustrated about their
inability to afford to move to a neighborhood with better schools. Doing the
best they can, they are highly aware that their son, Mathew, attends only "halfway decent schools" and is not getting the "best education." The Bryants' hopes
for Mathew are no different from the Conways' for Logan. What is different is
their capacity to follow through on their hopes and deliver opportunities. The
Conways are white and the Bryants are black. Because their incomes, professional status, and educations are nearly identical, conventional wisdom suggests

SOURCE: From Thomas M. Shapiro. 2004. *The Hidden Cost of Being African American:
How Wealth Perpetuates Inequality* (New York, Oxford University Press).

that race should be at most a minor factor in opportunities available to these two families, but we will see tangible connections between family assets and race. Differing family asset capacity, which has more to do with race than with merits or accomplishments, most likely will translate into different worlds for Mathew and Logan.

Demonstrating the unique and diverse social circumstances that blacks and whites face is the best way to understand racial differences in wealth holding. The ideas I develop... also push the sociology of wealth in another important direction, namely, an exploration of how the uses of wealth perpetuate inequality. Together, wealth accumulation and utilization highlight the ways in which the opportunity structure contributes to massive racial wealth inequality that worsens racial inequality.

My argument is grounded in three big ideas. First, I argue that family inheritance and continuing racial discrimination in crucial areas like homeownership are reversing gains earned in schools and on jobs and making racial inequality worse. Family inheritance is more encompassing than money passed at death, because for young adults it often includes paying for college, substantial down-payment assistance in buying a first home, and other continuing parental financial assistance. Consequently, it is virtually impossible for people of color to earn their way to equal wealth through wages. No matter how much blacks earn, they cannot preserve their occupational status for their children; they cannot outearn the wealth gap. Many believe that African Americans do not do as well as whites, other minorities, or immigrants because they spend too much money rather than save and invest in the future. They are unable to defer gratification, do not sacrifice for the future, and consume excessively. We will see how the facts speak otherwise. Second, these inheritances frequently amount to what I call transformative assets. This involves the capacity of unearned, inherited wealth to lift a family economically and socially beyond where their own achievements, jobs, and earnings would place them. These head-start assets set up different starting lines, establish different rules for success, fix different rewards for accomplishments, and ultimately perpetuate inequality. Third, the way families use head-start assets to transform their own lives—within current structures that reward them for doing so—has racial and class consequences for the homes they buy, the communities they live in, and the quality of schools their children attend. The same set of processes typically advantages whites while disadvantaging African Americans. My family interviews point to critical mechanisms of denial that insulate whites from privilege.

Homeownership is one of the bedrocks of the American Dream, and I explore homeownership as a prime way of delving into these big ideas. We are a nation of homeowners. In 2002 the homeownership rate was 68 percent, a historic high. Homeownership is by far the single most important way families accumulate wealth. Homeownership also is the way families gain access to the nicest communities, the best public services, and, most important for my argument, quality education. Homeownership is the most critical pathway for transformative assets; hence examining homeownership also keeps our eyes on contemporary discrimination in mortgage markets, the cost of home loans, residential segregation, and the way families accumulate wealth through home appreciation, all of which systematically disadvantage blacks. Homeownership appears critical to success in other areas of life as well, from how well a child does in school to better marital stability to positive civic participation to decreased domestic violence.[1] How young families acquire homes is one of the most tangible ways that the historical legacy of race plays out in the present generation and projects well into future. Understanding how young families can afford to buy homes and how this contributes greatly to the racial wealth gap brings us back full circle to the importance of family legacies.

These big ideas help us understand one of the most important issues facing America as we start the twenty-first century. African Americans were frozen out of the mainstream of American life over the first half of the last century, but since 1954 the civil rights movement has won many battles against racial injustice, and America has reached a broad national consensus in favor of a more tolerant, more inclusive society. Yet we live with a great paradox: Why is racial inequality increasing in this new era?

To fully appreciate the decisions American families like the Conways and Bryants face, we need to understand the extent, causes, and consequences of the vast increase in inequality that has taken place since the early 1970s. Inequality has increased during both Democratic and Republican administrations. Those at the top of the income distribution have increased their share the most. In fact, the slice of the income pie received by the top 1 percent of families is nearly twice as large as it was 30 years ago, and their share now is about as large as the share of the bottom 40 percent. This is not news. In *Nickel and Dimed*, liberal critic Barbara Ehrenreich tells her story of working at low-skill jobs in America's booming service sector, jobs like waitressing, cleaning houses, and retail sales. These are the fastest-growing jobs in America, and they highlight our current work-to-welfare reform strategy. Ehrenreich's experiences illustrate how hard it is to get by in America on poverty wages. More than anything else, perhaps, Ehrenreich's personal experiences demonstrate that in today's America more than hard work is necessary for economic success. I talked to many families who live these lives for real, and we will see how rising inequality makes assets even more critical for success.

In *Wealth and Democracy*, conservative strategist Kevin Phillips argues that current laissez-faire policies are pretenses to further enrich wealthy and powerful families. Rather than philosophical principles, conservative policies of tax cuts for the wealthy, gutting the inheritance tax, and less business regulation favor wealth and property at the expense of middle-class success. The Bush administration's gradual phase-out of the estate tax privileges unearned, inherited wealth over opportunity, hard work, and accomplishment. President Bush's 2003 tax stimulus package carved 39 percent of the benefits for the wealthiest 1 percent. I will broaden the discussion of rising inequality by bringing family wealth back into the picture. Phillips concludes his book with a dire warning: "Either democracy must be renewed, with politics brought back to life, or wealth is likely to cement a new and less democratic regime—plutocracy by some other name."

An ideology that equated personal gain with benefits to society accompanied the great economic boom of the last part of the twentieth century. Even though inequality increased in the past 20 years, despite loud words and little action, policies such as affordable housing and equitable school funding that challenged that mindset simply had no chance of getting off the ground. Ironically, historically low unemployment rates went hand-in-hand with rising inequality in an America where hard work no longer means economic success. Success includes harder work, less family time, and probably more stress. The average middle-income, two-parent family now works the equivalent of 16 more weeks than it did in 1979 due to longer hours, second jobs, and working spouses.[2] The years of economic stagnation subsequent to the boom produced a dramatic increase in the number of working poor, and working homeless families are a growing concern.[3] Since late 2001, in a period marked by a declining stock market and rising unemployment, an abundance of data has provided strong evidence that lower-income households are under severe economic stress: Personal bankruptcies, automobile repossessions, mortgage foreclosures, and other indicators of bad debt all reached records in 2002.[4]

What is the role of wealth and inheritance in rising inequality? The baby boom generation, which grew up during a long period of economic prosperity right after World War II, is in the midst of benefiting from the greatest inheritance of wealth in history. One reliable source estimates that parents will bequeath $9 trillion to their adult children between 1990 and 2030.[5] Given this fact, it is no wonder that an already ineffective estate tax (due to tax planning, family trusts, and loopholes), which takes 50 percent of estates worth more than $1 million, came under such ferocious political attack during the second Clinton administration and has been effectively repealed by the Bush administration.

This wealth inheritance will exacerbate already rising inequality. Economists Robert Avery and Michael Rendall presented a benchmark statistical study in 1993 showing that most inherited wealth will be pocketed by only a few.[6] According to the study, one-third of the money will go to 1 percent of the baby boomers, who will receive about $1.6 million apiece. Another third, representing an average bequest of $336,000, will go to the next 9 percent. The final slice, divided by the remaining 90 percent of the generation, will run about $40,000 apiece. We will see how this baby boomer inheritance not only fuels inequality but also intensifies racial inequality. Few people now talk about the profound effects—economic, social, and political—of that widening gap. We can argue for the privilege of passing along more unearned inequality, or we can take a stand for fairness and equality.

THE CONTEXT OF RACIAL INEQUALITY

Writing at the beginning of the twentieth century, historian W.E.B. Du Bois emphatically declared that the problem of the century was the problem of the color line. Writing again at midcentury, Du Bois reviewed what African Americans had accomplished in education, civil rights, voting rights, occupation, income, housing, literature and arts, and science. African Americans had made progress, he noted, although it was unequal, incomplete, and accompanied by wide gaps and temporary retreats. At about the same time that Du Bois was penning his assessment in a black newspaper, the *Pittsburgh Courier*, the Nobel economist Gunnar Myrdal published the widely read *An American Dilemma*. This influential and lengthy study documented the living conditions for African Americans during the first half of the century, revealing to many for the first time the impact of systematic discrimination in the United States. These two giants helped to define racial inequality in terms of equal opportunity and discrimination and to place these issues at the heart of a nation's concern. The twisted, politically narrow, and bureaucratically unfortunate notion of "affirmative action" substituted for equal opportunity by century's end, and affirmative action continues to frame our hopes and distrust regarding race. Even though the struggle for equal opportunity is far from completed, the single-minded and narrow focus on affirmative action forces compromises with our past, obscures our present understanding of racial inequality, and restricts policy in the future.

Du Bois and Myrdal correctly identified a color line of opportunity and discrimination at the core of the twentieth-century racial equality agenda in the United States. The agenda in the twenty-first century must go further to include the challenge of closing the wealth gap, which currently is 10 cents on the dollar, if we are to make real progress toward racial equality and democracy. Understanding the racial wealth gap is the key to understanding how racial inequality is passed along from generation to generation.

The enigma of racial inequality is still a festering public and private conversation in American society. After the country's dismantling of the most oppressive racist policies and practices of its past, many have come to believe that the United States has moved beyond race and that our most pressing racial concerns should center now on race-neutrality and color-blindness. Proclaiming the success of the civil rights agenda and the dawning of a postracial age in America, books by Shelby Steele, Abigail and Stephan Thernstrom, and others influenced not only the academic debates but elite and popular opinion as well.[7] Indeed, a review of the record shows impressive gains, most particularly in the areas of law, education, jobs, and earnings. Even though progress is real, this new political sensibility about racial progress and equality incorporates illusions that mask an enduring and robust racial hierarchy and continue to hinder efforts to achieve our ideals of democracy and justice.

In fact, we can consider seriously the declining economic significance of race because the measures we have traditionally used to gauge racial inequality focus almost exclusively on salaries. The black–white earnings gap narrowed considerably throughout the 1960s and 1970s. The earnings gap has remained relatively stable since then, with inequality rising again in the 1980s and closing once more during tight labor markets in the 1990s.[8] The average black family earned 55 cents for every dollar earned by the average white family in 1989; by 2000 it reached an all-time high of 64 cents on the dollar.[9] For black men working full-time, the gains are more impressive, as their wages reached 67 percent of those of fully employed white men, up from 62 percent in 1980 and only

50 percent in 1960.[10] How much the racial wage gap has closed, why it has closed, and what it means are the subjects of academic and political debate. One study, for example, argues that the racial wage gap is really 23 percent higher than the official figures because incarceration rates hide low wages and joblessness among blacks.[11] At comparable incomes, more African American family members work to earn the same money as white families. Working longer hours and more weeks per year means that middle-income black families worked the equivalent of 12 more weeks than white families to earn the same money in 2000.[12]

The tremendous growth of the black middle class often is cited as a triumphant sign of progress toward racial equality. Indeed, the raw numbers appear to justify celebration: In 1960 a little more than three-quarters of a million black men and women were employed in middle-class occupations; by 1980 the number increased to nearly three and a third million; and nearly seven million African Americans worked in middle-class jobs in 1995.[13] This impressive growth in achieving upward mobility, however, does not tell the whole story, as some argue that stagnating economic conditions and blacks' lower-middle-class occupational profile have stalled the march into the middle class since the mid-1970s.[14]

The real story of the meaning of race in modern America, however, must include a serious consideration of how one generation passes advantage and disadvantage to the next—how individuals' starting points are determined. While ending the old ways of outright exclusion, subjugation, segregation, custom, discrimination, racist ideology, and violence, our nation continues to reproduce racial inequality, racial hierarchy, and social injustice that is very real and formidable for those who experience it.

In law, in public policy, in custom, in education, in jobs, in health, indeed, in achievements, one could argue that America is more equal today than at any time in our past. Analysts and advocates scour the annual release of official government statistics on income to detect the latest trends in racial inequality. Traditional measures of economic well-being and inequality, such as income, education, and jobs, show authentic and impressive progress toward racial equality from the mid-1960s through the early 1980s and stagnation since.[15] This is not to suggest by any stretch of the imagination that we have seen the dawning of the age of racial parity in the United States, because, indeed, wide racial gaps and discrimination persist in all of these domains. Employment discrimination, educational discrimination, environmental discrimination, and discriminatory immigration, taxation, health, welfare, and transportation policies continue.[16] Despite the passage of major civil rights reforms, most whites and blacks continue to live in highly segregated communities. To achieve perfectly integrated communities, two-thirds of either all black or all white residents would have to move across racial boundaries. The same indicators show too that progress toward racial equality has halted since the early 1980s. Vast wealth differences and hence enormous disparities in opportunities remain between equally achieving and meritorious white and black families. Progress made since the early 1960s has stalled short of equality. Familiar for Du Bois and Myrdal is the dilemma that, despite narrowed gaps in so many important areas, new generations of whites and blacks still start with vastly different sets of options and opportunities. An asset perspective examines a modern element of the American dilemma: Similar achievements by people of similar abilities do not yield comparable results.

NOTES

1. Joint Center for Housing Studies of Harvard University (2002); Ross and Yinger (2002); Boshara (2001). Data for 2002 from the third-quarter report from the Census Bureau, Department of Commerce.

2. Data from Mishel, Bernstein, and Boushey (2003). The increase in work hours primarily comes from wives working more hours and more weeks per year. The actual figure is 660 more hours worked.

3. See Otto (1999).

4. See Wedner (2002) and Andrews (2003).

5. See Avery and Rendall (1993).

6. Avery and Rendall (1993).

7. Most particularly, Steele (1990) and Thernstrom and Thernstrom (1997).

8. The narrowing of the earnings gap during the 1990s appears to have centered on black men with relatively low educations. Altonji and Blank (1999) review the economics literature on the racial earning gap.

9. Data from Mishel, Bernstein, and Boushey (2003). The specific data refer to the ratio of black-to-white median family income.

10. Data from Shkury (2001);

11. Western (2001); Western and Pettit (2002).

12. Figure from Mishel, Bernstein, and Boushey (2003).

13. Patillo-McCoy (1999) contains a good discussion of the growth and condition of the black middle class. For a statistical profile, see Smith and Horton (1997).

14. Patillo-McCoy (1999) makes this case strongly.

15. In addition to the income literature already cited, see Jargowsky (1997) for increased African American labor force participation; Grodsky and Pager (2001) for African Americans' demonstrable gains in occupational attainment; and Hout (1984) on occupational mobility. Jencks and Phillips (1998) argue that the black-white education gap has narrowed since 1970, even though large differences remain. The black-white score gap at the end of high school, for example, has narrowed substantially since 1965.

16. On employment, see Kirschenman and Neckerman (1993); Moss and Tilly (2001); Hill (1993); Feagin and Sykes (1994). On education, see Orfield, Eaton, and Jones (1997). On environment, see Bullard (1997) and Denton.

REFERENCES

Altonji, Joseph, and Rebecca Blank. 1999. "Race and Gender in the Labor Market." In *Handbook of Labor Economics*, vol. 3, ed. Orley Ashenfelter and David Card. Amsterdam: Elsevier.

Avery, Robert, and Robert Rendall. 1993. "Estimating the Size and Distribution of the Baby Boomers' Prospective Inheritances." In *Proceedings of the Social Science Section of the American Statistical Association*. Alexandria, Va.: American Statistical Association.

Boshara, Ray. 2001. *Building Assets: A Report on the Asset-Development and IDA Field*. Washington, D.C.: Corporation for Economic Development.

Bullard, Robert, ed. 1997. *Unequal Protection: Environmental Justice and Communities of Color*. San Francisco: Sierra Club.

Feagin, Joe, and Melvin Sykes. 1994. *Living with Racism: The Black Middle-Class Experience*. Boston: Beacon.

Grodsky, Eric, and Devah Pager. 2001. "The Structure of Disadvantage: Individual and Occupational Determinants of the Black-White Wage Gap." *American Sociological Review* 66: 542–67.

Hill, Herbert. 1993. "Black Workers, Organized Labor, and Title VII of the 1964 Civil Rights Act." In *Race in America: The Struggle for Equality*, ed. Herbert Hill and James E. Hones Jr. Madison: University of Wisconsin Press.

Hout, Michael. 1984. "Occupational Mobility of Black Men, 1962 to 1973." *American Sociological Review* 49: 308–22.

Jargowsky, Paul. 1997. *Poverty and Place*. New York: Russell Sage Foundation.

Joint Center for Housing Studies of Harvard University. 2002. *The State of the Nation's Housing 2002*. Cambridge: JCHD.

Kirschenman, Joleen, and Kathryn M. Neckerman. 1991. "We'd Love to Hire Them, but...": The Meaning of Race for Employers." In *The Urban Underclass*, ed. Christopher Jencks and Paul E. Peterson.

Massey, Douglas, and Nancy Denton. 1993. *American Apartheid*. Cambridge: Harvard University Press.

Mishel, Lawrence, Jared Bernstein, and Heather Boushey. 2003. *The State of Working America, 2002–2003*. Ithaca: Cornell University Press.

Moss, Phillip, and Chris Tilly. 2001. *Stories Employers Tell: Race, Skill, and Hiring in America*. New York: Russell Sage Foundation.

Orfield, Gary, Susan Eaton, and Elaine Jones. 1997. *Dismantling Desegregation: The Quiet Reversal of Brown v. Board of Education*. New York: New Press.

Otto, Mary. 1999. "The Working Homeless is a Growing Reality." *Miami Herald*. December 26.

Patillo-McCoy, Mary. 1999. *Black Picket Fences*. Chicago: University of Chicago Press.

Phillips, Kevin. 2002. *Wealth and Democracy*. New York: Broadway Books.

Ross, Stephen, and John Yinger. 2002. *The Color of Credit: Mortgage Discrimination, Research Methodology, and Fair-Lending Enforcement*. Cambridge, Mass., and London: MIT Press.

Shkury, Shimon. 2001. "Wage Differences Between White Men and Black Men in the United States of America." Master's thesis, Department of Sociology, University of Pennsylvania.

Steele, Shelby. 1990. *The Content of Our Character: A New Vision of Race in America*. New York: St. Martin's.

Thernstrom, Stephan, and Abigail Thernstrom. 1997. *America in Black and White: One Nation, Indivisible*, New York: Simon & Schuster.

Wedner, Diane. 2002. "Mortgage Delinquency Rates Reach 30-Year High." *Los Angeles Times*. September 10.

Western, Bruce. 2002. "The Impact of Incarceration on Wage Mobility and Inequality." *American Sociological Review* 67: 526–46.

Western, Bruce, and Becky Pettit. 2002. "Black-White Wage Inequality, Employment Rates, and Incarceration." Department of Sociology, Princeton University. At www.princeton.edu/western/

17

Is Capitalism Gendered and Racialized?

JOAN ACKER

C apitalism is racialized and gendered in two intersecting historical processes. First, industrial capitalism emerged in the United States dominated by white males, with a gender- and race-segregated labor force, laced with wage inequalities, and a society-wide gender division of caring labor. The processes of reproducing segregation and wage inequality changed over time, but segregation and inequality were not eliminated. A small group of white males still dominate the capitalist economy and its politics. The society-wide gendered division of caring labor still exists. Ideologies of white masculinity and related forms of consciousness help to justify capitalist practices. In short, conceptual and material practices that construct capitalist production and markets, as well as beliefs supporting those practices, are deeply shaped through gender and race divisions of labor and power and through constructions of white masculinity.

Second, these gendered and racialized practices are embedded in and replicated through the gendered substructures of capitalism. These gendered substructures exist in ongoing incompatible organizing of paid production activities and

SOURCE: Joan Acker. 2006. *Class Questions, Feminist Answers*. Lanham, MD: Rowman and Littlefield.

unpaid domestic and caring activities. Domestic and caring activities are devalued and seen as outside the "main business" (Smith 1999) of capitalism. The commodification of labor, the capitalist wage form, is an integral part of this process, as family provisioning and caring become dependent upon wage labor. The abstract language of bureaucratic organizing obscures the ongoing impact on families and daily life. At the same time, paid work is organized on the assumption that reproduction is of no concern. The separations between paid production and unpaid life-sustaining activities are maintained by corporate claims that they have no responsibility for anything but returns to shareholders. Such claims are more successful in the United States, in particular, than in countries with stronger labor movements and welfare states. These often successful claims contribute to the corporate processes of establishing their interests as more important than those of ordinary people.

THE GENDERED AND RACIALIZED DEVELOPMENT OF U.S. CAPITALISM

Segregations and Wage Inequalities

Industrial capitalism is historically, and in the main continues to be, a white male project, in the sense that white men were and are the innovators, owners, and holders of power. Capitalism developed in Britain and then in Europe and the United States in societies that were already dominated by white men and already contained a gender-based division of labor. The emerging waged labor force was sharply divided by gender, as well as by race and ethnicity with many variations by nation and regions within nations. At the same time, the gendered division of labor in domestic tasks was reconfigured and incorporated in a gendered division between paid market labor and unpaid domestic labor. In the United States, certain white men, unburdened by caring for children and households and already the major wielders of gendered power, buttressed at least indirectly by the profits from slavery and the exploitation of other minorities, were, in the nineteenth century, those who built the U.S. factories and railroads, and owned and managed the developing capitalist enterprise. As far as we know, they were also heterosexual and mostly of Northern European heritage. Their wives and daughters benefited from the wealth they amassed and contributed in symbolic and social ways to the perpetuation of their class, but they were not the architects of the new economy.

Recruitment of the labor force for the colonies and then the United States had always been transnational and often coercive. Slavery existed prior to the development of industrialism in the United States: Capitalism was built partly on profits from that source. Michael Omi and Howard Winant (1994, 265) contend that the United States was a racial dictatorship for 258 years, from 1607 to 1865. After the abolition of slavery in 1865, severe exploitation, exclusion, and domination of blacks by whites perpetuated racial divisions cutting across gender and some class divisions, consigning blacks to the most menial, low-paying work in agriculture, mining, and domestic service. Early industrial workers were immigrants. For example, except for the brief tenure (twenty-five years) of young, native-born white women workers in the Lowell, Massachusetts, mills, immigrant women and children were the workers in the first mass production industry in the United States, the textile mills of Massachusetts and Philadelphia, Pennsylvania (Perrow 2002). This was a gender and racial/ethnic division of labor that still exists, but now on a global basis. Waves of European immigrants continued to come to the United States to work in factories and on farms. Many of these European immigrants, such as impoverished Irish, Poles, and eastern European Jews were seen as non-white or not-quite-white by white Americans and were used in capitalist production as low-wage workers, although some of them were actually skilled workers (Brodkin 1998). The experiences of racial oppression built into industrial capitalism varied by gender within these racial/ethnic groups.

Capitalist expansion across the American continent created additional groups of Americans who were segregated by race and gender into racial and ethnic enclaves and into low-paid and highly exploited work. This expansion included the extermination and expropriation of native peoples, the subordination of Mexicans in areas taken in the war with Mexico in 1845, and the recruitment of Chinese and other Asians as low-wage workers, mostly on the west coast (Amott and Matthaei 1996; Glenn 2002).

Women from different racial and ethnic groups were incorporated differently than men and differently than each other into developing capitalism in the late nineteenth and early twentieth centuries. White Euro-American men moved from farms into factories or commercial, business, and administrative jobs. Women aspired to be housewives as the male breadwinner family became the ideal. Married white women, working class and middle class, were housewives unless unemployment, low wages, or death of their husbands made their paid work necessary (Goldin 1990, 133). Young white women with some secondary education moved into the expanding clerical jobs and into elementary school teaching when white men with sufficient education were unavailable (Cohn 1985). African Americans, both women and men, continued to be confined to menial work, although some were becoming factory workers, and even teachers and professionals as black schools and colleges were formed (Collins 2000). Young women from first- and second-generation European immigrant families worked in factories and offices. This is a very sketchy outline of a complex process (Kessler-Harris 1982), but the overall point is that the capitalist labor force in the United States emerged as deeply segregated horizontally by occupation and stratified vertically by positions of power and control on the basis of both gender and race.

Unequal pay patterns went along with sex and race segregation, stratification, and exclusion. Differences in the earnings and wealth (Keister 2000) of women and men existed before the development of the capitalist wage (Padavic and Reskin 2002). Slaves, of course, had no wages and earned little after abolition. These patterns continued as capitalist wage labor became the dominant form and wages became the primary avenue of distribution to ordinary people. Unequal wages were justified by beliefs about virtue and entitlement. A living wage or a just wage for white men was higher than a living wage or a just wage for white women or for women and men from minority racial and ethnic groups (Figart, Mutari, and Power 2002). African-American women were at the bottom of the wage hierarchy.

The earnings advantage that white men have had throughout the history of modern capitalism was created partly by their organization to increase their wages and improve their working conditions. They also sought to protect their wages against the competition of others, women and men from subordinate groups (for example, Cockburn 1983, 1991). This advantage also suggests a white male coalition across class lines, (Connell 2000; Hartmann 1976), based at least partly in beliefs about gender and race differences and beliefs about the superior skills of white men. White masculine identity and self-respect were complexly involved in these divisions of labor and wages. This is another way in which capitalism is a gendered and racialized accumulation process (Connell 2000). Wage differences between white men and all other groups, as well as divisions of labor between these groups, contributed to profit and flexibility, by helping to maintain growing occupational areas, such as clerical work, as segregated and low paid. Where women worked in manufacturing or food processing, gender divisions of labor kept the often larger female work force in low-wage routine jobs, while males worked in other more highly paid, less routine, positions (Acker and Van Houten 1974). While white men might be paid more, capitalist organizations could benefit from this "gender/racial dividend." Thus, by maintaining divisions, employers could pay less for certain levels of skill, responsibility, and experience when the worker was not a white male.

This is not to say that getting a living wage was easy for white men, or that most white men achieved it. Labor-management battles, employers' violent tactics to prevent unionization, [and] massive unemployment during frequent economic depressions characterized the situation of white industrial workers as wage labor spread in the nineteenth and early twentieth centuries. During the same period, new white-collar jobs were created to manage, plan, and control the expanding industrial economy. This rapidly increasing middle class was also stratified by gender and race. The better-paid, more respected jobs went to white men; white women were secretaries and clerical workers; people of color were absent. Conditions and issues varied across industries and regions of the country. But, wherever you look, those variations contained underlying gendered and racialized divisions. Patterns of stratification and segregation were written into employment contracts in work content, positions in work hierarchies, and wage differences, as well as other forms of distribution.

These patterns persisted, although with many alterations, through extraordinary changes in production and social life. After World War II, white women, except for a brief period immediately after the war, went to work for pay in the expanding service sector, professional, and managerial fields. African Americans moved to the North in large numbers, entering industrial and service sector jobs. These processes accelerated after the 1960s, with the civil rights and women's movements, new civil rights laws, and affirmative action. Hispanics and Asian Americans, as well as other racial/ethnic groups, became larger proportions of the population, on the whole finding work in low-paid, segregated jobs. Employers continued, and still continue, to select and promote workers based on gender and racial identifications, although the processes are more subtle, and possibly less visible, than in the past (for example, Brown et al. 2003; Royster 2003). These processes continually recreate gender and racial inequities, not as cultural or ideological survivals from earlier times, but as essential elements in present capitalisms (Connell 1987, 103–106).

Segregating practices are a part of the history of white, masculine-dominated capitalism that establishes class as gendered and racialized. Images of masculinity support these practices, as they produce a taken-for-granted world in which certain men legitimately make employment and other economic decisions that affect the lives of most other people. Even though some white women and people from other-than-white groups now hold leadership positions, their actions are shaped within networks of practices sustained by images of masculinity (Wacjman 1998).

Masculinities and Capitalism

Masculinities are essential components of the ongoing male project, capitalism. While white men were and are the main publicly recognized actors in the history of capitalism, these are not just any white men. They have been, for example, aggressive entrepreneurs or strong leaders of industry and finance (Collinson and Hearn 1996). Some have been oppositional actors, such as self-respecting and tough workers earning a family wage, and militant labor leaders. They have been particular men whose locations within gendered and racialized social relations and practices can be partially captured by the concept of masculinity. "Masculinity" is a contested term. As Connell (1995, 2000), Hearn (1996), and others have pointed out, it should be pluralized as "masculinities," because in any society at any one time there are several ways of being a man. "Being a man" involves cultural images and practices. It always implies a contrast to an unidentified femininity.

Hegemonic masculinity can be defined as the taken-for-granted, generally accepted form, attributed to leaders and other influential figures at particular historical times. Hegemonic masculinity legitimates the power of those who embody it. More than one type of hegemonic masculinity may exist simultaneously, although they may share characteristics, as do the business leader and the sports star at the present time. Adjectives describing hegemonic masculinities closely follow those describing characteristics of successful business organizations, as Rosabeth Moss Kanter (1977) pointed out in the 1970s. The successful CEO and the successful organization are aggressive, decisive, competitive, focused on winning and defeating the enemy, taking territory from others. The ideology of capitalist markets is imbued with a masculine ethos. As R. W. Connell (2000, 35) observes, "The market is often seen as the antithesis of gender (marked by achieved versus ascribed status, etc.). But the market operates through forms of

rationality that are historically masculine and involve a sharp split between instru-
mental reason on the one hand, emotion and human responsibility on the other"
(Seidler 1989). Masculinities embedded in collective practices are part of the context within which certain men made and still make the decisions that drive and
shape the ongoing development of capitalism. We can speculate that how these
men see themselves, what actions and choices they feel compelled to make and
they think are legitimate, how they and the world around them define desirable
masculinity, enter into that decision making (Reed 1996). Decisions made at the
very top reaches of (masculine) corporate power have consequences that are experienced as inevitable economic forces or disembodied social trends. At the same
time, these decisions symbolize and enact varying hegemonic masculinities
(Connell 1995). However, the embeddedness of masculinity within the ideologies
of business and the market may become invisible, seen as just part of the way business is done. The relatively few women who reach the highest positions probably
think and act within these strictures.

Hegemonic masculinities and violence are deeply connected within capitalist
history: The violent acts of those who carried out the slave trade or organized
colonial conquests are obvious examples. Of course, violence has been an essential component of power in many other socioeconomic systems, but it continues
into the rational organization of capitalist economic activities. Violence is frequently a legitimate, if implicit, component of power exercised by bureaucrats
as well as "robber barons." Metaphors of violence, frequently military violence,
are often linked to notions of the masculinity of corporate leaders, as "defeating
the enemy" suggests. In contemporary capitalism, violence and its links to masculinity are often masked by the seeming impersonality of objective conditions.
For example, the masculinity of top managers, the ability to be tough, is
involved in the implicit violence of many corporate decisions, such as those cutting jobs in order to raise profits and, as a result, producing unemployment.
Armies and other organizations, such as the police, are specifically organized
around violence. Some observers of recent history suggest that organized violence, such as the use of the military, is still mobilized at least partly to reach
capitalist goals, such as controlling access to oil supplies. The masculinities of
those making decisions to deploy violence in such a way are hegemonic, in the
sense of powerful and exemplary. Nevertheless, the connections between masculinity, capitalism, and violence are complex and contradictory, as Jeff Hearn and
Wendy Parkin (2001) make clear. Violence is always a possibility in mechanisms
of control and domination, but it is not always evident, nor is it always used.

As corporate capitalism developed, Connell (1995) and others (for example,
Burris 1996) argue that a hegemonic masculinity based on claims to expertise
developed alongside masculinities organized around domination and control.
Hegemonic masculinity relying on claims to expertise does not necessarily lead
to economic organizations free of domination and violence, however (Hearn and
Parkin 2001). Hearn and Parkin (2001) argue that controls relying on both
explicit and implicit violence exist in a wide variety of organizations, including
those devoted to developing new technology.

Different hegemonic masculinities in different countries may reflect different national histories, cultures, and change processes. For example, in Sweden in the mid-1980s, corporations were changing the ways in which they did business toward a greater participation in the international economy, fewer controls on currency and trade, and greater emphasis on competition. Existing images of dominant masculinity were changing, reflecting new business practices. This seemed to be happening in the banking sector, where I was doing research on women and their jobs (Acker 1994a). The old paternalistic leadership, in which primarily men entered as young clerks expecting to rise to managerial levels, was being replaced by young, aggressive men hired as experts and managers from outside the banks. These young, often technically trained, ambitious men pushed the idea that the staff was there to sell bank products to customers, not, in the first instance, to take care of the needs of clients. Productivity goals were put in place; nonprofitable customers, such as elderly pensioners, were to be encouraged not to come into the bank and occupy the staff's attention. The female clerks we interviewed were disturbed by these changes, seeing them as evidence that the men at the top were changing from paternal guardians of the people's interests to manipulators who only wanted riches for themselves. The confirmation of this came in a scandal in which the CEO of the largest bank had to step down because he had illegally taken money from the bank to pay for his housing. The amount of money was small; the disillusion among employees was huge. He had been seen as a benign father; now he was no better than the callous young men on the way up who were dominating the daily work in the banks. The hegemonic masculinity in Swedish banks was changing as the economy and society were changing.

Hegemonic masculinities are defined in contrast to subordinate masculinities. White working class masculinity, although clearly subordinate, mirrors in some of its more heroic forms the images of strength and responsibility of certain successful business leaders. The construction of working class masculinity around the obligations to work hard, earn a family wage, and be a good provider can be seen as providing an identity that both served as a social control and secured male advantage in the home. That is, the good provider had to have a wife and probably children for whom to provide. Glenn (2002) describes in some detail how this image of the white male worker also defined him as superior to and different from black workers.

Masculinities are not stable images and ideals, but shifting with other societal changes. With the turn to neoliberal business thinking and globalization, there seem to be new forms. Connell (2000) identifies "global business masculinity," while Lourdes Beneria (1999) discusses the "Davos man," the global leader from business, politics, or academia who meets his peers once a year in the Swiss town of Davos to assess and plan the direction of globalization. Seeing masculinities as implicated in the ongoing production of global capitalism opens the possibility of seeing sexualities, bodies, pleasures, and identities as also implicated in economic relations.

In sum, gender and race are built into capitalism and its class processes through the long history of racial and gender segregation of paid labor and through the images and actions of white men who dominate and lead central capitalist endeavors. Underlying these processes is the subordination to production and the market of nurturing and caring for human beings, and the assignment of these responsibilities to women as unpaid work. Gender segregation that differentially affects women in all racial groups rests at least partially on the ideology and actuality of women as carers. Images of dominant masculinity enshrine particular male bodies and ways of being as different from the female and distanced from caring. . . . I argue that industrial capitalism, including its present neoliberal form, is organized in ways that are, at the same time, antithetical and necessary to the organization of caring or reproduction and that the resulting tensions contribute to the perpetuation of gendered and racialized class inequalities. Large corporations are particularly important in this process as they increasingly control the resources for provisioning but deny responsibility for such social goals.

REFERENCES

Acker, Joan. 1994a. The Gender Regime of Swedish Banks. *Scandinavian Journal of Management* 10, no. 2: 117–30.

Acker, Joan, and Donald Van Houten. 1974. Differential Recruitment and Control: The Sex Structuring of Organizations. *Administrative Science Quarterly* 19 (June, 1974): 152–63.

Amott, Teresa, and Julie Matthaei. 1996. *Race, Gender, and Work: A Multi-cultural Economic History of Women in the United States.* Revised edition. Boston: South End Press.

Beneria, Lourdes. 1999. Globalization, Gender and the Davos Man. *Feminist Economics* 5, no. 3: 61–83.

Brodkin, Karen. 1998. Race, Class, and Gender: The Metaorganization of American Capitalism. *Transforming Anthropology* 7, no. 2: 46–57.

Brown, Michael K., Martin Carnoy, Elliott Currie, Troy Duster, David B. Oppenheimer, Marjorie M. Shultz, and David Wellman. 2003. *White-Washing Race: The Myth of a Color-Blind Society.* Berkeley: University of California Press.

Burris, Beverly H. 1996. Technocracy, Patriarchy and Management. In *Men as Managers, Managers as Men,* ed. David L. Collinson and Jeff Hearn. London: Sage.

Cockburn, Cynthia. 1983. *Brothers.* London: Pluto Press.

———. 1991. *In the Way of Women: Men's Resistance to Sex Equality in Organization.* Ithaca, N.Y: ILR Press.

Cohn, Samuel. 1985. *The Process of Occupational Sex-Typing: The Femininization of Clerical Labor in Great Britain.* Philadelphia: Temple University Press.

Collins, Patricia Hill. 2000. *Black Feminist Thought,* second edition. New York and London: Routledge.

Collinson, David L., and Jeff Hearn. 1996. Breaking the Silence: On Men, Masculinities and Managements. In *Men as Managers, Managers as Men,* ed. David L. Collinson and Jeff Hearns. London: Sage.

Connell, R. W. 1987. *Gender & Power*. Stanford, Calif: Stanford University Press.

———. 1995. *Masculinities*. Berkeley: University of California Press.

———. 2000. *The Men and the Boys*. Berkeley: University of California Press.

Figart, Deborah M., Ellen Mutari, and Marilyn Power. 2002. *Living Wages, Equal Wages*. London and New York: Routledge.

Glenn, Evelyn Nakano. 2002. *Unequal Freedom: How Race and Gender Shaped American Citizenship and Labor*. Cambridge: Harvard University Press.

Goldin, Claudia. 1990. *Understanding the Gender Gap: An Economic History of American Women*. New York and Oxford: Oxford University Press.

Hearn, Jeff. 1996. Is Masculinity Dead? A Critique of the Concept of Masculinity/ Masculinities. In *Understanding Masculinities: Social Relations and Cultural Arenas*, ed. M. Mac an Ghaill. Buckingham: Oxford University Press.

———. 2004. From Hegemonic Masculinity to the Hegemony of Men. *Feminist Theory* 5, no. 1: 49–72.

Hearn, Jeff, and Wendy Parkin. 2001. *Gender, Sexuality and Violence in Organizations*. London: Sage.

Kanter, Rosabeth Moss. 1977. *Men and Women of the Corporation*. New York: Basic Books.

Keister, Lisa. 2000. *Wealth in America: Trends in Wealth Inequality*. Cambridge: Cambridge University Press.

Kessler-Harris, Alice. 1982. *Out to Work: A History of Wage-Earning Women in the United States*. New York: Oxford University Press.

Omi, Michael, and Howard Winant. 1994. *Racial Formation in the United States*. New York: Routledge.

Padavic, Irene, and Barbara Reskin. 2002. *Women and Men at Work*, second edition. Thousand Oaks, Calif.: Pine Forge Press.

Perrow, Charles. 2002. *Organizing America*. Princeton and Oxford: Princeton University Press.

Reed, Rosslyn. 1996. Entrepreneurialism and Paternalism in Australian Management: A Gender Critique of the "Self-Made" Man. In *Men as Managers, Managers as Men*, ed. David L. Collinson and Jeff Hearn. London: Sage.

Royster, Deirdre A. 2003. *Race and the Invisible Hand: How White Networks Exclude Black Men from Blue-Collar Jobs*. Berkeley: University of California Press.

Seidler, Victor J. 1989. *Rediscovering Masculinity: Reason, Language*, and *Sexuality*. London and New York: Routledge.

Smith, Dorothy. 1999. *Writing the Social: Critique, Theory, and Investigation*. Toronto: University of Toronto Press.

Wacjman, Judy. 1998. *Managing Like a Man*. Cambridge: Polity Press.

18

How the New Working Class Can Transform Urban America

ROBIN D. G. KELLEY

Corporate downsizing, deindustrialization, racist and sexist social policy, and
...the erosion of the welfare state do not respect the boundaries between
work and community, the household and public space. The battle for livable
wages and fulfilling jobs is inseparable from the fight for decent housing and
safe neighborhoods; the struggle to defuse cultural stereotypes of inner city residents cannot be easily removed from the intense fights for environmental justice.
Moreover, the struggle to remake culture itself, to develop new ideas, new relationships, and new values that place mutuality over materialism and collective
responsibility over "personal responsibility," and place greater emphasis on ending all forms of oppression rather than striving to become an oppressor, cannot
be limited to either home or work.

Standing in the eye of the storm are the new multiracial, urban working
classes. It is they, not the Democratic Party, not a bunch of smart policy analysts,
not corporate benevolence, who hold the key to transforming the city and the
nation. . . . But my point is very, very simple.

MAKING OF THE NEW URBAN WORKING CLASS

. . .The pervasive imagery of the "underclass" makes the very idea of a contemporary urban working class seem obsolete. Instead of hardworking urban residents, many of whom are Latino, Asian-Pacific Islanders, West Indian immigrants, and U.S.-born African Americans, the dominant image of the "ghetto"
is of idle black men drinking forty-ounce bottles of malt liquor and young
black women "with distended bellies, their youthful faces belying the fact that
they are often close to delivering their second or third child." White leftists nostalgic for the days before identity politics allegedly undermined the "class struggle" often have trouble seeing beyond the ruddy-faced hard hats or European
immigrant factory workers who populate social history's images of the
American working class. The ghetto is the last place to find American workers.

SOURCE: Robin D. G. Kelley. 1997. *Yo' Mama's Disfunktional: Fighting the Culture Wars
in Urban America*. (Boston: Beacon, 1997).

If you are looking for the American working class today, however, you will do just as well to look in hospitals and universities as in the sooty industrial suburbs and smokestack districts of days past. In Bethlehem, Pennsylvania, for example, once a stronghold of the steel industry, nursing homes have become the fastest growing source of employment, and the unions that set out to organize these workers have outgrown the steelworkers' union by leaps and bounds. The new working class is also concentrated in food processing, food services, and various retail establishments. In the world of manufacturing, sweatshops are making a huge comeback, particularly in the garment industry and electronics assembling plants, and homework (telephone sales, for example) is growing. These workers are more likely to be brown and female than the old blue-collar white boys we are so accustomed to seeing in popular culture. . . .

Organizing the new immigrant labor force is perhaps the fundamental challenge facing the labor movement. For one, a substantial proportion of immigrant workers is employed by small ethnic firms with little tolerance for labor unions. Besides obvious language and cultural barriers, union leaders are trying to tackle the herculean task of organizing thousands of tiny, independent, sometimes transient firms. Immigrants are also less represented in public sector jobs, which tend to have a much higher percentage of unionized employees. (Indeed, the heavy concentration of native-born black people in public sector jobs partly explains why African Americans have such a high unionization rate.) The most obvious barrier to organizing immigrant workers, however, has been discriminatory immigration policy. . . . The 1986 Immigration Reform and Control Act imposed legal sanctions against employers of "aliens" without proper documents. Thus, even when unions were willing to organize undocumented workers, fear of deportation kept many workers from joining the labor movement.

Unions are also partly to blame for the state of labor organizing among immigrant workers. Until recently, union leaders too often assumed that Latino and Asian workers were unorganizable or difficult to organize—arguments that have been made about women and African American workers in the past. . . . Even when union organizers were willing to approach undocumented workers, they often operated on the assumption that immigrants were easily manipulated by employers, willing to undercut prevailing wages, or were "target workers" whose goal was to make enough money to return to their place of origin. . . .

Several decades of grassroots, rank-and-file efforts on the part of workers of color [have tried to] re-orient unions toward issues of social justice, racism, sexism, and cultural difference within their ranks. Indeed, one of the most significant labor-based social justice movements emerged out of the Service Employees International Union (SEIU). . . . Launched in 1985, Justice for Janitors sought to build a mass movement to win union recognition and to address the needs of a workforce made up primarily of people of color, mainly immigrants. As Sweeney explained, "The strategy of Justice for Janitors was to build a mass movement, with workers making clear that they wanted union representation and winning 'voluntary recognition' from employers. The campaigns addressed the special needs of an immigrant workforce, largely from Latin America. In many cities, the janitors' cause became a civil rights movement—and a cultural crusade." Throughout the mid- to late 1980s, Justice for Janitors waged several successful "crusades" in Pittsburgh, Denver, San Diego, Los Angeles, and Washington, D.C. With support from Latino leaders and local church officials, for example, their Denver campaign yielded a sudden growth in unionization among janitors and wage increases of about ninety cents an hour.

From its inception, Justice for Janitors has been deeply committed to antiracism and mass mobilization through community-based organizing and civil disobedience. Heirs of the sit-down strikers of the 1930s and the Civil Rights movement of the 1950s and 1960s, they have waged militant, highly visible campaigns in major cities throughout the country. In Los Angeles, for example, Justice for Janitors is largely responsible for the dramatic increase in unionized custodial employees, particularly among workers contracted out by big firms to clean high-rise buildings. The percentage of janitors belonging to unions rose from 10 percent of the workforce in 1987 to 90 percent in 1995. Their success certainly did not come easily. Indeed, the turning point in their campaign began around 1990, when two hundred janitors struck International Service Systems (ISS), a Danish-owned company. A mass march in Century City, California, in support of the strikers generated enormous publicity after police viciously attacked the demonstrators. Overall, more than sixty people were hospitalized, including a pregnant woman who was beaten so severely she miscarried. An outpouring of sympathy for the strikers turned the tables, enabling them to win a contract with ISS covering some 2,500 janitors in Southern California. . . .

In Washington, D.C., Justice for Janitors led the struggle of Local 82 of the SEIU in its fight against U.S. Service Industries (USSI)—a private janitorial company that used nonunion labor to clean downtown office buildings. But because they conceived of themselves as a social movement, they did not stop with the protection of union jobs. In March 1995, Justice for Janitors organized several demonstrations in the district that led to over 200 arrests. Blocking traffic and engaging in other forms of civil disobedience, the protesters demanded an end to tax breaks to real estate developers as well as cutbacks in social programs for the poor. As union spokesperson Manny Pastreich put it, "This isn't just about 5,000 janitors; it's about issues that concern all D.C. residents—what's happening to their schools, their streets, their neighborhoods." Many people who participated in the March demonstrations came from all over the country, and not all were janitors. Greg Ceci, a longshoreman from Baltimore, saw the struggle as a general revitalization of the labor movement and a recognition, finally, that the unions need to lead a larger fight for social justice. "We need to reach out to the workers who have been ignored by mainstream unions. We need to fight back, and I want to be a part of it."

Indeed, D.C.'s Justice for Janitors is made up precisely of workers who have been ignored—poor women of color. Black and Latino women make up the majority of its membership, hold key leadership positions, and have put their bodies on the line for the SEIU as well as the larger cause of social justice. Twenty-four-year-old Dania Herring is an example of Justice for Janitors' new leadership cadre. A mother of four and resident of one of the poorest neighborhoods in the southeast section of the District, whose husband (at the time of the demonstrations) was an unemployed bricklayer, Herring had quit her job to become a full-time organizer for Local 82 and Justice for Janitors. Yet, these women were so militant that members of Washington's District Council dismissed them as "hooligans." "Essentially, they use anarchy as a means of organizing workers," stated African American councilman Harold Brazil. "And they do that under the mantle of justice—for janitors—or whoever else they want to organize."

In the end, their challenge to the city and to USSI paid off. In December 1995, the National Labor Relations Board (NLRB) concluded that USSI had "a history of pervasive illegal conduct" by threatening, interrogating, and firing employees they deemed unacceptable, especially those committed to union organizing. African American workers, in particular, had suffered most from the wave of firings, in part because USSI, like other employers, believed immigrant workers were more malleable and less committed to unionization because of their tenuous status as residents. The pattern was clear to Amy Parker, a janitor who had worked under USSI for three years. "Before USSI got the contract to clean my building, there were 18 African-Americans. Now, I am the only one." After three steady years cleaning the same building, she earned only $5.50 an hour. The NLRB's decision against USSI, therefore, was a substantial victory for Local 82 of the SEIU. It meant that USSI could no longer discriminate against the union and it generated at least a modicum of recognition for Justice for Janitors.

Such stories have been duplicated across the country. Justice for Janitors, in short, is more than a union. It is a dynamic social movement made up primarily of women and people of color, and it has the potential to redirect the entire labor movement. . . .

LABOR/COMMUNITY STRATEGIES AGAINST
CLASS-BASED RACISM

. . .Urban working people spend much if not most of their lives in their neighborhoods, in their homes, in transit, in the public spaces of the city, in houses of worship, in bars, clubs, barber-shops, hair and nail salons, in various retail outlets, in medical clinics, welfare offices, courtrooms, even jail cells. They create and maintain families, build communities, engage in local politics, and construct a sense of fellowship that is sometimes life sustaining. These community ties are crucial to the success of any labor movement. . . .

Working people live in communities that are as embattled as the workplace itself. Black and Latino workers, for example, must contend with issues of police brutality and a racist criminal justice system, housing discrimination, lack of city services, toxic waste, inadequate health care facilities, sexual assault and domestic violence, and crime and neighborhood safety. And at the forefront of these community-based movements have been women, usually struggling mothers of all ages dedicated to making life better for themselves and their children— mothers who, as we have seen, have become the scapegoats for virtually everything wrong with the "inner city." In cities across the United States, working-class black and Latina women built and sustained community organizations that registered voters, patrolled the streets, challenged neighborhood drug dealers, defended the rights of prisoners, and fought vigorously for improvements in housing, city services, health care, and public assistance. . . .

Whereas most community- and labor-based organizations limit their focus to an issue or set of issues, even when they are able to see the bigger picture, once in a while there are movements that attempt to fight on all fronts. Such organizations, where they do exist, are often products of the best elements of Third World, feminist, and Black Liberation movements. Rather than see race, gender, and sexuality as "problems," they are, instead, pushing working-class politics in new directions. . . .

One of the most visible and successful examples of such a broad-based radical movement is the Labor/Community Strategy Center based in Los Angeles. The leaders of the Strategy Center have deep roots in social movements that go back to Black Liberation and student activism of the 1960s, urban antipoverty programs, farmworkers' movements, organized labor, and popular left movements in El Salvador and Mexico. They have been at the forefront of the struggle for clean air in the Harbor area of Los Angeles—a region with a high concentration of poor communities of color. They have worked closely with Justice for Janitors, providing crucial support to Local 660 of the SEIU. . . .

The Strategy Center's most important campaign during the last few years has been the Bus Riders Union (BRU), a multiracial organization of transit-dependent working people who have declared war on race and class inequality in public transportation. . . . Public transportation is one of the few issues that touches the lives of many urban working people across race, ethnic, and gender lines. And ultimately, equal access to affordable transportation is tied to equal access to employment opportunities, especially now that manufacturing and other medium- to high-wage jobs have migrated to suburban rings or industrial parks. The combination of rising fares and limits on services has sharply curtailed the ability of workers to get to work. The cost of transportation for many families is a major part of their monthly budget. The average working-class commuter in Los Angeles rides the bus between sixty and eighty times a month, including for work, family visits, medical care, and shopping. For transit-dependent wage earners, bus fare could potentially comprise up to one-fourth or more of their total income! For many poor people even a small increase makes riding prohibitive. They might visit family less often, skip grocery shopping, or even miss work if they do not have bus fare. Forcing low-income riders off the buses with higher fares moves the Los Angeles County Metropolitan Transit Authority (MTA) out of the business of transporting the urban poor.

After three years of conducting research and riding the buses to recruit members, in 1994 the Strategy Center and the Bus Riders Union, along with the Southern Christian Leadership Conference and the Korean Immigrant Workers' Advocates as coplaintiffs, filed a class action suit against the MTA on behalf of 350,000 bus riders, the vast majority of whom are poor people of color. Represented by the National Association for the Advancement of Colored People's Legal Defense and Educational Fund, the plaintiffs challenged proposals to raise bus fares from $1.10 to $1.35; eliminate monthly bus passes, which cost $42 for unlimited use; and introduce a zone system on the blue line commuter rail that would increase fares by more than 100 percent for half of the passengers. . . . The plaintiffs argued that the MTA's policies violate Title VI of the Civil Rights Act of 1964, which provides that no person "shall on the ground of race, color, or national origin, be excluded from participation in, be denied the benefits of, or be subjected to discrimination under any program or activity receiving Federal financial assistance."

The evidence that the MTA *has* created a separate and unequal transit system is overwhelming. On one side is the underfunded, overcrowded bus system which carries 94 percent of the MTA's passengers (80 percent of whom are African American, Latino, Asian-Pacific Islander, and Native American) yet receives less than one-third of the MTA's total expenditures. As a result, riders of color experience greater peak-hour overcrowding and tend to wait at neighborhood bus stops that are benchless and unsheltered. On the other side is the lavish commuter rail system connecting (or planning to connect) the predominantly white suburbs to the L.A. downtown business district. Although commuter line passengers make up 6 percent of the ridership, the system receives 70.9 percent of the MTA's resources. To make matters worse, the fare hike and proposed elimination of the monthly bus passes were to be accompanied by massive cutbacks in bus service.

. . .The proposed rail system has already created a two-tiered riding public—luxury trains for the white middle class, buses for the colored poor. Furthermore, the MTA paid for its Metrolink (train) lines with revenues earned from poor and working-class bus riders. Monies that should have been earmarked to improve bus service instead went to a partially finished, impractical rail system that will cost approximately $183 billion and only serve 11 percent of the population. It doesn't take a high IQ to realize that poor black and brown riders who have to crowd onto South Central, Inglewood, and East Los Angeles buses have been subsidizing the commuter rail. The average commuter rail rider is a white male professional with a household income of $64,450, who receives a $58 transportation subsidy from his employer and owns a car. By contrast, a typical bus rider is a person of color with no car and whose household income falls below $15,000. Put another way, the MTA's subsidy for bus #204, which runs through predominantly black and Latino neighborhoods along Vermont Avenue, will remain 34 cents per passenger, while the projected subsidy for the proposed Azusa and Torrance rail lines will be $55.64 and $52.87, respectively. Of course, the MTA will use revenues from sales tax and fare-box receipts to pay for this, but it is also dependent on federal funding: hence the violation of Title VI of the Civil Rights Act.

Like the Student Non-Violent Coordinating Committee in the 1960s, Strategy Center and BRU organizers take litigation *and* mass organizing seriously, but they are very clear which component drives the process. They have held legal workshops for members, conducted hours and hours of background research, prepared the initial strategy document on which the case is based, and carefully examined and re-examined all legal documentation pertaining to the suit. At the same time, they have worked hard to ensure that "the class" has a voice in the class action suit. In addition to collecting detailed declarations from riders, their main core of organizers are also plaintiffs—that is, bus riders. . . .

The Bus Riders Union has been incredibly successful. Besides building a truly multicultural, multiracial mass movement, the BRU...wrested a settlement from the MTA that imposed a two-year freeze on bus fares, a substantial reduction in the price of monthly and biweekly bus passes, a 5-percent increase in the number of buses in the MTA fleet in order to reduce overcrowding, the creation of new bus lines with the intention of eradicating "transportation segregation," and the creation of a joint working group in which the BRU would play a major role in drafting a five-year plan to improve bus service for all. Hence a solidly antiracist movement resulted in a victory for all riders, including transit-dependent white workers, the disabled, the elderly, and students.

While the settlement marks a critical landmark in the history of civil rights struggle, Strategy Center organizers understand that the specific legal challenge is less important than the social movement that grows out of it.... Through their own struggles, in the courts and city council chambers, on the streets and buses of Los Angeles, the Bus Riders Union has advanced a radical vision of social justice.... Lacking such a vision, many of us have not been able to see why the struggle over public transportation is so important for working people, or how antiracism benefits the entire working class.... The bus campaigns powerfully demonstrate that the issues facing the vast majority of people of color, for the most part, are working-class issues. Thus, unless we understand the significance of the "billions for buses" campaign or the historic class action suit against the MTA, we will never understand the deeper connections between welfare, workfare, warfare, and bus fare...or bus riders and unions, for that matter.

In this tragic era of pessimism and defeat, these grassroots, working-class radical movements go forward as if they might win.... The women and men who have built and sustained these organizations do not have the luxury not to fight back. In most cases, they are battling over issues basic to their own survival—decent wages, healthy environment, essential public services. At the heart of these movements are folks like my mother and the many other working-class women of color... who have borne the brunt of the material and ideological war against the poor. They understand, better than anyone, the necessity of fighting back. And they also understand that change does not come on its own. As Dr. Martin Luther King, Jr., so eloquently explained:

A solution of the present crisis will not take place unless men and women work for it. Human progress is neither automatic nor inevitable. Even a superficial look at history reveals that no social advance rolls in on the wheels of inevitability. Every step toward the goal of justice requires sacrifice, suffering, and struggle; the tireless exertions and passionate concern of dedicated individuals. Without persistent effort, time itself becomes an ally of the insurgent and primitive forces of irrational emotionalism and social destruction. This is no time for apathy or complacency. This is the time for vigorous and positive action.

The time is now.

19

Sex and Gender Through the Prism of Difference

MAXINE BACA ZINN, PIERRETTE HONDAGNEU-SOTELO,
AND MICHAEL MESSNER

"Men can't cry." "Women are victims of patriarchal oppression." "After divorces, single mothers are downwardly mobile, often moving into poverty." "Men don't do their share of housework and child care." "Professional women face barriers such as sexual harassment and a 'glass ceiling' that prevent them from competing equally with men for high-status positions and high salaries." "Heterosexual intercourse is an expression of men's power over women." Sometimes, the students in our sociology and gender studies courses balk at these kinds of generalizations. And they are right to do so. After all, some men are more emotionally expressive than some women, some women have more power and success than some men, some men do their share—or more—of housework and child care, and some women experience sex with men as both pleasurable and empowering. Indeed, contemporary gender relations are complex and changing in various directions, and as such, we need to be wary of simplistic, if handy, slogans that seem to sum up the essence of relations between women and men.

On the other hand, we think it is a tremendous mistake to conclude that "all individuals are totally unique and different," and that therefore all generalizations about social groups are impossible or inherently oppressive. In fact, we are convinced that it is this very complexity, this multifaceted nature of contemporary gender relations, that fairly begs for a sociological analysis of gender. We use the image of "the prism of difference" to illustrate our approach to developing this sociological perspective on contemporary gender relations. The *American Heritage Dictionary* defines "prism," in part, as "a homogeneous transparent solid, usually with triangular bases and rectangular sides, used to produce or analyze a continuous spectrum." Imagine a ray of light—which to the naked eye appears to be only one color—refracted through a prism onto a white wall. To the eye, the result is not an infinite, disorganized scatter of individual colors. Rather, the refracted light displays an order, a structure of relationships among the different colors—a rainbow. Similarly, we propose to use the "prism of difference" ... to analyze a

SOURCE: From Maxine Baca Zinn, Pierrette Hondagneu-Sotelo, and Michael Messner,
Gender through the Prism of Difference 3rd ed. (New York: Oxford University Press, 2005)
Copyright © by Pearson Education. Reprinted by permission.

continuous spectrum of people, in order to show how gender is organized and experienced differently when refracted through the prism of sexual, racial/ethnic, social class, physical abilities, age, and national citizenship differences.

EARLY WOMEN'S STUDIES: CATEGORICAL VIEWS OF "WOMEN" AND "MEN"

...It is possible to make good generalizations about women and men. But these generalizations should be drawn carefully, by always asking the questions "*which* women?" and "*which* men?" Scholars of sex and gender have not always done this. In the 1960s and 1970s, women's studies focused on the differences *between* women and men rather than *among* women and men. The very concept of gender, women's studies scholars demonstrated, is based on socially defined difference between women and men. From the macro level of social institutions such as the economy, politics, and religion, to the micro level of interpersonal relations, distinctions between women and men structure social relations. Making men and women *different* from one another is the essence of gender. It is also the basis of men's power and domination. Understanding this was profoundly illuminating. Knowing that difference produced domination enabled women to name, analyze, and set about changing their victimization.

In the 1970s, riding the wave of a resurgent feminist movement, colleges and universities began to develop women's studies courses that aimed first and foremost to make women's lives visible. The texts that were developed for these courses tended to stress the things that women shared under patriarchy—having the responsibility for housework and child care, the experience or fear of men's sexual violence, a lack of formal or informal access to education, and exclusion from high-status professional and managerial jobs, political office, and religious leadership positions (Brownmiller, 1975; Kanter, 1977).

The study of women in society offered new ways of seeing the world. But the 1970s approach was limited in several ways. Thinking of gender primarily in terms of differences *between* women and men led scholars to overgeneralize about both. The concept of patriarchy led to a dualistic perspective of male privilege and female subordination. Women and men were cast as opposites. Each was treated as a homogeneous category with common characteristics and experiences. This approach *essentialized* women and men. Essentialism, simply put, is the notion that women's and men's attributes and indeed women and men themselves are categorically different. From this perspective, male control and coercion of women produced conflict between the sexes. The feminist insight originally introduced by Simone De Beauvoir in 1953—that women, as a group, had been socially defined as the "other" and that men had constructed themselves as the subjects of history, while constructing women as their objects—fueled an energizing sense of togetherness among many women. As college students read books such as *Sisterhood Is Powerful* (Morgan, 1970), many of them joined organizations that fought—with some success—for equality and justice for women.

THE VOICES OF "OTHER" WOMEN

Although this view of women as an oppressed "other" was empowering for certain groups of women, some women began to claim that the feminist view of universal sisterhood ignored and marginalized their major concerns. It soon became apparent that treating women as a group united in its victimization by patriarchy was biased by too narrow a focus on the experiences and perspectives of women from more privileged social groups. "Gender" was treated as a generic category, uncritically applied to women. Ironically, this analysis, which was meant to unify women, instead produced divisions between and among them. The concerns projected as "universal" were removed from the realities of many women's lives. For example, it became a matter of faith in second-wave feminism that women's liberation would be accomplished by breaking down the "gendered public-domestic split." Indeed, the feminist call for women to move out of the kitchen and into the workplace resonated in the experiences of many of the college-educated white women who were inspired by Betty Friedan's 1963 book, *The Feminine Mystique.* But the idea that women's movement into workplaces was itself empowering or liberating seemed absurd or irrelevant to many working-class women and women of color. They were already working for wages, as had many of their mothers and grandmothers, and did not consider access to jobs and public life "liberating." For many of these women, liberation had more to do with organizing in communities and workplaces—often alongside men—for better schools, better pay, decent benefits, and other policies to benefit their neighborhoods, jobs, and families. The feminism of the 1970s did not seem to address these issues.

As more and more women analyzed their own experiences, they began to address the power relations that created differences among women and the part that privileged women played in the oppression of others. For many women of color, working-class women, lesbians, and women in contexts outside the United States (especially women in non-Western societies), the focus on male domination was a distraction from other oppressions. Their lived experiences could support neither a unitary theory of gender nor an ideology of universal sisterhood. As a result, finding common ground in a universal female victimization was never a priority for many groups of women.

Challenges to gender stereotypes soon emerged. Women of varied races, classes, national origins, and sexualities insisted that the concept of gender be broadened to take their differences into account (Baca Zinn et al., 1986; Hartmann, 1976; Rich, 1980; Smith, 1977). Many women began to argue that their lives were affected by their location in a number of different hierarchies: as African Americans, Latinas, Native Americans, or Asian Americans in the race hierarchy; as young or old in the age hierarchy; as heterosexual, lesbian, or bisexual in the sexual orientation hierarchy; and as women outside the Western industrialized nations, in subordinated geopolitical contexts. These arguments made it clear that women were not victimized by gender alone but by the historical and systematic denial of rights and privileges based on other differences as well.

MEN AS GENDERED BEINGS

As the voices of "other" women in the mid- to late 1970s began to challenge and expand the parameters of women's studies, a new area of scholarly inquiry was beginning to stir—a critical examination of men and masculinity. To be sure, in those early years of gender studies, the major task was to conduct studies and develop courses about the lives of women in order to begin to correct centuries of scholarship that rendered invisible women's lives, problems, and accomplishments. But the core idea of feminism—that "femininity" and women's subordination is a social construction—logically led to an examination of the social construction of "masculinity" and men's power. Many of the first scholars to take on this task were psychologists who were concerned with looking at the social construction of "the male sex role" (e.g., Pleck, 1976). By the late 1980s, there was a growing interdisciplinary collection of studies of men and masculinity, much of it by social scientists (Brod, 1987; Kaufman, 1987; Kimmel, 1987; Kimmel & Messner, 1989).

Reflecting developments in women's studies, the scholarship on men's lives tended to develop three themes: First, what we think of as "masculinity" is not a fixed, biological essence of men, but rather is a social construction that shifts and changes over time as well as between and among various national and cultural contexts. Second, power is central to understanding gender as a relational construct, and the dominant definition of masculinity is largely about expressing difference from—and superiority over—anything considered "feminine." And third, there is no singular "male sex role." Rather, at any given time there are various masculinities. R. W. Connell (1987; 1995; 2002) has been among the most articulate advocates of this perspective. Connell argues that hegemonic masculinity (the dominant form of masculinity at any given moment) is constructed in relation to femininities *as well as* in relation to various subordinated or marginalized masculinities. For example, in the United States, various racialized masculinities (e.g., as represented by African American men, Latino immigrant men, etc.) have been central to the construction of hegemonic (white middle-class) masculinity. This "othering" of racialized masculinities helps to shore up the privileges that have been historically connected to hegemonic masculinity. When viewed this way, we can better understand hegemonic masculinity as part of a system that includes gender as well as racial, class, sexual, and other relations of power.

The new literature on men and masculinities also begins to move us beyond the simplistic, falsely categorical, and pessimistic view of men simply as a privileged sex class. When race, social class, sexual orientation, physical abilities, immigrant, or national status are taken into account, we can see that in some circumstances, "male privilege" is partly—sometimes substantially—muted (Kimmel & Messner, 2004). Although it is unlikely that we will soon see a "men's movement" that aims to undermine the power and privileges that are connected with hegemonic masculinity, when we begin to look at "masculinities" through the prism of difference, we can begin to see similarities and possible points of coalition between and among certain groups of women and men

(Messner, 1998). Certain kinds of changes in gender relations—for instance, a national family leave policy for working parents—might serve as a means of uniting particular groups of women and men.

GENDER IN INTERNATIONAL CONTEXTS

It is an increasingly accepted truism that late twentieth-century increases in transnational trade, international migration, and global systems of production and communication have diminished both the power of nation-states and the significance of national borders. A much more ignored issue is the extent to which gender relations—in the United States and elsewhere in the world—are increasingly linked to patterns of global economic restructuring. Decisions made in corporate headquarters located in Los Angeles, Tokyo, or London may have immediate repercussions on how women and men thousands of miles away organize their work, community, and family lives (Sassen, 1991). It is no longer possible to study gender relations without giving attention to global processes and inequalities. . . .

Around the world, women's paid and unpaid labor is key to global development strategies. Yet it would be a mistake to conclude that gender is molded from the "top down." What happens on a daily basis in families and workplaces simultaneously constitutes and is constrained by structural transnational institutions. For instance, in the second half of the twentieth century young, single women, many of them from poor rural areas, were (and continue to be) recruited for work in export assembly plants along the U.S.-Mexico border, in East and Southeast Asia, in Silicon Valley, in the Caribbean, and in Central America. While the profitability of these multinational factories depends, in part, on management's ability to manipulate the young women's ideologies of gender, the women. . . do not respond passively or uniformly, but actively resist, challenge, and accommodate. At the same time, the global dispersion of the assembly line has concentrated corporate facilities in many U.S. cities, making available myriad managerial, administrative, and clerical jobs for college educated women. Women's paid labor is used at various points along this international system of production. Not only employment but also consumption embodies global interdependencies. There is a high probability that the clothing you are wearing and the computer you use originated in multinational corporate headquarters and in assembly plants scattered around third world nations. And if these items were actually manufactured in the United States, they were probably assembled by Latin American and Asian-born women.

Worldwide, international labor migration and refugee movements are creating new types of multiracial societies. While these developments are often discussed and analyzed with respect to racial differences, gender typically remains absent. As several commentators have noted, the white feminist movement in the United States has not addressed issues of immigration and nationality. Gender, however, has been fundamental in shaping immigration policies (Chang, 1994; Hondagneu-Sotelo, 1994). Direct labor recruitment programs generally solicit either male or female labor (e.g., Filipina nurses and Mexican male farm workers), national disenfranchisement has particular repercussions for women and men, and current immigrant laws are based on very gendered notions of what constitutes "family unification." As Chandra Mohanty suggests, "analytically these issues are the contemporary metropolitan counterpart of women's struggles against colonial occupation in the geographical third world" (1991:23). Moreover, immigrant and refugee women's daily lives often challenge familiar feminist paradigms. The occupations in which immigrant and refugee women concentrate—paid domestic work, informal sector street vending, assembly or industrial piece work performed in the home—often blur the ideological distinction between work and family and between public and private spheres (Hondagneu-Sotelo, 2001; Parrenas, 2001).

FROM PATCHWORK QUILT TO PRISM

All of these developments—the voices of "other" women, the study of men and masculinities, and the examination of gender in transnational contexts—have helped redefine the study of gender. By working to develop knowledge that is inclusive of the experiences of all groups, new insights about gender have begun to emerge. Examining gender in the context of other differences makes it clear that nobody experiences themselves as solely gendered. Instead, gender is configured through cross-cutting forms of difference that carry deep social and economic consequences.

By the mid-1980s, thinking about gender had entered a new stage, which was more carefully grounded in the experiences of diverse groups of women and men. This perspective is a general way of looking at women and men and understanding their relationships to the structure of society. Gender is no longer viewed simply as a matter of two opposite categories of people, males and females, but a range of social relations among differently situated people. Because centering on difference is a radical challenge to the conventional gender framework, it raises several concerns. If we think of all the systems that converge to simultaneously influence the lives of women and men, we can imagine an infinite number of effects these interconnected systems have on different women and men. Does the recognition that gender can be understood only contextually (meaning that there is no singular "gender" per se) make women's studies and men's studies newly vulnerable to critics in the academy? Does the immersion in difference throw us into a whirlwind of "spiraling diversity" (Hewitt, 1992:316) whereby multiple identities and locations shatter the categories "women" and "men"? . . .

We take a position directly opposed to an empty pluralism. Although the categories "woman" and "man" have multiple meanings, this does not reduce gender to a "postmodern kaleidoscope of lifestyles. Rather, it points to the *relational* character of gender" (Connell, 1992:736). Not only are masculinity and femininity relational, but different *masculinities* and *femininities* are interconnected through other social structures such as race, class, and nation. The concept of relationality suggests that "the lives of different groups are interconnected even without face-to-face relations (Glenn, 2002:14). The meaning of "woman" is defined by the existence of women of different races and classes. Being a white woman in the United States is meaningful only insofar as it is set apart from and in contradistinction to women of color.

Just as masculinity and femininity each depend on the definition of the other to produce domination, differences *among* women and *among* men are also created in the context of structured relationships. Some women derive benefits from their race and class position and from their location in the global economy, while they are simultaneously restricted by gender. In other words, such women are subordinated by patriarchy, yet their relatively privileged positions within hierarchies of race, class, and the global political economy intersect to create for them an expanded range of opportunities, choices, and ways of living. They may even use their race and class advantage to minimize some of the consequences of patriarchy and/or to oppose other women. Similarly, one can become a man in opposition to other men. For example, "the relation between heterosexual and homosexual men is central, carrying heavy symbolic freight. To many people, homosexuality is the *negation* of masculinity.... Given that assumption, antagonism toward homosexual men may be used to define masculinity" (Connell, 1992:736).

In the past decade, viewing gender through the prism of difference has profoundly reoriented the field (Acker, 1999; Glenn, 1999, 2002; Messner, 1996; West & Fenstermaker, 1995). Yet analyzing the multiple constructions of gender does not just mean studying groups of women and groups of men as different. It is clearly time to go beyond what we call the "patchwork quilt" phase in the study of women and men—that is, the phase in which we have acknowledged the importance of examining differences within constructions of gender, but do so largely by collecting together a study here on African American women, a study there on gay men, a study on working-class Chicanas, and so on. This patchwork quilt approach too often amounts to no more than "adding difference and stirring." The result may be a lovely mosaic, but like a patchwork quilt, it still tends to overemphasize boundaries rather than to highlight bridges of interdependency. In addition, this approach too often does not explore the ways that social constructions of femininities and masculinities are based on and reproduce relations of power. In short, we think that the substantial quantity of research that has now been done on various groups and subgroups needs to be analyzed within a framework that emphasizes differences and inequalities not as discrete areas of separation, but as interrelated bands of color that together make up a spectrum....

REFERENCES

Acker, Joan. 1999. "Rewriting Class, Race and Gender: Problems in Feminist Rethinking" Pp. 44–69 in Myra Marx Ferree, Judith Lorber, and Beth B. Hess (eds.), *Revisioning Gender.* Thousand Oaks, CA: Sage Publications.

Baca Zinn, M., L., Weber Cannon, E., Higgenbotham, & B., Thornton Dill. 1986. "The Costs of Exclusionary Practices in Women's Studies," *Signs: Journal of Women in Culture and Society* 11: 290–303.

Brod, Harry (ed.). 1987. *The Making of Masculinities: The New Men's Studies.* Boston: Allen & Unwin.

Brownmiller, Susan. 1975. *Against Our Will: Men, Women, and Rape.* New York: Simon & Schuster.

Chang, Grace. 1994. "Undocumented Latinas: The New 'Employable Mothers.'" Pp. 259–285 in Evelyn Nakano Glenn, Grace Chang, and Linda Rennie Forcey (eds.), *Mothering, Ideology, Experience, and Agency.* New York and London: Routledge.

Connell, R. W. 1987. *Gender and Power.* Stanford, CA: Stanford University Press.

Connell, R. W. 1992. "A Very Straight Gay: Masculinity, Homosexual Experience, and the Dynamics of Gender," *American Sociological Review* 57: 735–751.

Connell, R. W. 1995. *Masculinities.* Berkeley: University of California Press.

Connell, R. W. 2002. *Gender.* Cambridge: Polity.

De Beauvoir, Simone. 1953. *The Second Sex.* New York: Knopf.

Glenn, Evelyn Nakano. 1999. "The Social Construction and Institutionalization of Gender and Race: An Integrative Framework," Pp. 3–43 in Myra Marx Ferree, Judith Lorber, and Beth B. Hess (eds.), *Revisioning Gender.* Thousand Oaks. CA: Sage Publications.

Glenn, Evelyn Nakano. 2002. *Unequal Sisterhood: How Race and Gender Shaped American Citizenship and Labor.* Cambridge, MA: Harvard University Press.

Hartmann, Heidi. 1976. "Capitalism, Patriarchy, and Job Segregation by Sex," *Signs: Journal of Women in Culture and Society* 1(3), part 2, spring: 137–167.

Hewitt, Nancy A. 1992. "Compounding Differences," *Feminist Studies* 18: 313–326.

Hondagneu-Sotelo, Pierrette. 1994. *Gendered Transitions: Mexican Experiences of Immigration.* Berkeley: University of California Press.

Hondagneu-Sotelo, Pierrette. 2001. *Doméstica: Immigrant Workers Cleaning and Caring in the Shadows of Affluence.* Berkeley: University of California Press.

Kanter, Rosabeth Moss. 1977. *Men and Women of the Corporation.* New York: Basic Books.

Kaufman, Michael. 1987. *Beyond Patriarchy: Essays by Men on Pleasure, Power, and Change.* Toronto and New York: Oxford University Press.

Kimmel, Michael S. (ed.). 1987. *Changing Men: New Directions in Research on Men and Masculinity.* Newbury Park, CA: Sage.

Kimmel, Michael S. 1996. *Manhood in America: A Cultural History.* New York: Free Press.

Kimmel, Michael S. & Michael A. Messner (eds.). 1989. *Men's Lives.* New York: Macmillan.

Kimmel, Michael S. & Michael A. Messner (eds.). 2004. *Men's Lives,* 6th ed. Boston: Pearson.

Messner, Michael A. 1996. "Studying Up on Sex," *Sociology of Sport Journal* 13: 221–237.

Messner, Michael A. 1998. *Politics of Masculinities: Men in Movements.* Thousand Oaks, CA: Sage Publications.

Mohanty, Chandra Talpade. 1991. "Cartographies of Struggle: Third World Women and the Politics of Feminism." Pp. 51–80 in Chandra Talpade Mohanty, Ann Russo, and Lourdes Torres, (eds.), *Third World Women and the Politics of Feminism.* Bloomington: Indiana University Press.

Morgan, Robin. 1970. *Sisterhood Is Powerful: An Anthology of Writing from the Women's Liberation Movement.* New York: Vintage Books.

Parrenas, Rhacel Salazar. 2001. *Servants of Globalization: Women, Migration and Domestic Work.* Stanford: Stanford University Press.

Pleck, J. H. 1976. "The Male Sex Role: Definitions, Problems, and Sources of Change," *Journal of Social Issues* 32: 155–164.

Rich, Adrienne. 1980. "Compulsory Heterosexuality and the Lesbian Experience," *Signs: Journal of Women in Culture and Society* 5: 631–660.

Sassen, Saskia. 1991. *The Global City: New York, London, Tokyo.* Princeton: Princeton University Press.

Smith, Barbara. 1977. *Toward a Black Feminist Criticism.* Freedom, CA: Crossing Press.

West, Candace & Sarah Fenstermaker. 1995. "Doing Difference," *Gender & Society* 9: 8–37.

20

The Myth of the Latin Woman
I Just Met a Girl Named María

JUDITH ORTIZ COFER

On a bus trip to London from Oxford University, where I was earning some graduate credits one summer, a young man, obviously fresh from a pub, spotted me and as if struck by inspiration went down on his knees in the aisle. With both hands over his heart he broke into an Irish tenor's rendition of "María" from *West Side Story*. My politely amused fellow passengers gave his lovely voice the round of gentle applause it deserved. Though I was not quite

SOURCE: From Judith Ortiz Cofer, *The Latin Deli: Prose & Poetry* (Athens: University of Georgia Press, 1993). Copyright © 1993 by Judith Ortiz Cofer. Reprinted by permission of the publisher.

as amused, I managed my version of an English smile: no show of teeth, no extreme contortions of the facial muscles—I was at this time of my life practicing reserve and cool. Oh, that British control, how I coveted it. But María had followed me to London, reminding me of a prime fact of my life: you can leave the Island, master the English language, and travel as far as you can, but if you are a Latina, especially one like me who so obviously belongs to Rita Moreno's gene pool, the Island travels with you.

This is sometimes a very good thing—it may win you that extra minute of someone's attention. But with some people, the same things can make *you* an island—not so much a tropical paradise as an Alcatraz, a place nobody wants to visit. As a Puerto Rican girl growing up in the United States and wanting like most children to "belong," I resented the stereotype that my Hispanic appearance called forth from many people I met.

Our family lived in a large urban center in New Jersey during the sixties, where life was designed as a microcosm of my parents' casas on the island. We spoke in Spanish, we ate Puerto Rican food bought at the bodega, and we practiced strict Catholicism complete with Saturday confession and Sunday mass at a church where our parents were accommodated into a one-hour Spanish mass slot, performed by a Chinese priest trained as a missionary for Latin America.

As a girl I was kept under strict surveillance, since virtue and modesty were, by cultural equation, the same as family honor. As a teenager I was instructed on how to behave as a proper señorita. But it was a conflicting message girls got, since the Puerto Rican mothers also encouraged their daughters to look and act like women and to dress in clothes our Anglo friends and their mothers found too "mature" for our age. It was, and is, cultural, yet I often felt humiliated when I appeared at an American friend's party wearing a dress more suitable to a semiformal than to a playroom birthday celebration. At Puerto Rican festivities, neither the music nor the colors we wore could be too loud. I still experience a vague sense of letdown when I'm invited to a "party" and it turns out to be a marathon conversation in hushed tones rather than a fiesta with salsa, laughter, and dancing—the kind of celebration I remember from my childhood.

I remember Career Day in our high school, when teachers told us to come dressed as if for a job interview. It quickly became obvious that to the barrio girls, "dressing up" sometimes meant wearing ornate jewelry and clothing that would be more appropriate (by mainstream standards) for the company Christmas party than as daily office attire. That morning I had agonized in front of my closet, trying to figure out what a "career girl" would wear because, essentially, except for Marlo Thomas on TV, I had no models on which to base my decision. I knew how to dress for school: at the Catholic school I attended we all wore uniforms; I knew how to dress for Sunday mass, and I knew what dresses to wear for parties at my relatives' homes. Though I do not recall the precise details of my Career Day outfit, it must have been a composite of the above choices. But I remember a comment my friend (an Italian-American) made in later years that coalesced my impressions of that day. She said that at the business school she was attending the Puerto Rican girls always stood out for wearing "everything at once." She meant, of course, too much jewelry, too many accessories. On that day at school, we were simply made the negative models by the nuns who were themselves not credible fashion experts to any of us. But it was painfully obvious to me that to the others, in their tailored skirts and silk blouses, we must have seemed "hopeless" and "vulgar." Though I now know that most adolescents feel out of step much of the time, I also know that for the Puerto Rican girls of my generation that sense was intensified. The way our teachers and classmates looked at us that day in school was just a taste of the culture clash that awaited us in the real world, where prospective employers and men on the street would often misinterpret our tight skirts and jingling bracelets as a come-on.

Mixed cultural signals have perpetuated certain stereotypes—for example, that of the Hispanic woman as the "Hot Tamale" or sexual firebrand. It is a one-dimensional view that the media have found easy to promote. In their special vocabulary, advertisers have designated "sizzling" and "smoldering" as the adjectives of choice for describing not only the foods but also the women of Latin America. From conversations in my house I recall hearing about the harassment that Puerto Rican women endured in factories where the "boss men" talked to them as if sexual innuendo was all they understood and, worse, often gave them the choice of submitting to advances or being fired.

It is custom, however, not chromosomes, that leads us to choose scarlet over pale pink. As young girls, we were influenced in our decisions about clothes and colors by the women—older sisters and mothers who had grown up on a tropical island where the natural environment was a riot of primary colors, where showing your skin was one way to keep cool as well as to look sexy. Most important of all, on the island, women perhaps felt freer to dress and move more provocatively, since, in most cases, they were protected by the traditions, mores, and laws of a Spanish/Catholic system of morality and machismo whose main rule was: *You may look at my sister, but if you touch her I will kill you.* The extended family and church structure could provide a young woman with a circle of safety in her small pueblo on the Island; if a man "wronged" a girl, everyone would close in to save her family honor.

This is what I have gleaned from my discussions as an adult with older Puerto Rican women. They have told me about dressing in their best party clothes on Saturday nights and going to the town's plaza to promenade with their girlfriends in front of the boys they liked. The males were thus given an opportunity to admire the women and to express their admiration in the form of *piropos:* erotically charged street poems they composed on the spot. I have been subjected to a few piropos while visiting the Island, and they can be outrageous, although custom dictates that they must never cross into obscenity. This ritual, as I understand it, also entails a show of studied indifference on the woman's part; if she is "decent," she must not acknowledge the man's impassioned words. So I do understand how things can be lost in translation. When a Puerto Rican girl dressed in her idea of what is attractive meets a man from the mainstream culture who has been trained to react to certain types of clothing as a sexual signal, a clash is likely to take place. The line I first heard based on this aspect of the myth happened when the boy who took me to my first formal dance leaned over to plant a sloppy overeager kiss painfully on my mouth, and when I didn't respond with sufficient passion said in a resentful tone: "I thought you Latin girls were supposed to mature early"—my first instance of being thought of as a fruit or vegetable—I was supposed to *ripen*, not just grow into womanhood like other girls.

It is surprising to some of my professional friends that some people, including those who should know better, still put others "in their place." Though rarer, these incidents are still commonplace in my life. It happened to me most recently during a stay at a very classy metropolitan hotel favored by young professional couples for their weddings. Late one evening after the theater, as I walked toward my room with my new colleague (a woman with whom I was coordinating an arts program), a middle-aged man in a tuxedo, a young girl in satin and lace on his arm, stepped directly into our path. With his champagne glass extended toward me, he exclaimed, "Evita!"

Our way blocked, my companion and I listened as the man half-recited, half-bellowed "Don't Cry for Me, Argentina." When he finished, the young girl said: "How about a round of applause for my daddy?" We complied, hoping this would bring the silly spectacle to a close. I was becoming aware that our little group was attracting the attention of the other guests. "Daddy" must have perceived this too, and he once more barred the way as we tried to walk past him. He began to shout-sing a ditty to the tune of Bamba"—except the lyrics were about a girl named María whose exploits all rhymed with her name and gonorrhea. The girl kept saying "Oh, Daddy" and looking at me with pleading eyes. She wanted me to laugh along with the others. My companion and I stood silently waiting for the man to end his offensive song. When he finished, I looked not at him but at his daughter. I advised her calmly never to ask her father what he had done in the army. Then I walked between them and to my room. My friend complimented me on my cool handling of the situation. I confessed to her that I really had wanted to push the jerk into the swimming pool. I knew that this same man—probably a corporate executive, well educated, even worldly by most standards—would not have been likely to regale a white woman with a dirty song in public. He would perhaps have checked his impulse by assuming that she could be somebody's wife or mother, or at least *somebody* who might take offense. But to him, I was just an Evita or a María: merely a character in his cartoon-populated universe.

Because of my education and my proficiency with the English language, I have acquired many mechanisms for dealing with the anger I experience. This was not true for my parents, nor is it true for the many Latin women working at menial jobs who must put up with stereotypes about our ethnic group such as: "They make good domestics." This is another facet of the myth of the Latin women in the United States. Its origin is simple to deduce. Work as domestics, waitressing, and factory jobs are all that's available to women with little English and few skills. The myth of the Hispanic menial has been sustained by the same media phenomenon that made "Mammy" from *Gone with the Wind* America's idea of the black woman for generations; María, the housemaid or counter girl, is now indelibly etched into the national psyche. The big and the little screens have presented us with the picture of the funny Hispanic maid, mispronouncing words and cooking up a spicy storm in a shiny California kitchen.

This media-engendered image of the Latina in the United States has been documented by feminist Hispanic scholars, who claim that such portrayals are partially responsible for the denial of opportunities for upward mobility among Latinas in the professions. I have a Chicana friend working on a Ph.D. in philosophy at a major university. She says her doctor still shakes his head in puzzled amazement at all the "big words" she uses. Since I do not wear my diplomas around my neck for all to see, I too have on occasion been sent to that "kitchen," where some think I obviously belong.

One such incident that has stayed with me, though I recognize it as a minor offense, happened on the day of my first public poetry reading. It took place in Miami in a boat-restaurant where we were having lunch before the event. I was nervous and excited as I walked in with my notebook in hand. An older woman motioned me to her table. Thinking (foolish me) that she wanted me to autograph a copy of my brand new slender volume of verse, I went over. She ordered a cup of coffee from me, assuming that I was the waitress. Easy enough to mistake my poems for menus, I suppose. I know that it wasn't an intentional act of cruelty, yet with all the good things that happened that day, I remember that scene most clearly, because it reminded me of what I had to overcome before anyone would take me seriously. In retrospect I understand that my anger gave my reading fire, that I have almost always taken doubts in my abilities as a challenge—and that the result is, most times, a feeling of satisfaction at having won a convert when I see the cold, appraising eyes warm to my words, the body language change, the smile that indicates that I have opened some avenue for communication. That day I read to that woman and her lowered eyes told me that she was embarrassed at her little faux pas, and when I willed her to look up to me, it was my victory, and she graciously allowed me to punish her with my full attention. We shook hands at the end of the reading, and I never saw her again. She has probably forgotten the whole thing but maybe not.

Yet I am one of the lucky ones. My parents made it possible for me to acquire a stronger footing in the mainstream culture by giving me the chance at an education. And books and art have saved me from the harsher forms of ethnic and racial prejudice that many of my Hispanic *compañeras* have had to endure. I travel a lot around the United States, reading from my books of poetry and my novel, and the reception I most often receive is one of positive interest by people who want to know more about my culture. There are, however, thousands of Latinas without the privilege of an education or the entrée into society that I have. For them life is a struggle against the misconceptions perpetuated by the myth of the Latina as whore, domestic, or criminal. We cannot change this by legislating the way people look at us. The transformation, as I see it, has to occur at a much more individual level. My personal goal in my public life is to try to replace the old pervasive stereotypes and myths about Latinas with a much more interesting set of realities. Every time I give a reading, I hope the stories I tell, the dreams and fears I examine in my work, can achieve some universal truth which will get my audience past the particulars of my skin color, my accent, or my clothes.

I once wrote a poem in which I called us Latinas "God's brown daughters." This poem is really a prayer of sorts, offered upward, but also, through the human-to-human channel of art, outward. It is a prayer for communication, and for respect. In it, Latin women pray "in Spanish to an Anglo God/with a Jewish heritage," and they are "fervently hoping/that if not omnipotent,/at least He be bilingual."

21

The Bachelor: Whiteness in the Harem

RACHEL E. DUBROFSKY

Over the course of the first season of *American Idol* in the summer of 2002, more people voted by phone to help select a winner than voted in the 2000 U.S. presidential election (Albiniak, 2002, p. 22). Contemporary reality-based programming has captured the attention of TV viewers in the United States, and television scholars have responded accordingly. However, little work on reality-based shows featuring the activities of everyday people focuses on women, although women figure centrally in the reality-based romance genre, or on the intersection of women and race, though people of color figure more prominently in reality-based shows than in scripted shows. This article outlines how *The Bachelor* is "raced": The series is a context in which only white people find romantic partners, while women of color work to facilitate the coupling of white people. I examine racial stereotypes, racial references, and the ways that narrative structures oppressively regulate the television text.

I base my noting of race on visible racial markers and comments by participants about their racial background. Some women are marked by their dark skin or physical features as women of color and are treated as such on the series. Other women who are not explicitly marked physically as women of color but are described with a specific ethnic heritage—for instance, Latin American descent—are marked by the series sometimes as women of color, and sometimes not. I do not put the word "white" in quotation marks here, although quotation marks would properly emphasize the term's instability.

I argue that the structure, mise-en-scène, and setting of *The Bachelor* evoke the Westernized trope of the Eastern harem (Ahmed, 1982; Shohat & Stam, 1994), enforcing timeworn racist gender dynamics and duplicating the trope's imperialist, Orientalist, and oppressive racist structures (Said, 1978). The very

SOURCE: *Critical Studies in Media Communication* 23 (March 2006): 39–56.

form of *The Bachelor* naturalizes the desire of white men for women of color as a means of preparing for union with their ultimate partners, white women. In *The Bachelor*, whiteness is an implicit prerequisite for finding a mate. While many of the white women do not find love with the bachelor, they may be the center of the storyline for one or more episodes. In fact, the more spectacularly the white women fail to become the bachelor's partner, the more screen time they get. This is not the case for women of color, who work only to frame the narrative about white people forming a romantic union.

THE BACHELOR'S SUCCESS

Originally aired in March 2002, *The Bachelor* was one of the first reality-based shows to focus on romance. It has proved the most enduring, with eight eight-week seasons aired to date. In *The Bachelor,* a man (a different white man each season) selects one woman from among 25 eligible women to be his potential bride. Based on information given in the first episode of seasons one, two, and six, the bachelors for those seasons were selected over other applicants because they were well-rounded, wanted marriage, had an established career, and boasted a great personality. Bachelor Bob Guiney was selected for season three because he was the most popular (among audiences) of the men rejected on the first season of *The Bachelorette*. Other bachelors were approached by producers to star on the series: Andrew Firestone from season four is heir to the Firestone fortune; Jesse Palmer from season five is a football player for the New York Giants; Charlie O'Connell from season seven is a minor Hollywood actor; and Travis Stork from season eight is an emergency doctor. In season two's special *The Bachelor Revealed* (no specials of this kind aired during other seasons), Chris Harrison, the host (who is white), revealed that when selecting from female applicants who applied for the series, producers "scrutinize" every aspect: "She must be single and between the ages of 21 and 35. She must be adventurous, ready for marriage. She should be intelligent. She should be ambitious."

In each season, the bachelor goes on a series of dates with the women (some one-on-one dates with "lucky" contestants and some group dates, during which individual women vie for his attention). At the end of each episode, at least one participant is eliminated during a "rose ceremony," during which the bachelor presents a rose to each of the women he wants to keep for the next week. In the first few rose ceremonies, the bachelor eliminates several women at once; when the group has been narrowed to four, he eliminates the women one at a time, until he selects his final choice. Thus far none of the bachelors have married the woman [he] selected; only two of the seven couples remain together at the time of this writing (at the time of this writing season eight had not ended yet).

Each season also includes two specials. In the *The Women Tell All* special, the second-to-last episode aired in the season, before the finale, the eliminated women discuss their experiences on the show. When this special airs, the audience does not yet know whom the bachelor has chosen. The *After the Final Rose* special is the last episode in the season, airing after the finale (where the bachelor makes his final selection). In this special, the couple is reunited. . . .

Because the series uses the rhetoric of realism (through its form and content, the activities of "real" people doing "real" things), it naturalizes the constructions of race and romance it promotes to its audience. This article breaks down some of this "common-sense" rhetoric to ask how TV naturalizes the process of a white man and woman as the final pair. How is it that none of the ways women of color behave provides access to the central narrative? How does the harem structure situate race and experiences with racial "others" as both a necessary part of a white person's journey to finding a mate and a challenge to be overcome?

REALITY-BASED PROGRAMMING

The term "reality-based television" refers broadly to shows that are unscripted, though most have a very specific structure (with set tasks and events for each episode). The term purposely implies that the shows are based on reality without suggesting that they *are* reality—emphasizing the constructedness not only of reality-based programming but also of TV representations more generally. My analysis assumes that what occurs on reality-based shows is a constructed fiction, like the action on scripted shows, with the twist that real people create the fiction of the series. In other words, reality-based shows use footage of "real" people in "real" situations to create a fictional text, while scripted shows use a script to create the action. What happens on reality-based shows is not, of course, a representation of what "really" happened. The narrative is constructed by TV workers, sometimes using a tiny percentage of the footage actually shot. Like Stam (2000), I approach characters not as "real" people but rather as discursive constructions, and I look at "accents" and "intonations" ... discernible in the televisual voice, examining which ambient ethnic voices are "heard" and which are elided or distorted. I assume, therefore, that the story about race in *The Bachelor* is assembled through the editing process. The production process shapes the final product, the "voices," "accents," and "intonations."

WOMEN OF COLOR IN "THE BACHELOR"

While women of color appear on the show, they do not thrive. In the first season, all four women of color in the initial pool were eliminated by the third week. In the second season, the last two of the three women of color voluntarily left during the second week. The only woman of color on the third season was eliminated in the first week. All three woman of color on the fourth season were eliminated by the sixth week. On the fifth season, all four women of color were eliminated by the second week. On the sixth season, the last of the three women of color in the original pool was eliminated by the fourth week.

However, a Cuban American woman, Mary, joined the series on the third episode of the sixth season, and she was the woman bachelor Byron Velvick finally chose as his mate. This was the first and only time the series added women in the middle of a season (a white woman was added along with Mary). Mary had appeared in season four, when the series marked her explicitly as Cuban American. In season six her ethnicity was not mentioned until the second-to-last episode. I argue that while Mary was marked as a woman of color in season four, she was effectively "whitened" in season six. Because she was not marked physically as a woman of color, the series could represent her ethnicity in a mutable fashion. The only woman of color on the seventh season was eliminated at the first rose ceremony. The three women of color on the eighth season were eliminated in the first episode.

MAKING TOO MUCH OF TOO LITTLE?

Focusing on women of color in *The Bachelor* is tricky: Because they figure so little, I risk seeming to be making "too much" of "too little." Indeed, nearly all women are eventually eliminated each season. Perhaps it is statistically insignificant that only one woman who was marked as a woman of color (and not marked as such in the season when the bachelor proposed to her) ended up with a bachelor. Possibly most of the women of color never "clicked" romantically with the bachelors.

The series invites us to consider race within the logic of relational choice, rather than within the logic of representation or of production. What is crucial is how consistently the series marks the presence of the women of color. Appadurai suggests that universities often encourage diversity on "the principle that more difference is better" but frequently fail to create "a habitus where diversity is at the heart of the apparatus itself" (1996, p. 26). This is also true of television. *The Bachelor* habitus is such that women of color exist but are mostly irrelevant to the dominant narrative, except to the extent that their actions work to frame the white women's access to this narrative, or to frame the white bachelor's journey to finding his ideal mate. In other words, the show's racism is not overt. At times it is difficult to pin down. This is what Hall calls "inferential racism": when racist representations are unspoken and naturalized, making the racist premises upon which the representations rely difficult to bring to the surface (2003, p. 91).

FRAMING WHITE WOMEN'S ILLEGITIMATE
BEHAVIOR

Women of color on *The Bachelor* verify the behavior of white women and thus re-center whiteness. Women of color generally frame the white women's illegitimate behavior—their excessive emotional behavior—thus highlighting that these particular women are unsuitable matches for the bachelor. For example, Anindita, a South Asian American woman who voluntarily left at the second rose ceremony on the second season, caused quite a stir during a group date with bachelor Aaron Buerge and four white women by confronting Christi about her negative feelings about Suzanne. The central focus of this scene, however, is on how hurt Christi was by the day's events. In fact, during the confrontation, Anindita's voice was heard but she appeared only in the corner of the frame. What was apparent was Christi's emotion, given the close-ups of her teary face. While Anindita exposed the tension between the two women—even inciting it—she was never central to this narrative, much less to the romance narrative with the bachelor. Anindita's role was to highlight Christi's excessive emotionality.

Karin, an African American woman, acted to frame and re-center the presence of a white woman, Lee-Ann, in the fourth season. Although Karin made it to the fourth rose ceremony, she was never seen interacting with the bachelor. That she remained on the show for so long perhaps indicates some interaction with Bob, yet the focus was on Bob's interactions with the other women, never with her. In fact, the series referred to her in only three ways: as beautiful (by Bob and the other women); as high-maintenance (by the other women); and as a great friend (by Lee-Ann).

Karin had little or no screen time. Although the other women referred to Karin as high-maintenance and commented on her lack of enthusiasm for the day's group date at a water-park, this received little attention. Nor did her friendship with Lee-Ann, a much disliked fellow-participant, emerge as an issue. The absence of any fuss over Karin's marked lack of enthusiasm or her closeness to a detested participant is curious, considering that whenever any of the (white) women set themselves apart from the main action of the series (by being disdainful of activities, aloof, self-absorbed, and so forth), this became a focus of at least one episode and thereafter marked the woman negatively. In essence, although the series presented Karin behaving like some of the white women, she did not signify in the same ways.

Karin's most notable moment was when she tried to comfort Lee-Ann, who was furious with the other women and with the bachelor. At this point, we heard Karin speak for longer (and not very long at that) than she had at any other time. Lee-Ann had been the dramatic center of much of the last four episodes (because she set herself apart from the other women and incurred their disdain). Karin's interactions with her foregrounded this drama. To some extent,... Karin was very visible but had no voice. Comments about Karin repeatedly referenced her beauty, yet she herself rarely spoke and little attention was paid to her behavior.

Frances, an Asian American woman on the second season (who left voluntarily), figured only nominally on the show, but she framed the over-emotional behavior of a white woman, Heather. The series did not show Frances interacting with the bachelor on her only date with him (a group date), although she did narrate a lot of the action on the date. However, in *The Women Tell All* special,

Frances became the catalyst for an emotional outburst from Heather. Frances joked about how Heather cooked lots of good food for the women in an attempt to fatten them up. While all the women laughed, the next shot showed Heather bursting into tears. When the host asked what was wrong, Heather claimed Frances's comment hurt her. Frances looked confused but apologized. Again, a woman of color's fleeting presence served to frame the characterization of a white woman as excessively emotional.

If the explicit aim of the series is to promote romance (ideally leading to marriage, or at least to a long-term commitment) and winners are those who succeed in becoming the star of the romance narrative, then the overriding message is that women of color do not count. They are positioned as neither legitimate nor illegitimate romantic partners for the bachelor. Yet the women of color are a vital part of the story of two white people finding a partner: They verify the bachelor's choices by highlighting the unsuitability of certain white women; they serve as window dressing for the white women, giving them a "special flavor, an added spice" (Hooks, 1992, p. 157).

CHOOSING TO LEAVE THE SHOW

Despite the show's avowed attempt to offer *all* the women the opportunity for love with the bachelor and the power to make choices, these are seriously circumscribed when it comes to women of color. Gray (1995) explains that what he calls assimilationist programs "celebrate racial invisibility and color blindness... [by integrating] individual black characters into hegemonic white worlds void of any hint of African American traditions, social struggle, racial conflicts, and cultural difference" (p. 85). In these texts, in contrast to the hegemonic status of whiteness, Gray adds: "Blackness simply works to reaffirm, shore up, and police the cultural and moral boundaries of the existing racial order. From the privileged angle of their normative race and class positions, whites are portrayed as sympathetic advocates for the elimination of prejudice" (1995, p. 87). The overriding assimilationist paradigm of *The Bachelor* is apparent in the very set-up of the series: By pairing white men with women of color as open to the possibility of romance, the series constructs them as implicitly willing to engage in interracial relationships. On the surface, the series operates as if color does not matter, as if people in the series (and implicitly the makers of the show) are neutral when it comes to racial differences, or cultural differences read as racial ones, and will treat everyone as if these differences do not exist. Everyone, ostensibly, can compete to win the rewards of the show—finding a romantic partner. The suggestion is that white women and women of color have access to the same choices, will benefit from the same rewards, and suffer the same consequences for the choices they make. However, the choices and opportunities afforded women of color do not allow access to the central romance narrative, as they do for white women. The actions of women of color do not bring them closer to winning the bachelor's heart, though it is often through the actions of women of color that white women lose access to the bachelor's heart. . . .

HAREM STRUCTURE

Shohat suggests that one way to look at texts is to explore their *"inferential ethnic presences,* that is, the various ways in which ethnic cultures penetrate the screen without always literally being represented by ethnic and racial themes or even characters" (1991, p. 223). While the premise of the series, to find the bachelor a long-term romantic partner, may not immediately bring to mind the concept of the harem, the very set-up of *The Bachelor* implicitly references the harem: one man with 25 beautiful women who live in the same quarters and are always at the bachelor's disposal. In fact, the women have little to do but lounge around and wait to share time with the bachelor. . . .

The women on *The Bachelor* are often so similar in appearance (e.g., body size and skin color) that, at least visually, they may at times appear to be interchangeable. The structure of the show is such that the supply of willing women—willing to make themselves accessible to this one man, and no other man, for the duration of the show—is endless. The series adds to the harem structure the notion of choice: The women's willingness is bolstered by the paradigm of choice since the women are told that they can choose either to leave or to stay and accept the will of the bachelor (who will decide if they stay or leave). . . .

The décor of the bachelor pad mimics this construction of the harem with its sumptuous, boudoir-like furniture; the array of sitting rooms with stuffed couches, throw rugs, and oversized pillows; and wall hangings in rich dark colors. Of course, the requisite hot-tub and swimming pool are nestled in a lush garden, with many verdant private settings for midnight trysts. The harem décor persists throughout the show. Almost every date is located near a pool and a hot tub (or this is the final destination of the date), and many private rooms (with candles, carpets, pillows, and a fireplace) are conducive to intimacy.

The bachelor sometimes has the opportunity to share intimacies with several women in a single night. For example, on a group date he may spend some time with all the women together and then divide his time between kissing one woman in a secluded garden area, "making-out" with another in an intimate boudoir-like room, and embracing yet a third in a hot tub. The sexually charged atmosphere is enhanced by the "far-Eastern," "Orientalist" themes of the activities as well as the setting, many requiring participants to stay low to the ground, ever ready for a "tumble in the hay." In season one, for example, bachelor Alex Michael and Amanda enjoyed sushi in a private room with vaguely "Oriental" red tapestries on the wall; they sat at a low table with a carved-out floor for their feet (it looked like they were sitting on the floor). After dinner, they rolled over onto the floor and "made out." Then they moved into another private room, donned kimonos, and gave each other full-body massages. In season four, bachelor

Bob and three of the women went to a karaoke pajama party. Wearing negligée-like attire, the women sang to Bob and lounged together on deep plush red cushions. Bob "made out" with one of the women in the more private, curtained-off rooms in the back, adorned with a bed and pillows in deep shades of red and an "Oriental"-looking red tapestry on one wall. On another date, Bob and several women went to a private cabaret club where the women went on stage and danced for him in erotic poses. In the third season, bachelor Andrew took a few of the women to share Ethiopian food, which they ate with their hands off a low table, sitting on the floor in a room decorated with "Middle-Eastern"-looking tapestries. At the end of the meal, they lounged on cushions on the floor and were entertained by belly dancers who encouraged the women to dance for Andrew. In the fifth season, bachelor Jesse took three women on a date to an opulent tent set outdoors where they were greeted by a live elephant. The set designers decked out the tent in "Middle-Eastern"-looking décor, with gilded red pillows strewn across the floor, gilded red tapestries hanging from the walls, a low table at the center of the room, and "Oriental" rugs layered on the ground. "Middle-Eastern"-sounding music played in the background. Jesse spent part of the date doing what one might expect in this setting: rolling around and "making out" with one of the women. Over the course of the date, he kissed all three women, either in the tent or in secluded areas outside.

The harem structure was even more explicit in the sixth season, when bachelor Byron moved into a guest house in the garden behind the bachelorette pad where the women lived, giving him unlimited access to the women. In episode four, for example, he organized a pajama party at the women's house, providing the lingerie for the women to wear. While the women had access to the bachelor, Byron decided just how much: in the fifth episode, for example, Byron insisted that Jayne, who came to visit him, leave his quarters because, as he told us in voice-over, he did not want to "complicate" things by having any of the women spend the night. When the bachelor is not present, the activities of the women are staged for the pleasure of "the scopic privilege of the master in an exclusionary space inaccessible to other men" (Shohat & Stam, 1994, p. 164): We witness them navigating the difficulties of living together while they frolic about in bikinis, prance from hot tub to pool, play lawn games, drink lots of alcohol, and, of course, gossip and fight with one another.

The harem décor accentuates the sexual possibilities at the bachelor's disposal and emphasizes that the women are to be ever-ready and willing to make use of the cushy privacy (cameras notwithstanding) afforded by this setting. The space of *The Bachelor* is one where sexual dilettantism is to be celebrated and explored, part of the Western myth about the Eastern harem, a place of sexual excess, of limitless pleasure for Western men (Ahmed, 1982; Alloula, 1986). The bachelor is the Sheik of this realm, the white playboy who has arrived in a dark land to frolic before returning to more serious ventures and to his white leading lady. On the road to finding a mate, the bachelor has numerous opportunities for lusty forays with many women who await the pleasure of being conquered by him.

EXOTIC STOPS EN ROUTE TO FINDING A MATE

The permeation of the "Eastern" via the implicit (never explicitly referenced) trope of the harem is emblematic of the way *The Bachelor* works through both implicit and explicit issues having to do with race: Racial diversity frames the central narrative of two white people forming a romantic union. Diversity ultimately works to maintain pedagogies of whiteness, while situating diversity as essential in defining and highlighting whiteness.... The white couple can fully enjoy and partake in the sensual rituals of the East (when in Rome do as the Romans do), while in the end distancing themselves from the very same eroticized culture that enabled them to find each other. People of color in *The Bachelor* are part of the backdrop, the setting, of the otherwise white narrative. While the bachelor may frolic with women of color during his sojourn in the harem, in the end, he leaves the harem with his chosen white partner. . . .

THE WHITE HERO AND HIS MISTRESS

The sexual gallivanting of the bachelor is "otherized" within the larger romance narrative of the series; surely the bachelor is to use this time to sow his wild oats with the express purpose of settling down with one woman at the end, presumably in a home with a white picket fence and very few Persian carpets or red pillows and certainly no elephant grazing on the front lawn. Male protagonists in *The Bachelor,* like the protagonists in classic Hollywood harem films, explore the evils of the Westernized version of the Eastern harem before overcoming these excesses in the interest of sustaining a monogamous union. The bachelors transgress Western monogamous norms via the trope of the Western version of the Eastern harem in order to reaffirm Western norms. A romantic union for a white man can only be found with one white woman; being in love means not lusting after others. The harem experience and the women who are part of this process are, to some extent, the grotesque but necessary other (Stallybrass and White, 1986).

Just as the harem narrative is embedded in the Western narrative of monogamous love, portrayals of people of color are embedded in the defining of whiteness. Through the trope of the Eastern harem, *The Bachelor*'s racist, imperialist structure tells a story about the romantic heterosexual union of two white people; it relies on the myth of white man's ability to conquer the dark Other to find his way towards the ideal white mate. Women of color and white women serve different purposes in this Orientalist set-up. White women can rise above the harem structure to become the bachelor's chosen woman, for whom he will forsake the harem's pleasures. Unless they can be whitened, however, women of color cannot.

By locating the harem structure in the United States (in Hollywood, the harem remains in the East but at the end the white couple returns to the West), the series reinscribes not only imperialist desires, but also the historical power dynamic of slavery in the United States. Given the history of slavery in the United States, and the relationship of women of color to their white masters—as sexual slaves, as mothers to their illegitimate children, but never as legitimate romantic partners (wives) or as equal citizens—the positioning of women of color in *The Bachelor* is particularly disturbing. Here again women of color fulfill the time worn roles of satisfying the sexual desires of white men and taking care of white men (by highlighting the inappropriate and potentially dangerous behaviors of white women who want to marry white men). *The Bachelor* tells a very specific story about whiteness, where whiteness is essential to finding a romantic partner. Once women of color help the white heroes find one another, they must disappear into the background—just as the harem structure disappears to let the white master and his chosen mistress take center stage.

REFERENCES

Ahmed, L. (1982). Western ethnocentrism and perceptions of the harem. *Feminist Studies, 8,* 521–534.

Albiniak, P. (2002, September 9). Ideal, not idle summer for Fox. *Broadcasting and Cable,* p. 22.

Alloula, M. (1986). *The colonial harem.* Minneapolis, MN: University of Minnesota Press.

Appadurai, A. (1996). Diversity and disciplinarity as cultural artifacts. In C. Nelson & D. Parameshwar Gaonkar (Eds.), *Disciplinarity and dissent in cultural studies* (pp. 23–36). New York: Routledge.

Gray, H. (1995). *Watching race: Television and the struggle for "blackness."* Minneapolis, MN: University of Minnesota Press.

Hall, S. (2003). The whites of their eyes. In G. Dines & J. M. Humez (Eds.), *Gender, race, and class in media: A text-reader* (pp. 89–93) (2nd ed.). London: Sage Publications.

Hooks, B. (1992). Madonna: Plantation mistress or soul sister. *Black looks: Race and representation* (pp. 157–164). Toronto: Between the Lines.

Said, E. W. (1978). *Orientalism.* New York: Vintage Books.

Shaheen, J. G. (2001). *Reel bad Arabs: How Hollywood vilifies a people.* New York: Olive Branch Press.

Shohat, E. (1991). Ethnicities-in-relation: Toward a multicultural reading of American cinema. In L. D. Friedman (Ed.), *Unspeakable images: Ethnicity and the American cinema* (pp. 215–250). Chicago: University of Illinois Press.

Shohat, E., & Stam, R. (1994). *Unthinking Eurocentrism: Multiculturalism and the media.* New York: Routledge.

Stallybrass, P., & White, A. (1986). *Transgression: The politics and poetics of transgression.* Ithaca, NY: Cornell University Press.

Stam, R. (2000). Bakhtin, polyphony, and ethnic/racial representation. In L. D. Friedman (Ed.), *Unspeakable images: Ethnicity and the American cinema* (pp. 251–276). Chicago: University of Illinois Press.

22

Masculinities and Athletic Careers

MICHAEL MESSNER

It is now widely accepted in sport sociology that social institutions such as the
...media, education, the economy, and (a more recent and controversial addition to the list) the black family itself all serve to systematically channel disproportionately large numbers of young black men into football, basketball, boxing,
and baseball, where they are subsequently "stacked" into low-prestige and high-
risk positions, exploited for their skills, and finally, when their bodies are used
up, excreted from organized athletics at a young age with no transferable skills
with which to compete in the labor market (Edwards 1984; Eitzen and Purdy
1986; Eitzen and Yetman 1977).

While there are racial differences in involvement in sports, class, age, and
educational differences seem more significant. Rudman's (1986) initial analysis
revealed profound differences between whites' and blacks' orientations to sports.
Blacks were found to be more likely than whites to view sports favorably, to
incorporate sports into their daily lives, and to be affected by the outcome of
sporting events. However, when age, education, and social class were factored
into the analysis, Rudman found that race did not explain whites' and blacks'
different orientations. Blacks' affinity to sports is best explained by their tendency
to be clustered disproportionately in lower-income groups.

The 1980s has ushered in what Wellman (1986, p. 43) calls a "new political
linguistics of race," which emphasizes cultural rather than structural causes (and
solutions) to the problems faced by black communities. The advocates of the
cultural perspective believe that the high value placed on sports by black communities has led to the development of unrealistic hopes in millions of black
youths. They appeal to family and community to bolster other choices based
upon a more rational assessment of "reality." Visible black role models in many
other professions now exist, they say, and there is ample evidence which proves
that sports careers are, at best, a bad gamble.

Critics of the cultural perspective have condemned it as conservative and
victim blaming. But it can also be seen as a response to the view of black athletes

SOURCE: From *Gender & Society* 3 (March 1989): 71–88. Reprinted by permission of
Sage Publications, Inc.

as little more than unreflexive dupes of an all-powerful system, which ignores the importance of agency. Gruneau (1983) has argued that sports must be examined within a theory that views human beings as active subjects who are operating within historically constituted structural constraints. Gruneau's reflexive theory rejects the simplistic views of sports as either a realm of absolute oppression or an arena of absolute freedom and spontaneity. Instead, he argues, it is necessary to construct an understanding of how and why participants themselves actively make choices and construct and define meaning and a sense of identity within the institutions [in which] they find themselves.

None of these perspectives considers the ways that gender shapes men's definitions of meaning and choices. Within the sociology of sport, gender as a process that interacts with race and class is usually ignored or taken for granted—except when it is *women* athletes who are being studied. Sociologists who are attempting to come to grips with the experiences of black men in general, and in organized sports in particular, have almost exclusively focused their analytic attention on the variable "black," while uncritically taking "men" as a given. Hare and Hare (1984), for example, view masculinity as a biologically determined tendency to act as a provider and protector that is thwarted for black men by socioeconomic and racist obstacles. Staples (1982) does view masculinity largely as a socially produced script, but he accepts this script as a given, preferring to focus on black men's blocked access to male role fulfillment. These perspectives on masculinity fail to show how the male role itself, as it interacts with a constricted structure of opportunity, can contribute to locking black men into destructive relationships and lifestyles (Franklin 1984; Majors 1986).

This article will examine the relationships among male identity, race, and social class by listening to the voices of former athletes. I will first briefly describe my research. Then I will discuss the similarities and differences in the choices and experiences of men from different racial and social class backgrounds. Together, these choices and experiences help to construct what Connell (1987) calls "the gender order." Organized sports, it will be suggested, is a practice through which men's separation from and power over women is embodied and naturalized at the same time that hegemonic (white, heterosexual, professional-class) masculinity is clearly differentiated from marginalized and subordinated masculinities.

DESCRIPTION OF RESEARCH

Between 1983 and 1985, I conducted 30 open-ended, in-depth interviews with male former athletes. My purpose was to add a critical understanding of male gender identity to Levinson's (1978) conception of the "individual life-course"—specifically, to discover how masculinity develops and changes as a man interacts with the socially constructed world of organized sports. Most of the men I interviewed had played the U.S. "major sports"—football, basketball, baseball, track. At the time of the interview, each had been retired from playing organized sports for at least 5 years. Their ages ranged from 21 to 48, with the median, 33. Fourteen were black, 14 were white, and 2 were Hispanic. Fifteen of the 16 black and Hispanic men had come from poor or working-class families, while the majority (9 of 14) of the white men had come from middle-class or professional families. Twelve had played organized sports through high school, 11 through college, and 7 had been professional athletes. All had at some time in their lives based their identities largely on their roles as athletes and could therefore be said to have had athletic careers.

MALE IDENTITY AND ORGANIZED SPORTS

. . . For the men in my study, the rule-bound structure of organized sports became a context in which they struggled to construct a masculine positional identity.

All of the men in this study described the emotional salience of their earliest experiences in sports in terms of relationships with other males. It was not winning and victories that seemed important at first; it was something "fun" to do with fathers, older brothers or uncles, and eventually with same-aged peers. As a man from a white, middle-class family said, "The most important thing was just being out there with the rest of the guys—being friends." A 32-year-old man from a poor Chicano family, whose mother had died when he was 9 years old, put it more succinctly:

> What I think sports did for me is it brought me into kind of an instant family. By being on a Little League team, or even just playing with kids in the neighborhood, it brought what I really wanted, which was some kind of closeness.

Though sports participation may have initially promised "some kind of closeness," by the ages of 9 or 10, the less skilled boys were already becoming alienated from—or weeded out of—the highly competitive and hierarchical system of organized sports. Those who did experience some early successes received recognition from adult males (especially fathers and older brothers) and held higher status among peers. As a result, they began to pour more and more of their energies into athletic participation. It was only after they learned that they would get recognition from other people for being a good athlete—indeed, that this attention was contingent upon *being a winner*—that performance and winning (the dominant values of organized sports) became extremely important. For some, this created pressures that served to lessen or eliminate the fun of athletic participation (Messner 1987a, 1987b).

While feminist psychoanalytic and developmental theories of masculinity are helpful in explaining boys' early attraction and motivations in organized sports, the imperatives of core gender identity do not fully determine the contours and directions of the life course. As Rubin (1985) and Levinson (1978) have pointed out, an understanding of the lives of men must take into account the processual nature of male identity as it unfolds through interaction between the internal (psychological ambivalences) and the external (social, historical, and institutional) contexts.

To examine the impact of the social contexts, I divided my sample into two comparison groups. In the first group were 10 men from higher-status backgrounds, primarily white, middle-class, and professional families. In the second group were 20 men from lower-status backgrounds, primarily minority, poor, and working-class families. While my data offered evidence for the similarity of experiences and motivations of men from poor backgrounds, independent of race, I also found anecdotal evidence of a racial dynamic that operates independently of social class. However, my sample was not large enough to separate race and class, and so I have combined them to make two status groups.

In discussing these two groups, I will focus mainly on the high school years. During this crucial period, the athletic role may become a master status for a young man, and he is beginning to make assessments and choices about his future. It is here that many young men make a major commitment to—or begin to back away from—athletic careers.

Men from Higher-Status Backgrounds

The boyhood dream of one day becoming a professional athlete—a dream shared by nearly all the men interviewed in this study—is rarely realized. The sports world is extremely hierarchical. The pyramid of sports careers narrows very rapidly as one climbs from high school, to college, to professional levels of competition (Edwards 1984; Harris and Eitzen 1978; Hill and Lowe 1978). In fact, the chances of attaining professional status in sports are approximately 4/100,000 for a white man, 2/100,000 for a black man, and 3/100,000 for a Hispanic man in the United States (Leonard and Reyman 1988). For many young athletes, their dream ends early when coaches inform them that they are not big enough, strong enough, fast enough, or skilled enough to compete at the higher levels. But six of the higher-status men I interviewed did not wait for coaches to weed them out. They made conscious decisions in high school or in college to shift their attentions elsewhere—usually toward educational and career goals. Their decision not to pursue an athletic career appeared to them in retrospect to be a rational decision based on the growing knowledge of how very slim their chances were to be successful in the sports world. For instance, a 28-year-old white graduate student said:

> By junior high I started to realize that I was a good player—maybe even one of the best in my community—but I realized that there were all these people all over the country and how few will get to play pro sports. By high school, I still dreamed of being a pro—I was a serious athlete, I played hard—but I knew it wasn't heading anywhere. I wasn't going to play pro ball.

A 32-year-old white athletic director at a small private college had been a successful college baseball player. Despite considerable attention from professional scouts, he had decided to forgo a shot at a baseball career and to enter graduate school to pursue a teaching credential. As he explained this decision:

> At the time I think I saw baseball as pissing in the wind, really. I was married, I was 22 years old with a kid. I didn't want to spend 4 or 5 years in the minors with a family. And I could see I wasn't a superstar; so it wasn't really worth it. So I went to grad school. I thought that would be better for me.

Perhaps most striking was the story of a high school student body president and top-notch student who was also "Mr. Everything" in sports. He was named captain of his basketball, baseball, and football teams and achieved All-League honors in each sport. This young white man from a middle-class family received attention from the press and praise from his community and peers for his athletic accomplishments, as well as several offers of athletic scholarships from universities. But by the time he completed high school, he had already decided to quit playing organized sports. As he said:

> I think in my own mind I kind of downgraded the stardom thing. I thought that was small potatoes. And sure, that's nice in high school and all that, but on a broad scale, I didn't think it amounted to all that much. So I decided that my goal's to be a dentist, as soon as I can.

In his sophomore year of college, the basketball coach nearly persuaded him to go out for the team, but eventually he decided against it:

> I thought, so what if I can spend two years playing basketball? I'm not going to be a basketball player forever and I might jeopardize my chances of getting into dental school if I play.

He finished college in three years, completed dental school, and now, in his mid-30s, is again the epitome of the successful American man: a professional with a family, a home, and a membership in the local country club.

How and why do so many successful male athletes from higher-status backgrounds come to view sports careers as "pissing in the wind," or as "small potatoes"? How and why do they make this early assessment and choice to shift from sports and toward educational and professional goals? The white, middle-class institutional context, with its emphasis on education and income, makes it clear to them that choices exist and that the pursuit of an athletic career is not a particularly good choice to make. Where the young male once found sports to be a convenient institution within which to construct masculine status, the postadolescent and young adult man from a higher-status background simply *transfers* these same strivings to other institutional contexts: education and careers.

For the higher-status men who had chosen to shift from athletic careers, sports remained important on two levels. First, having been a successful high school or college athlete enhances one's adult status among other men in the community—but only as a badge of masculinity that is *added* to his professional status. In fact, several men in professions chose to be interviewed in their offices, where they publicly displayed the trophies and plaques that attested to their earlier athletic accomplishments. Their high school and college athletic careers may have appeared to them as "small potatoes," but many successful men speak of their earlier status as athletes as having "opened doors" for them in their present professions and in community affairs. Similarly, Farr's (1988) research on "Good Old Boys Sociability Groups" shows how sports, as part of the glue of masculine culture, continues to facilitate "dominance bonding" among privileged men long after active sports careers end. The college-educated, career-successful men in Farr's study rarely express overtly sexist, racist, or classist attitudes; in fact, in their relationships with women, they "often engage in expressive intimacies" and "make fun of exaggerated "'machismo'" (p. 276). But though they outwardly conform more to what Pleck (1982) calls "the modern male role," their informal relationships within their sociability groups, in effect, affirm their own gender and class status by constructing and clarifying the boundaries between themselves and women and lower-status men. This dominance bonding is based largely upon ritual forms of sociability (camaraderie, competition), "the superiority of which was first affirmed in the exclusionary play activities of young boys in groups" (Farr 1988, p. 265).

In addition to contributing to dominance bonding among higher-status adult men, sports remains salient in terms of the ideology of gender relations. Most men continued to watch, talk about, and identify with sports long after their own disengagement from athletic careers. Sports as a mediated spectacle provides an important context in which traditional conceptions of masculine superiority—conceptions recently contested by women—are shored up. As a 32-year-old white professional-class man said of one of the most feared professional football players today:

> A woman can do the same job as I can do—maybe even be my boss. But I'll be damned if she can go out on the football field and take a hit from Ronnie Lott.

Violent sports as spectacle provide linkages among men in the project of the domination of women, while at the same time helping to construct and clarify differences among various masculinities. The statement above is a clear identification with Ronnie Lott *as a man,* and the basis of the identification is the violent male body. As Connell (1987, p. 85) argues, sports is an important organizing institution for the embodiment of masculinity. Here, men's power over women becomes naturalized and linked to the social distribution of violence. Sports, as a practice, suppresses natural (sex) similarities, constructs differences, and then, largely through the media, weaves a structure of symbol and interpretation around these differences that naturalizes them (Hargreaves 1986, p. 112). It is also significant that the man who made the above statement about Ronnie Lott was quite aware that he (and perhaps 99 percent of the rest of the U.S. male population) was probably as incapable as most women of taking a "hit" from someone like

Lott and living to tell of it. For middle-class men, the "tough guys" of the culture industry—the Rambos, the Ronnie Lotts who are fearsome "hitters," who "play hurt"—are the heroes who "prove" that "we men" are superior to women. At the same time, they play the role of the "primitive other," against whom higher-status men define themselves as "modern" and "civilized."

Sports, then, is important from boyhood through adulthood for men from higher-status backgrounds. But it is significant that by adolescence and early adulthood, most of these young men have concluded that sports *careers* are not for them. Their middle-class cultural environment encourages them to decide to shift their masculine strivings in more "rational" directions: education and non-sports careers. Yet their previous sports participation continues to be very important to them in terms of constructing and validating their status within privileged male peer groups and within their chosen professional careers. And organized sports, as a public spectacle, is a crucial locus around which ideologies of male superiority over women, as well as higher-status men's superiority over lower-status men, are constructed and naturalized.

Men from Lower-Status Backgrounds

For the lower-status young men in this study, success in sports was not an added proof of masculinity; it was often their only hope of achieving public masculine status. A 34-year-old black bus driver who had been a star athlete in three sports in high school had neither the grades nor the money to attend college, so he accepted an offer from the U.S. Marine Corps to play on their baseball team. He ended up in Vietnam, where a grenade blew four fingers off his pitching hand. In retrospect, he believed that his youthful focus on sports stardom and his concomitant lack of effort in academics made sense:

> You can go anywhere with athletics—you don't have to have brains. I mean, I didn't feel like I was gonna go out there and be a computer expert, or something that was gonna make a lot of money. The only thing I could do and live comfortably would be to play sports—just to get a contract—doesn't matter if you play second or third team in the pros, you're gonna make big bucks. That's all I wanted, a confirmed livelihood at the end of my ventures, and the only way I could do it would be through sports. So I tried. It failed, but that's what I tried.

Similar, and even more tragic, is the story of a 34-year-old black man who is now serving a life term in prison. After a career-ending knee injury at the age of 20 abruptly ended what had appeared to be a certain road to professional football fame and fortune, he decided that he "could still be rich and famous" by robbing a bank. During his high school and college years, he said, he was nearly illiterate:

> I'd hardly ever go to classes and they'd give me Cs. My coaches taught some of the classes. And I felt, "So what? They owe me that! I'm an athlete! I

thought that was what I was born to do—to play sports—and everybody understood that.

Are lower-status boys and young men simply duped into putting all their eggs into one basket? My research suggested that there was more than "hope for the future" operating here. There were also immediate psychological reasons that they chose to pursue athletic careers. By the high school years, class and ethnic inequalities had become glaringly obvious, especially for those who attended socioeconomically heterogeneous schools. Cars, nice clothes, and other signs of status were often unavailable to these young men, and this contributed to a situation in which sports took on an expanded importance for them in terms of constructing masculine identities and status. A white, 36-year-old man from a poor, single-parent family who later played professional baseball had been acutely aware of his low-class status in his high school:

> I had one pair of jeans, and I wore them every day. I was always afraid of what people thought of me—that this guy doesn't have anything, that he's wearing the same Levi's all the time, he's having to work in the cafeteria for his lunch. What's going on? I think that's what made me so shy. . . . But boy, when I got into sports, I let it all hang out—[laughs]—and maybe that's why I became so good, because I was frustrated, and when I got into that element, they gave me my uniform in football, basketball, and baseball, and I didn't have to worry about how I looked, because then it was me who was coming out, and not my clothes or whatever. And I think that was the drive.

Similarly, a 41-year-old black man who had a 10-year professional football career described his insecurities as one of the few poor blacks in a mostly white, middle-class school and his belief that sports was the one arena in which he could be judged solely on his merit:

> I came from a poor family, and I was very sensitive about that in those days. When people would say things like "Look at him—he has dirty pants on," I'd think about it for a week. [But] I'd put my pants on and I'd go out on the football field with the intention that I'm gonna do a job. And if that calls on me to hurt you, I'm gonna do it. It's as simple as that. I demand respect just like everybody else.

"Respect" was what I heard over and over when talking with the men from lower-status backgrounds, especially black men. I interpret this type of respect to be a crystallization of the masculine quest for recognition through public achievement, unfolding within a system of structured constraints due to class and race inequities. The institutional context of education (sometimes with the collusion of teachers and coaches) and the constricted structure of opportunity in the economy made the pursuit of athletic careers appear to be the most rational choice to these young men.

The same is not true of young lower-status women. Dunkle (1985) points out that from junior high school through adulthood, young black men are far more likely to place high value on sports than are young black women, who are more likely to value academic achievement. There appears to be a gender dynamic operating in adolescent male peer groups that contributes toward their valuing sports more highly than education. Franklin (1986, p. 161) has argued that many of the normative values of the black male peer group (little respect for nonaggressive solutions to disputes, contempt for nonmaterial culture) contribute to the constriction of black men's views of desirable social positions, especially through education. In my study, a 42-year-old black man who did succeed in beating the odds by using his athletic scholarship to get a college degree and eventually becoming a successful professional said:

> By junior high, you either got identified as an athlete, a thug, or a bookworm. It's very important to be seen as somebody who's capable in some area. And you don't want to be identified as a bookworm. I was very good with books, but I was kind of covert about it. I was a closet bookworm. But with sports, I was somebody; so I worked very hard at it.

For most young men from lower-status backgrounds, the poor quality of their schools, the attitudes of teachers and coaches, as well as the antieducation environment within their own male peer groups, made it extremely unlikely that they would be able to succeed as students. Sports, therefore, became *the* arena in which they attempted to "show their stuff." For these lower-status men, as Baca Zinn (1982) and Majors (1986) argued in their respective studies of Chicano men and black men, when institutional resources that signify masculine status and control are absent, physical presence, personal style, and expressiveness take on increased importance. What Majors (1986, p. 6) calls "cool pose" is black men's expressive, often aggressive, assertion of masculinity. This self-assertion often takes place within a social context in which the young man is quite aware of existing social inequities. As the black bus driver, referred to above, said of his high school years:

> See, the rich people use their money to do what they want to do. I use my ability. If you wanted to be around me, if you wanted to learn something about sports, I'd teach you. But you're gonna take me to lunch. You're gonna let me use your car. See what I'm saying? In high school I'd go where I wanted to go. I didn't have to be educated. I was well-respected. I'd go somewhere, and they'd say, "Hey, that's Mitch Harris,[1] yeah, that's a bad son of a bitch!"

Majors (1986) argues that although "cool pose" represents a creative survival technique within a hostile environment, the most likely long-term effect of this masculine posturing is educational and occupational dead ends. As a result, we can conclude, lower-status men's personal and peer-group responses to a constricted structure of opportunity—responses that are rooted, in part, in the developmental insecurities and ambivalences of masculinity—serve to lock many of these young men into limiting activities such as sports.

SUMMARY AND CONCLUSIONS

This research has suggested that within a social context that is stratified by social class and by race, the choice to pursue—or not to pursue—an athletic career is explicable as an individual's rational assessment of the available means to achieve a respected masculine identity. For nearly all of the men from lower-status backgrounds, the status and respect that they received through sports was temporary—it did not translate into upward mobility. Nonetheless, a strategy of discouraging young black boys and men from involvement in sports is probably doomed to fail, since it ignores the continued existence of structural constraints. Despite the increased number of black role models in nonsports professions, employment opportunities for young black males have actually deteriorated. . . .

But it would be a mistake to conclude that we simply need to breed socioeconomic conditions that make it possible for poor and minority men to mimic the "rational choices" of white, middle-class men. If we are to build an appropriate understanding of the lives of all men, we must critically analyze white middle-class masculinity, rather than uncritically taking it as a normative standard. To fail to do this would be to ignore the ways in which organized sports serve to construct and legitimate gender differences and inequalities among men and women.

Feminist scholars have demonstrated that organized sports give men from all backgrounds a means of status enhancement that is not available to young women. Sports thus serve the interests of all men in helping to construct and legitimize their control of public life and their domination of women (Bryson 1987; Hall 1987; Theberge 1987). Yet concrete studies are suggesting that men's experiences within sports are not all of a piece. Brian Pronger's (1990) research suggests that gay men approach sports differently than straight men do, with a sense of "irony." And my research suggests that although sports are important for men from both higher- and lower-status backgrounds, there are crucial differences. In fact, it appears that the meaning that most men give to their athletic strivings has more to do with competing for status among men than it has to do with proving superiority over women. How can we explain this seeming contradiction between the feminist claim that sports links all men in the domination of women and the research findings that different groups of men relate to sports in very different ways?

. . . Connell's (1987) concept of the "gender order" is useful. The gender order is a dynamic process that is constantly in a state of play. Moving beyond static gender-role theory and reductionist concepts of patriarchy that view men as an undifferentiated group which oppresses women, Connell argues that at any given historical moment, there are competing masculinities—some hegemonic, some marginalized, some stigmatized. Hegemonic masculinity (that definition of masculinity which is culturally ascendant) is constructed in relation to various subordinated masculinities as well as in relation to femininities. The project of male domination of women may tie all men together, but men share very unequally in the fruits of this domination.

These are key insights in examining the contemporary meaning of sports. Utilizing the concept of the gender order, we can begin to conceptualize how hierarchies of race, class, age, and sexual preference among men help to construct and legitimize men's overall power and privilege over women. And how, for some black, working-class, or gay men, the false promise of sharing in the fruits of hegemonic masculinity often ties them into their marginalized and subordinate statuses within hierarchies of intermale dominance. For instance, black men's development of what Majors (1986) calls "cool pose" within sports can be interpreted as an example of creative resistance to one form of social domination (racism); yet it also demonstrates the limits of an agency that adopts other forms of social domination (masculinity) as its vehicle. As Majors (1990) points out:

> Cool Pose demonstrates black males' potential to transcend oppressive conditions in order to express themselves as men. [Yet] it ultimately does not put black males in a position to live and work in more egalitarian ways with women, nor does it directly challenge male hierarchies.

Indeed, as Connell's (1990) analysis of an Australian "Iron Man" shows, the commercially successful, publicly acclaimed athlete may embody all that is valued in present cultural conceptions of hegemonic masculinity—physical strength, commercial success, supposed heterosexual virility. Yet higher-status men, while they admire the public image of the successful athlete, may also look down on him as a narrow, even atavistic, example of masculinity. For these higher-status men, their earlier sports successes are often status enhancing and serve to link them with other men in ways that continue to exclude women. Their decisions not to pursue athletic careers are equally important signs of their status vis-à-vis other men. Future examinations of the contemporary meaning and importance of sports to men might take as a fruitful point of departure that athletic participation, and sports as public spectacle, serve to provide linkages among men in the project of the domination of women, while at the same time helping to construct and clarify differences and hierarchies among various masculinities.

NOTE

1.   "Mitch Harris" is a pseudonym.

REFERENCES

Baca Zinn, M. 1982. "Chicano Men and Masuclinity." *Journal of Ethnic Studies* 10:29–44.

Bryson, L. 1987. "Sport and the Maintenance of Masculine Hegemony." *Women's Studies International Forum* 10:349–60.

Connell, R. W. 1987. *Gender and Power.* Stanford, CA: Stanford University Press.

——. 1990. "An Iron Man: The Body and Some Contradictions of Hegemonic Masculinity." In *Sport, Men, and the Gender Order: Critical Feminist Perspectives,* edited by M. A. Messner and D. S. Sabo. Champaign, IL: Human Kinetics.

Dunkle, M. 1985. "Minority and Low-Income Girls and Young Women in Athletics." *Equal Play* 5 (Spring-Summer): 12–13.

Edwards, H. 1984. "The Collegiate Athletic Arms Race: Origins and Implications of the "Rule 48' Controversy." *Journal of Sport and Social Issues* 8:4–22.

Eitzen, D. S. and D. A. Purdy. 1986. "The Academic Preparation and Achievement of Black and White College Athletes." *Journal of Sport and Social Issues* 10:15–29.

Eitzen, D. S. and N. B. Yetman. 1977. "Immune From Racism?" *Civil Rights Digest* 9:3–13.

Farr, K. A. 1988. "Dominance Bonding Through the Good Old Boys Sociability Group." *Sex Roles* 18:259–77.

Franklin, C. W. II. 1984. *The Changing Definition of Masculinity.* New York: Plenum.

——. 1986. "Surviving the Institutional Decimation of Black Males: Causes, Consequences, and Intervention." In *The Making of Masculinities: The New Men's Studies,* edited by H. Brod, pp. 155–70. Winchester, MA: Allen & Unwin.

Gruneau, R. 1983. *Class, Sports, and Social Development.* Amherst: University of Massachusetts Press.

Hall, M. A. (ed.). 1987. "The Gendering of Sport, Leisure, and Physical Education." *Women's Studies International Forum* 10:361–474.

Hare, N. and J. Hare. 1984. *The Endangered Black Family: Coping with the Unisexualization and Coming Extinction of the Black Race.* San Francisco, CA: Black Think Tank.

Hargreaves, J. A. 1986. "Where's the Virtue? Where's the Grace? A Discussion of the Social Production of Gender Through Sport." *Theory, Culture and Society* 3:109–21.

Harris, D. S. and D. S. Eitzen. 1978. "The Consequences of Failure in Sport." *Urban Life* 7:177–88.

Hill, P. and B. Lowe. 1978. "The Inevitable Metathesis of the Retiring Athlete." *International Review of Sport Sociology* 9:5–29.

Leonard, W. M. II and J. M. Reyman. 1988. "The Odds of Attaining Professional Athlete Status: Refining the Computations." *Sociology of Sport Journal* 5:162–69.

Levinson, D. J. 1978. *The Seasons of a Man's Life.* New York: Ballantine.

Majors, R. 1986. "Cool Pose: The Proud Signature of Black Survival." *Changing Men: Issues in Gender, Sex, and Politics* 17:5–6.

——. 1990. "Cool Pose: Black Masculinity in Sports." In *Sport, Men, and the Gender Order: Critical Feminist Perspectives,* edited by M. A. Messner and D. S. Sabo. Champaign, IL: Human Kinetics.

Messner, M. 1987a. "The Meaning of Success: The Athletic Experience and the Development of Male Identity." In *The Making of Masculinities: The New Men's Studies,* edited by H. Brod, pp. 193–209. Winchester, MA: Allen & Unwin.

———. 1987b. "The Life of a Man's Seasons: Male Identity in the Lifecourse of the Athlete." In *Changing Men: New Directions in Research on Men and Masculinity,* edited by M. S. Kimmel, pp. 53–67. Newbury Park, CA: Sage.

Pleck, J. H. 1982. *The Myth of Masculinity.* Cambridge: MIT Press.

Pronger, B. 1990. "Gay Jocks: A Phenomenology of Gay Men in Athletics." In *Sport, Men, and the Gender Order: Critical Feminist Perspectives,* edited by M. A. Messner and D. S. Sabo. Champaign, IL: Human Kinetics.

Rubin, L. B. 1985. *Just Friends: The Role of Friendship in Our Lives.* New York: Harper & Row.

Rudman, W. J. 1986. "The Sport Mystique in Black Culture." *Sociology of Sport Journal* 3:305–19.

Staples, Robert. 1982. *Black Masculinity.* New York: Black Scholar Press.

Theberge, N. 1987. "Sport and Women's Empowerment." *Women's Studies International Forum* 10:387–93.

Wellman, D. 1986. "The New Political Linguistics of Race." *Socialist Review* 87/88: 43–62.

23

Gladiators, Gazelles, and Groupies
Basketball Love and Loathing

JULIANNE MALVEAUX

The woman in me loves the sheer physical impact of a basketball game. I don't dwell on the fine points, the three-point shot or the overtime, and can't even tell you which teams make me sweat (though I could have told you twenty years ago). I can tell you, though, how much I enjoy watching mostly black men engage in an elegantly (and sometimes inelegantly) skilled game, and how engrossing the pace and physicality can be. In my younger days, I spent hours absorbed by street basketball games, taking the subway to watch the Rucker Pros in upper Manhattan, hanging on a bench near the basketball courts on West Fourth Street in the Village. I didn't consider myself a basketball groupie— I was a nerd who liked the game. Still, there is something about muscled, scantily clad men (the shorts have gotten longer in the 1990s) that spoke to me. Once I

SOURCE: From Todd Boyd and Kenneth L. Shropshire, eds., *Basketball Jones: America Above the Rim* (New York: New York University Press, 2000), pp. 51–58. Reprinted by permission of New York University Press.

even joked that watching basketball games was like safe sex in the age of AIDS. Yes, the woman in me, even in my maturity, loves the physical impact of a basketball game, a game where muscles strain, sweat flies, and intensity demands attention.

The feminist in me abhors professional basketball and the way it reinforces gender stereotypes. Men play, women watch. Men at the center, women at the periphery, and eagerly so. Men making millions, women scheming to get to some of the millions through their sex and sexuality. The existence of the Women's National Basketball Association hardly ameliorates my loathing for the patriarchal underbelly of the basketball sport, since women players are paid a scant fraction of men's pay and attract a fraction of the live and television audience. To be sure, women's basketball will grow and develop and perhaps even provide men with role models of what sportsmanship should be. I am also emphatically clear that basketball isn't the only patriarchy in town. Despite women's participation in politics it, too, is a patriarchy, with too many women plying sex and sexuality as their stock in trade, as the impeachment imbroglio of 1998–99 reminds us.

The race person in me has mixed feelings about basketball. On one hand, I enjoy seeing black men out there making big bank, the kind of bank they can't make in business, science, or more traditional forms of work. On the other hand, I'm aware of the minuscule odds that any high school hoopster will become a Michael Jordan. If some young brothers spent less time on the hoops and more in the labs, perhaps the investment of time in chemistry would yield the same mega-millions that professional basketball does. This is hardly the sole fault of the youngsters with NBA aspirations. It is also the responsibility of coaches to introduce a reality check to those young people whose future focus is exclusively on basketball. And it is society's responsibility to make sure there are opportunities for young men, especially young African American men, outside basketball. The athletic scholarship should not be the sole passport to college for low-income young men. Broader access to quality education must be a societal mandate; encouraging young African American men to focus on higher education is critical to our nation's fullest development.

Professional basketball bears an unfortunate similarity to an antebellum plantation, with some coaches accustomed to barking orders and uttering racial expletives to get maximum performance from their players. While few condone the fact that Latrell Sprewell, then of the Golden State Warriors, choked his coach P. J. Carlesimo in 1997, many understand that rude, perhaps racist, but certainly dismissive treatment motivated his violent action. The race of coaches, writers, and commentators, in contrast to that of the players, often has broader racial implications. White players whose grammar needs a boost often get it from sportswriters, who make them sound far more erudite than they are. The same writers quote black players with exaggerated, cringe-producing broken English—"dis," "dat," "dese," and "dose"—almost a parody of language. White commentators frequently remark on the "natural talent" of black players, compared to the skill of white ones. Retired white players are more likely to get invitations to coach and manage than are their black counterparts.

Despite this plantation tension, there are those who tout the integration in basketball as something lofty and desirable and the sport itself as one that teaches discipline, teamwork, and structure. In *Values of the Game,* former New Jersey senator and 2000 presidential candidate Bill Bradley describes the game as one of passion, discipline, selflessness, respect, perspective, courage, and other virtues. Though Bradley took a singular position as a senator in lifting up the issue of police brutality around the Rodney King beating, his record on race matters is otherwise relatively unremarkable. But Bradley gets credit for being far more racially progressive than he actually is because he, a white man, played for the New York Knicks for a decade, living at close quarters with African American men.

As onerous as I find the basketball plantation, the race person in me wonders why I am so concerned about the gladiators. Few if any of them exhibit anything that vaguely resembles social consciousness. Michael Jordan, for example, passed on the opportunity to make a real difference in the 1990 North Carolina Senate race between the former Charlotte mayor and Democratic candidate, the African American architect Harvey Gantt, and the ultra conservative and racially manipulative Republican senator Jesse Helms, preferring to save his endorsements for Nike. Charles Barkley once arrogantly crowed that he was not a role model, ignoring the biblical adage that much is expected of those to whom much is given. Few of the players, despite their millions, invest in the black community and in black economic development. Magic Johnson is a notable exception, with part ownership in a Los Angeles–based black bank and development of a theater chain that has the potential for revitalizing otherwise abandoned inner-city communities. Johnson's example notwithstanding, basketball players are more likely to make headlines for their shenanigans than for giving back to the community.

With my mixed feelings, love and loathing, why pay attention to basketball at all? Why not tune it out as forcefully as I tune out anything else I'm disinterested in? The fact is that it is nearly impossible to tune out, turn off, or ignore basketball. It is a cultural delimiter, a national export, a medium through which messages about race, gender, and power are transmitted not only nationally but also internationally. As Walter LaFeber observes in his book, *Michael Jordan and the New Global Capitalism,* Mr. Jordan's imagery has been used not only to sell the basketball game and Nike shoes but to make hundreds of millions of dollars for both Mr. Jordan and the companies he represents. Jordan is not to be faulted for brokering his skill into millions of dollars. Still, the mode and methods of his enrichment make the game of basketball a matter of intellectual curiosity as one explores the way in which capitalism and popular culture intersect. Jordan's millions, interesting as they may be, are less interesting than the way in which media have created his iconic status in both domestic and world markets and offered him up as a role model of the reasonable, affable African American man, the antithesis of "bad boy" trash talkers who so frequently garner headlines.

From a gender perspective, too, there is another set of questions. What would a woman have to do to achieve the same influence, iconic status, and bankability as Jordan? Is it even conceivable, in a patriarchal world, that a woman could earn, gain, or be invented in as iconic a status as a Michael Jordan? A woman might sing—but songbirds come and go and aren't often connected with international marketing of consumer products. She might dance—but that, too, would not turn into international marketability. She might possess the dynamic athleticism of the tennis-playing Williams sisters, Venus and Serena, or she might have the ethereal beauty of a Halle Berry. In either case, her product identification would solely be connected with "women's" products—hair, beauty, and women's sports items. The commercial heavy lifting has been left to the big boys, or to one big boy in particular, Michael Jordan.

To be sure, the public acceptance of women's sports has changed. Thanks to Title IX, women's sports get better funding and more attention at the college level than they did only decades ago. After several fits and starts, the professional women's basketball league seems to be doing well, although not as well as the NBA. In time this may change, but my informal survey of male basketball fans suggests that some find women's basketball simply less entertaining than the NBA. "When women start dunking and trash talking, more people will start watching," said a man active in establishing Midnight Basketball teams in urban centers. If this is the case, then it raises questions both about women's basketball and about the tension between male players and officials about on-court buffoonery. Do officials kill the goose that laid the golden egg when they ask a Dennis Rodman to "behave himself"?

My ire is not about gender envy; it is about imagery and the replication of gender-oppressive patterns by using basketball, especially, to connect an image of sportsmanship, masculinity, with product identification. In a commercial context that promotes this connection, there is both a subtle and a not-so-subtle message about the status of women.

There is also a set of subtle messages about race and race relations that emerges when African American men are used in the same way that African American women were used in the "trade cards" of the early twentieth century. In the patriarchal context, hoopsters become hucksters while the invisible (white) male role as power broker is reinforced. In too many ways the rules and profit of the game reinforce the rules, profit, and history of American life, with iconic gladiators serving as symbols and servants of multinational interests. Black men who entertain serve as stalkers for white men who measure profits. Neutered black men can join their white colleagues in chachinging cash registers but can never unlock the golden handcuffs and the platinum muzzles that limit their ability to generate independent opinions. Women? Always seen, never heard. Ornamental or invisible assets. Pawns in an unspoken game.

A subtext of the basketball culture relegates women, especially African American women, to a peripheral, dependent, and soap-operatic role. These women are sometimes seen as sources of trouble (as in the imbroglio with the Washington then Bullets, now Wizards, where apparently false rape charges spotlighted the lives of two players and perhaps changed the long-term composition of the team). Or they are depicted (for example, in a 1998 *Sports Illustrated* feature) as predatory sperm collectors, whose pregnancies are part of a "plot" to collect child support from high-earning hoopsters.

There are aspects to some women's behavior in relationship with basketball stars that are hardly laudable. (Some of this is detailed in the fictional *Homecourt*

Advantage, written by two basketball significant others.) At the same time, it is clear that this behavior is more a symptom than a cause of predatory patriarchy. After all, no one forces a player into bed or into a relationship with an unscrupulous woman. But conditions often make a player irresistible to women whose search for legitimacy comes through attachment to a man.

The disproportionate attention and influence that the basketball lifestyle gives some men makes them thoroughly irresistible to some women. This begs the question: In a patriarchy, just what can women do to attain the same influence and irresistibility? In a patriarchy, male games are far more intriguing than female games. Women, spectators, can watch and attach themselves to high-achieving gladiators. There is little that they can do, in the realm of sport, capitalism, and imagery, to gain the same access that men have. Again, this is not restricted to basketball but is a reflection of gender roles in our society. One might describe this as the "Hillary Rodham Clinton dilemma." Does a woman gain more power and influence by developing her own career or by attaching herself to a powerful man?

This dilemma may be a short-term one, given changes in the status of women in our society. Still, a range of scholars have noted that while women have come a long way, with higher incomes, higher levels of labor-force participation, and increased representation in executive ranks, at the current pace, income equality (which may not be the equivalent of the breakdown of patriarchy) will occur in the middle of the twenty-first century.

Those who depict the basketball culture have been myopic in their focus on the men who shoot hoops, because their coverage has ignored the humanity of the women who hover around hoopsters like moths drawn to a flame. Those writers who are eager to write about the "paternity ward" or about "Darwin's athletes" might also ask how the gender relations in basketball replicate or differ from gender relations in our society. Unfortunately, gender relations in basketball are too often the norm in society. Men play, women watch. To the extent that the basketball culture is elevated, women's roles are denigrated. Why are we willing to endow male hoopsters with a status that no woman can attain in our society? Sedentary men sing the praises of male hoopsters, sit enthralled and engrossed by their games, thus elevating a certain form of achievement in our society. This behavior is seen as normal, even laudable. Yet it takes women, often the daughters or sisters of these enthralled men, out of the iconic, high-achievement mix and pushes them to the periphery of a culture that reveres the athletic achievement derived from the combination of basketball prowess and patriarchy.

The combination of prowess and patriarchy denigrates the groupie who is attracted by the bright lights, but it also dehumanizes the swift gazelle, the gladiator, and the hoopster, whose humanity is negated by his basketball identity. Media coverage depicts young men out of control, trash-talking, wild-walking, attention-grabbing icons. Men play, women watch. Were there other center stages, this would be of limited interest. But the fact is that this stage is one from which other stages reverberate. Men play, women watch, in politics, economics, technology, and sports. And it is "blown up" into international cultural supremacy in basketball.

This woman watches, with love and loathing, the way basketball norms reinforce those that exist in our society. It is a triumph of patriarchy, this exuberance of masculine physicality. It is a reminder to women that feminism notwithstanding, we have yet to gain economic and cultural equivalency with Michael Jordan's capitalist dominance. While we should not, perhaps, ask men to walk away from arenas in which they can dominate, we must ask ourselves why there is no equivalent space for us; why the basketball tenet that men play, women watch, reverberates in so many other sectors of our society.

REFERENCES

Bradley, Bill. *Values of the Game.* New York: Artisan Press, 1998.

Ewing, Rita, and Crystal McCrary. *Homecourt Advantage.* New York: Avon Books, 1998.

Frey, Darcy. *The Last Shot: City Street, Basketball Dreams.* Boston: Houghton Mifflin, 1994.

Hoberman, John. *Darwin's Athletes: How Sport Has Damaged Black America and Preserved the Myth of Race.* New York: Mariner Books, 1997.

LaFeber, Walter. *Michael Jordan and the New Global Capitalism.* New York: W. W. Norton & Co., 1999.

Wahl, Grant, and L. Jon Wertheim. "Paternity Ward." *Sports Illustrated,* May 4, 1998, at 62.

24

"Is This a White Country, or What?"

LILLIAN B. RUBIN

"They're letting all these coloreds come in and soon there won't be any place left for white people," broods Tim Walsh, a thirty-three-year-old white construction worker. "It makes you wonder: Is this a white country, or what?"

It's a question that nags at white America, one perhaps that's articulated most often and most clearly by the men and women of the working class. For it's they who feel most vulnerable, who have suffered the economic contractions of recent decades most keenly, who see the new immigrants most clearly as direct competitors for their jobs.

It's not whites alone who stew about immigrants. Native-born blacks, too, fear the newcomers nearly as much as whites—and for the same economic

SOURCE: © Lillian B. Rubin. Originally published by HarperCollins. Reprinted by permission of Dunham Literary as agent for the author.

reasons. But for whites the issue is compounded by race, by the fact that the newcomers are primarily people of color. For them, therefore, their economic anxieties have combined with the changing face of America to create a profound uneasiness about immigration—a theme that was sounded by nearly 90 percent of the whites I met, even by those who are themselves first-generation, albeit well-assimilated, immigrants.

Sometimes they spoke about this in response to my questions; equally often the subject of immigration arose spontaneously as people gave voice to their concerns. But because the new immigrants are dominantly people of color, the discourse was almost always cast in terms of race as well as immigration, with the talk slipping from immigration to race and back again as if these are not two separate phenomena. "If we keep letting all them foreigners in, pretty soon there'll be more of them than us and then what will this country be like?" Tim's wife, Mary Anne, frets. "I mean, this is *our* country, but the way things are going, white people will be the minority in our own country. Now does that make any sense?"

Such fears are not new. Americans have always worried about the strangers who came to our shores, fearing that they would corrupt our society, dilute our culture, debase our values. So I remind Mary Anne, "When your ancestors came here, people also thought we were allowing too many foreigners into the country. Yet those earlier immigrants were successfully integrated into the American society. What's different now?"

"Oh, it's different, all right," she replies without hesitation. "When my people came, the immigrants were all white. That makes a big difference.". . .

Listening to Mary Anne's words I was reminded again how little we Americans look to history for its lessons, how impoverished is our historical memory.

For, in fact, being white didn't make "a big difference" for many of those earlier immigrants. The dark-skinned Italians and the eastern European Jews who came in the late nineteenth and early twentieth centuries didn't look very white to the fair-skinned Americans who were here then. Indeed, the same people we now call white—Italians, Jews, Irish—were seen as another race at that time. Not black or Asian, it's true, but an alien other, a race apart, although one that didn't have a clearly defined name. Moreover, the racist fears and fantasies of native-born Americans were far less contained then than they are now, largely because there were few social constraints on their expression.

When, during the nineteenth century, for example, some Italians were taken for blacks and lynched in the South, the incidents passed virtually unnoticed. And if Mary Anne and Tim Walsh, both of Irish ancestry, had come to this country during the great Irish immigration of that period, they would have found themselves defined as an inferior race and described with the same language that was used to characterize blacks: "low-browed and savage, grovelling and bestial, lazy and wild, simian and sensual."[1] Not only during that period but for a long time afterward as well, the U.S. Census Bureau counted the Irish as a distinct and separate group, much as it does today with the category it labels "Hispanic."

But there are two important differences between then and now, differences that can be summed up in a few words: the economy and race. Then, a growing industrial economy meant that there were plenty of jobs for both immigrant and native workers, something that can't be said for the contracting economy in which we live today. True, the arrival of the immigrants, who were more readily exploitable than native workers, put Americans at a disadvantage and created discord between the two groups. Nevertheless, work was available for both.

Then, too, the immigrants—no matter how they were labeled, no matter how reviled they may have been—were ultimately assimilable, if for no other reason than that they were white. As they began to lose their alien ways, it became possible for native Americans to see in the white ethnics of yesteryear a reflection of themselves. Once this shift in perception occurred, it was possible for the nation to incorporate them, to take them in, chew them up, digest them, and spit them out as Americans—with subcultural variations not always to the liking of those who hoped to control the manners and mores of the day, to be sure, but still recognizably white Americans.

Today's immigrants, however, are the racial other in a deep and profound way. . . . And integrating masses of people of color into a society where race conciousness lies at the very heart of our central nervous system raises a whole new set of anxieties and tensions. . . .

The increased visibility of other racial groups has focused whites more self-consciously than ever on their own racial identification. Until the new immigration shifted the complexion of the land so perceptibly, whites didn't think of themselves as white in the same way that Chinese know they're Chinese and African-Americans know they're black. Being white was simply a fact of life, one that didn't require any public statement, since it was the definitive social value against which all others were measured. "It's like everything's changed and I don't know what happened," complains Marianne Bardolino. "All of a sudden you have to be thinking all the time about these race things. I don't remember growing up thinking about being white like I think about it now. I'm not saying I didn't know there was coloreds and whites; it's just that I didn't go along thinking, *Gee, I'm a white person.* I never thought about it at all. But now with all the different colored people around, you have to think about it because they're thinking about it all the time."

"You say you feel pushed now to think about being white, but I'm not sure I understand why. What's changed?" I ask.

"I told you," she replies quickly, a small smile covering her impatience with my question. "It's because they think about what they are, and they want things their way, so now I have to think about what I am and what's good for me and my kids." She pauses briefly to let her thoughts catch up with her tongue, then continues. "I mean, if somebody's always yelling at you about being black or Asian or something, then it makes you think about being white. Like, they want the kids in school to learn about their culture, so then I think about being white and being Italian and say: What about my culture? If they're going to teach about theirs, what about mine?"

To which America's racial minorities respond with bewilderment. "I don't understand what white people want," says Gwen Tomalson. "They say if black kids are going to learn about black culture in school, then white people want their kids to learn about white culture. I don't get it. What do they think kids have been learning about all these years? It's all about white people and how they live and what they accomplished. When I was in school you wouldn't have thought black people existed for all our books ever said about us."

As for the charge that they're "thinking about race all the time," as Marianne Bardolino complains, people of color insist that they're forced into it by a white world that never lets them forget. "If you're Chinese, you can't forget it, even if you want to, because there's always something that reminds you," Carol Kwan's husband, Andrew, remarks tartly. "I mean, if Chinese kids get good grades and get into the university, everybody's worried and you read about it in the papers."

While there's little doubt that racial anxieties are at the center of white concerns, our historic nativism also plays a part in escalating white alarm. The new immigrants bring with them a language and an ethnic culture that's vividly expressed wherever they congregate. And it's this also, the constant reminder of an alien presence from which whites are excluded, that's so troublesome to them.

The nativist impulse isn't, of course, given to the white working class alone. But for those in the upper reaches of the class and status hierarchy—those whose children go to private schools, whose closest contact with public transportation is the taxi cab—the immigrant population supplies a source of cheap labor, whether as nannies for their children, maids in their households, or workers in their businesses. They may grouse and complain that "nobody speaks English anymore," just as working-class people do. But for the people who use immigrant labor, legal or illegal, there's a payoff for the inconvenience—a payoff that doesn't exist for the families in this study but that sometimes costs them dearly. For while it may be true that American workers aren't eager for many of the jobs immigrants are willing to take, it's also true that the presence of a large immigrant population—especially those who come from developing countries where living standards are far below our own—helps to make these jobs undesirable by keeping wages depressed well below what most American workers are willing to accept. . . .

It's not surprising, therefore, that working-class women and men speak so angrily about the recent influx of immigrants. They not only see their jobs and their way of life threatened, they feel bruised and assaulted by an environment that seems suddenly to have turned color and in which they feel like strangers in their own land. So they chafe and complain: "They come here to take advantage of us, but they don't really want to learn our ways," Beverly Sowell, a thirty-three-year old white electronics assembler, grumbles irritably. "They live different than us; it's like another world how they live. And they're so clannish. They keep to themselves, and they don't even *try* to learn English. You go on the bus these days and you might as well be in a foreign country; everybody's talking some other language, you know, Chinese or Spanish or something. Lots of them have been here a long time, too, but they don't care; they just want to take what they can get."

But their complaints reveal an interesting paradox, an illuminating glimpse into the contradictions that beset native-born Americans in their relations with those who seek refuge here. On the one hand, they scorn the immigrants; on the other, they protest because they "keep to themselves." It's the same contradiction that dominates black-white relations. Whites refuse to integrate blacks but

are outraged when they stop knocking at the door, when they move to sustain
the separation on their own terms'—in black theme houses on campuses, for
example, or in the newly developing black middle-class suburbs.

I wondered, as I listened to Beverly Sowell and others like her, why the
same people who find the lifeways and languages of our foreign-born population
offensive also care whether they "keep to themselves."

"Because like I said, they just shouldn't, that's all," Beverly says stubbornly.
"If they're going to come here, they should be willing to learn our ways—you
know what I mean, be real Americans. That's what my grandparents did, and
that's what they should do."

"But your grandparents probably lived in an immigrant neighborhood when
they first came here, too," I remind her.

"It was different," she insists. "I don't know why; it was. They wanted to be
Americans; these here people now, I don't think they do. They just want to take
advantage of this country. . . .

"Everything's changed, and it doesn't make sense. Maybe you get it, but I
don't. We can't take care of our own people and we keep bringing more and
more foreigners in. Look at all the homeless. Why do we need more people here
when our own people haven't got a place to sleep?"

"Why do we need more people here?"—a question Americans have asked
for two centuries now. Historically, efforts to curb immigration have come during economic downturns, which suggests that when times are good, when
American workers feel confident about their future, they're likely to be more
generous in sharing their good fortune with foreigners. But when the economy
falters, as it did in the 1990s, and workers worry about having to compete for
jobs with people whose standard of living is well below their own, resistance to
immigration rises. "Don't get me wrong; I've got nothing against these people,"
Tim Walsh demurs. "But they don't talk English, and they're used to a lot less,
so they can work for less money than guys like me can. I see it all the time; they
get hired and some white guy gets left out."

It's this confluence of forces—the racial and cultural diversity of our new
immigrant population; the claims on the resources of the nation now being
made by those minorities who, for generations, have called America their
home; the failure of some of our basic institutions to serve the needs of our people; the contracting economy, which threatens the mobility aspirations of
working-class families—all these have come together to leave white workers
feeling as if everyone else is getting a piece of the action while they get nothing.
"I feel like white people are left out in the cold," protests Diane Johnson, a
twenty-eight-year-old white single mother who believes she lost a job as a bus
driver to a black woman. "First it's the blacks; now it's all those other colored
people, and it's like everything always goes their way. It seems like a white person doesn't have a chance anymore. It's like the squeaky wheel gets the grease,
and they've been squeaking and we haven't," she concludes angrily.

Until recently, whites didn't need to think about having to "squeak"—at
least not specifically as whites. They have, of course, organized and squeaked at

various times in the past—sometimes as ethnic groups, sometimes as workers. But not as whites. As whites they have been the dominant group, the favored ones, the ones who could count on getting the job when people of color could not. Now suddenly there are others—not just individual others but identifiable groups, people who share a history, a language, a culture, even a color—who lay claim to some of the rights and privileges that formerly had been labeled "for whites only." And whites react as if they've been betrayed, as if a sacred promise has been broken. They're white, aren't they? They're *real* Americans, aren't they? This is their country, isn't it?

The answers to these questions used to be relatively unambiguous. But not anymore. Being white no longer automatically assures dominance in the politics of a multiracial society. Ethnic group politics, however, has a long and fruitful history. As whites sought a social and political base on which to stand, therefore, it was natural and logical to reach back to their ethnic past. Then they, too, could be "something"; they also would belong to a group; they would have a name, a history, a culture, and a voice. "Why is it only the blacks or Mexicans or Jews that are 'something'?" asks Tim Walsh. "I'm Irish, isn't that something, too? Why doesn't that count?"

In reclaiming their ethnic roots, whites can recount with pride the tribulations and transcendence of their ancestors and insist that others take their place in the line from which they have only recently come. "My people had a rough time, too. But nobody gave us anything, so why do we owe them something? Let them pull their share like the rest of us had to do," says Al Riccardi, a twenty-nine-year-old white taxi driver.

From there it's only a short step to the conviction that those who don't progress up that line are hampered by nothing more than their own inadequacies or, worse yet, by their unwillingness to take advantage of the opportunities offered them. "Those people, they're hollering all the time about discrimination," Al continues, without defining who "those people" are. "Maybe once a long time ago that was true, but not now. The problem is that a lot of those people are lazy. There's plenty of opportunities, but you've got to be willing to work hard."

He stops a moment, as if listening to his own words, then continues, "Yeah, yeah, I know there's a recession on and lots of people don't have jobs. But it's different with some of those people. They don't really want to work, because if they did, there wouldn't be so many of them selling drugs and getting in all kinds of trouble."

"You keep talking about 'those people' without saying who you mean," I remark.

"Aw c'mon, you know who I'm talking about," he says, his body shifting uneasily in his chair. "It's mostly the black people, but the Spanish ones, too."

In reality, however, it's a no-win situation for America's people of color, whether immigrant or native born. For the industriousness of the Asians comes in for nearly as much criticism as the alleged laziness of other groups. When blacks don't make it, it's because, whites like Al Riccardi insist, their culture doesn't teach respect for family; because they're hedonistic, lazy, stupid, and/or criminally inclined. But when Asians demonstrate their ability to overcome the obstacles of an alien language and culture, when the Asian family seems to be the repository of our most highly regarded traditional values, white hostility doesn't disappear. It just changes its form. Then the accomplishments of Asians, the speed with which they move up the economic ladder, aren't credited to their superior culture, diligence, or intelligence—even when these are granted—but to the fact that they're "single minded," "untrustworthy," "clannish drones," "narrow people" who raise children who are insufficiently "well rounded."[2]. . .

Not surprisingly, as competition increases, the various minority groups often are at war among themselves as they press their own particular claims, fight over turf, and compete for an ever-shrinking piece of the pie. In several African-American communities, where Korean shopkeepers have taken the place once held by Jews, the confrontations have been both wrenching and tragic. A Korean grocer in Los Angeles shoots and kills a fifteen-year-old black girl for allegedly trying to steal some trivial item from the store.[3] From New York City to Berkeley, California, African-Americans boycott Korean shop owners who, they charge, invade their neighborhoods, take their money, and treat them disrespectfully.[4] But painful as these incidents are for those involved, they are only symptoms of a deeper malaise in both communities—the contempt and distrust in which the Koreans hold their African-American neighbors, and the rage of blacks as they watch these new immigrants surpass them.

Latino-black conflict also makes headlines when, in the aftermath of the riots in South Central Los Angeles, the two groups fight over who will get the lion's share of the jobs to rebuild the neighborhood. Blacks, insisting that they're being discriminated against, shut down building projects that don't include them in satisfactory numbers. And indeed, many of the jobs that formerly went to African-Americans are now being taken by Latino workers. In an article entitled "Black vs. Brown," Jack Miles, an editorial writer for the *Los Angeles Times*, reports that "janitorial firms serving downtown Los Angeles have almost entirely replaced their unionized black work force with non-unionized immigrants."[5]. . .

But the disagreements among America's racial minorities are of little interest or concern to most white working-class families. Instead of conflicting groups, they see one large mass of people of color, all of them making claims that endanger their own precarious place in the world. It's this perception that has led some white ethnics to believe that reclaiming their ethnicity alone is not enough, that so long as they remain in their separate and distinct groups, their power will be limited. United, however, they can become a formidable countervailing force, one that can stand fast against the threat posed by minority demands. But to come together solely as whites would diminish their impact and leave them open to the charge that their real purpose is simply to retain the privileges of whiteness. A dilemma that has been resolved, at least for some, by the birth of a new entity in the history of American ethnic groups—the "European-Americans."[6]. . .

At the University of California at Berkeley, for example, white students and their faculty supporters insisted that the recently adopted multicultural curriculum include a unit of study of European-Americans. At Queens College in New York City, where white ethnic groups retain a more distinct presence, Italian-American students launched a successful suit to win recognition as a disadvantaged minority and gain the entitlements accompanying that status, including special units of Italian-American studies.

White high school students, too, talk of feeling isolated and, being less sophisticated and wary than their older sisters and brothers, complain quite openly that there's no acceptable and legitimate way for them to acknowledge a white identity. "There's all these things for all the different ethnicities, you know, like clubs for black kids and Hispanic kids, but there's nothing for me and my friends to join," Lisa Marshall, a sixteen-year-old white high school student, explains with exasperation. "They won't let us have a white club because that's supposed to be racist. So we figured we'd just have to call it something else, you know, some ethnic thing, like Euro-Americans. Why not? They have African-American clubs."

Ethnicity, then, often becomes a cover for "white," not necessarily because these students are racist but because racial identity is now such a prominent feature of the discourse in our social world. In a society where racial consciousness is so high, how else can whites define themselves in ways that connect them to a community and, at the same time, allow them to deny their racial antagonisms?

Ethnicity and race—separate phenomena that are now inextricably entwined. Incorporating newcomers has never been easy, as our history of controversy and violence over immigration tells us.[7] But for the first time, the new immigrants are also people of color, which means that they tap both the nativist and racist impulses that are so deeply a part of American life. As in the past, however, the fear of foreigners, the revulsion against their strange customs and seemingly unruly ways, is only part of the reason for the anti-immigrant attitudes that are increasingly being expressed today. For whatever xenophobic suspicions may arise in modern America, economic issues play a critical role in stirring them up.

REFERENCES

Alba, Richard D. *Ethnic Identity*. New Haven: Yale University Press, 1990.
Roediger, David R. *The Wages of Whiteness*. New York: Verso, 1991.

NOTES

1. David R. Roediger, *The Wages of Whiteness* (New York: Verso, 1991), p. 133.
2. These were, and often still are, the commonly held stereotypes about Jews. Indeed, the Asian immigrants are often referred to as "the new Jews."
3. Soon Ja Du, the Korean grocer who killed fifteen-year-old Latasha Harlins, was found guilty of voluntary manslaughter and sentenced to four hundred hours of community service, a $500 fine, reimbursement of funeral costs to the Harlins family, and five years' probation.

4. The incident in Berkeley didn't happen in the black ghetto, as most of the others did. There, the Korean grocery store is near the University of California campus, and the woman involved in the incident is an African-American university student who was maced by the grocer after an argument over a penny.

5. Jack Miles, "Blacks vs. Browns," *Atlantic Monthly* (October 1992), pp. 41–68.

6. For an interesting analysis of what he calls "the transformation of ethnicity," see Richard D. Alba, *Ethnic Identity* (New Haven, CT: Yale University Press, 1990).

7. In the past, many of those who agitated for a halt to immigration were immigrants or native-born children of immigrants. The same often is true today. As anti-immigrant sentiment grows, at least some of those joining the fray are relatively recent arrivals. One man in this study, for example—a fifty-two-year-old immigrant from Hungary—is one of the leaders of an anti-immigration group in the city where he lives.

25

Optional Ethnicities

For Whites Only?

MARY C. WATERS

What does it mean to talk about ethnicity as an option for an individual? To argue that an individual has some degree of choice in their ethnic identity flies in the face of the commonsense notion of ethnicity many of us believe in— that one's ethnic identity is a fixed characteristic, reflective of blood ties and given at birth. However, social scientists who study ethnicity have long concluded that while ethnicity is based on a *belief* in a common ancestry, ethnicity is primarily a *social* phenomenon, not a biological one (Alba 1985, 1990; Barth 1969; Weber [1921] 1968, p. 389). The belief that members of an ethnic group have that they share a common ancestry may not be a fact. There is a great deal of change in ethnic identities across generations through intermarriage, changing allegiances, and changing social categories. There is also a much larger amount of change in the identities of individuals over their lives than is commonly believed. While most people are aware of the phenomenon known as "passing"—people raised as one race who change at some point and claim a different race as their

SOURCE: From Silvia Pedraza and Rubén G. Rumbaut, eds., *Origins and Destinies: Immigration, Race and Ethnicity in America* (Belmont, CA: Wadsworth, 1996), pp. 444–54. Reprinted by permission.

identity—there are similar life course changes in ethnicity that happen all the time and are not given the same degree of attention as "racial passing."

White Americans of European ancestry can be described as having a great deal of choice in terms of their ethnic identities. The two major types of options White Americans can exercise are (1) the option of whether to claim any specific ancestry, or to just be "White" or American, [Lieberson (1985) called these people "unhyphenated Whites"] and (2) the choice of which of their European ancestries to choose to include in their description of their own identities. In both cases, the option of choosing how to present yourself on surveys and in everyday social interactions exists for Whites because of social changes and societal conditions that have created a great deal of social mobility, immigrant assimilation, and political and economic power for Whites in the United States. Specifically, the option of being able to not claim any ethnic identity exists for Whites of European background in the United States because they are the majority group—in terms of holding political and social power, as well as being a numerical majority. The option of choosing among different ethnicities in their family backgrounds exists because the degree of discrimination and social distance attached to specific European backgrounds has diminished over time. . . .

SYMBOLIC ETHNICITIES FOR WHITE AMERICANS

What do these ethnic identities mean to people and why do they cling to them rather than just abandoning the tie and calling themselves American? My own field research with suburban Whites in California and Pennsylvania found that later-generation descendants of European origin maintain what are called "symbolic ethnicities." Symbolic ethnicity is a term coined by Herbert Gans (1979) to refer to ethnicity that is individualistic in nature and without real social cost for the individual. These symbolic identifications are essentially leisure-time activities, rooted in nuclear family traditions and reinforced by the voluntary enjoyable aspects of being ethnic (Waters 1990). Richard Alba (1990) also found later-generation Whites in Albany, New York, who chose to keep a tie with an ethnic identity because of the enjoyable and voluntary aspects to those identities, along with the feelings of specialness they entailed. An example of symbolic ethnicity is individuals who identify as Irish, for example, on occasions such as Saint Patrick's Day, on family holidays, or for vacations. They do not usually belong to Irish American organizations, live in Irish neighborhoods, work in Irish jobs, or marry other Irish people. The symbolic meaning of being Irish American can be constructed by individuals from mass media images, family traditions, or other intermittent social activities. In other words, for later-generation White ethnics, ethnicity is not something that influences their lives unless they want it to. In the world of work and school and neighborhood, individuals do not have to admit to being ethnic unless they choose to. And for an increasing number of European-origin individuals whose parents and grandparents have intermarried, the ethnicity they claim is largely a matter of personal choice as they sort through all of the possible combinations of groups in their genealogies. . . .

RACE RELATIONS AND SYMBOLIC ETHNICITY

However much symbolic ethnicity is without cost for the individual, there is a cost associated with symbolic ethnicity for the society. That is because symbolic ethnicities of the type described here are confined to White Americans of European origin. Black Americans, Hispanic Americans, Asian Americans, and American Indians do not have the option of a symbolic ethnicity at present in the United States. For all of the ways in which ethnicity does not matter for White Americans, it does matter for non-Whites. Who your ancestors are does affect your choice of spouse, where you live, what job you have, who your friends are, and what your chances are for success in American society, if those ancestors happen not to be from Europe. The reality is that White ethnics have a lot more choice and room for maneuver than they themselves think they do. The situation is very different for members of racial minorities, whose lives are strongly influenced by their race or national origin regardless of how much they may choose not to identify themselves in terms of their ancestries.

When White Americans learn the stories of how their grandparents and great-grandparents triumphed in the United States over adversity, they are usually told in terms of their individual efforts and triumphs. The important role of labor unions and other organized political and economic actors in their social and economic successes are left out of the story in favor of a generational story of individual Americans rising up against communitarian, Old World intolerance, and New World resistance. As a result, the "individualized" voluntary, cultural view of ethnicity for Whites is what is remembered.

One important implication of these identities is that they tend to be very individualistic. There is a tendency to view valuing diversity in a pluralist environment as equating all groups. The symbolic ethnic tends to think that all groups are equal; everyone has a background that is their right to celebrate and pass on to their children. This leads to the conclusion that all identities are equal and all identities in some sense are interchangeable—"I'm Italian American, you're Polish American. I'm Irish American, you're African American." The important thing is to treat people as individuals and all equally. However, this assumption ignores the very big difference between an individualistic symbolic ethnic identity and a socially enforced and imposed racial identity.

My favorite example of how this type of thinking can lead to some severe misunderstandings between people of different backgrounds is from the *Dear Abby* advice column. A few years back a person wrote in who had asked an acquaintance of Asian background where his family was from. His acquaintance answered that this was a rude question and he would not reply. The bewildered White asked Abby why it was rude, since he thought it was a sign of respect to

wonder where people were from, and he certainly would not mind anyone asking HIM about where his family was from. Abby asked her readers to write in to
say whether it was rude to ask about a person's ethnic background. She reported
that she got a large response, that most non-Whites thought it was a sign of dis-
respect, and Whites thought it was flattering:

> Dear Abby,
> I am 100 percent American and because I am of Asian ancestry I am often
> asked "What are you?" It's not the personal nature of this question that
> bothers me, it's the question itself. This query seems to question my very
> humanity. "What am I? Why I am a person like everyone else!"
> Signed, A REAL AMERICAN

> Dear Abby,
> Why do people resent being asked what they are? The Irish are so
> proud of being Irish, they tell you before you even ask. Tip O'Neill has
> never tried to hide his Irish ancestry.
> Signed, JIMMY.

> (Reprinted by permission of Universal Press Syndicate)

In this exchange Jimmy cannot understand why Asians are not as happy to
be asked about their ethnicity as he is, because he understands his ethnicity and
theirs to be separate but equal. Everyone has to come from somewhere—his
family from Ireland, another's family from Asia—each has a history and each
should be proud of it. But the reason he cannot understand the perspective of
the Asian American is that all ethnicities are not equal; all are not symbolic, cost-
less, and voluntary. When White Americans equate their own symbolic ethnicities with the socially enforced identities of non-White Americans, they obscure
the fact that the experiences of Whites and non-Whites have been qualitatively
different in the United States and that the current identities of individuals partly
reflect that unequal history.

In the next section I describe how relations between Black and White students
on college campuses reflect some of these asymmetries in the understanding of
what a racial or ethnic identity means. While I focus on Black and White students
in the following discussion, you should be aware that the myriad other groups in
the United States—Mexican Americans, American Indians, Japanese Americans—
all have some degree of social and individual influences on their identities, which
reflect the group's social and economic history and present circumstance.

RELATIONS ON COLLEGE CAMPUSES

Both Black and White students face the task of developing their race and ethnic
identities. Sociologists and psychologists note that at the time people leave home
and begin to live independently from their parents, often ages eighteen to

twenty-two, they report a heightened sense of racial and ethnic identity as they sort through how much of their beliefs and behaviors are idiosyncratic to their families and how much are shared with other people. It is not until one comes in close contact with many people who are different from oneself that individuals realize the ways in which their backgrounds may influence their individual personality. This involves coming into contact with people who are different in terms of their ethnicity, class, religion, region, and race. For White students, the ethnicity they claim is more often than not a symbolic one—with all of the voluntary, enjoyable, and intermittent characteristics I have described above.

Black students at the university are also developing identities through interactions with others who are different from them. Their identity development is more complicated than that of Whites because of the added element of racial discrimination and racism, along with the "ethnic" developments of finding others who share their background. Thus Black students have the positive attraction of being around other Black students who share some cultural elements, as well as the need to band together with other students in a reactive and oppositional way in the face of racist incidents on campus.

Colleges and universities across the country have been increasing diversity among their student bodies in the last few decades. This has led in many cases to strained relations among students from different racial and ethnic backgrounds. The 1980s and 1990s produced a great number of racial incidents and high racial tensions on campuses. While there were a number of racial incidents that were due to bigotry, unlawful behavior, and violent or vicious attacks, much of what happens among students on campuses involves a low level of tension and awkwardness in social interactions.

Many Black students experience racism personally for the first time on campus. The upper-middle-class students from White suburbs were often isolated enough that their presence was not threatening to racists in their high schools. Also, their class background was known by their residence and this may have prevented attacks being directed at them. Often Black students at the university who begin talking with other students and recognizing racial slights will remember incidents that happened to them earlier that they might not have thought were related to race.

Black college students across the country experience a sizeable number of incidents that are clearly the result of racism. Many of the most blatant ones that occur between students are the result of drinking. Sometimes late at night, drunken groups of White students coming home from parties will yell slurs at single Black students on the street. The other types of incidents that happen include being singled out for special treatment by employees, such as being followed when shopping at the campus bookstore, or going to the art museum with your class and the guard stops you and asks for your I.D. Others involve impersonal encounters on the street—being called a nigger by a truck driver while crossing the street, or seeing old ladies clutch their pocketbooks and shake in terror as you pass them on the street. For the most part these incidents are not specific to the university environment, they are the types of incidents middle-class Blacks face every day throughout American society, and they have been documented by sociologists (Feagin 1991).

In such a climate, however, with students experiencing these types of incidents and talking with each other about them, Black students do experience a tension and a feeling of being singled out. It is unfair that this is part of their college experience and not that of White students. Dealing with incidents like this, or the ever-present threat of such incidents, is an ongoing developmental task for Black students that takes energy, attention, and strength of character. It should be clearly understood that this is an asymmetry in the "college experience" for Black and White students. It is one of the unfair aspects of life that results from living in a society with ongoing racial prejudice and discrimination. It is also very understandable that it makes some students angry at the unfairness of it all, even if there is no one to blame specifically. It is also very troubling because, while most Whites do not create these incidents, some do, and it is never clear until you know someone well whether they are the type of person who could do something like this. So one of the reactions of Black students to these incidents is to band together.

In some sense then, as Blauner (1992) has argued, you can see Black students coming together on campus as both an "ethnic" pull of wanting to be together to share common experiences and community, and a "racial" push of banding together defensively because of perceived rejection and tension from Whites. In this way the ethnic identities of Black students are in some sense similar to, say, Korean students wanting to be together to share experiences. And it is an ethnicity that is generally much stronger than, say, Italian Americans. But for Koreans who come together there is generally a definition of themselves as "different from" Whites. For Blacks reacting to exclusion, there is a tendency for the coming together to involve both being "different from" but also "opposed to" Whites.

The anthropologist John Ogbu (1990) has documented the tendency of minorities in a variety of societies around the world, who have experienced severe blocked mobility for long periods of time, to develop such oppositional identities. An important component of having such an identity is to describe others of your group who do not join in the group solidarity as devaluing and denying their very core identity. This is why it is not common for successful Asians to be accused by others of "acting White" in the United States, but it is quite common for such a term to be used by Blacks and Latinos. The oppositional component of a Black identity also explains how Black people can question whether others are acting "Black enough." On campus, it explains some of the intense pressures felt by Black students who do not make their racial identity central and who choose to hang out primarily with non-Blacks. This pressure from the group, which is partly defining itself by not being White, is exacerbated by the fact that race is a physical marker in American society. No one immediately notices the Jewish students sitting together in the dining hall, or the one Jewish student sitting surrounded by non-Jews, or the Texan sitting with the Californians, but everyone notices the Black student who is or is not at the "Black table" in the cafeteria.

An example of the kinds of misunderstandings that can arise because of different understandings of the meanings and implications of symbolic versus oppositional identities concerns questions students ask one another in the dorms about personal appearances and customs. A very common type of interaction in the dorm concerns questions Whites ask Blacks about their hair. Because Whites tend to know little about Blacks, and Blacks know a lot about Whites, there is a general asymmetry in the level of curiosity people have about one another. Whites, as the numerical majority, have had little contact with Black culture; Blacks, especially those who are in college, have had to develop bicultural skills—knowledge about the social worlds of both Whites and Blacks. Miscommunication and hurt feelings about White students' questions about Black students' hair illustrate this point. One of the things that happens freshman year is that White students are around Black students as they fix their hair. White students are generally quite curious about Black students' hair—they have basic questions such as how often Blacks wash their hair, how they get it straightened or curled, what products they use on their hair, how they comb it, etc. Whites often wonder to themselves whether they should ask these questions. One thought experiment Whites perform is to ask themselves whether a particular question would upset them. Adopting the "do unto others" rule, they ask themselves, "If a Black person was curious about my hair would I get upset?" The answer usually is "No, I would be happy to tell them." Another example is an Italian American student wondering to herself, "Would I be upset if someone asked me about calamari?" The answer is no, so she asks her Black roommate about collard greens, and the roommate explodes with an angry response such as, "Do you think all Black people eat watermelon too?" Note that if this Italian American knew her friend was Trinidadian American and asked about peas and rice the situation would be more similar and would not necessarily ignite underlying tensions.

Like the debate in *Dear Abby,* these innocent questions are likely to lead to resentment. The issue of stereotypes about Black Americans and the assumption that all Blacks are alike and have the same stereotypical cultural traits has more power to hurt or offend a Black person than vice versa. The innocent questions about Black hair also bring up a number of asymmetries between the Black and White experience. Because Blacks tend to have more knowledge about Whites than vice versa, there is not an even exchange going on, the Black freshman is likely to have fewer basic questions about his White roommate than his White roommate has about him. Because of the differences historically in the group experiences of Blacks and Whites there are some connotations to Black hair that don't exist about White hair. (For instance, is straightening your hair a form of assimilation, do some people distinguish between women having "good hair" and "bad hair" in terms of beauty and how is that related to looking "White"?) Finally, even a Black freshman who cheerfully disregards or is unaware that there are these asymmetries will soon slam into another asymmetry if she willingly answers every innocent question asked of her. In a situation where Blacks make up only 10 percent of the student body, if every non-Black needs to be educated about hair, she will have to explain it to nine other students. As one Black student explained to me, after you've been asked a couple of times about something so personal you begin to feel like you are an attraction in a zoo, that you are at the university for the education of the White students.

INSTITUTIONAL RESPONSES

Our society asks a lot of young people. We ask young people to do something that no one else does as successfully on such a wide scale—that is to live together with people from very different backgrounds, to respect one another, to appreciate one another, and to enjoy and learn from one another. The successes that occur every day in this endeavor are many, and they are too often overlooked. However, the problems and tensions are also real, and they will not vanish on their own. We tend to see pluralism working in the United States in much the same way some people expect capitalism to work. If you put together people with various interests and abilities and resources, the "invisible hand" of capitalism is supposed to make all the parts work together in an economy for the common good.

There is much to be said for such a model—the invisible hand of the market can solve complicated problems of production and distribution better than any "visible hand" of a state plan. However, we have learned that unequal power relations among the actors in the capitalist marketplace, as well as "externalities" that the market cannot account for, such as long-term pollution, or collusion between corporations, or the exploitation of child labor, means that state regulation is often needed. Pluralism and the relations between groups are very similar. There is a lot to be said for the idea that bringing people who belong to different ethnic or racial groups together in institutions with no interference will have good consequences. Students from different backgrounds will make friends if they share a dorm room or corridor, and there is no need for the institution to do any more than provide the locale. But like capitalism, the invisible hand of pluralism does not do well when power relations and externalities are ignored. When you bring together individuals from groups that are differentially valued in the wider society and provide no guidance, there will be problems. In these cases the "invisible hand" of pluralist relations does not work, and tensions and disagreements can arise without any particular individual or group of individuals being "to blame." On college campuses in the 1990s some of the tensions between students are of this sort. They arise from honest misunderstandings, lack of a common background, and very different experiences of what race and ethnicity mean to the individual.

The implications of symbolic ethnicities for thinking about race relations are subtle but consequential. If your understanding of your own ethnicity and its relationship to society and politics is one of individual choice, it becomes harder to understand the need for programs like affirmative action, which recognize the ongoing need for group struggle and group recognition, in order to bring about social change. It also is hard for a White college student to understand the need that minority students feel to band together against discrimination. It also is easy, on the individual level, to expect everyone else to be able to turn their ethnicity on and off at will, the way you are able to, without understanding that ongoing discrimination and societal attention to minority status makes that impossible for individuals from minority groups to do. The paradox of symbolic ethnicity is that it depends upon the ultimate goal of a pluralist society, and at the same time makes it more difficult to achieve that ultimate goal. It is dependent upon the concept that all ethnicities mean the same thing, that enjoying the traditions of one's heritage is an option available to a group or an individual, but that such a heritage should not have any social costs associated with it.

As the Asian Americans who wrote to *Dear Abby* make clear, there are many societal issues and involuntary ascriptions associated with non-White identities. The developments necessary for this to change are not individual but societal in nature. Social mobility and declining racial and ethnic sensitivity are closely associated. The legacy and the present reality of discrimination on the basis of race or ethnicity must be overcome before the ideal of a pluralist society, where all heritages are treated equally and are equally available for individuals to choose or discard at will, is realized.

REFERENCES

Alba, Richard D. 1985. *Italian Americans: Into the of Twilight Ethnicity*. Englewood Cliffs, NJ: Prentice-Hall.

Alba, Richard D. 1990. *Ethnic Identity: The Transformation of White America*. New Haven: Yale University Press.

Barth, Frederick. 1969. *Ethnic Groups and Boundaries*. Boston: Little, Brown.

Blauner, Robert. 1992. "Talking Past Each Other: Black and White Languages of Race." *American Prospect* (Summer): 55–64.

Feagin, Joe R. 1991. "The Continuing Significance of Race: Anti-Black Discrimination in Public Places." *American Sociological Review* 56: 101–17.

Gans, Herbert. 1979. "Symbolic Ethnicity: The Future of Ethnic Groups and Cultures in America." *Ethnic and Racial Studies* 2: 1–20.

Lieberson, Stanley. 1985. *Making It Count: The Improvement of Social Research and Theory*. Berkeley: University of California Press.

Ogbu, John. 1990. "Minority Status and Literacy in Comparative Perspective." *Daedalus* 119: 141–69.

Waters, Mary C. 1990. *Ethnic Options: Choosing Identities in America*. Berkeley: University of California Press.

Weber, Max. [1921]/1968. *Economy and Society: An Outline of Interpretive Sociology*. Eds. Guenther Roth and Claus Wittich, trans. Ephraim Fischoff. New York: Bedminister Press.

26

Global Woman

BARBARA EHRENREICH AND ARLIE RUSSELL HOCHSCHILD

"Whose baby are you?" Josephine Perera, a nanny from Sri Lanka, asks Isadora, her pudgy two-year-old charge in Athens, Greece.

Thoughtful for a moment, the child glances toward the closed door of the next room, in which her mother is working, as if to say, "That's my mother in there."

"No, you're *my* baby," Josephine teases, tickling Isadora lightly. Then, to settle the issue, Isadora answers, "Together!" She has two mommies—her mother and Josephine. And surely a child loved by many adults is richly blessed.

In some ways, Josephine's story—which unfolds in an extraordinary documentary film, *When Mother Comes Home for Christmas,* directed by Nilita Vachani—describes an unparalleled success. Josephine has ventured around the world, achieving a degree of independence her mother could not have imagined, and amply supporting her three children with no help from her ex-husband, their father. Each month she mails a remittance check from Athens to Hatton, Sri Lanka, to pay the children's living expenses and school fees. On her Christmas visit home, she bears gifts of pots, pans, and dishes. While she makes payments on a new bus that Suresh, her oldest son, now drives for a living, she is also saving for a modest dowry for her daughter, Norma. She dreams of buying a new house in which the whole family can live. In the meantime, her work as a nanny enables Isadora's parents to devote themselves to their careers and avocations.

But Josephine's story is also one of wrenching global inequality. While Isadora enjoys the attention of three adults, Josephine's three children in Sri Lanka have been far less lucky. According to Vachani, Josephine's youngest child, Suminda, was two—Isadora's age—when his mother first left home to work in Saudi Arabia. Her middle child, Norma, was nine; her oldest son, Suresh, thirteen. From Saudi Arabia, Josephine found her way first to Kuwait, then to Greece. Except for one two-month trip home, she has lived apart from her children for ten years. She writes them weekly letters, seeking news of relatives, asking about school, and complaining that Norma doesn't write back.

Although Josephine left the children under her sister's supervision, the two youngest have shown signs of real distress. Norma has attempted suicide three times. Suminda, who was twelve when the film was made, boards in a grim, Dickensian orphanage that forbids talk during meals and showers. He visits his aunt on holidays. Although the oldest, Suresh, seems to be on good terms with

SOURCE: Ehrenreich, Barbara, and Arlie Russell Hochschild eds. 2003. *Global Woman: Nannies, Maids, and Sex Workers in the New Economy.* New York: Metropolitan.

his mother, Norma is tearful and sullen, and Suminda does poorly in school, picks quarrels, and otherwise seems withdrawn from the world. Still, at the end of the film, we see Josephine once again leave her three children in Sri Lanka to return to Isadora in Athens. For Josephine can either live with her children in desperate poverty or make money by living apart from them. Unlike her affluent First World employers, she cannot both live with her family and support it.

Thanks to the process we loosely call "globalization," women are on the move as never before in history. In images familiar to the West from television commercials for credit cards, cell phones, and airlines, female executives jet about the world, phoning home from luxury hotels and reuniting with eager children in airports. But we hear much less about a far more prodigious flow of female labor and energy: the increasing migration of millions of women from poor countries to rich ones, where they serve as nannies, maids, and sometimes sex workers. In the absence of help from male partners, many women have succeeded in tough "male world" careers only by turning over the care of their children, elderly parents, and homes to women from the Third World. This is the female underside of globalization, whereby millions of Josephines from poor countries in the south migrate to do the "women's work" of the north—work that affluent women are no longer able or willing to do. These migrant workers often leave their own children in the care of grandmothers, sisters, and sisters-in-law. Sometimes a young daughter is drawn out of school to care for her younger siblings.

This pattern of female migration reflects what could be called a worldwide gender revolution. In both rich and poor countries, fewer families can rely solely on a male breadwinner. In the United States, the earning power of most men has declined since 1970, and many women have gone out to "make up the difference." By one recent estimate, women were the sole, primary, or coequal earners in more than half of American families. So the question arises: Who will take care of the children, the sick, the elderly? Who will make dinner and clean house?

While the European or American woman commutes to work an average twenty-eight minutes a day, many nannies from the Philippines, Sri Lanka, and India cross the globe to get to their jobs. Some female migrants from the Third World do find something like "liberation," or at least the chance to become independent breadwinners and to improve their children's material lives. Other, less fortunate migrant women end up in the control of criminal employers— their passports stolen, their mobility blocked, forced to work without pay in brothels or to provide sex along with cleaning and child-care services in affluent homes. But even in more typical cases, where benign employers pay wages on time, Third World migrant women achieve their success only by assuming the cast-off domestic roles of middle- and high-income women in the First World— roles that have been previously rejected, of course, by men. And their "commute" entails a cost we have yet to fully comprehend.

The migration of women from the Third World to do "women's work" in affluent countries has so far received little scholarly or media attention—for reasons that are easy enough to guess. First, many, though by no means all, of the new female migrant workers are women of color, and therefore subject to the racial "discounting" routinely experienced by, say, Algerians in France, Mexicans in the United States, and Asians in the United Kingdom. Add to racism the private "indoor" nature of so much of the new migrants' work. Unlike factory workers, who congregate in large numbers, or taxi drivers, who are visible on the street, nannies and maids are often hidden away, one or two at a time, behind closed doors in private homes. Because of the illegal nature of their work, most sex workers are even further concealed from public view.

At least in the case of nannies and maids, another factor contributes to the invisibility of migrant women and their work—one that, for their affluent employers, touches closer to home. The Western culture of individualism, which finds extreme expression in the United States, militates against acknowledging help or human interdependency of nearly any kind. Thus, in the time-pressed upper middle class, servants are no longer displayed as status symbols, decked out in white caps and aprons, but often remain in the background, or disappear when company comes. Furthermore, affluent career women increasingly earn their status not through leisure, as they might have a century ago, but by apparently "doing it all"—producing a full-time career, thriving children, a contented spouse, and a well-managed home. In order to preserve this illusion, domestic workers and nannies make the house hotel-room perfect, feed and bathe the children, cook and clean up—and then magically fade from sight.

The lifestyles of the First World are made possible by a global transfer of the services associated with a wife's traditional role—child care, home-making, and sex—from poor countries to rich ones. To generalize and perhaps oversimplify: in an earlier phase of imperialism, northern countries extracted natural resources and agricultural products—rubber, metals, and sugar, for example—from lands they conquered and colonized. Today, while still relying on Third World countries for agricultural and industrial labor, the wealthy countries also seek to extract something harder to measure and quantify, something that can look very much like love. Nannies like Josephine bring the distant families that employ them real maternal affection, no doubt enhanced by the heartbreaking absence of their own children in the poor countries they leave behind. Similarly, women who migrate from country to country to work as maids bring not only their muscle power but an attentiveness to detail and to the human relationships in the household that might otherwise have been invested in their own families. Sex workers offer the simulation of sexual and romantic love, or at least transient sexual companionship. It is as if the wealthy parts of the world are running short on precious emotional and sexual resources and have had to turn to poorer regions for fresh supplies.

There are plenty of historical precedents for this globalization of traditional female services. In the ancient Middle East, the women of populations defeated in war were routinely enslaved and hauled off to serve as household workers and concubines for the victors. Among the Africans brought to North America as slaves in the sixteenth through nineteenth centuries, about a third were women and children, and many of those women were pressed to be concubines, domestic servants, or both. Nineteenth-century Irishwomen—along with many rural Englishwomen—migrated to English towns and cities to work as domestics in the homes of the growing upper middle class. Services thought to be innately feminine—child care, housework, and sex—often win little recognition or pay. But they have always been sufficiently in demand to transport over long distances if necessary. What is new today is the sheer number of female migrants and the very long distances they travel. Immigration statistics show huge numbers of women in motion, typically from poor countries to rich. Although the gross statistics give little clue as to the jobs women eventually take, there are reasons to infer that much of their work is "caring work," performed either in private homes or in institutional settings such as hospitals, hospices, child-care centers, and nursing homes. . . .

Most women, like men, migrate from the south to the north and from poor countries to rich ones. Typically, migrants go to the nearest comparatively rich country, preferably one whose language they speak or whose religion and culture they share. There are also local migratory flows: from northern to southern Thailand, for instance, or from East Germany to West. But of the regional or cross-regional flows, four stand out. One goes from Southeast Asia to the oil-rich Middle and Far East—from Bangladesh, Indonesia, the Philippines, and Sri Lanka to Bahrain, Oman, Kuwait, Saudi Arabia, Hong Kong, Malaysia, and Singapore. Another stream of migration goes from the former Soviet bloc to western Europe—from Russia, Romania, Bulgaria, and Albania to Scandinavia, Germany, France, Spain, Portugal, and England. A third goes from south to north in the Americas, including the stream from Mexico to the United States, which scholars say is the longest-running labor migration in the world. A fourth stream moves from Africa to various parts of Europe. France receives many female migrants from Morocco, Tunisia, and Algeria. Italy receives female workers from Ethiopia, Eritrea, and Cape Verde.

Female migrants overwhelmingly take up work as maids or domestics. As women have become an ever greater proportion of migrant workers, receiving countries reflect a dramatic influx of foreign-born domestics. In the United States, African-American women, who accounted for 60 percent of domestics in the 1940s, have been largely replaced by Latinas, many of them recent migrants from Mexico and Central America. In England, Asian migrant women have displaced the Irish and Portuguese domestics of the past. In French cities, North African women have replaced rural French girls. In western Germany, Turks and women from the former East Germany have replaced rural native-born women. . . .

Why this transfer of women's traditional services from poor to rich parts of the world? The reasons are, in a crude way, easy to guess. Women in Western countries have increasingly taken on paid work, and hence need other—paid domestics and caretakers for children and elderly people—to replace them. For their part, women in poor countries have an obvious incentive to migrate: relative and absolute poverty. The "care deficit" that has emerged in the wealthier countries as women enter the workforce *pulls* migrants from the Third World and post-communist nations; poverty *pushes* them.

In broad outline, this explanation holds true. Throughout western Europe, Taiwan, and Japan, but above all in the United States, England, and Sweden, women's employment has increased dramatically since the 1970s. In the United States, for example, the proportion of women in paid work rose from 15 percent of mothers of children six and under in 1950 to 65 percent today. Women now make up 46 percent of the U.S. labor force. Three-quarters of mothers of children eighteen and under and nearly two-thirds of mothers of children age one and younger now work for pay. Furthermore, according to a recent International Labor Organization study, working Americans averaged longer hours at work in the late 1990s than they did in the 1970s. By some measures, the number of hours spent at work have increased more for women than for men, and especially for women in managerial and professional jobs.

Meanwhile, over the last thirty years, as the rich countries have grown much richer, the poor countries have become—in both absolute and relative terms—poorer. Global inequalities in wages are particularly striking. In Hong Kong, for instance, the wages of a Filipina domestic are about fifteen times the amount she could make as a schoolteacher back in the Philippines. In addition, poor countries turning to the IMF or World Bank for loans are often forced to undertake measures of so-called structural adjustment, with disastrous results for the poor and especially for poor women and children. To qualify for loans, governments are usually required to devalue their currencies, which turns the hard currencies of rich countries into gold and the soft currencies of poor countries into straw. Structural adjustment programs also call for cuts in support for "noncompetitive industries" and for the reduction of public services such as health care and food subsidies for the poor. Citizens of poor countries, women as well as men, thus have a strong incentive to seek work in more fortunate parts of the world.

But it would be a mistake to attribute the globalization of women's work to a simple synergy of needs among women—one group, in the affluent countries, needing help and the other, in poor countries, needing jobs. For one thing, this formulation fails to account for the marked failure of First World governments to meet the needs created by its women's entry into the workforce. The downsized American—and to a lesser degree, western European—welfare state has become a "deadbeat dad." Unlike the rest of the industrialized world, the United States does not offer public child care for working mothers, nor does it ensure paid family and medical leave. Moreover, a series of state tax revolts in the 1980s reduced the number of hours public libraries were open and slashed school-enrichment and after-school programs. Europe did not experience anything comparable. Still, tens of millions of western European women are in the workforce who were not before—and there has been no proportionate expansion in public services.

Secondly, any view of the globalization of domestic work as simply an arrangement among women completely omits the role of men. Numerous studies, including some of our own, have shown that as American women took on paid employment, the men in their families did little to increase their contribution to the work of the home. For example, only one out of every five men among the working couples whom Hochschild interviewed for *The Second Shift* in the 1980s shared the work at home, and later studies suggest that while working mothers are doing somewhat less housework than their counterparts twenty years ago, most men are doing only a little more. With divorce, men frequently abdicate their child-care responsibilities to their ex-wives. In most cultures of the First World outside the United States, powerful traditions even more firmly discourage husbands from doing "women's work." So, strictly speaking, the presence of immigrant nannies does not enable affluent women to enter the workforce; it enables affluent *men* to continue avoiding the second shift.

The men in wealthier countries are also, of course, directly responsible for the demand for immigrant sex workers—as well as for the sexual abuse of many migrant women who work as domestics. Why, we wondered, is there a particular demand for "imported" sexual partners? Part of the answer may lie in the fact that new immigrants often take up the least desirable work, and, thanks to the AIDS epidemic, prostitution has become a job that ever fewer women deliberately choose. But perhaps some of this demand ... grows out of the erotic lure of the "exotic." Immigrant women may seem desirable sexual partners for the same reason that First World employers believe them to be especially gifted as caregivers: they are thought to embody the traditional feminine qualities of nurturance, docility, and eagerness to please. Some men feel nostalgic for these qualities, which they associate with a bygone way of life. Even as many wage-earning Western women assimilate to the competitive culture of "male" work and ask respect for making it in a man's world, some men seek in the "exotic Orient" or "hot-blooded tropics" a woman from the imagined past.

Of course, not all sex workers migrate voluntarily. An alarming number of women and girls are trafficked by smugglers and sold into bondage. Because trafficking is illegal and secret, the numbers are hard to know with any certainty. Kevin Bales estimates that in Thailand alone, a country of 60 million, half a million to a million women are prostitutes, and one out of every twenty of these is enslaved. . . . Many of these women are daughters whom northern hill-tribe families have sold to brothels in the cities of the south. Believing the promises of jobs and money, some begin the voyage willingly, only to discover days later that the "arrangers" are traffickers who steal their passports, define them as debtors, and enslave them as prostitutes. Other women and girls are kidnapped, or sold by their impoverished families, and then trafficked to brothels. Even worse fates befall women from neighboring Laos and Burma, who flee crushing poverty and repression at home only to fall into the hands of Thai slave traders.

If the factors that pull migrant women workers to affluent countries are not as simple as they at first appear, neither are the factors that push them. Certainly relative poverty plays a major role, but, interestingly, migrant women often do not come from the poorest classes of their societies. In fact, they are typically more affluent and better educated than male migrants. Many female migrants from the Philippines and Mexico, for example, have high school or college diplomas and have held middle-class—albeit low-paid—jobs back home. One study of Mexican migrants suggests that the trend is toward increasingly better-educated female migrants. Thirty years ago, most Mexican-born maids in the United States had been poorly educated maids in Mexico. Now a majority have high school degrees and have held clerical, retail, or professional jobs before leaving for the United States. Such women are likely to be enterprising and adventurous enough to resist the social pressures to stay home and accept their lot in life.

Noneconomic factors—or at least factors that are not immediately and directly economic—also influence a woman's decision to emigrate. By migrating, a woman may escape the expectation that she care for elderly family members, relinquish her paycheck to a husband or father, or defer to an abusive husband. Migration may also be a practical response to a failed marriage and the need to provide for children without male help. . . . And there are forces at work that may be making the men of poor countries less desirable as husbands. Male unemployment runs high in the countries that supply female domestics to the First World. Unable to make a living, these men often grow demoralized and cease contributing to their families in other ways. . . .

To an extent then, the globalization of child care and housework brings the ambitious and independent women of the world together: the career-oriented upper-middle-class woman of an affluent nation and the striving woman from a crumbling Third World or postcommunist economy. Only it does not bring them together in the way that second-wave feminists in affluent countries once liked to imagine—as sisters and allies struggling to achieve common goals. Instead, they come together as mistress and maid, employer and employee, across a great divide of privilege and opportunity.

27

The Contested Meanings of "Asian American"

Racial Dilemmas in the Contemporary U.S.

NAZLI KIBRIA

In the social and political discourse of the U.S. today, the idea of race is one . . . that is prominent, and yet highly ambiguous in meaning. Among the developments that both contribute to and reflect the current uncertainties of race is the emergence of such racialized constructs and groups as Asian American, Latino/Hispanic and Native American. Bringing together distinct ethnic or national groups, these pan-ethnic collectivities constitute an increasingly important and visible aspect of racial politics in the U.S. Omi and Winant (1996) argue that these pan-ethnic formations are the basis for new forms of racialization or the development of new racial subjects.

SOURCE: *Ethnic and Racial Studies*, Volume 21, Number 5, September 1998. Copyright Routledge.

In this article I explore the contests of racial meaning that are part of the contemporary pan-ethnic formation of "Asian American." There are many ways in which the concept of "Asian American" has become institutionalized in U.S. life. "Asian American" is evident, for example, in classifications of race in the census and other bureaucratic forms. Since the late 1960s, we have also seen the emergence of a vast array of groups and organizations under the banner of "Asian American." Central to these developments has been the definition of "Asian American" as signifying a racial minority group bound by common racial interests. This conceptualization is, however, an increasingly uncertain one, in ways that reflect such forces as the 1965 Immigration Act and its effects on the Asian American population. But also implicated are more general uncertainties about the meaning of race in the U.S. today. Embedded in the contestations of Asian American meaning nowadays are such larger questions as "What constitutes a racial group?" and "What is the character of racial disadvantage in the U.S. today?"

DILEMMAS OF RACIAL MEANING

Race is a system of power, one that draws on physical differences to construct and give meaning to racial groups and the hierarchy in which they are embedded (Miles 1989; Sanjek 1994). The focus of this study is on the conditions and processes that produce racial categories and their meanings: the social construction of race. Drawing on constructionist perspectives of race and ethnicity, I suggest that the production of racial meaning is a dynamic and ongoing matter (Omi and Winant 1986; Nagel 1994). By definition, given their embeddedness in a larger system of power, racial categories reflect the externally imposed designations or assignments of dominant groups upon others. But racial definition is also shaped by the actions of the categorized group itself. That is, within the limits of prevailing structures of opportunity and constraint, racialized groups work to shape their own identities. Cornell and Hartmann (1998) discuss this interaction of structure and agency in the formation of identities:

> Identities are made, but by an interaction between circumstantial or human assignment on the one hand and assertion on the other. Construction involves both the passive experience of being "made" by external forces, including not only material circumstances but the claims that other persons or groups make about the group in question, and the active process by which the group "makes" itself. . . . This interaction is continuous, and it involves all those processes through which identities are made and remade, from the initial formation of a collective identity through its maintenance, reproduction, transformation, and even repudiation over time. Construction refers not to a one-time event but an ongoing project. (p. 80)

There are several thematic issues that are often apparent in the dynamics of racial meaning. I suggest that it is useful to conceptualize these issues as ongoing dilemmas or core sets of questions about racial categories. These dilemmas operate as points of focus or conceptual and thematic guideposts in the production of racial meanings. The specific questions that flow from these dilemmas will vary for racialized groups, in ways that reflect their particular circumstances. At the same time, the questions that arise will also reflect larger ongoing debates on race in the society in question. But whatever their specific content, I suggest that these emergent questions become the lightning-rod for contests about the racialized group and the meanings that surround it. Thus, for example, within institutional forums of various sorts (for example, political groups, newspapers), these questions may generate discussion and debate about the racialized group. For individual members of the racialized group, the questions may provide a focus for reflections on identity.

Among the dilemmas of racial meaning are issues of *boundary*. What are the criteria for membership in this racial category? To put it simply, what are the grounds for being identified by it? The distinguishing feature of racial categories is that they involve what Miles (1989) describes as "the signification of human biological characteristics in such a way as to define and construct differentiated social collectivities" (p. 75). The dilemma of racial boundaries involves questions about this signification, in particular the interpretation of physical characteristics.

Racial boundaries reflect relations of power, in particular the ability of the dominant group to construct and impose definitions upon others. At the same time, those who are categorized cannot simply be understood as passive recipients of externally imposed definitions. For example, in *Who Is Black?* Davis (1991) describes the development of the "one-drop rule" or the rule of hypodescent for defining black identity in the U.S. Imposed by white people as a means of maintaining power and enforcing segregation, the "one-drop rule" has also been widely accepted and enforced by African Americans as a basis for defining membership. It is of note, however, that the definition of black identity has become a growing focus of debate in the U.S., reflecting such conditions as a rising rate of black outmarriage as well as the growing presence of immigrants from the Caribbean and Africa who sometimes resist incorporation into a black American identity (Waters 1994; Mathews 1996).

Closely related to issues of boundary are those of *community*. How do the racial boundaries suggest membership in a community? That is, how do the boundaries mark a community of persons who share more than simply an externally imposed designation? These issues are linked to processes of ethnicization or the making of an ethnic group. That is, questions of community are embedded in the dynamics by which a racialized group comes to define itself in terms of common ancestry, history and a set of cultural symbols that they see as capturing the essence of their peoplehood (Cornell and Hartmann 1998, p. 19). Reflecting their particular histories of incorporation and community-building in the U.S., racialized groups vary in the extent of their ethnicization. These differences can be expected to shape the specific questions of community that surround them. . . .

Intrinsic to racial construction is the locating of groups within a racial hierarchy. The dilemmas of racial meaning thus include questions of *positioning*—what is the position of the group within the racial hierarchy? In short, where and how do they fit in? Stereotypes and images of racial groups, with their presumptions about characteristic racial behaviour and temperament, play an important part in this process of hierarchical location. These racial notions are developed and affirmed as they become embedded in social institutions such as the law, economy and education. Here they may take the form of systematic and cross-cutting institutional patterns or dynamics that shape the access of racial groups to societal resources. Because racial membership is widely *believed* to be a given, biological matter, the presumed traits of race, and the institutional conditions and inequalities with which they are intertwined, can also be seen as "natural," inherent.

As part of a system of power, racial categories and their meanings reflect the ability of dominant groups to impose their designations upon others. Thus, in the U.S. white people have exerted control over the production of racial images and stereotypes. Scholars further observe that white dominance has marked the very terms or terrain on which race definition has unfolded (Frankenberg 1993). That is, race definition in the U.S. has occurred in relation to "whiteness," which has been defined as the centre or norm—the standard against which the racial identities of others are defined and measured. . . .

Thus in a variety of ways, racial groups can work to challenge and shape their racial location. For example, an important activity of contemporary racial minority movements has been the challenging and subverting of stereotypes. The movements have therefore been an important force in contesting such stereotypes and more generally, expanding and shaping the institutional and conceptual space within which they are discussed.

THE ASIAN AMERICAN CONSTRUCT AND POST-1965 DEVELOPMENTS

"Asian American" is a contemporary, post–Civil Rights construct. While the Asian settlers of the late nineteenth and early twentieth centuries often found themselves categorized or lumped together as "Asiatics" or "Orientals" by the dominant U.S. society, they did not respond by banding together. Instead, they maintained separate institutions and identities and at certain times engaged in strategic acts of distancing or disidentification from one another (Hayano 1981). But in the 1960s, young U.S.-born Asian American activists on college campuses, inspired by the civil rights struggles of the time, organized the Asian American movement as a means of political empowerment and mobilization. Rejecting the then common term "Oriental," they coined "Asian American," a term that has since gained currency.

Like the other pan-ethnic collectivities promoted by the movements of the time (for example, Native American, Latino), the Asian American concept has become an institutionalized dimension of the contemporary U.S. racial system. Like the others, it has been a basis for monitoring antidiscrimination efforts and of organizing access to government resources. Further reinforcing its significance has been the multiculturalist movement, which often draws on the pan-ethnic categories to organize the promotion of cultural diversity (Hollinger 1995).

Accompanying these developments has been the mushrooming of Asian American groups and organizations which aim at mobilizing and advancing pan-Asian interests. The institutionalization of the Asian American movement is evident not only in the presence of an extensive network of pan-Asian associations of various sorts, but also in a series of pan-Asian publications (for example, *Asian Week, A. Magazine, Amerasia Journal*) whose stated goal is to serve a pan-Asian audience. One of the most important of the institutional developments has been the growth of Asian American Studies programmes and classes on college campuses. Among other things, Asian American Studies has provided an important forum for reflection and debate on the meaning of "Asian American.". . .

Critical to understanding the questions that surround the concept of Asian American today is the 1965 Immigration Reform Act, which lifted race restrictions and so opened the doors to Asian immigration. Prior to the late 1960s, the Asian American population was small in number. The flow of Asian immigration, most notably of Chinese and Japanese immigrants to the U.S. that had commenced in the mid-1800s, had been effectively halted with a series of restrictive immigration laws passed in the late nineteenth and early twentieth centuries. The 1970 census showed Asian Pacific Americans to be 0.7 per cent of the total U.S. population. But in 1990, reflecting the 1965 Act, their numbers had risen to 2.9 per cent (Min 1997a, p. 15). The rise in numbers, along with other developments, has affected the collective character of the Asian American population in important ways. For one thing, the foreign-born comprise a higher percentage of the Asian American population in comparison to the period preceding the 1965 Act (Espiritu 1992, p. 26; Barringer, Gardner and Levin 1993, p. 43). In addition, the high proportion of immigrants, along with a global context that in many instances facilitates "host" and "homeland" linkages, suggests processes of transnationalism to be an important feature of Asian American life today.

Concurrent with the expansion in Asian American numbers has been its growing ethnic diversity. In contrast to the largely Japanese and Chinese origins of Asian Americans in the first half of the century, today Asian Americans are highly diverse in their national origins. Asian Indians, Koreans and Filipinos, for example, are among the fastest growing segments of the Asian American population. There has also been a general shift away from the largely working-class origins of the late nineteenth– early twentieth–century Asian immigrants. While there is considerable socio-economic diversity within the Asian American population today, it is also the case that many post-1965 Asian immigrants come from professional, white-collar and highly educated backgrounds (Min 1997a, p. 17). The middle-class background of many Asian immigrants, in conjunction with the phenomenal growth and success of some Asian economies, has lent support and credence to the popular image of Asians as a "model minority": a group that is culturally programmed for economic success.

The middle-class image does not, however, do justice to the realities of socio-economic diversity within the new Asian immigrant stream. In fact, many analysts characterize the Asian American population today as polarized, consisting of two sharply disparate socio-economic segments (Mar and Kim

1994; Ong and Hee 1994). In contrast to those admitted in the 1970s, recent Asian immigrants have been less select and more diverse in their socio-economic origins. Further contributing to a movement away from a purely middle-class profile is the entry since the mid-1970s of refugees from Vietnam, Cambodia and Laos whose levels of education and training are, on average, lower than that of the other post-1965 Asian immigrant groups (Barringer, Gardner and Levin 1993, pp. 31–32). Thus, while economic polarization is apparent *within* Asian ethnic groups, it is also one that can coincide with ethnic boundaries. In other words, there are important differences in the socio-economic profiles of Asian-origin groups. For example, the income levels and poverty rates for Vietnamese Americans contrast sharply and negatively with those of Japanese or Filipino Americans.

The growth of the Asian American population has occurred at a time of complex transformation in the racial environment of the U.S. In the aftermath of the civil rights struggles, institutionalized racial barriers have been challenged in the U.S. But there has also been racial polarization and backlash, especially apparent in the last two decades. Feeding resentment and retreat from progressive policy has been the structural transformation of the U.S. economy, most notably the decline suffered by the U.S. manufacturing sector (Mar and Kim 1994). In recent years affirmative action programmes that were instituted in the 1960s and 1970s have come under attack. There has also been a rising tide of anti-immigrant sentiment, as expressed in calls for various restrictive measures, such as the curtailing of government welfare benefits and social services to immigrants. . . .

By increasing the size and visibility of the Asian American population, post-1965 Asian immigration has lent vigour and urgency to issues of the Asian American community. But the new immigration has also posed challenges to the very notion of "Asian American," especially to its meaning and viability as a basis of community. Here it is important to point out that, as during the movement's inception, pan-Asian organizations today continue to involve primarily U.S.-born Asians (Espiritu and Ong 1994). The relative absence in these forums of immigrants, who constitute a significant proportion of the Asian American population, gives rise to questions about the possibility of the pan-Asian community.

Central to these questions is the meaningfulness of the notion of shared racial interest and location that has guided pan-Asian activity. . . . There is, for one thing, the element of ambiguity that surrounds Asian racial boundaries in the U.S. today. That is, the fact that not all of those who are institutionally or officially Asian American are widely seen or labelled as racially "Asian" during informal social encounters generates doubts about racial interest and location as bases of unity. The important role of an Asian American historical past in creating a sense of shared racial fate is also made uncertain by the post-1965 transformation of the Asian American population into a largely immigrant one. The U.S. roots of many Asian Americans are recent, not extending back to the late nineteenth–early twentieth–century Asian immigrations that are emphasized in Asian American history.

Uncertainties about shared racial location and interest also reflect the complexities of class that have been part of the post-1965 transformation of the Asian American population. As I have described, many post-1965 Asian immigrants have been highly educated and held white-collar and professional occupations in their countries of origin. For such persons, the working-class origins of Asian immigration that is emphasized in Asian American history as well as the agenda of solidarity with other racially oppressed groups may seem distant and irrelevant. Also important are the divisions of class as well as the occupational niche within the new Asian immigrant stream. These divisions are likely to feed into different experiences and perceptions of racism and thus the understanding of racial location and interest. . . .

It is important to point out that there are ways in which contemporary conditions *do* support the vision of Asian American community as a racial and political one. In the current tide of racial resentment and backlash, Asian Americans often find themselves to be the target of fears about an invasion by foreign economic interests and by immigrants. These attacks tend to be indiscriminate, without regard to national origin and involving a wide range of Asian groups. More generally, the political advantages that can be derived from banding together still work to support a racial interest vision of the Asian American community. At the same time, as I have described here, the meaning of Asian American community is a highly contested matter. In ways that reflect post-1965 forces, a specific focus of contest is the vision of race-centred political community as it has been defined by the Asian American movement. This contest reflects larger uncertainties about the unitary nature of racial interests in the face of class and other divisions. . . .

Positioning Dilemmas: Asians as a "Model Minority"

In affirming the identity of Asian Americans as a racial minority, commonality and solidarity with other oppressed groups of colour has been a central aspect of the ideology of pan-Asianism. Post-1965 conditions have, however, worked to generate questions about what Asian Americans share with other racial minorities by challenging the notion of Asian racial disadvantage. Among the relevant developments is the growth of the East Asian and Southeast Asian economies in recent decades. By the early 1980s, talk of the "Asian tiger economies" and the "Asian miracle," often attributed to "Confucian" values, had become commonplace in the U.S. Also relevant, of course, are the middle-class and professional backgrounds and socio-economic achievements of a substantial segment of the post-1965 Asian immigrants as I have described earlier. Even prior to the 1965 Act, in the years after World War II, there are indications of socio-economic mobility into the middle class among U.S.-born Asian Americans (Mar and Kim 1994).

The stereotype of Asians as a model minority has, I suggest, provided an important ideological focus for interpretation and debate of the uncertainties of Asian racial disadvantage and position raised by the[se] developments. In other words, the model minority stereotype and the debates surrounding it have offered a forum for understanding and considering the Asian racial position today. On the one hand, the model minority stereotype offers a way of making sense of the general indicators of Asian achievement. In essence, it does so by suggesting that Asians are a minority group endowed with cultural values such as a strong work ethic and devotion to education that predispose them to economic and educational achievement. Worth noting here is the radical contrast of this image to earlier ones—of Asians as [u]nassimilable, inscrutable, tricky and immoral heathens (Hurh and Kim 1989). This late nineteenth– early twentieth–century imagery provided the ideological backdrop to the exclusion acts that were to bar Asian immigration to the U.S. But the mid-1960s seem to have marked a shift in the U.S. imagery of Asians. Since that time, media reports on the educational and economic achievements of Asian Americans have become commonplace. . . .

The model minority stereotype appears to situate Asians in a privileged position within the racial hierarchy of the U.S. The stereotype suggests the transient nature of racial disadvantage for Asian Americans. Simultaneously, it evokes the culturally rooted and ultimately essentialist nature of the chronic disadvantage of other racial minorities. A critical dimension of the stereotype is the suggestion that Asians are different from other racial minorities. Their cultural values, in particular their commitment to work, self-sufficiency, family and education, allow them to overcome disadvantage, in contrast to other minorities whose predisposition causes them to remain permanently mired in poverty and disability. As an explanatory variable, "culture" here takes on an essential, biological quality. Through this comparison, Asians emerge as a group for whom racial disadvantage, if it exists, is a temporary condition. In this sense they are heirs to the European immigrant story of successful triumph over adversity and, eventually, assimilation into the "mainstream." It is not surprising that media reports on Asian successes often draw parallels between the Asian and Jewish American experiences.

The Asian American movement has been a major force in challenging the model minority stereotype. Like other minority movements, it has approached the exposing of group stereotypes as an important political activity. But a critique of the model minority stereotype has also been important to the movement because of the questions that the stereotype can raise about the ideology of pan-Asianism. When left unpacked, the model minority stereotype brings into question the movement's core ideas of Asian racial disadvantage and commonality with other groups of colour. Critique of the stereotype has also been driven by a recognition of the ways in which its connotation of self-sufficiency and achievement work to undercut the access of Asian American service organizations to funding that targets disadvantaged populations. . . .

Critique of the model minority stereotype has also focused on its larger implications for race relations and inequality in the U.S. Here the stereotype is seen as an instrument of white supremacy, used to pit Asians and other minorities against each other and thus weaken minority solidarity and power. Also pointed out are the ways in which the stereotype legitimates the absence or failure of programmes and measures to address racial inequality (Osajima 1988). That is, since Asian

Americans are able to pull themselves up with their cultural bootstraps, racial disadvantage is not a structural aspect of U.S. life that requires active intervention.

CONCLUSION

Like the other pan-ethnic movements that emerged in the 1960s, "Asian American" was conceived by its founders as a strategy of political empowerment. Shared racial identity and interests provided the central rationale for pan-Asian organizing. The need to sustain and preserve this strategy is perhaps more pressing than ever, given the anti-immigrant sentiment and fears of Asian economic competition that mark the racial environment of the U.S. today. But this task is complicated by the sharp contests of meaning that surround the concept of "Asian American" today.

Post-1965 conditions have exacerbated the inherent contradictions and ambiguities of the racial interests framework that has guided pan-Asian activity. Among those who are potentially encompassed by the Asian American concept, divisions of class, ethnicity and ascribed racial identity have become prominent. These divisions complicate efforts to articulate and forge a pan-Asian agenda based on common racial location and interests. I suggest that what can help to clarify the potential bases of solidarity among Asian Americans today are studies that actively compare racial experiences across different groups of Asian Americans, instead of simply assuming a common one. At present we know little about how the rank and file of those who are encompassed by the Asian American umbrella actually see and relate to the Asian American concept. Involvement in the Asian American movement and its institutions has been the province of primarily U.S.-born and middle-class Asians. The question how other Asian Americans relate to the political and strategic understanding of "Asian American" that have been advanced by the Asian American movement is one that needs investigation.

Contestations of Asian American meaning are embedded in the larger uncertainties that mark the racial environment of the U.S. today. As I have discussed, questions about the identity of Asian Americans as a racial minority group emerge in the context of increased concern and confusion regarding the nature of racial disadvantage; in particular, whether it is reflective of essential inferiority. Also apparent is heightened uncertainty about the fundamental substance of racial boundaries. In established and popular U.S. understanding, racial boundaries are pure, discrete, immutable and easily defined by physical characteristics. Among the conditions that now explicitly challenge these ideas is the growth of a self-identified "multiracial" population that refuses to be racially categorized in unidimensional terms, as reflected in their demands for a "multiracial" category in the census (Marriot 1996). Also relevant are the racial ambiguities of the increasingly prominent Hispanic population who are identified as a racial group in some contexts, even though individuals within it experience different racial labels, ranging from "white" to "black" (Oboler 1995). In ways that reflect the more fluid character of racial classifications in Latin America, this population is also more likely to resist and thus to challenge the singular and discrete categories of the U.S. racial system (Fernandez 1992). The ambiguities of "Asian American" as a signifier of a racial group are part of these emergent challenges.

REFERENCES

Barringer, Herbert, Gardner, Robert and Levin, Michael. 1993. *Asians and Pacific Islanders in the United States*, New York: Russell Sage Foundation.

Cornell, Stephen and Hartmann, Douglas. 1998. *Ethnicity and Race: Making Identities in a Changing World*, Thousand Oaks, CA: Pine Forge Press.

Davis, F. James. 1991. *Who Is Black? One Nation's Definition*, University Park, PA: Pennsylvania State University Press.

Espiritu, Yen L. 1992. *Asian American Panethnicity: Bridging Institutions and Identities*, Philadelphia, PA: Temple University Press.

Espiritu, Yen L. and Ong, Paul. 1994. "Class constraints on racial solidarity among Asian American," in Paul Ong, Edna Bonacich and Lucie Cheng (eds.), *The New Asian Immigration in Los Angeles and Global Restructuring*, Philadelphia, PA: Temple University Press, pp. 295–321.

Fernandez, Carlos. 1992. "La Raza and the melting pot: a comparative look at multi-ethnicity," in Maria Root (ed.) *Racially Mixed People in America*, Newbury Park, CA: Sage Publications, pp. 126–43.

Frankenburg, Ruth. 1993. *White Women, Race Matters: The Social Construction of Whiteness*, Minneapolis, MN: University of Minnesota Press.

Hayano, David. 1981. "Ethnic identification and disidentification: Japanese-Americans view of Chinese-Americans," *Ethnic Groups*, vol. 3, no. 2, pp. 157–71.

Hollinger, David. 1995. *Postethnic America: Beyond Multiculturalism*, New York: Basic Books.

Hurh, Won Moo and Kim, Kwang Chung. 1989. "The 'success' image of Asian Americans: its validity, and its practical and theoretical implications," *Ethnic and Racial Studies*, vol. 12, no. 4: 512–38.

Mar, Don and Kim, Marlene. 1994. "Historical trends" in *The State of Asian Pacific America: Economic Diversity, Issues and Policies*, Los Angeles, CA: Asian Pacific American Public Policy Institute, pp. 13–31.

Marriot, Michel. 1996. "Multiracial Americans ready to claim their own identity," *The New York Times*, 20 July.

Mathews, Linda. 1996. "More than identity ride on a new racial category," *The New York Times*, 6 July.

Miles, Robert. 1989. *Racism*, London: Routledge.

Min, Pyong Gap. 1997a. "Introduction," in Pyong Gap Min (ed.), *Asian Americans: Contemporary Trends and Issues*, Thousand Oaks, CA: Sage Publications, pp. 1–9.

Nagel, Joane. 1994. Constructing ethnicity: creating and recreating ethnic identity and culture," *Social Problems*, vol. 41, no. 1, pp. 152–76.

Oboler, Suzanne. 1995. *Ethnic Labels, Latino Lives,* Minneapolis, MN: University of Minnesota Press.

Omi, Michael and Winant, Howard. 1986. "Contesting the meaning of race in the post–civil rights movement era," in Silvia Pedraza and Ruben G. Rumbaut (eds.), *Origins and Destinies: Immigration, Race, and Ethnicity in America,* Belmont, CA: Wadsworth Publishing Co, pp. 470–78.

Ong, Paul and Hee, Suzanne. 1994. "Economic diversity" in *The State of Asian Pacific America: Economic Diversity, Issues and Policies,* Los Angeles, CA: Asian Pacific American Public Policy Institute, pp. 31–56.

Osajima, Keith. 1988. "Asian Americans as the model minority: an analysis of the popular press image in the 1960s and 1980s," in Gary Okihiro, Shirley Hune, Arthur Hansen and John Liu (eds.), *Reflections on Shattered Windows,* Pullman, WA: Washington State University Press, pp. 165–74.

Sanjek, Roger. 1994. "The enduring inequalities of race," in Steven Gregory and Roger Sanjek (eds), *Race,* New Brunswick, NJ: Rutgers University Press, pp. 1–17.

Waters, Mary. 1994. "Ethnic and racial identities of second-generation black immigrants in New York City," *International Migration Review,* vol. 28, no. 4, pp. 795–820.

28

"No Lattés Here"

Asian American Youth and the Cyber Café Obsession

MARY YU DANICO AND LINDA TRINH VÕ

In urbanized suburbs—where there are limited alternative recreational spaces or activities for youth—cyber cafés attract elementary- to college-age Asian American youth. "PC rooms," "PC Bangs," "Internet Cafés," and "Cyber Cafés" have surfaced across the country, but southern California is home to the nation's largest cluster of these cafés. Located in mini malls or strip malls, cyber cafés have become the fastest-growing business in Asian ethnic enclaves, particularly among Koreans, Vietnamese, and Chinese. Using the latest in interactive technology and state-of-the-art computers, youth compete in computer games with individuals who sit beside them, or even in other cafés. Although these cafés offer an affordable and accessible form of entertainment, some critics argue that video games encourage violence and are addictive, referring to them

SOURCE: Lee, Jennifer, and Min Zhou, eds. 2004. *Asian American Youth.* New York: Routledge, pp. 177–189.

as "on-line heroin." Our research, however, finds that cyber cafés not only provide Asian American youth with a social outlet but also offer a safe space where they can form bonds of friendship and reclaim their masculinity.

In southern California, recently enacted laws have restricted youths' freedom to drive, imposed curfews in some areas, and provided few accessible social outlets, leaving teenagers only limited options to establish their own leisure space. Asian American youth—most of whom are immigrants, refugees, or the children of immigrants or refugees—have even fewer spaces to "kick back." In this article, we discuss the development of the cyber café culture in ethnic communities and explain why Asian American youth are drawn to these social outlets. We examine the evolution of cyber cafés from places where youths play virtual games to alternative social outlets for them to hang out after school, into the night and on weekends, sometimes even luring them away from school. In particular, we focus on the activities that take place in cyber cafés and explore why Asian American youth, in particular, are drawn to the cyber community.

Although some sites are known to be hangouts for "deviant behavior" such as gang havens for drug dealing and other vices, for the purposes of this article, we focus on the economic, social, and cultural reasons that attract Asian American youth to cyber cafés. This research is based on ethnographic observation of cyber cafés and their surrounding areas, formal and informal interviews with cyber café owners, managers, and patrons, and focus groups with cyber café youth patrons.

THE TRANSNATIONAL AND DOMESTIC DEVELOPMENT OF CYBER CAFÉS

Cyber cafés are part of the global Internet computer revolution and have become a profitable niche for ethnic entrepreneurs. Like other trendy Asian predecessors—such as Pokémon, karaoke, and anime—the cyber gaming industry is another imported "craze" from Asia. The cyber café phenomenon first surfaced in South Korea along with the term PC "Bang" (meaning "room" in Korean). Following the 1997 Asian financial crisis, which drastically devalued its currency, Korea turned to the Internet to help reinvigorate its fallen economy. Borrowing $57 billion from the International Monetary Fund, Korea set its sights on becoming technologically competitive with First World countries like the United States (Sullivan, 2001). Today, South Korea, with its 48.6 million population, leads the world in high-speed home access. Although many South Koreans have personal computers and therefore need not frequent cafés to play video games, the country boasts more than 20,000 cyber cafés that are dedicated to gaming.

Although high-speed home access is readily available in their homes, Koreans go to cafés to escape their notoriously tight living quarters and to meet, socialize, and play computer games with their friends (Graham, 2002). Gaming is serious business in South Korea; players compete in nationally televised tournaments, win cash prizes, are offered movie roles and product endorsements, and receive

Hollywood-style celebrity status and treatment (Baker, 2000). Other Asian countries, such as Taiwan, Malaysia, the Philippines, and China, quickly followed suit, and café mania has now spread across Asia and beyond (Baguioro, 2002; Bangkok Post, 2001; Ni, 2002). By the late 1990s, PC Bangs made a trans-Pacific move to the United States, finding their way into predominantly Asian communities throughout the nation.

Cyber cafés are different from traditional Internet cafés where people drink coffee, check their e-mail, or browse the web. In the past, because of the high price of purchasing a computer and costly Internet connections, people flocked to these Internet cafés where computers were readily available for $7 to $14 an hour. In metropolitan areas like New York, San Francisco, and Honolulu, patrons of Internet cafés came to hang out, drink cappuccino, check their e-mail, and chat on-line. The cost to "hang out" in Internet cafés limited its access to the larger public, and, therefore, during the initial period of the Internet craze, middle- and upper-middle-class yuppies, or DINKs (couples with Double Income No Kids), were their primary patrons. By the mid-1990s, however, personal computers and Internet services became more accessible and affordable and, consequently, made them available to a much wider population. As a result, by 1998, Internet cafés in metropolitan cities like New York and San Francisco went bankrupt (Marriot, 1998). However, as Internet cafés disappeared from upscale neighborhoods, PC Bangs gained widespread popularity in urban, low-income Asian immigrant communities.

According to the web site www.cybercafes.com, southern California alone is home to 395 cyber cafés. However, this list includes cafés that serve drinks and food, so it is difficult to determine exactly how many strictly offer game playing. These businesses—owned mainly by Vietnamese, Chinese-Vietnamese, or Koreans—have been noted for "spreading like viruses" throughout Orange and Los Angeles counties, with some owners operating businesses in multiple locations. Community groups estimate that there are more than twenty cyber cafés in Los Angeles's Koreatown, over ten in the West Los Angeles area, and many more have been sprouting up in the northern part of the San Fernando Valley and the eastern part of downtown in cities like Diamond Bar. Although most communities do not keep an accurate count of cyber cafés, Garden Grove, which is part of Little Saigon and home to 135,548 Vietnamese in Orange County, has been keeping track of these businesses since 1998 (Song, 2001). In 2000, there were only two cafés, but two years later, there were about twenty, with additional clusters of cafés in surrounding cities such as Santa Ana and Westminster.

Located in nondescript strip malls that dot southern California's commercial landscape, many cyber cafés are barely noticeable and tucked amid other businesses. For instance, having visited one cyber café in a strip mall that intersects Little Saigon and Orange County Koreatown, we found that it was located in a two-story mall amid dentist, real estate, insurance, and social service offices as well as an auto parts and paint store, judging from the Korean and/or Vietnamese language signs, the businesses cater to a predominantly ethnic clientele. The strip mall also includes entertainment businesses such as a dance studio, a Vietnamese restaurant, and a Vietnamese café. In the second-story cyber café, Vietnamese and Korean teens sit quietly in clusters playing games in the airy and spacious room, surrounded by anime posters. Other cyber café sites we visited in Los Angeles and Orange counties were similarly intermixed with other ethnic businesses.

In these cafés, visible signs—like those marking parking regulations on the street—are posted listing rules such as no yelling and no cursing. In all the sites, glass windows may cover some sidewalls letting in natural light and allowing pedestrians to see the activities inside; however, large portions of the windows were covered to prevent glare on the computer screens. On average, the cafés have 40 to 70 cubicles, each with a state-of-the-art computer and a plush executive-style office chair. Just a few years ago, the price for playing on the computers was $5 an hour, but it has since dropped to only $2 or less owing to increased competition.

Depending on the location, some cyber cafés are racially diverse. For example, an owner of a cyber café in Orange County's Garden Grove area estimates that his business is one-third Vietnamese, one-third Korean, one-fourth Mexican, and the rest "American," meaning white. In these cafés, a range of languages is spoken with English, Korean, Vietnamese, and Chinese dialects. Some 1.5- and second-generation youth "code-switch," and combine English with an ethnic language, symbolizing generational as well as ethnic bonds that form in the cafés.

During our visits to various cyber cafés, we observed individuals with their headsets on, playing intently on their computers, while hip-hop music played in the background. Some youth sat in clusters, casually chatting and joking around with one another, and, in most cases, the employees are individuals who are gaming fans and play at the terminals when customers do not occupy them. Although they are called cafés, they are not coffeehouses in the traditional sense; the refreshments served in cyber cafés are typically cold drinks such as soda and a few snacks.

THE SOCIAL AND ECONOMIC REALITIES OF IMMIGRANT AND REFUGEE YOUTH

The cyber cafés that have sprouted up across the southern California region cluster in middle- and working-class neighborhoods in which immigrants and refugees reside. Although some immigrant and refugee families lead seemingly middle-class lives, family members often work long hours at multiple jobs to maintain a middle-class lifestyle. Still others struggle to live on meager incomes or depend on welfare services to make ends meet. The busy workday leaves parents with little time to closely monitor their children, which leaves latchkey immigrant or refugee children with hours of unsupervised time after school and on weekends. As a result, cyber cafés serve as quasi-after-school programs for many Asian American working-class male youth. Not surprisingly, the busiest time for the cyber cafés is after 3:00 in the afternoon, when school lets out. Capitalizing on the popularity, some cafés lure youth by promoting "happy hours" where they reduce costs even further between the hours of 3:00 and 5:00 p.m. Although many youths inform their parents that they are at the cyber cafés, others do not tell their parents their whereabouts and sometimes skip school to play. Some youths reported that as long as they were attending school and receiving decent grades, their parents did not question their activities or have any control over them.

Cyber cafés are popular not just because they serve as convenient after-school diversions for immigrant youth but because of the digital divide that separates those who can afford household technological advances (such as home computers and Internet access) from those who cannot. Children of working-class immigrant families who do not have the luxury of home computers turn to cyber cafés to fill this void; at a cost of $1.50 to $2.00 an hour, this form of entertainment is an affordable way to play the most technologically advanced games. Immigrant youth often use their allowance money or the money they earn from their minimum-wage jobs to play for hours at these cafés. For working-class youth who have few economic resources, cyber cafés provide an affordable means of entertainment that is a lot cheaper than going to the movies, eating out at fast-food restaurants, or even going to a mall. Spending hours at cafés is even more appealing when managers and owners provide discounts to repeat customers, which, in turn, keeps the youth coming back. Moreover, the vast landscape and the nonpedestrian friendly environment in southern California make it difficult for youth to navigate without a mode of transportation. Many youths who do not have cars or access to them must rely on public transportation, which is limited in these areas. Others get rides from family members and friends, while some ride their bikes to these sites. Hence, the proximity of the cafés to their homes makes them convenient.

Problems and pressures at home also make cyber cafés a welcome diversion for immigrant youth. Immigrant children of working-class parents often live in cramped, multigenerational households, and in some cases, with multiple families, so the cyber cafés provide a refuge for youth. Furthermore, children of immigrants often have adult responsibilities at home; they act as translators for their first-generation parents who have no or limited English-language ability and must therefore rely on their children to assist them with complicated medical, housing, or financial matters (Park, 2002).

Given the linguistic and cultural barriers between the two generations, parents and their 1.5- and second-generation children often have difficulty communicating and relating to one another. The barriers make it difficult for youth to confide in their parents about their daily struggles at school and at home (Zhou and Bankston, 1998). Consequently, these young adults feel disconnected from their parents and look for a second "family" with their peers. Ultimately, many find a second "home" at the cafés. Some youth even have computers of their own, but they frequent cyber cafés because they provide a sense of community and belonging that keeps them coming back for more. Anthropologist Tim Tangherlini, who has researched cyber cafés, elaborates, "It's a multiethnic community that really speaks to the alienation of a lot of the big cities.... It's a place where kids can come together and make connections instead of playing by themselves in their parents' homes" (quoted in Song, 2001).

In addition, because parents work long hours, immigrant youth must often fend for themselves and look after their younger siblings, which explains the presence of elementary school-aged children at cyber cafés. Asian American males often bring their siblings with them to the cafés when their parents saddle them with babysitting responsibilities. Females, by contrast, are expected to assume the gendered responsibilities of doing household chores and taking care of younger children. Compared to their male counterparts, females have less leeway and leisure time to spend on their own, which explains, in part, the smaller presence of females at these cafés. Although Asian American females may have more adult and household responsibilities than males, both often complain that they do not have a childhood (Danico, 2004).

CYBER CAFÉS AS SAFE SOCIAL OUTLETS
FOR YOUTH

Asian immigrant youth have difficulty locating safe social spaces where they can hang out and establish friendships while also forging a sense of community. Entertainment and social sites in ethnic communities often exclude youth who are under the age of 18 or 21. For instance, pool halls often cater to an older crowd, and ethnic cafés with hostesses and karaoke bars are expensive and serve alcohol, and therefore, exclude youth. And while extremely popular, the import car racing scene, which gathers informally and periodically, is not easily accessible for those without cars or connections to people with cars. . . .

It is difficult for Asian American youth to just "hang out" in public urban spaces in groups without arousing suspicion from law enforcement officials. Asian American high school and college-age youth in southern California often complain about profiling and harassment by police officers who stop them because of the import cars they drive and the type of clothing they wear, automatically treating them as "gang" members. Given the difficulties of finding a space to just hang out, Asian American youth gravitate to cyber cafés because they provide a safe and comfortable space to socialize. . . .

Cyber cafés are also appealing for youth who hold similar interests in the subculture of video gaming.

> Even if you're sitting at home playing by yourself, it's not as fun [as] playing with someone else. You can't say things like, "Oh my God, you just died!" or "Go get the last guy!" It's like renting versus seeing a movie at a theater. When you're alone you don't laugh at things. When everyone is laughing, you're laughing too.

. . .Players, along with some parents, consider computer game rooms a safer alternative than hanging out on the streets. Although the games they play depict violent acts, in a number of these settings, there is a sense of camaraderie and bonding among the young patrons and employees who maintain a relaxed attitude. . . .

Unlike ethnic pool halls or other ethnic leisure sites that are typically thought of as exclusively male domains, cyber cafés welcome women, although, as noted earlier, few females come. Cyber café owners, managers, and players all concur that female customers are rare, and, by some estimates, the male to female ratio is twenty to one. Our observation confirmed that young women come in far less frequently than males, and when they do, they often arrive in pairs, although this varies on the time and the cafés. Whereas most males come to play the computer games, the young women come mainly to use the Internet and check and send e-mail. However, a few play similar games to their male counterparts. In some cases, boyfriends or brothers introduce the young women to the games and often accompany them to the cafés. Males certainly respect the skills of good girl players, but the stereotype holds that only "unattractive" girls are serious players. . . .

ASIAN AMERICAN MEN AND MASCULINITY

One late afternoon in Orange County's Koreatown, we visited a café where the patrons explained that, although camaraderie and having a place to go after school were the initial reasons for coming to the café, the games appealed to their "basic male instincts." In our focus groups, Asian American youth remarked that they were attracted to Counter-Strike because of the "guns and violence." For example, when asked what was appealing about Counter-Strike, a Korean American youth replied: "You're killing other people. Every time it's different. You have skills at certain things. When you first start, you have an adrenaline rush killing people, especially when you have your first kill." And as a Chinese American youth reiterated, "It's pretty much what guys are raised to [do]; you see things on TV and the media, especially in the U.S. . . . Women play with Barbie, men play with guns, it's natural."

Cyber cafés have become a social and cultural outlet where young Asian American men can feel a sense of achievement and assert their masculinity in a society that often demasculinizes them. In a culture in which Asian American male youths see few positive images of themselves in the media, computer games are a safe means by which they can gain a sense of empowerment. In some cases, these youth may not excel at athletics or school; however, they are good at these games. They are not the "model minority" whiz kids, nor are they the trouble-making "gang bangers"—extreme stereotypical labels often ascribed to Asian American males. They struggle with school and realize that they are not living up to the educational expectations of their immigrant parents, who have made incredible sacrifices for them to have educational opportunities unavailable in their homeland. For instance, when asked about the most challenging aspect of his life, a Korean American male responded, "The most challenging thing is school. The fact that I go to a JC (junior college) . . . there's a pride factor. My girlfriend goes to Berkeley." This Korean American youth had internalized the parental pressure of getting into a "top" college, and his failure to do so had affected his sense of identity as an Asian American male, especially when he compared himself to his "smarter" girlfriend.

The youth we interviewed were also cognizant of the racialized and gendered constructions of Asian American males. In our focus groups, for example, participants identified themselves ethnically as Chinese American or Korean American or Asian American but not as "American." When we asked, "Why not just say you're American?" they seemed to concur with the statement made by one individual, "Because we look different." A Chinese American youth in his early twenties in one of our focus groups expressed his perception of Asian American men and women: "Asian women are more favored than Asian men in Hollywood. Asian men are known for stereotypes such as kung fu.... We have it better than [they did in] my parents' time, but obviously, we're still a minority, I guess."

When asked about his ethnic preferences in dating, he stated, "I'm open to dating Asian women as long as they're not the stereotypical 'Koreatown' girl, like they have to dress up every time you go out and wear a lot of make-up." He did not want to date a "stereotypical Koreatown girl," who could be read as a female who wanted a guy with a lot of money (not necessarily referring to a Korean American female). His aversion to dating stereotypical Koreatown girls, however, appeared more like a defensive reaction to his own class status. He also remarked with some disdain that it was easier for Asian women to date non-Asian men and explained, "My parents would want me to date Chinese." However, he added, "I wouldn't mind dating a white woman because it's not typical" and "You have more status [as an Asian male] if you date white women."

Like the other Asian American males in the study, this male recognized how racial, gender, class, and status dynamics played out in the "real world," and he also understood his position in it. Asian American male youth gravitate toward the cafés where they can hang out with others like them and act out their aggression through the video games. The violent computer games allow players to be judged by their gaming skills rather than their race, class, or status, and provide a way for young men to display their virtual masculine abilities. They may not be able to control their social environment or the direction of their real lives, but through their "virtual lives," they can acquire and maintain a sense of status and power. . . .

CONCLUSION

Like all youth, Asian American youth seek alternative social outlets where they can hang out with others like them. For low-income Asian American teenagers who reside in southern California's urbanized suburbs, finding social outlets that are both accessible and affordable presents a challenge. They want a diversion from the boredom, stresses, and problems of everyday life but have limited economic resources to entertain themselves. Cyber cafés provide an affordable space for them to fill some of these needs and offer a venue for Asian American youth to channel their energies. The cafés also provide a space where young Asian American men can form bonds of friendship and reclaim their masculinity in a society in which they sometimes feel demasculinized and marginalized. The video games allow them to become someone who has skills and status, and give them an opportunity to demonstrate their gaming prowess. At the same time, the cyber community offers them an opportunity to participate in a network that encourages teamwork and alliances to defeat "an enemy" in their virtual lives. Cyber café players have created a growing youth subculture with their own language, rituals, and protocol on how to be an "engaged citizen" of this community. . . .

With few after-school programs, organized activities, or youth centers in their communities that cater to the interests of high school or college age immigrant and refugee youths, cyber cafés provide an attractive option for them. The cyber café culture is likely to grow in popularity as long as technological gaming innovations continue, as long as officials allow the businesses to stay open, and, most importantly, as long as new generations of Asian American youths remain attracted to this form of leisure.

REFERENCES

Baguioro, Luz, "Manila Kids Skip Class to Play Counter-Strike," *Straits Times* (Singapore), January 27, 2002.

Baker, Michael, "Internet Fad Grips Korean Youth Culture," *Courier Mail*, May 6, 2000, p. 21.

Bangkok Post [online], "Internet: Youths Lost in Profitable World of Cyber Games; Café Owners Rush to Lure More Children," Copyright FT Asia Africa Intelligence Wire, August 20, 2001.

Danico, Mary Yu, *The 1.5 Generation: Becoming Korean American in Hawaii* (Honolulu: University of Hawaii Press, 2004).

Graham, Jefferson, "Cyber Cafés Serve an Explosive Brew," *USA Today*, February 7, 2002, p. 10.

Marriot, Michel. "The Sad Ballad of the Cyber Café," *New York Times on the Web*, April 16, 1998.

Ni, Ching-Ching, "Dens of the Cyber Addicts," *Los Angeles Times*, June 28, 2002, pp. A1, A4, and A5.

Park, Lisa Sun-Hee, "Asian Immigrant Entrepreneurial Children" in *Contemporary Asian American Communities: Intersections and Divergences*, ed. Linda Trinh Võ and Rick Bonus (Philadelphia: Temple University Press, 2002), 161–177.

Song, Jason, "Fast Times at the Cybercafe," *Los Angeles Times*, September 10, 2001, p. E1.

Sullivan, John, "PC Bangs a Hit with Local Kids," *Newsday New York*, August 7, 2001, p. A13.

29

Prisons for Our Bodies, Closets for Our Minds

Racism, Heterosexism, and Black Sexuality

PATRICIA HILL COLLINS

White fear of black sexuality is a basic ingredient of white racism.

CORNEL WEST

For African Americans, exploring how sexuality has been manipulated in defense of racism is not new. Scholars have long examined the ways in which "white fear of black sexuality" has been a basic ingredient of racism. For example, colonial regimes routinely manipulated ideas about sexuality in order to maintain unjust power relations. Tracing the history of contact between English explorers and colonists and West African societies, historian Winthrop Jordan contends that English perceptions of sexual practices among African people reflected preexisting English beliefs about Blackness, religion, and animals. American historians point to the significance of sexuality to chattel slavery. In the United States, for example, slaveowners relied upon an ideology of Black sexual deviance to regulate and exploit enslaved Africans. Because Black feminist analyses pay more attention to women's sexuality, they too identify how the sexual exploitation of women has been a basic ingredient of racism. For example, studies of African American slave women routinely point to sexual victimization as a defining feature of American slavery. Despite the important contributions of this extensive literature on race and sexuality, because much of the literature assumes that sexuality means *hetero*sexuality, it ignores how racism and heterosexism influence one another.

In the United States, the assumption that racism and heterosexism constitute two separate systems of oppression masks how each relies upon the other for meaning. Because neither system of oppression makes sense without the other, racism and heterosexism might be better viewed as sharing one history with similar yet disparate effects on all Americans differentiated by race, gender, sexuality, class, and nationality. People who are positioned at the margins of both systems and who are harmed by both typically raise questions about the intersections of racism and heterosexism much earlier and/or more forcefully than those people

SOURCE: Collins, Patricia Hill. 2004. *Black Sexual Politics: African Americans and the New Racism.* New York: Routledge.

who are in positions of privilege. In the case of intersections of racism and heterosexism, Black lesbian, gay, bisexual, and transgendered (LGBT) people were among the first to question how racism and heterosexism are interconnected. As African American LGBT people point out, assuming that all Black people are heterosexual and that all LGBT people are White distorts the experiences of LGBT Black people. Moreover, such comparisons misread the significance of ideas about sexuality to racism and race to heterosexism.

Until recently, questions of sexuality in general, and homosexuality in particular, have been treated as crosscutting, divisive issues within antiracist African American politics. The consensus issue of ensuring racial unity subordinated the allegedly crosscutting issue of analyzing sexuality, both straight and gay alike. This suppression has been challenged from two directions. Black women, both heterosexual and lesbian, have criticized the sexual politics of African American communities that leave women vulnerable to single motherhood and sexual assault. Black feminist and womanist projects have challenged Black community norms of a sexual double standard that punishes women for behaviors in which men are equally culpable. Black gays and lesbians have also criticized these same sexual politics that deny their right to be fully accepted within churches, families, and other Black community organizations. Both groups of critics argue that ignoring the heterosexism that underpins Black patriarchy hinders the development of a progressive Black sexual politics. . . .

Developing a progressive Black sexual politics requires examining how racism and heterosexism mutually construct one another.

MAPPING RACISM AND HETEROSEXISM: THE
PRISON AND THE CLOSET

. . . Racism and heterosexism, the prison and the closet, appear to be separate systems, but LGBT African Americans point out that *both* systems affect their everyday lives. If racism and heterosexism affect Black LGBT people, then these systems affect *all* people, including heterosexual African Americans. Racism and heterosexism certainly converge on certain key points. For one, both use similar state-sanctioned institutional mechanisms to maintain racial and sexual hierarchies. For example, in the United States, racism and heterosexism both rely on segregating people as a mechanism of social control. For racism, segregation operates by using race as a visible marker of group membership that enables the state to relegate Black people to inferior schools, housing, and jobs. Racial segregation relies on enforced membership in a visible community in which racial discrimination is tolerated. For heterosexism, segregation is enforced by pressuring LGBT individuals to remain closeted and thus segregated from one another. Before social movements for gay and lesbian liberation, sexual segregation meant that refusing to claim homosexual identities virtually eliminated any group-based political action to resist heterosexism. For another, the state has played a very important role in sanctioning both forms of oppression. In support of racism, the state sanctioned laws that regulated where Black people could live, work, and attend school. In support of heterosexism, the state maintained laws that refused to punish hate crimes against LGBT people, that failed to offer protection when LGBT people were stripped of jobs and children, and that generally sent a message that LGBT people who came out of the closet did so at their own risk.

Racism and heterosexism also share a common set of practices that are designed to discipline the population into accepting the status quo. These disciplinary practices can best be seen in the enormous amount of attention paid both by the state and organized religion to the institution of marriage. If marriage were in fact a natural and normal occurrence between heterosexual couples and if it occurred naturally within racial categories, there would be no need to regulate it. People would naturally choose partners of the opposite sex and the same race. Instead, a series of laws have been passed, all designed to regulate marriage. For example, for many years, the tax system has rewarded married couples with tax breaks that have been denied to single taxpayers or unmarried couples. The message is clear—it makes good financial sense to get married. Similarly, to encourage people to marry within their assigned race, numerous states passed laws banning interracial marriage. These restrictions lasted until the landmark Supreme Court decision in 1967 that overturned state laws. The state has also passed laws designed to keep LGBT people from marrying. In 1996, the U.S. Congress passed the Federal Defense of Marriage Act that defined marriage as a "legal union between one man and one woman." In all of these cases, the state perceives that it has a compelling interest in disciplining the population to marry and to marry the correct partners.

Racism and heterosexism also manufacture ideologies that defend the status quo. When ideologies that defend racism and heterosexism become taken-for-granted and appear to be natural and inevitable, they become hegemonic. Few question them and the social hierarchies they defend. Racism and heterosexism both share a common cognitive framework that uses binary thinking to produce hegemonic ideologies. Such thinking relies on oppositional categories. It views race through two oppositional categories of Whites and Blacks, gender through two categories of men and women, and sexuality through two oppositional categories of heterosexuals and homosexuals. A master binary of normal and deviant overlays and bundles together these and other lesser binaries. In this context, ideas about "normal" race (whiteness, which ironically, masquerades as racelessness), "normal" gender (using male experiences as the norm), and "normal" sexuality (heterosexuality, which operates in a similar hegemonic fashion) are tightly bundled together. In essence, to be completely "normal," one must be White, masculine, and heterosexual, the core hegemonic White masculinity. This mythical norm is hard to see because it is so taken-for-granted. Its antithesis, its Other, would be Black, female, and lesbian, a fact that Black lesbian feminist Audre Lorde pointed out some time ago.

Within this oppositional logic, the core binary of normal/deviant becomes ground zero for justifying racism and heterosexism. The deviancy assigned to race and that assigned to sexuality becomes an important point of contact between the two systems. Racism and heterosexism both require a concept of sexual deviancy for meaning, yet the form that deviance takes within each system differs. For racism, the point of deviance is created by a *normalized White heterosexuality* that depends on a *deviant Black heterosexuality* to give it meaning. For heterosexism, the point of deviance is created by this very same *normalized White heterosexuality* that now depends on a *deviant White homosexuality*. Just as racial normality requires the stigmatization of the sexual practices of Black people, heterosexual normality relies upon the stigmatization of the sexual practices of homosexuals. In both cases, installing White heterosexuality as normal, natural, and ideal requires stigmatizing alternate sexualities as abnormal, unnatural, and sinful.

The purpose of stigmatizing the sexual practices of Black people and those of LGBT people may be similar, but the content of the sexual deviance assigned to each differs. Black people carry the stigma of *promiscuity* or excessive or unrestrained heterosexual desire. This is the sexual deviancy that has both been assigned to Black people and been used to construct racism. In contrast, LGBT people carry the stigma of *rejecting* heterosexuality by engaging in unrestrained homosexual desire. Whereas the deviancy associated with promiscuity (and, by implication, with Black people as a race) is thought to lie in an *excess* of heterosexual desire, the pathology of homosexuality (the invisible, closeted sexuality that becomes impossible within heterosexual space) seemingly resides in the *absence* of it.

While analytically distinct, in practice, these two sites of constructed deviancy work together and both help create the "sexually repressive culture" in America. . . . Despite their significance for American society overall, here I confine my argument to the challenges that confront Black people. Both sets of ideas frame a hegemonic discourse of *Black sexuality* that has at its core ideas about an assumed promiscuity among heterosexual African American men and women and the impossibility of homosexuality among Black gays and lesbians. How have African Americans been affected by and reacted to this racialized system of heterosexism (or this sexualized system of racism)?

AFRICAN AMERICANS AND THE RACIALIZATION OF PROMISCUITY

Ideas about Black promiscuity that produce contemporary sexualized spectacles such as Jennifer Lopez, Destiny's Child, Ja Rule, and the many young Black men on the U.S. talk show circuit have a long history. Historically, Western science, medicine, law, and popular culture reduced an African-derived aesthetic concerning the use of the body, sensuality, expressiveness, and spirituality to an ideology about *Black sexuality*. The distinguishing feature of this ideology was its reliance on the idea of Black promiscuity. The possibility of distinctive and worthwhile African-influenced worldviews on anything, including sexuality, as well as the heterogeneity of African societies expressing such views, was collapsed into an imagined, pathologized Western discourse of what was thought to be essentially African. To varying degrees, observers from England, France, Germany, Belgium, and other colonial powers perceived African sensuality, eroticism, spirituality, and/or sexuality as deviant, out of control, sinful, and as an essential feature of racial difference. . . .

With all living creatures classified in this way, Western scientists perceived African people as being more natural and less civilized, primarily because African people were deemed to be closer to animals and nature, especially the apes and monkeys whose appearance most closely resembled humans. Like African people, animals also served as objects of study for Western science because understanding the animal kingdom might reveal important insights about civilization, culture, and what distinguished the human "race" from its animal counterparts as well as the human "races" from one another. . . .

Those most proximate to animals, those most lacking civilization, also were those humans who came closest to having the sexual lives of animals. Lacking the benefits of Western civilization, people of African descent were perceived as having a biological nature that was inherently more sexual than that of Europeans. The primitivist discourse thus created the category of "beast" and the sexuality of such beasts as "wild." The legal classification of enslaved African people as chattel (animal-like) under American slavery that produced controlling images of bucks, jezebels, and breeder women drew meaning from this broader interpretive framework.

Historically, this ideology of Black sexuality that pivoted on a Black heterosexual promiscuity not only upheld racism but it did so in gender-specific ways. In the context of U.S. society, beliefs in Black male promiscuity took diverse forms during distinctive historical periods. For example, defenders of chattel slavery believed that slavery safely domesticated allegedly dangerous Black men because it regulated their promiscuity by placing it in the service of slave owners. Strategies of control were harsh and enslaved African men who were born in Africa or who had access to their African past were deemed to be the most dangerous. In contrast, the controlling image of the rapist appeared after emancipation because Southern Whites' feared that the unfettered promiscuity of Black freedmen constituted a threat to the Southern way of life. . . .

The events themselves may be over, but their effects persist under the new racism. This belief in an inherent Black promiscuity reappears today. For example, depicting poor and working-class African American inner-city neighborhoods as dangerous urban jungles where SUV-driving White suburbanites come to score drugs or locate prostitutes also invokes a history of racial and sexual conquest. Here sexuality is linked with danger, and understandings of both draw upon historical imagery of Africa as a continent replete with danger and peril to the White explorers and hunters who penetrated it. Just as contemporary safari tours in Africa create an imagined Africa as the "White man's playground" and mask its economic exploitation, jungle language masks social relations of hyper-segregation that leave working-class Black communities isolated, impoverished, and dependent on a punitive welfare state and an illegal international drug trade. Under this logic, just as wild animals (and the proximate African natives)

belong in nature preserves (for their own protection), unassimilated, undomesti-
cated poor and working-class African Americans belong in racially segregated
neighborhoods. . . .

African American women also live with ideas about Black women's promiscuity and lack of sexual restraint. Reminiscent of concerns with Black women's
fertility under slavery and in the rural South, contemporary social welfare policies
also remain preoccupied with Black women's fertility. In prior eras, Black
women were encouraged to have many children. Under slavery, having many
children enhanced slave owners' wealth and a good "breeder woman" was less
likely to be sold. In rural agriculture after emancipation, having many children
ensured a sufficient supply of workers. But in the global economy of today, large
families are expensive because children must be educated. Now Black women
are seen as producing too many children who contribute less to society than
they take. Because Black women on welfare have long been seen as undeserving,
long-standing ideas about Black women's promiscuity become recycled and
redefined as a problem for the state. . . .

RACISM AND HETEROSEXISM REVISITED

On May 11, 2003, a stranger killed fifteen-year-old Sakia Gunn who, with four
friends, was on her way home from New York's Greenwich Village. Sakia and
her friends were waiting for the bus in Newark, New Jersey, when two men got
out of a car, made sexual advances, and physically attacked them. The women
fought back, and when Gunn told the men that she was a lesbian, one of them
stabbed her in the chest.

Sakia Gunn's murder illustrates the connections among class, race, gender,
sexuality, and age. Sakia lacked the protection of social class privilege. She and
her friends were waiting for the bus in the first place because none had access to
private automobiles that offer protection for those who are more affluent. In
Gunn's case, because her family initially did not have the money for her funeral,
she was scheduled to be buried in a potter's grave. Community activists took up
a collection to pay for her funeral. She lacked the gendered protection provided
by masculinity. Women who are perceived to be in the wrong place at the
wrong time are routinely approached by men who feel entitled to harass and
proposition them. Thus, Sakia and her friends share with all women the vulnerabilities that accrue to women who negotiate public space. She lacked the protection of age—had Sakia and her friends been middle-aged, they may not have
been seen as sexually available. Like African American girls and women, regardless of sexual orientation, they were seen as approachable. Race was a factor, but
not in a framework of interracial race relations. Sakia and her friends were
African American, as were their attackers. In a context where Black men are
encouraged to express a hyper-heterosexuality as the badge of Black masculinity,
women like Sakia and her friends can become important players in supporting

patriarchy. They challenged Black male authority, and they paid for the transgression of refusing to participate in scripts of Black promiscuity. But the immediate precipitating catalyst for the violence that took Sakia's life was her openness about her lesbianism. Here, homophobic violence was the prime factor. Her death illustrates how deeply entrenched homophobia can be among many African American men and women, in this case, beliefs that resulted in an attack on a teenaged girl.

How do we separate out and weigh the various influences of class, gender, age, race, and sexuality in this particular incident? Sadly, violence against Black girls is an everyday event. What made this one so special? Which, if any, of the dimensions of her identity got Sakia Gunn killed? There is no easy answer to this question, because *all* of them did. More important, how can any Black political agenda that does not take *all* of these systems into account, including sexuality, ever hope adequately to address the needs of Black people as a collectivity? One expects racism in the press to shape the reports of this incident. In contrast to the 1998 murder of Matthew Shepard, a young, White, gay man in Wyoming, no massive protests, nationwide vigils, and renewed calls for federal hate crimes legislation followed Sakia's death. But what about the response of elected and appointed officials? The African American mayor of Newark decried the crime, but he could not find the time to meet with community activists who wanted programmatic changes to retard crimes like Sakia's murder. The principal of her high school became part of the problem. As one activist described it, "students at Sakia's high school weren't allowed to hold a vigil. And the kids wearing the rainbow flag were being punished like they had on gang colors."

Other Black leaders and national organizations spoke volumes through their silence. The same leaders and organizations that spoke out against the police beating of Rodney King by Los Angeles area police, the rape of immigrant Abner Louima by New York City police, and the murder of Timothy Thomas by Cincinnati police said nothing about Sakia Gunn's death. Apparently, she was just another unimportant little Black girl to them. But to others, her death revealed the need for a new politics that takes the intersections of racism and heterosexism as well as class exploitation, age discrimination, and sexism into account. Sakia was buried on May 16 and a crowd of approximately 2,500 people attended her funeral. The turnout was unprecedented: predominantly Black, largely high school students, and mostly lesbians. Their presence says that as long as African American lesbians like high school student Sakia Gunn are vulnerable, then every African American woman is in danger; and if all Black women are at risk, then there is no way that any Black person will ever be truly safe or free.

30

The Invention of Heterosexuality

JONATHAN NED KATZ

H eterosexuality is old as procreation, ancient as the lust of Eve and Adam. That first lady and gentleman, we assume, perceived themselves, behaved, and felt just like today's heterosexuals. We suppose that heterosexuality is unchanging, universal, essential: ahistorical.

Contrary to that common sense conjecture, the concept of heterosexuality is only one particular historical way of perceiving, categorizing, and imagining the social relations of the sexes. Not ancient at all, the idea of heterosexuality is a modern invention, dating to the late nineteenth century. The heterosexual belief, with its metaphysical claim to eternity, has a particular, pivotal place in the social universe of the late nineteenth and twentieth centuries that it did not inhabit earlier. This essay traces the historical process by which the heterosexual idea was created as ahistorical and taken-for-granted. . . .

By not studying the heterosexual idea in history, analysts of sex, gay and straight, have continued to privilege the "normal" and "natural" at the expense of the "abnormal" and "unnatural." Such privileging of the norm accedes to its domination, protecting it from questions. By making the normal the object of a thoroughgoing historical study we simultaneously pursue a pure truth and a sex-radical and subversive goal: we upset basic preconceptions. We discover that the heterosexual, the normal, and the natural have a history of changing definitions. Studying the history of the term challenges its power.

Contrary to our usual assumption, past Americans and other peoples named, perceived, and socially organized the bodies, lusts, and intercourse of the sexes in ways radically different from the way we do. If we care to understand this vast past sexual diversity, we need to stop promiscuously projecting our own hetero and homo arrangement. Though lip service is often paid to the distorting, ethnocentric effect of such conceptual imperialism, the category heterosexuality continues to be applied uncritically as a universal analytical tool. Recognizing the time-bound and culturally specific character of the heterosexual category can help us begin to work toward a thoroughly historical view of sex. . . .

SOURCE: From Jonathan Ned Katz. Copyright © 1990.
Reprinted by permission of the author.

BEFORE HETEROSEXUALITY: EARLY VICTORIAN TRUE LOVE, 1820–1860

In the early nineteenth-century United States, from about 1820 to 1860, the heterosexual did not exist. Middle-class white Americans idealized a True Womanhood, True Manhood, and True Love, all characterized by "purity"—the freedom from sensuality.[1] Presented mainly in literary and religious texts, this True Love was a fine romance with no lascivious kisses. This ideal contrasts strikingly with late nineteenth- and twentieth-century American incitements to a hetero sex.[2]

Early Victorian True Love was only realized within the mode of proper procreation, marriage, the legal organization for producing a new set of correctly gendered women and men. Proper womanhood, manhood, and progeny—not a normal male-female eros—was the main product of this mode of engendering and of human reproduction.

The actors in this sexual economy were identified as manly men and womanly women and as procreators, not specifically as erotic beings or heterosexuals. Eros did not constitute the core of a heterosexual identity that inhered, democratically, in both men and women. True Women were defined by their distance from lust. True Men, though thought to live closer to carnality, and in less control of it, aspired to the same freedom from concupiscence.

Legitimate natural desire was for procreation and a proper manhood or womanhood; no heteroerotic desire was thought to be directed exclusively and naturally toward the other sex; lust in men was roving. The human body was thought of as a means towards procreation and production; penis and vagina were instruments of reproduction, not of pleasure. Human energy, thought of as a closed and severely limited system, was to be used in producing children and in work, not wasted in libidinous pleasures.

The location of all this engendering and procreative labor was the sacred sanctum of early Victorian True Love, the home of the True Woman and True Man—a temple of purity threatened from within by the monster masturbator, an archetypal early Victorian cult figure of illicit lust. The home of True Love was a castle far removed from the erotic exotic ghetto inhabited most notoriously then by the prostitute, another archetypal Victorian erotic monster....

LATE VICTORIAN SEX-LOVE: 1860–1892

"Heterosexuality" and "homosexuality" did not appear out of the blue in the 1890s. These two eroticisms were in the making from the 1860s on. In late Victorian America and in Germany, from about 1860 to 1892, our modern idea of an eroticized universe began to develop, and the experience of a hetero-lust began to be widely documented and named....

In the late nineteenth-century United States, several social factors converged to cause the eroticizing of consciousness, behavior, emotion, and identity that became typical of the twentieth-century Western middle class. The transformation of the family from producer to consumer unit resulted in a change in family members' relation to their own bodies; from being an instrument primarily of work, the human body was integrated into a new economy, and began more commonly to be perceived as a means of consumption and pleasure. Historical work has recently begun on how the biological human body is differently integrated into changing modes of production, procreation, engendering, and pleasure so as to alter radically the identity, activity, and experience of that body.[3]

The growth of a consumer economy also fostered a new pleasure ethic. This imperative challenged the early Victorian work ethic, finally helping to usher in a major transformation of values. While the early Victorian work ethic had touted the value of economic production, that era's procreation ethic had extolled the virtues of human reproduction. In contrast, the late Victorian economic ethic hawked the pleasures of consuming, while its sex ethic praised an erotic pleasure principle for men and even for women.

In the late nineteenth century, the erotic became the raw material for a new consumer culture. Newspapers, books, plays, and films touching on sex, "normal" and "abnormal," became available for a price. Restaurants, bars, and baths opened, catering to sexual consumers with cash. Late Victorian entrepreneurs of desire incited the proliferation of a new eroticism, a commoditized culture of pleasure.

In these same years, the rise in power and prestige of medical doctors allowed these upwardly mobile professionals to prescribe a healthy new sexuality. Medical men, in the name of science, defined a new ideal of male-female relationships that included, in women as well as men, an essential, necessary, normal eroticism. Doctors, who had earlier named and judged the sex-enjoying woman a "nymphomaniac," now began to label women's *lack* of sexual pleasure a mental disturbance, speaking critically, for example, of female "frigidity" and "anesthesia."[4]

By the 1880s, the rise of doctors as a professional group fostered the rise of a new medical model of Normal Love, replete with sexuality. The new Normal Woman and Man were endowed with a healthy libido. The new theory of Normal Love was the modern medical alternative to the old Cult of True Love. The doctors prescribed a new sexual ethic as if it were a morally neutral, medical description of health. The creation of the new Normal Sexual had its counterpart in the invention of the late Victorian Sexual Pervert. The attention paid the sexual abnormal created a need to name the sexual normal, the better to distinguish the average him and her from the deviant it.

HETEROSEXUALITY: THE FIRST YEARS, 1892–1900

In the periodization of heterosexual American history suggested here, the years 1892 to 1900 represent "The First Years" of the heterosexual epoch, eight key years in which the idea of the heterosexual and homosexual were initially and tentatively formulated by U.S. doctors. The earliest-known American use of the word "heterosexual" occurs in a medical journal article by Dr. James G. Kiernan of Chicago, read before the city's medical society on March 7, 1892, and published that May—portentous dates in sexual history.[5] But Dr. Kiernan's heterosexuals were definitely not exemplars of normality. Heterosexuals, said Kiernan, were defined by a mental condition, "psychical hermaphroditism." Its symptoms were "inclinations to both sexes." These heterodox sexuals also betrayed inclinations "to abnormal methods of gratification," that is, techniques to insure pleasure without procreation. Dr. Kiernan's heterogeneous sexuals did demonstrate "traces of the normal sexual appetite" (a touch of procreative desire). Kiernan's normal sexuals were implicitly defined by a monolithic other-sex inclination and procreative aim. Significantly, they still lacked a name.

Dr. Kiernan's article of 1892 also included one of the earliest-known uses of the word "homosexual" in American English. Kiernan defined "Pure homosexuals" as persons whose "general mental state is that of the opposite sex." Kiernan thus defined homosexuals by their deviance from a gender norm. His heterosexuals displayed a double deviance from both gender and procreative norms.

Though Kiernan used the new words heterosexual and homosexual, an old procreative standard and a new gender norm coexisted uneasily in his thought. His word heterosexual defined a mixed person and compound urge, abnormal because they wantonly included procreative and non-procreative objectives, as well as same-sex and different-sex attractions.

That same year, 1892, Dr. Krafft-Ebing's influential *Psychopathia Sexualis* was first translated and published in the United States.[6] But Kiernan and Krafft-Ebing by no means agreed on the definition of the heterosexual. In Krafft-Ebing's book, "hetero-sexual" was used unambiguously in the modern sense to refer to an erotic feeling for a different sex. "Homosexual" referred unambiguously to an erotic feeling for a "same sex." In Krafft-Ebing's volume, unlike Kiernan's article, heterosexual and homosexual were clearly distinguished from a third category, a "psychosexual hermaphroditism," defined by impulses toward both sexes.

Krafft-Ebing hypothesized an inborn "sexual instinct" for relations with the "opposite sex," the inherent "purpose" of which was to foster procreation. Krafft-Ebing's erotic drive was still a reproductive instinct. But the doctor's clear focus on a different-sex versus same-sex sexuality constituted a historic, epochal move from an absolute procreative standard of normality toward a new norm. His definition of heterosexuality as other-sex attraction provided the basis for a revolutionary, modern break with a centuries-old procreative standard.

It is difficult to overstress the importance of that new way of categorizing. The German's mode of labeling was radical in referring to the biological sex, masculinity or femininity, and the pleasure of actors (along with the procreant purpose of acts). Krafft-Ebing's heterosexual offered the modern world a new norm that came to dominate our idea of the sexual universe, helping to change it from a mode of human reproduction and engendering to a mode of pleasure. The heterosexual category provided the basis for a move from a production-oriented, procreative imperative to a consumerist pleasure principle—an institutionalized pursuit of happiness. . . .

Only gradually did doctors agree that heterosexual referred to a normal, "other-sex" eros. This new standard-model heterosex provided the pivotal term for the modern regularization of eros that paralleled similar attempts to standardize masculinity and femininity, intelligence, and manufacturing.[7] The idea of heterosexuality as the master sex from which all others deviated was (like the idea of the master race) deeply authoritarian. The doctors' normalization of a

sex that was hetero proclaimed a new heterosexual separatism—an erotic apartheid that forcefully segregated the sex normals from the sex perverts. The new,
strict boundaries made the emerging erotic world less polymorphous—safer for
sex normals. However, the idea of such creatures as heterosexuals and homosexuals emerged from the narrow world of medicine to become a commonly
accepted notion only in the early twentieth century. In 1901, in the comprehensive *Oxford English Dictionary,* "heterosexual" and "homosexual" had not yet
made it.

THE DISTRIBUTION OF THE HETEROSEXUAL
MYSTIQUE: 1900–1930

In the early years of this heterosexual century the tentative hetero hypothesis was
stabilized, fixed, and widely distributed as the ruling sexual orthodoxy: The
Heterosexual Mystique. Starting among pleasure-affirming urban working-class
youths, southern blacks, and Greenwich Village bohemians as defensive subculture, heterosex soon triumphed as dominant culture.[8]

In its earliest version, the twentieth-century heterosexual imperative usually
continued to associate heterosexuality with a supposed human "need," "drive,"
or "instinct" for propagation, a procreant urge linked inexorably with carnal lust
as it had not been earlier. In the early twentieth century, the falling birth rate,
rising divorce rate, and "war of the sexes" of the middle class were matters of
increasing public concern. Giving vent to heteroerotic emotions was thus praised
as enhancing baby-making capacity, marital intimacy, and family stability. (Only
many years later, in the mid-1960s, would heteroeroticism be distinguished
completely, in practice and theory, from procreativity and male-female pleasure
sex justified in its own name.)

The first part of the new sex norm—hetero—referred to a basic gender diver-
gence. The "oppositeness" of the sexes was alleged to be the basis for a universal,
normal, erotic attraction between males and females. The stress on the sexes'
"oppositeness," which harked back to the early nineteenth century, by no means
simply registered biological differences of females and males. The early twentieth-
century focus on physiological and gender dimorphism reflected the deep anxieties
of men about the shifting work, social roles, and power of men over women, and
about the ideals of womanhood and manhood. That gender anxiety is documented, for example, in 1897, in *The New York Times'* publication of the Reverend
Charles Parkhurst's diatribe against female "andromaniacs," the preacher's derogatory, scientific-sounding name for women who tried to "minimize distinctions by
which manhood and womanhood are differentiated."[9] The stress on gender difference was a conservative response to the changing social-sexual division of activity
and feeling which gave rise to the independent "New Woman" of the 1880s and
eroticized "Flapper" of the 1920s.

The second part of the new hetero norm referred positively to sexuality. That novel upbeat focus on the hedonistic possibilities of male-female conjunctions also reflected a social transformation—a revaluing of pleasure and procreation, consumption and work in commercial, capitalist society. The democratic attribution of a normal lust to human females (as well as males) served to authorize women's enjoyment of their own bodies and began to undermine the early Victorian idea of the pure True Woman—a sex-affirmative action still part of women's struggle. The twentieth-century Erotic Woman also undercut the nineteenth-century feminist assertion of women's moral superiority, cast suspicions of lust on women's passionate romantic friendships with women, and asserted the presence of a menacing female monster, "the lesbian."[10] . . .

In the perspective of heterosexual history, this early twentieth-century struggle for the more explicit depiction of an "opposite-sex" eros appears in a curious new light. Ironically, we find sex-conservatives, the social purity advocates of censorship and repression, fighting against the depiction not just of sexual perversity but also of the new normal hetero-sexuality. That a more open depiction of normal sex had to be defended against forces of propriety confirms the claim that heterosexuality's predecessor, Victorian True Love, had included no legitimate eros. . . .

THE HETEROSEXUAL STEPS OUT: 1930–1945

In 1930, in *The New York Times,* heterosexuality first became a love that dared to speak its name. On April 20th of that year, the word "heterosexual" is first known to have appeared in *The New York Times Book Review*. There, a critic described the subject of André Gide's *The Immoralist* proceeding "from a heterosexual liaison to a homosexual one." The ability to slip between sexual categories was referred to casually as a rather unremarkable aspect of human possibility. This is also the first known reference by *The Times* to the new hetero/homo duo.[11]

In September the second reference to the hetero/homo dyad appeared in *The New York Times Book Review,* in a comment on Floyd Dell's *Love in the Machine Age*. This work revealed a prominent antipuritan of the 1930s using the dire threat of homosexuality as his rationale for greater heterosexual freedom. *The Times* quoted Dell's warning that current abnormal social conditions kept the young dependent on their parents, causing "infantilism, prostitution and homosexuality." Also quoted was Dell's attack on the "inculcation of purity" that "breeds distrust of the opposite sex." Young people, Dell said, should be "permitted to develop normally to heterosexual adulthood." "But," *The Times* reviewer emphasized, "such a state already exists, here and now." And so it did. Heterosexuality, a new gender-sex category, had been distributed from the narrow, rarified realm of a few doctors to become a nationally, even internationally, cited aspect of middle-class life.[12] . . .

HETEROSEXUAL HEGEMONY: 1945–1965

The "cult of domesticity" following World War II—the reassociation of women with the home, motherhood, and child-care; men with fatherhood and wage work outside the home—was a period in which the predominance of the hetero norm went almost unchallenged, an era of heterosexual hegemony. This was an age in which conservative mental-health professionals reasserted the old link between heterosexuality and procreation. In contrast, sex-liberals of the day strove, ultimately with success, to expand the heterosexual ideal to include within the boundaries of normality a wider-than-ever range of nonprocreative, premarital, and extramarital behaviors. But sex-liberal reform actually helped to extend and secure the dominance of the heterosexual idea, as we shall see when we get to Kinsey.

The postwar sex-conservative tendency was illustrated in 1947, in Ferdinand Lundberg and Dr. Marynia Farnham's books, *Modern Woman: The Lost Sex.* Improper masculinity and femininity was exemplified, the authors decreed, by "engagement in heterosexual relations...with the complete intent to see to it that they do not eventuate in reproduction."[13] Their procreatively defined heterosex was one expression of a postwar ideology of fecundity that, internalized and enacted dutifully by a large part of the population, gave rise to the postwar baby boom.

The idea of the feminine female and masculine male as prolific breeders was also reflected in the stress, specific to the late 1940s, on the homosexual as sad symbol of "sterility"—that particular loaded term appears incessantly in comments on homosex dating to the fecund forties.

In 1948, in *The New York Times Book Review,* sex liberalism was in ascendancy. Dr. Howard A. Rusk declared that Alfred Kinsey's just published report on *Sexual Behavior in the Human Male* had found "wide variations in sex concepts and behavior." This raised the question: "What is 'normal' and 'abnormal'?" In particular, the report had found that "homosexual experience is much more common than previously thought," and "there is often a mixture of both homo and hetero experience."[14]

Kinsey's counting of orgasms indeed stressed the wide range of behaviors and feelings that fell within the boundaries of a quantitative, statistically accounted heterosexuality. Kinsey's liberal reform of the hetero/homo dualism widened the narrow, old hetero category to accord better with the varieties of social experience. He thereby contradicted the older idea of a monolithic, qualitatively defined, natural procreative act, experience, and person.[15]

Though Kinsey explicitly questioned "whether the terms 'normal' and 'abnormal' belong in a scientific vocabulary," his counting of climaxes was generally understood to define normal sex as majority sex. This quantified norm constituted a final, society-wide break with the old qualitatively defined reproductive standard. Though conceived of as purely scientific, the statistical definition of the normal as the-sex-most-people-are-having substituted a new, quantitative moral standard for the old, qualitative sex ethic—another triumph for the spirit of capitalism.

Kinsey also explicitly contested the idea of an absolute, either/or antithesis between hetero and homo persons. He denied that human beings "represent two discrete populations, heterosexual and homosexual." The world, he ordered, "is not to be divided into sheep and goats." The hetero/homo division was not nature's doing: "Only the human mind invents categories and tries to force facts into separated pigeon-holes. The living world is a continuum."[16]

With a wave of the taxonomist's hand, Kinsey dismissed the social and historical division of people into heteros and homos. His denial of heterosexual and homosexual personhood rejected the social reality and profound subjective force of a historically constructed tradition which, since 1892 in the United States, had cut the sexual populaton in two and helped to establish the social reality of a heterosexual and homosexual identity.

On the one hand, the social construction of homosexual persons has led to the development of a powerful gay liberation identity politics based on an ethnic group model. This has freed generations of women and men from a deep, painful, socially induced sense of shame, and helped to bring about a society-wide liberalization of attitudes and responses to homosexuals.[17] On the other hand, contesting the notion of homosexual and heterosexual persons was one early, partial resistance to the limits of the hetero/homo construction. Gore Vidal, rebel son of Kinsey, has for years been joyfully proclaiming:

> there is no such thing as a homosexual or a heterosexual person. There are only homo- or heterosexual acts. Most people are a mixture of impulses if not practices, and what anyone does with a willing partner is of no social or cosmic significance.
>
> So why all the fuss? In order for a ruling class to rule, there must be arbitrary prohibitions. Of all prohibitions, sexual taboo is the most useful because sex involves everyone.... We have allowed our governors to divide the population into two teams. One team is good, godly, straight; the other is evil, sick, vicious.[18]

HETEROSEXUALITY QUESTIONED:1965–1982

By the late 1960s, anti-establishment counterculturalists, fledgling feminists, and homosexual-rights activists had begun to produce an unprecedented critique of sexual repression in general, of women's sexual repression in particular, of marriage and the family—and of some forms of heterosexuality....

Heterosexual History: Out of The Shadows

Our brief survey of the heterosexual idea suggests a new hypothesis. Rather than naming a conjunction old as Eve and Adam, heterosexual designates a word and concept, a norm and role, an individual and group identity, a behavior and feeling, and a peculiar sexual-political institution particular to the late nineteenth and twentieth centuries.

Because much stress has been placed here on heterosexuality as word and concept, it seems important to affirm that heterosexuality (and homosexuality) came into existence before it was named and thought about. The formulation of the heterosexual idea did not create a heterosexual experience or behavior; to suggest otherwise would be to ascribe determining power to labels and concepts. But the titling and envisioning of heterosexuality did play an important role in consolidating the construction of the heterosexual's social existence. Before the wide use of the word "heterosexual," I suggest, women and men did not mutually lust with the same profound, sure sense of normalcy that followed the distribution of "heterosexual" as universal sanctifier.

According to this proposal, women and men make their own sexual histories. But they do not produce their sex lives just as they please. They make their sexualities within a particular mode of organization given by the past and altered by their changing desire, their present power and activity, and their vision of a better world. That hypothesis suggests a number of good reasons for the immediate inauguration of research on a historically specific heterosexuality.

The study of the history of the heterosexual experience will forward a great intellectual struggle still in its early stages. This is the fight to pull heterosexuality, homosexuality, and all the sexualities out of the realm of nature and biology [and] into the realm of the social and historical. Feminists have explained to us that anatomy does not determine our gender destinies (our masculinities and femininities). But we've only recently begun to consider that *biology does not settle our erotic fates.* The common notion that biology determines the object of sexual desire, or that physiology and society together cause sexual orientation, are determinisms that deny the break existing between our bodies and situations and our desiring. Just as the biology of our hearing organs will never tell us why we take pleasure in Bach or delight in Dixieland, our female or male anatomies, hormones, and genes will never tell us why we yearn for women, men, both, other, or none. That is because desiring is a self-generated project of individuals within particular historical cultures. Heterosexual history can help us see the place of values and judgments in the construction of our own and others' pleasures, and to see how our erotic tastes—our aesthetics of the flesh—are socially institutionalized through the struggle of individuals and classes.

The study of heterosexuality in time will also help us to recognize the *vast historical diversity of sexual emotions and behaviors*—a variety that challenges the monolithic heterosexual hypothesis. John D'Emilio and Estelle Freedman's *Intimate Matters: A History of Sexuality in America* refers in passing to numerous substantial changes in sexual activity and feeling: for example, the widespread use of contraceptives in the nineteenth century, the twentieth-century incitement of the female orgasm, and the recent sexual conduct changes by gay men in response to the AIDS epidemic. It's now a commonplace of family history that people in particular classes feel and behave in substantially different ways under different, historical conditions. Only when we stop assuming an invariable essence of heterosexuality will we begin the research to reveal the full variety of sexual emotions and behaviors.

The historical study of the heterosexual experience can help us *understand the erotic relationships of women and men in terms of their changing modes of social organization*. Such model analysis actually characterizes a sex history well underway. This suggests that the eros-gender-procreation system (the social ordering of lust, femininity and masculinity, and baby-making) has been linked closely to a society's particular organization of power and production. To understand the subtle history of heterosexuality we need to look carefully at correlations between (1) society's organization of eros and pleasure; (2) its mode of engendering persons as feminine or masculine (its making of women and men); (3) its ordering of human reproduction; and (4) its dominant political economy. This General Theory of Sexual Relativity proposes that substantial historical changes in the social organization of eros, gender, and procreation have basically altered the activity and experience of human beings within those modes.

A historical view locates heterosexuality and homosexuality in time, helping us distance ourselves from them. This distancing can help us formulate new questions that clarify our long-range sexual-political goals: What has been and is the social function of sexual categorizing? Whose interests have been served by the division of the world into heterosexual and homosexual? Do we dare not draw a line between those two erotic species? Is some sexual naming socially necessary? Would human freedom be enhanced if the sex-biology of our partners in lust was of no particular concern, and had no name? In what kind of society could we all more freely explore our desire and our flesh?

As we move [into the year 2000], a new sense of the historical making of the heterosexual and homosexual suggests that these are ways of feeling, acting, and being with each other that we can together unmake and radically remake according to our present desire, power, and our vision of a future political-economy of pleasure.

NOTES

1. Barbara Welter, "The Cult of True Womanhood: 1820–1860," *American Quarterly*, vol. 18 (Summer 1966); Welter's analysis is extended here to include True Men and True Love.

2. Some historians have recently told us to revise our idea of sexless Victorians: their experience and even their ideology, it is said, were more erotic than we previously thought. Despite the revisionists, I argue that "purity" was indeed the dominant, early Victorian, white middle-class standard. For the debate on Victorian sexuality see John D'Emilio and Estelle Freedman, *Intimate Matters: A History of Sexuality in America* (New York: Harper & Row, 1988), p. xii.

3. See, for example, Catherine Gallagher and Thomas Laqueur, eds., "The Making of the Modern Body: Sexuality and Society in the Nineteenth Century," *Representations*, no. 14 (Spring 1986) (republished, Berkeley: University of California Press, 1987).

4. This reference to females reminds us that the invention of heterosexuality had vastly different impacts on the histories of women and men. It also differed in its impact

on lesbians and heterosexual women, homosexual and heterosexual men, the middle class and working class, and on different religious, racial, national, and geographic groups.

5. Dr. James G. Kieman, "Responsibility in Sexual Perversion," *Chicago Medical Recorder*, vol. 3 (May 1892), pp. 185–210.

6. R. von Krafft-Ebing, *Psychopathia Sexualis, with Especial Reference to Contrary Sexual Instinct: A Medico-Legal Study*, trans. Charles Gilbert Chaddock (Philadelphia: F. A. Davis, 1892), from the 7th and revised German ed. Preface, November 1892.

7. For the standardization of gender see Lewis Terman and C. C. Miles, *Sex and Personality, Studies in Femininity and Masculinity* (New York: McGraw Hill, 1936). For the standardization of intelligence see Lewis Terman, *Stanford-Binet Intelligence Scale* (Boston: Houghton Mifflin, 1916). For the standardization of work, see "scientific management" and "Taylorism" in Harry Braverman, *Labor and Monopoly Capital: The Degradation of Work in the Twentieth Century* (New York: Monthly Review Press, 1974).

8. See D'Emilio and Freedman, *Intimate Matters*, pp. 194–201, 231, 241, 295–96; Ellen Kay Trimberger, "Feminism, Men, and Modern Love: Greenwich Village, 1900–1925," in *Powers of Desire: The Politics of Sexuality*, ed. Ann Snitow, Christine Stansell, Sharon Thompson (New York: Monthly Review Press, 1983), pp. 131–52; Kathy Peiss, "'Charity Girls' and City Pleasures: Historical Notes on Working Class Sexuality, 1880–1920," in *Powers of Desire*, pp. 74–87; and Mary P. Ryan, "The Sexy Saleslady: Psychology, Heterosexuality, and Consumption in the Twentieth Century," in her *Womanhood in America*, 2nd ed. (New York: Franklin Watts, 1979), pp. 151–82.

9. [Rev. Charles Parkhurst], "Woman. Calls Them Andromaniacs. Dr. Parkhurst So Characterizes Certain Women Who Passionately Ape Everything That Is Mannish. Woman Divinely Preferred. Her Supremacy Lies in Her Womanliness, and She Should Make the Most of It—Her Sphere of Best Usefulness the Home," *The New York Times*, May 23, 1897, p. 16:1.

10. See Lisa Duggan, "The Social Enforcement of Heterosexuality and Lesbian Resistance in the 1920s," in *Class, Race, and Sex: The Dynamics of Control*, ed. Amy Swerdlow and Hanah Lessinger (Boston: G. K. Hall, 1983), pp. 75–92; Rayna Rapp and Ellen Ross, "The Twenties Backlash: Compulsory Heterosexuality, the Consumer Family, and the Waning of Feminism," in *Class, Race, and Sex;* Christina Simmons, "Companionate Marriage and the Lesbian Threat," *Frontiers*, vol. 4, no. 3 (Fall 1979), pp. 54–59; and Lillian Faderman, *Surpassing the Love of Men* (New York: William Morrow, 1981).

11. Louis Kronenberger, review of André Gide, *The Immoralist, New York Times Book Review*, April 20, 1930, p. 9.

12. Henry James Forman, review of Floyd Dell, *Love in the Machine Age* (New York: Farrar & Rinehart), *New York Times Book Review*, September 14, 1930, p. 9.

13. Ferdinand Lundberg and Dr. Marynia F. Farnham, *Modern Woman: The Lost Sex* (New York: Harper, 1947).

14. Dr. Howard A. Rusk, *New York Times Book Review*, January 4, 1948, p. 3.

15. Alfred Kinsey, Wardell B. Pomeroy, Clyde E. Martin, *Sexual Behavior in the Human Male* (Philadelphia: W. B. Saunders, 1948), pp. 199–200.

16. Kinsey, *Sexual Behavior*, pp. 637, 639.

17. See Steven Epstein, "Gay Politics, Ethnic Identity: The Limits of Social Constructionism," *Socialist Review* 93/94 (1987), pp. 9–54.

18. Gore Vidal, "Someone to Laugh at the Squares With" [Tennessee Williams], *New York Review of Books*, June 13, 1985; reprinted in his *At Home: Essays, 1982–1988* (New York: Random House, 1988), p. 48.

31

Get a Life, Girls

ARIEL LEVY

Some version of a sexy, scantily clad temptress has been around through the ages, and there has always been a demand for smut. But whereas this was once a guilty pleasure on the margins—on the almost entirely male margins— now, strippers, porn stars and Playboy bunnies have gone mainstream, writing bestsellers, starring in reality television shows, living a life we're all encouraged to emulate. Prepubescent girls wear "thong" underpants; their mothers drive off to the gym for pole-dancing classes after lunch.

Last week Anita Roddick, founder of the Body Shop, hit out at what she called "pimp and ho chic." "A lot of people seem to think that it's cool to be a pimp or a ho," she said. "But it's not cool. The reality is dark, evil, appalling and unregulated. There are thousands of ads, mostly focused on women and young girls, that say you are not attractive, you are not sexy, you are not intelligent unless you look like this. Something has gone very wrong."

What Dame Anita called "ho chic" I call "raunch culture," and it's every-where. Men and women alike have developed a taste for kitschy, slutty stereo-types of female sexuality—we don't even think about it any more, we just expect to see women flashing and stripping and groaning everywhere we look.

Not so long ago the revelation that a woman in the public eye had appeared in any kind of pornography would have destroyed her reputation. Think of Vanessa Williams, crowned the first black Miss America in 1983, and how quickly she was dethroned after her nude photos surfaced in *Penthouse*. She managed to make a comeback as a singer, but the point is that being exposed as a

SOURCE: *The Spectator*, March 4, 2006, p. 22.

porn star then was something you needed to come back from. Now, it's the comeback itself.

Paris Hilton was just a normal, blonde New York socialite, an heiress with a taste for table-dancing, before she and her former boyfriend Rick Solomon made a video of themselves having sex. Somehow the footage found its way on to the internet and was distributed worldwide, after which Paris Hilton became one of the most recognizable and marketable female celebrities in the world. Since the advent of the sex tapes, Hilton has become famous enough to warrant a slew of endorsement deals. There is a Paris Hilton jewelry line (bellybutton rings feature prominently), a perfume, and a string of nightclubs called Club Paris set to open in London, New York, Atlanta, Madrid, Miami and Las Vegas. She also has a modelling contract for Guess Jeans, and her book, *Confessions of an Heiress,* was a bestseller. Her debut CD—the first single is entitled "Screwed"—is about to be released. Paris Hilton isn't some disgraced exile of our society. On the contrary, she has become our mascot, the embodiment of our collective fixations— blondness, hotness, richness, anti–intellectualism and exhibitionism.

The rise of raunch culture in the West seems counterintuitive to some. What about conservative values and evangelical Christianity? they wonder. But raunch culture transcends political parties both in America and the UK because the values people vote for are not necessarily the same values they live by. Even if people consider themselves conservative, their political ideals may be a reflec- tion of the way they wish things were, rather than an indication of how they plan to lead their lives.

Raunch culture is not essentially progressive; it is essentially commercial. It isn't about opening our minds to the possibilities and mysteries of sexuality. If we were to acknowledge that sexuality is personal and unique, it would become unwieldy. Making sexiness into something simple and quantifiable makes it easier to explain and to market. If you remove the human factor from sex and make it about stuff—big fake boobs, bleached blonde hair, long nails, poles, thongs— then you can sell it. Suddenly, sex requires shopping: you need plastic surgery, peroxide, a manicure, a mall.

There is a disconnection between sexiness, or "hotness," and sex itself. As Paris Hilton told *Rolling Stone,* "My boyfriends always tell me I'm not sexual. Sexy, but not sexual." And any 14-year-old who has downloaded her sex tapes can tell you that Hilton looks excited when she is posing for the camera, bored when she is engaged in actual sex. (In one tape, Hilton took a cellphone call during intercourse.) She is the perfect sexual celebrity for this moment, because our interest is in the appearance of sexiness, not the existence of sexual pleasure.

Passion isn't the point. The glossy, overheated thumping of sexuality in our culture is less about "connection" than consumption. "Hotness" has become our cultural currency, and a lot of people spend a lot of time and a lot of money trying to acquire it. Hotness is not the same thing as beauty, which has been valued throughout history. Hot can mean popular. Hot can mean talked about. But when it pertains to women, hot means two things in particular: fuckable and saleable. These are the literal job criteria for our role models: strippers and porn stars. These are women whose profession is based on faking lust, imitating actual female sexual pleasure and power. If we're all trying to look like porn stars these days (and we are), we're imitating an imitation of arousal. It's a long way from sexual liberation.

This is not a situation foisted upon women. In the West, in the 21st century, we have opportunities and expectations that our mothers never had. We have attained a degree of hard-won (and still threatened) freedom in our personal lives; we are gradually penetrating the highest levels of the workforce; we get to go to college and play sports and be secretaries of state. But to look around, you'd think that all any of us women want to do is to rip off our clothes and shake our butts in men's faces.

So why do we go in for raunch culture? Why, when the feminist movement was supposed to have freed us from stereotypes, have we deliberately embraced them again? The freedom to be sexually provocative or promiscuous is not enough freedom; it is not the only "women's issue" worth paying attention to. And we are not even free in the sexual arena. We have simply adopted a new norm, a new role to play: lusty, busty exhibitionist. There are other choices. It's ironic that we call this "adult" entertainment, when reducing sexuality to implants and polyester underpants is really pretty adolescent.

32

Darker Shades of Queer: Race and Sexuality at the Margins

CHONG-SUK HAN

By now, I've listened with a mild sense of amusement as countless gay leaders, an overwhelmingly white group, attempt to explain that their oppressed status as sexual minorities provides them with an enlightened sense of social justice that enables them to understand the plight of those who are racially oppressed. As sexual minorities, they say they share a history of oppression. This "shared history of oppression," they explain, provides them with exceptional insight and somehow absolves them of blame when it comes to the racial social hierarchy. Certainly, this view is not unique. People of color also believe that they possess special insights when it comes to social justice. Yet, two things are bitterly clear about our "shared" American experiences. One, a shared history of

SOURCE: Tarrant, Shera, ed. 2008. *Men Speak Out: Views on Gender, Sex, and Power.* New York: Routledge.

oppression rarely leads to coalition building among those who have been system-atically denied their rights. More devastatingly, such shared experiences of oppression rarely lead to sympathy for others who are also marginalized, trauma-tized, and minimized by the dominant society. Rather, all too miserably, those who should naturally join in fighting discrimination find it more comforting to join their oppressors in oppressing others. In fact, they trade in one oppressed status for the other. Doing so, many gay white folks become more racist than non-gay folks while many people of color become more homophobic. As a gay man of color, I see this on a routine basis, whether it be racism in the gay community or homophobia in communities of color. And it pisses me off.

Psychologists have theories, I'm sure, about why such things happen. Perhaps some of us feel some comfort in the ability to claim at least a small share of the privileges and benefits of belonging to the so-called majority group. Maybe gay white folks use their race to buy into the mainstream while straight people of color use their sexuality for the same purpose. If they have something in common with those with power, perhaps some of that power will trickle down or rub off on them. I doubt it, but it's easy to believe. It's even comforting. What easier way to make oneself feel better than to marginal-ize others. But for now, it doesn't really matter why they do what they do. I'm not interested in why it happens. Rather, I'm interested in exposing it, con-demning it, shaming it, and stopping it. Many gay activists want to believe that there aren't issues of racism within the gay community. As members of an oppressed group, they like to think that they are above oppressing others. Yet, looking around any gayborhood, something becomes blatantly clear to those of us on the outside looking in. Within the queer spaces that have sprung up in once neglected and forgotten neighborhoods, inside the slick new storefronts, in trendy restaurants, and on magazine covers, gay America has given a whole new meaning to the term "whitewash."

Whiteness in the gay community is everywhere, from what we see, what we experience, and more importantly, what we desire. Media images now popular in television and film such as *Queer as Folk, Queer Eye for the Straight Guy, The L-Word,* and the like promote a monolithic image of the gay community as being overwhelmingly upper-middle class—if not simply rich—and white. These images aren't new. Flipping through late night television, I'm often struck by how white and rich gay characters are whether they are on reruns of the *Golden Girls* or rebroadcasts of *Making Love.* The only difference now is that there are more of them, and they aren't so angst ridden about their sexuality. We now live in a country where gay TV characters are out and proud instead of languishing in perpetual shame, hiding in the shadows. They revel in their sexuality, at least on the screen. I can't help but think, though, that their revelation comes largely from their privilege of whiteness, the privilege to "be like everyone else," with only one minor difference.

Even the most perfunctory glance through gay publications exposes the pau-city of non-white gay images. It's almost as if no gay men or women of color exist outside of fantasy cruises to Jamaica, Puerto Rico, or the "Orient." To the larger gay community, our existence, as gay men and women of color, is merely a footnote, an inconvenient fact that is addressed in the most insignificant and patronizing way. Sometime between Stonewall and *Will and Grace,* gay leaders decided that the best way to be accepted was to mimic upper-middle-class white America. As a sexually marginalized group, the idea is to take what they can from the dominant group and claim it as their own. Rarely a day goes by when I don't read something, by some gay "leader," about the purchasing power of the gay "community."

Sometimes, racism in the gay community takes on a more explicit form aimed at excluding men and women of color from gay institutions. All over the country, gay people of color are routinely asked for multiple forms of identification to enter the most basic of gay premises, the gay bar. Some might argue that this is a minor irritant. But in a country where gays and lesbians are expected to hide their sexuality while in public spaces, gay bars and other gay businesses provide us with the few opportunities to freely be ourselves. Denied access to the bars, gay men and women of color often lose the ability to see and socialize with others like us who also turn to these allegedly safe places for not only their social aspects but for their affirming aspects, as well. Isolated incidents might be easily forgotten, but news reports and buzz on various online forums expose such practices as endemic in gay communities. From New York to Los Angeles, from Seattle to Miami, the borders of gayness are patrolled by those who deny the existence of gay men and women of color. And much like a neighborhood with more than six black families is quickly labeled "black" and, hence, avoided by many white buyers, a bar with more than six non-white folks is quickly labeled "ethnic" and avoided by the white clientele. Except, of course, those on the prowl for something "different" to wet their sexual appetite. In so many ways, their actions are nothing more than a cheap version of sexual tourism.

More importantly, gay men and women of color are routinely denied leadership roles in gay organizations that purport to speak for all of us. In effect, it is the needs and concerns of a largely middle-class gay white community that come to the forefront of what is thought to be a gay cause. Interjecting race in these community organizations is no easy task. On too many occasions, gay men and women of color have been told not to muddy the waters of the primary goal by bringing in concerns that might be addressed elsewhere. When mainstream gay organizations actually address issues of race, gay white men and women continue to set the agenda for what is and is not considered appropriate for discussion. During one community forum on race, the organizers, again an overwhelmingly white bunch, informed the audience that we would not actually talk about racism, as "everyone is capable of racism." With one sweeping generalization, this group of white men trivialized the everyday experiences of the men, and a few women, of color in the audience by denying our personal experiences and by turning the accusation back on us. It's a funny feeling to sit in an audience as an Asian gay man and watch as middle-class white men claim victimhood in racist America. Ultimately, isn't that what their claim that "everyone is capable of racism" boils down to? Isn't what they're really saying that they are also victims of racism?

Sadly, racism isn't just about how we are treated. It's also about how we treat ourselves. The primacy of whiteness in the gay community often manifests as internalized racism. In his 2002 online essay, "No Blacks Allowed," Keith Boykin argues that "in a culture that devalues black males and elevates white males," black men deal with issues of self-hatred that white men do not. Boykin argues that this racial self-hatred makes gay black men see other gay black men as unsuitable sexual partners and white males as the ultimate sexual partners.

This desire for white male companionship is not limited just to black men, and neither is racial self-hatred. Rather, it seems to be pandemic among many gay men of color. Even the briefest visit to a gay bar betrays the dirty secret that gay men of color don't see each other as potential life partners. Rather, we see each other as competitors for the few white men who might be willing to date someone considered lower on the racial hierarchy. We spend our energy and time contributing to the dominance of whiteness by putting white men on the pedestal while ignoring those who would otherwise be our natural allies.

The primacy of white masculinity in the gay community is no accident. It is a carefully choreographed racialization of men of color that mimics the masculine hierarchy in straight communities. Related to the "middle-classing" and "whitening" of gay America, images of gay men have mirrored the mainstream. No longer the Nellie queens of the 1960s, stereotypical images of gay men have changed from swivel-hipped sissies to muscle-bound he-men. So ingrained is the image of the "ideal" man within the gay community, "straight-acting" is a selling point in gay personal ads.

But masculinity doesn't exist in a vacuum. Masculinity is built upon the femininity of Others. Lacking female femininity, the gay community thrusts this role upon Asian men. Given the way that Asian men are gendered in the mainstream, this isn't a difficult task. If gay white men are masculine, they are masculine compared to gay Asian men. If masculinity is desirable and femininity is not, then clearly white men are desirable but Asian men are not.

While black men have escaped the feminization inflicted on Asian men, their masculinity is also heavily gendered. Rather than lacking masculinity, they are hyper-masculine, outside of the norm and, thus, to be feared. Gay Latino men and gay Native American men fare no better in the gay imagination. In the gay white mind, they are exotic beings who exist for the pleasure of white male consumption. Much like heterosexual male claims to the right of sexual consumption of women, many gay white men have claimed the right to pick and choose what they want from their sexual partners by positioning whiteness as a bargaining tool in sexual encounters.

Ironically, we strive for the attention of the very same white men who view us as nothing more than an inconvenience. "No femmes, no fats, and no Asians" is a common quote found in many gay personal ads, both in print and in cyberspace. Gay white men routinely tell us that we are lumped with the very least of desirable men within the larger gay community. In this way, we are reduced to no more than one of many characteristics that are considered undesirable. Rather than confronting this racism, many of my gay Asian brothers have become apologists for this outlandish racist behavior. We damage ourselves by not only allowing it, but actively participating in it. We excuse their racist behavior because we engage in the same types of behavior. When seeking sexual partners for ourselves, we also exclude "femmes, fats, and Asians."

The rationale we use, largely to fool ourselves, to justify the inability of seeing each other as potential partners and allies, is laughable at best. Many Asian guys have told me that dating other Asians would be like "dating [their] brother, father, uncle, etc." Yet, we never hear white men argue that dating other white men would be like dating their brothers or fathers. This type of logic grants individuality to white men while feeding into the racist stereotype that all of "us" are indistinguishable from one another and therefore easily interchangeable.

Some of us rely on tired stereotypes. Boykin writes about the professional gay black man who degrades other black men as being of a lower social class while thinking nothing of dating blue-collar white men. The Asian version is that they are "Americanized" and looking for "American" men. In effect, we help white men oppress other men of color by buying into the racial social structure that the former group has created.

This self-hatred and willingness to join our oppressors in oppressing others is clearly evident in an April 4, 2006, column posted on Advocate.com. Jasmyne Cannick, a self-described black lesbian, argues that we should not extend any rights to "illegal" immigrants until gays are granted full rights. When did equality become a zero-sum game? Does extending human rights to immigrants somehow limit the amount of equal rights left for us? This is as absurd as claiming that extending marriage rights to gays and lesbians would somehow weaken heterosexual marriages. Extending rights to immigrants does not limit our rights in any way. To the contrary, it reinforces the commitment to equality and fairness for everyone, including gays and lesbians. As for Cannick's argument that these immigrants are "illegal," I might remind her that until the Supreme Court ruled on *Lawrence v. Texas* in 2003 millions of us were engaging in the horribly illegal activity of loving someone of the same sex. Thousands of us also got married "illegally." Why does Cannick believe that we have the right to challenge laws that brand us or our actions illegal, while others do not share this right? Clearly, those with the real power at the *Advocate* wanted to make a statement and found a pawn in Cannick. Her willingness to stand so firmly against granting rights to "illegal" immigrants simply reinforces the erroneous belief that it is acceptable for some members of society not to have the exact same rights enjoyed by others. Shamefully and unapologetically, Cannick quotes Audre Lorde at the very same time she uses the master's tools to build a fence around another oppressed group.

If we are invisible in the dominant gay community, perhaps we are doubly so in our own communities of color. If we are a footnote in the gay community, we are an endnote in communities of color—an inconvenient fact that is buried in the back out of view. We are told by family and friends that being gay is a white problem. We are told, early in life, that we must avoid such stigma at all costs. When we try to interject issues of sexuality, we are told that there is precious little time to waste on trivial needs while we pursue racial justice. Cannick writes that "lesbians and gays should not be second-class citizens. Our issues should not get bumped to the back of the line in favor of extending rights to people who have entered this country illegally." I've heard that exact sentence elsewhere. In fact, nearly verbatim, except "entered this country illegally" was replaced with "chosen an immoral lifestyle." See how that works? Master's tools.

I've seen those who are marginalized use the master's tools in numerous instances, now too legion to list. Citing Leviticus, some people of color who are also members of the clergy have vehemently attacked homosexuality as an abomination. This is the same Leviticus that tells us that wearing cloth woven of two fabrics and eating pork or shrimp is an abomination punishable by death. Yet not surprisingly, rarely do Christian fundamentalists picket outside of a Gap or a Red Lobster. If hypocrisy has a border, those yielding Leviticus as their weapon of choice must have crossed it by now. It must be convenient to practice a religion with such disdain that the word of God need only be obeyed when it reinforces one's own hatred and bigotry. How else do we explain those who condemn *Brokeback Mountain* based on their religious views while, in the same breath, praise *Walk the Line,* a movie about two adulterous country singers?

More problematic is that we choose to practice historic amnesia by ignoring the fact that Leviticus was used by slave owners to justify slavery by arguing that God allowed the owning of slaves and selling of daughters. Anti-miscegenation laws, too, were justified using the Bible. In 1965, Virginia trial court judge Leon Bazile sentenced an interethnic couple who were married in Washington, D.C., to a jail term using the Bible as his justification. In his ruling, he wrote, "Almighty God created the races white, black, yellow, malay and red, and he placed them on separate continents. The fact that he separated the races shows that he did not intend for the races to mix." Scores of others also used the story of Phinehas, who distinguished himself in the eyes of God by murdering an interracial couple, thereby preventing a plague to justify their own bigotry. Have we forgotten that the genocide and removal of Native Americans was also largely justified on biblical grounds?

Have we simply decided to pick and choose the parts of the Bible that reinforce our own prejudices and use it against others in the exact same way that it has been used against us? Have we really gotten so adept at using the master's tools that he no longer needs to use them himself to keep us all in our place?

Given the prevalence of negative racial attitudes in the larger gay community and the homophobia in communities of color, gay people of color have to begin building our own identities. For gay people of color to be truly accepted by both the gay community and communities of color, we must form connections with each other first and build strong and lasting coalitions with each other rather than see each other as competitors for the attention of potential white partners. We must begin confronting whiteness where it stands while simultaneously confronting homophobia. More importantly, we must begin doing this within our own small circle of gay people of color. We must confront our own internalized racism that continues to put gay white people on a pedestal while devaluing other gays and lesbians of color. Certainly, this is easier said than done. The task at hand seems insurmountable. In Seattle, a group of gay, lesbian, and transgendered social activists from various communities of color have launched the Queer People of Color Liberation Project. Through a series of live performances, they plan on telling their own stories to counter the master narratives found within the larger gay community and within communities of color.

Certainly, gay people of color have allies both in the mainstream gay community and in our communities of color. Recently, Khalil Hassam, a high school student in Seattle, won a national ACLU scholarship for opposing prejudice. Hassam, the only Muslim student at University Prep High School, decided to fight for justice after a Muslim speaker made derogatory comments about homosexuals. Despite his own marginalized status as a Muslim American, Hassam confronted the homophobia found within his own community. Examples such as these are scattered throughout the country. Nonetheless, there is much more that allies, both straight and gay, can do to promote social justice. We must see gay rights and civil rights not as exclusive, but as complementary. All too often, even those on the left support "other" causes out of a Niemoellerian fear of having no one left to speak up for us if the time should come. I propose that the motivation to join in political efforts should come not from such fears, but from the belief that there are no such "other" causes. Rather, as Martin Luther King, Jr. reminded us, "an injustice anywhere is a threat to justice everywhere." We must remind ourselves, contrary to what Cannick may want us to believe, that social justice is not a zero–sum game. Granting rights to others does not diminish our rights. It is the exact opposite. Ensuring that rights are guaranteed to others ensures that they are guaranteed to us.

Ultimately, the crisis for gay men of color is one of masculinity. The centrality of masculinity within the gay community leads to rejecting all those outside of the masculine norm. Likewise, the centrality of masculinity in communities of color leads to stripping away masculinity based on stereotypical perceptions of homosexuality. Gay men of color are told, by both communities, that we are somehow not masculine enough to be full members in either community. Some of us have tried to attain this mythical masculine norm. We spend hours at the gym toning our bodies and building our muscles to fit with the gay masculine norm. Some of us disguise our speech, alter our style, and watch our steps in an effort to appear more straight to the untrained eye. And while a few of us may escape the cage of the masculine abnormality, we leave our brothers behind. "You're pretty masculine for an Asian guy," we are told. "You don't act like a black guy," they say. Ironically, both of these betray the cage of acceptable masculinity that binds us to the mythical norm. Asian men are not masculine enough, black men are too masculine. The narrow range of acceptable masculinity is reserved for white men—gay or straight—because, ultimately, it benefits them. Rather than see the explicit racist statement embedded in this compliment, we secretly blush and giggle at the attention. I can't speak eloquently or competently about lesbian experiences with the expectations of femininity they experience. Yet, whenever I listen to my gay Asian sisters speak of needing to be more lipstick to attract the butch white woman, I wonder what's happening over there in the lesbian community. It sounds vaguely familiar.

The real solution lies not in mimicking the masculinity found in the larger society, but in abolishing it. We need to think about re-envisioning what it means to be masculine and, for that matter, what it means to be feminine and the social value we place on each of these categories. Sadly, I don't have the answers, at least not yet. But I can't help yet feel that feminist scholars who attack the gendered hierarchy have a point. Perhaps one of them will find the answer someday. Until then, I can continue to make sure that concerns regarding race are brought up in the gay community and homophobia is confronted in communities of color. Perhaps one day, I won't need to do either.

33

Selling Sex for Visas

Sex Tourism as a Stepping-stone to International Migration

DENISE BRENNAN

On the eve of her departure for Germany to marry her German client–turned-boyfriend, Andrea, a Dominican sex worker, spent the night with her Dominican boyfriend. When I dropped by the next morning to wish her well, her Dominican boyfriend was still asleep. She stepped outside, onto her porch. She could not lie about her feelings for her soon-to-be husband. "No," she said, "it's not love." But images of an easier life for herself and her two daughters compelled her to migrate off the island and out of poverty. She put love aside—at least temporarily.

Andrea, like many Dominican sex workers in Sosúa, a small town on the north coast of the Dominican Republic, makes a distinction between marriage *por amor* (for love) and marriage *por residencia* (for visas). After all, why waste a marriage certificate on romantic love when it can be transformed into a visa to a new land and economic security?

Since the early 1990s, Sosúa has been a popular vacation spot for male European sex tourists, especially Germans. Poor women migrate from throughout the Dominican Republic to work in Sosúa's sex trade; there, they hope to

SOURCE: Ehrenreich, Barbara and Arlie Russell Hochschild, eds. 2003. *Global Woman: Nannies, Maids, and Sex Workers in the New Economy.* New York: Metropolitan, pp. 154–168.

meet and marry foreign men who will sponsor their migration to Europe. By migrating to Sosúa, these women are engaged in an economic strategy that is both familiar and altogether new: they are attempting to capitalize on the very global linkages that exploit them. These poor single mothers are not simply using sex work in a tourist town with European clients as a survival strategy; they are using it as an *advancement* strategy.

The key aims of this strategy are marriage and migration off the island. But even short of these goals, Sosúa holds out special promise to its sex workers, who can establish ongoing transnational relationships with the aid of technologies such as fax machines at the phone company in town (the foreign clients and the women communicate about the men's return visits in this manner) and international money wires from clients overseas. Sosúa's sex trade also stands apart from that of many other sex-tourist destinations in the developing world in that it does not operate through pimps, nor is it tied to the drug trade; young women are not trafficked to Sosúa, and as a result they maintain a good deal of control over their working conditions.

Certainly, these women still risk rape, beatings, and arrest; the sex trade is dangerous, and Sosúa's is no exception. Nonetheless, Dominican women are not coerced into Sosúa's trade but rather end up there through networks of female family members and friends who have worked there. Without pimps, sex workers keep all their earnings; they are essentially working freelance. They can choose the bars and nightclubs in which to hang out, the number of hours they work, the clients with whom they will work, and the amount of money to charge.

There has been considerable debate over whether sex work can be anything but exploitative. The stories of Dominican women in Sosúa help demonstrate that there is a wide range of experiences within the sex trade, some of them beneficial, others tragic.... I have been particularly alarmed at the media's monolithic portrayal of sex workers in sex-tourist destinations, such as Cuba, as passive victims easily lured by the glitter of consumer goods. These overly simplistic and implicitly moralizing stories deny that poor women are capable of making their own labor choices. The women I encountered in Sosúa had something else to say.

SEX WORKERS AND SEX TOURISTS

Sex workers in Sosúa are at once independent and dependent, resourceful and exploited. They are local agents caught in a web of global economic relations. To the extent that they can, they try to take advantage of the men who are in Sosúa to take advantage of them. The European men who frequent Sosúa's bars might see Dominican sex workers as exotic and erotic because of their dark skin color; they might pick one woman over another in the crowd, viewing them all as commodities for their pleasure and control. But Dominican sex workers often see the men, too, as readily exploitable—potential dupes, walking visas, means by which the women might leave the island, and poverty, behind.

Even though only a handful of women have actually married European men and migrated off the island, the possibility of doing so inspires women to move to Sosúa from throughout the island and to take up sex work. Once there, however,

Dominican sex workers are beholden to their European clients to deliver visa sponsorships, marriage proposals, and airplane tickets. Because of the differential between sex workers and their clients in terms of mobility, citizenship, and socioeconomic status, these Dominican sex workers might seem to occupy situations parallel to those that prevail among sex workers throughout the developing world. Indeed, I will recount stories here of disappointment, lies, and unfulfilled dreams. Yet some women make modest financial gains through Sosúa's sex trade—gains that exceed what they could achieve working in export-processing zones or domestic service, two common occupations among poor Dominican women. These jobs, on average, yield fewer than 1,000 pesos ($100) a month, whereas sex workers in Sosúa charge approximately 500 pesos for each encounter with a foreign client.

Sex tourism, it is commonly noted, is fueled by the fantasies of white, First World men who exoticize dark-skinned "native" bodies in the developing world, where they can buy sex for cut-rate prices. These two components—racial stereotypes and the economic disparity between the developed and the developing worlds—characterize sex-tourist destinations everywhere. But male sex tourists are not the only ones who travel to places like Sosúa to fulfill their fantasies. Many Dominican sex workers look to their clients as sources not only of money, marriage, and visas, but also of greater gender equity than they can hope for in the households they keep with Dominican men. Some might hope for romance and love, but most tend to fantasize about greater resources and easier lives.

Yet even for the women with the most pragmatic expectations, there are few happy endings. During the time I spent with sex workers in Sosúa, I, too, became invested in the fantasies that sustained them through their struggles. Although I learned to anticipate their return from Europe, disillusioned and divorced, I continued to hope that they would find financial security and loving relationships. Similarly, Sosúa's sex workers built their fantasies around the stories of their few peers who managed to migrate as the girlfriends or wives of European tourists—even though nearly all of these women returned, facing downward mobility when they did so. Though only a handful of women regularly receive money wires from clients in Europe, the stories of those who do circulate among sex workers like Dominicanized versions of Hollywood's *Pretty Woman*.

The women who pursue these fantasies in Sosúa tend to be pushed by poverty and single motherhood. Of the fifty women I interviewed and the scores of others I met, only two were not mothers. The practice of consensual unions (of not marrying but living together), common among the poor in the Dominican Republic, often leads to single motherhood, which then puts women under significant financial pressure. Typically, these women receive no financial assistance from their children's fathers. I met very few sex workers who had sold sex before migrating to Sosúa, and I believe that the most decisive factor propelling these women into the sex trade is their status as single mothers. Many women migrated to Sosúa within days of their partners' departure from the household and their abandonment of their financial obligations to their children.

Most women migrated from rural settings with meager job opportunities, among them sporadic agricultural work, low-wage hairstyling out of one's home, and waitressing. The women from Santo Domingo, the nation's capital, had also held low-paying jobs, working in domestic service or in *zonas francas* (export-processing zones). Women who sell sex in Sosúa earn more money, more quickly than they can in any other legal job available to poor women with limited educations (most have not finished school past their early teens) and skill bases. These women come from *los pobres,* the poorest class in the Dominican Republic, and they simply do not have the social networks that would enable them to land work, such as office jobs, that offer security or mobility. Rather, their female-based social networks can help them find factory jobs, domestic work, restaurant jobs, or sex work.

Sex work offers women the possibility of making enough money to start a savings account while covering their own expenses in Sosúa and their children's expenses back home. These women tend to leave their children in the care of female family members, but they try to visit and to bring money at least once a month. If their home communities are far away and expensive to get to, they return less frequently. Those who manage to save money use it to buy or build homes back in their home communities. Alternatively, they might try to start small businesses, such as *colmados* (small grocery stores), out of their homes.

While saving money is not possible in factory or domestic work, sex workers, in theory at least, make enough money to build up modest savings. In practice, however, it is costly to live in Sosúa. Rooms in boardinghouses rent for 30 to 50 pesos a day, while apartments range from 1,500 to 3,000 pesos a month, and also incur start-up costs that most women cannot afford (such as money for a bed and cooking facilities). Since none of the boardinghouses have kitchens, women must spend more for take-out or restaurant meals. On top of these costs, they must budget for bribes to police officers (for release from jail), since sex workers usually are arrested two to five times a month. To make matters worse, the competition for clients is so fierce, particularly during the low-volume tourist seasons, that days can go by before a woman finds a client. Many sex workers earn just enough to cover their daily expenses in Sosúa while sending home modest remittances for their children. Realizing this, and missing their children, most women return to their home communities in less than a year, just as poor as when they first arrived. . . .

The sex workers I interviewed, who generally have no immediate family members abroad, have never had reliable transnational resources available to them. Not only do they not receive remittances but they cannot migrate legally through family sponsorship. Sex workers' transnational romantic ties act as surrogate family-migration networks. Consequently, migration to Sosúa from other parts of the Dominican Republic can be seen as both internal and international, since Sosúa is a stepping-stone to migration to other countries. For some poor young women, hanging out in the tourist bars of Sosúa is a better use of their time than waiting in line at the United States embassy in Santo Domingo. Carla, a first-time sex worker, explained why Sosúa draws women from throughout the country: "We come here because we dream of a ticket," she said, referring to an airline ticket. But without a visa—which they can obtain through marriage—that airline ticket is of little use.

If sex workers build their fantasies around their communities' experiences of migration, the fantasies sex tourists hope to enact in Sosúa are often first suggested through informal networks of other sex tourists. Sosúa first became known among European tourists by word of mouth. Most of the sex tourists I met in Sosúa had been to other sex-tourist destinations as well. These seasoned sex tourists, many of whom told me that they were "bored" with other destinations, decided to try Sosúa and Dominican women based on the recommendations of friends. This was the case for a group of German sex tourists who were drinking at a bar on the beach. They nodded when the German bar owner explained, "Dominican girls like to fuck." One customer chimed in, "With German women it's over quickly. But Dominican women have fiery blood. . . . When the sun is shining it gives you more hormones."

The Internet is likely to increase the traffic of both veteran and first-time sex tourists to previously little-known destinations like Sosúa. Online travel services provide names of "tour guides" and local bars in sex-tourism hot spots. On the World Sex Guide, a Web site on which sex tourists share information about their trips, one sex tourist wrote that he was impressed by the availability of "dirt cheap colored girls" in Sosúa, while another gloated, "When you enter the discos, you feel like you're in heaven! A tremendous number of cute girls and something for everyone's taste (if you like colored girls like me)!"

As discussions and pictures of Dominican women proliferate on the Internet sites—for "travel services" for sex tourists, pen-pal services, and even cyber classified advertisements in which foreign men "advertise" for Dominican girlfriends or brides—Dominican women are increasingly often associated with sexual availability. A number of articles in European magazines and newspapers portray Dominican women as sexually voracious. The German newspaper *Express* even published a seven-day series on the sex trade in Sosúa, called "Sex, Boozing, and Sunburn," which included this passage: "Just going from the street to the disco—there isn't any way men can take one step alone. Prostitutes bend over, stroke your back and stomach, and blow you kisses in your ear. If you are not quick enough, you get a hand right into the fly of your pants. Every customer is fought for, by using every trick in the book." A photo accompanying one of the articles in this series shows Dieter, a sex tourist who has returned to Sosúa nine times, sitting at a German-owned bar wearing a T-shirt he bought in Thailand; the shirt is emblazoned with the words SEX TOURIST.

With all the attention in the European press and on the Internet associating Dominican women with the sex industry, fear of a stigma has prompted many Dominican women who never have been sex workers to worry that the families and friends of their European boyfriends or spouses might wonder if they once were. And since Dominican women's participation in the overseas sex trade has received so much press coverage in the Dominican Republic, women who have lived or worked in Europe have become suspect at home. "I know when I tell people I was really with a folk-dance group in Europe, they don't believe me," a former dancer admitted. When Sosúans who were not sex workers spoke casually among themselves of a woman working overseas as a domestic, waitress, or dancer, they inevitably would raise the possibility of sex work, if only to rule it out explicitly. One Dominican café owner cynically explained why everyone assumes that Dominican women working overseas must be sex workers: "Dominican women have become known throughout the world as prostitutes. They are one of our biggest exports.". . .

Marginalized women in a marginalized economy can and do fashion creative strategies to control their economic lives. Globalization and the accompanying transnational phenomena, including sex tourism, do not simply shape everything in their paths. Individuals react and resist. Dominican sex workers use sex, romance, and marriage as means of turning Sosúa's sex trade into a site of opportunity and possibility, not just exploitation and domination. But exits from poverty are rarely as permanent as the sex workers hope; relationships sour, and subsequently, an extended family's only lifeline from poverty disintegrates. For every promise of marriage a tourist keeps, there are many more stories of disappointment. Dominican women's attempts to take advantage of these "walking visas" call attention, however, to the savviness and resourcefulness of the so-called powerless.

PART III

The Structure of Social Institutions

MARGARET L. ANDERSEN AND PATRICIA HILL COLLINS

In the United States and globally, social institutions exert a powerful influence on everyday life. Social institutions are also important channels for societal penalties and privileges. The type of work you do, the structure of your family, whether your religion will be recognized or suppressed, the kind of education you receive, the images you see of yourself in the media, and how you are treated by the state are all shaped by the social institutions of the society where you live. People rely on institutions to meet their needs, although social institutions treat some groups better than others. When a specific institution (such as the economy) fails them, people often appeal to another institution (such as the state) for redress. In this sense, institutions are both sources of support and sources of repression.

The concept of an institution is abstract because there is no thing or object that one can point to as an institution. *Social institutions* are the established societal patterns of behavior organized around particular purposes. The economy is an institution, as are the family, the media, education, health care, and the state—important societal institutions examined here. Each is organized around a specific purpose; in the case of the economy, for example, its purpose is the production, distribution, and consumption of goods and services. Within a given institution, there may be various patterns, such as different family structures, but as a whole, institutions are general patterns of behavior that emerge because of the specific societal conditions in which groups live. Institutions do change over time because societal conditions evolve and groups challenge specific institutional structures. Yet institutions are also enduring and persistent, even in the face of active efforts to change them. Institutions confront people from birth and live on after people die.

Across all societies, institutions are patterned by intersections of race, class, gender, sexuality, age, ethnicity, and disability (among others), and, as a result, the effects of these systems of power on social institutions differ from one society to the next. Each society has a distinctive history and institutional configuration of social inequalities. In the United States, race, class, and gender constitute fundamental categories that shape American social institutions that serve as basic conduits for social inequalities. Moreover, social institutions are interconnected. We think the interrelationships among institutions are especially apparent when studied in the context of race, class, and gender. Race, class, and gender oppression rests on a network of interconnected social institutions. Understanding the interconnections among institutions helps us see that we are all part of one historically created system that finds structural form in interrelated social institutions.

Despite their significance, dominant American ideology portrays institutions as neutral in their treatment of different groups; indeed, the liberal framework of the law allegedly makes access to public institutions (such as education and work) gender and race blind. Still, institutions differentiate on the basis of race, class, and gender. As the readings included here show, institutions are actually structured on the basis of race, class, and gender relations.

As an example, think of the economy. Economic institutions in American society are founded on capitalism—an economic system based on the pursuit of profit and the principle of private ownership. Such a system creates class inequality because, in simple terms, the profits of some stem from the exploitation of the labor of others. Race and gender further divide the U.S. capitalist economy, resulting in labor and consumer markets that routinely advantage some and disadvantage others. In particular, corporate and government structures create jobs for some while leaving others underemployed or without work.

Those with jobs encounter a dual labor market that includes: (1) a primary labor market characterized by relatively high wages, opportunities for advancement, employee benefits, and rules of due process that protect workers' rights; and, (2) a secondary labor market (where most women and minorities are located) characterized by low wages, little opportunity for advancement, few benefits, and little protection for workers. One result of the dual labor market is the persistent wage gap between men and women (see Figure 1) and between Whites and people of color—even when they have the same level of education. Women tend to be clustered in jobs that employ mostly women workers; such jobs have been economically devalued as a result. Gender segregation and race segregation intersect in the dual labor market, so women of color are most likely to be working in occupations where most of the other workers are also women of color. At the same time, men of

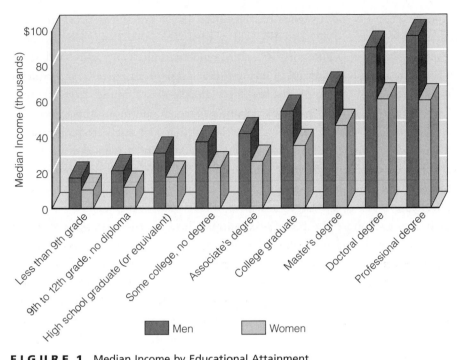

FIGURE 1   Median Income by Educational Attainment
SOURCE: U.S. Census Bureau. 2005. *Historical Income Tables*. www.census.gov.

color are also segregated into particular segments of the market. Indeed, there is a direct connection between gender and race segregation and wages because wages are lowest in occupations where women of color predominate (U.S. Department of Labor 2005). This is what it means to say that institutions are structured by intersections of race, class, and gender: institutions are built from and then reflect the historical and contemporary patterns of race, class, and gender relations in society.

As a second example, think of the state. The state refers to the organized system of power and authority in society. This includes the government, the police, the military, and the law (as reflected in both social policy and the civil and criminal justice system). The state is supposed to protect all citizens, regardless of their race, class, and gender (as well as other characteristics, such as disability and age); yet state policies routinely privilege men. This means not only that the majority of powerful people in the state are men (such as elected officials, judges, police, and the military) but, just as important, that the state works to protect men's interests. Policies about reproductive rights provide a good example. They are largely enacted by men but have an important effect on women. Similarly, welfare policies designed to encourage people to work are based on the model of men's experiences because they presume that staying home to care for one's children is not working. In this sense, the state is a gendered institution.

Most people do not examine the institutional structure of society when thinking about intersections of race, class, and gender. The individualist framework of dominant American culture sees race, class, gender, ethnicity, sexuality, and age as attributes of individuals, instead of seeing them as systems of power and inequality that are embedded in institutional structures. People do, of course, have identities of race, class, and gender, and these identities have an enormous impact on individual experience. However, seeing race, class, and gender solely from an individual viewpoint overlooks how profoundly embedded these identities are in the structure of American institutions. Moving historically marginalized groups to the center of analysis clarifies the importance of social institutions as links between individual experience and larger structures of race, class, and gender.

In this section of the book, we examine how social institutions structure systems of power and inequality of race, class, gender, ethnicity, and sexuality that, in turn, structure those very same social institutions. We look at five important social institutions: work and the economic system; families; cultural institutions such as the media that reproduce ideas; education and health; and the state and social welfare policies. Each institution can be shown to have a unique impact on different groups (such as the discriminatory treatment of African American men by the criminal justice system). At the same time, each institution and its interrelationship with other institutions can be seen as specifically structured through intersecting dynamics of race, class, and gender.

WORK AND ECONOMIC TRANSFORMATION

Structural transformations in the global economy have dramatically changed the conditions under which all people work. Three basic transformations currently affect the character of work. First, employers have turned to new technologies and job export to cut the costs of labor and to increase profitability. In some cases, new technologies have made some jobs obsolete. In other cases, where technology could not replace workers, jobs have been exported to countries where workers work for lower wages. Race, class, and gender frame how individual workers encounter these dual processes of "de-skilling" and job export.

Second, the new global economy has spurred the growth of the service sector. Here the dual economy operates by creating high-paying service work for skilled, college-educated workers (accountants, marketing representatives, etc.) and low-paying service work for everyone else (food service workers, nursing home aides, child care workers, and domestic workers, for example). With this growing service sector comes a shift in the kinds of skills required of workers. Here, too, race, class, and gender intersect in privileging and penalizing groups of workers. In "Racism in Toyland," Christine L. Williams shows how the labor market within this new service section reflects intersecitons of race, class, and gender. In "'Soft' Skills and Race," Philip Moss and Chris Tilly show how the "soft skills" of having the ability to interact with a range of people and demonstrating motivation to employers are increasingly important skills in the service economy. African American men, in particular, are perceived as lacking these skills and as a result are less likely to be hired for certain types of jobs.

Finally, global economic interdependence means that national borders no longer match up with corporations and the products they make. Instead, products may be manufactured and assembled in a variety of places. This has fostered major population migrations; workers migrate in search of jobs and economic opportunities. The influx of new immigrant workers is nothing new in the United States. Domestic service captures these trends of "de-skilling," the growth of the service sector, and migration in search of economic opportunity. Service work has long served as an entry point to the labor market for poor women and women of color, and immigrant women from a few non-European nations are the newest generation of domestic workers (Hondagneu-Sotelo 2001).

These changes have brought new opportunities for some groups—those positioned to benefit from the changes—while, for others, they have meant massive economic dislocation. The 30 million people who make up the working poor are the one group most affected by these economic trends. Katherine S. Newman points out in "The Invisible Poor" that powerful media images erroneously equate poverty with Black, undeserving welfare mothers. Instead, people on welfare are not the largest group of poor people in the United States. Rather, the working poor whose earnings do not allow them to afford decent housing, health care, child care, and nutrition constitute the largest group. Many are young workers or mothers with children, or they are African Americans, Latinos, or immigrants. Moreover, economic dislocation is not confined to workers in bad jobs. Kenneth W. Brown's experiences with "The Indignities of Unemployment" show that even workers in good jobs can be negatively affected by these economic trends.

FAMILIES

Families are another primary social institution profoundly influenced by intersecting systems of race, class, and gender. Historically, the family has been presumed to be the world of women: the ideology of the family (i.e., dominant belief systems about the family) purports that families are places for nurturing, love, and support—characteristics that have been associated with women. This ideal identifies women with the private world of the family and men with the public sphere of work. Family ideology, of course, only projects an ideal because the large number of working mothers shows that few families actually fit the presumed ideal. Nonetheless, the ideology of the family provides a standard against which all families are judged. Moreover, the ideology of the family is class and race specific; that is, it ignores and distorts the family experiences of African Americans, Latinos, and most Whites.

Bonnie Thornton Dill's essay, "Our Mothers' Grief: Racial-Ethnic Women and the Maintenance of Families," shows that for African American, Chinese American, and Mexican American women, family structure is deeply affected by the relationships of families to the structures of race, class, and gender. Dill's historical analysis of racial-ethnic women and their families in "Our Mothers' Grief" examines diverse patterns of family organization directly influenced by a group's placement in the larger political economy. Just as the political economy of the nineteenth century affected women's experience in families, the political economy of the late twentieth century shapes family relations for women and men of all races.

The articles in this section illustrate several points that have emerged from feminist studies of families (Thorne and Yalom 1992). First, the family is not monolithic. The now widely acknowledged diversity among families refutes the idea that there is a normative family: White, middle-class, with children, organized around a heterosexual married couple (preferably with the wife not employed), needing little support from relatives or neighbors. As the experiences of African Americans, Latinos, Asian Americans, lesbians, gays, and others reveal, this so-called "normal" family actually represents a minority perspective. Traditional myths about the American family mask the diversity of families in the United States. For example, one important myth about all families is that they are racially homogeneous. As Erica Chito Childs points out in "Navigating Interracial Borders: Black-White Couples and Their Social Worlds," many individuals challenge the myth by marrying across racial boundaries. At the same time, Childs notes that navigating the Black/White border can encounter far more resistance than expected (Childs 2005).

Second, in this context of the idealized family, the best way to analyze families is to look at their underlying structures. For example, many presume that families are formed around a heterosexual norm. But as Kath Weston points out in "Straight Is to Gay as Family Is to No Family," gays and lesbians also participate in their families of origin and create families of their own. Claiming sexual identities as gay, lesbian, or bisexual does not mean relinquishing family or, worse yet, being defined as anti-family or as being a threat to family. Gay and lesbian partners often form permanent relationships, often including children. Although such partnerships are not uniformly legally recognized, they are part of the diversity among families. Raising children in diverse family environments so that they appreciate not only their own families, but also diverse ways of forming families, is a challenge in a society that devalues gays and lesbians and people of color.

Third, the dominant ideology that has glorified the family routinely distorts the actual behavior of women, people of color, children, and the elderly. Annette Lareau's study ("Unequal Childhoods") of how race and class foster unequal childhoods among America's children examines one such distortion. Lareau shows how children from different social classes and racial groups encounter different opportunity structures. The very concept of American childhood varies tremendously by race, class, and gender.

A fourth theme that has emerged from new studies of families concerns family boundaries. The boundaries separating family, work, and other institutions are more fluid than was previously believed. No institution is isolated from another. By detailing patterns of domestic work done by new immigrant women of color, Pierrette Hondagneu-Sotelo's article "Doméstica" illustrates the permeable boundaries between work and family. In essence, the belief in the idealized family form achived by middle-class families is possible only through the domestic work of women of color. Changes in the economy, for example, affect family structures; moreover, this effect is reciprocal because changes in the family generate changes in the economy. The situation of working mothers challenges this assumption of the separation of work and family. The behavior of employed women of color, that the presumed dichotomy between the public and private spheres is false, is especially evident in light of the experience of people of color. Racial-ethnic families have rarely been provided the protection and privacy that the alleged split between public and private assumes. From welfare policy to reproductive rights, this dichotomy is based on the myth that families are insulated from the society around them.

Recentering one's thinking about the family by understanding the interconnections of race, class, and gender reveals new understandings about families and the myths that have pervaded assumptions about family experience. In particular, ideology that presents families as peaceful, loving, nurturing spaces masks the underlying conflicts that are embedded in family systems. Although we do not examine family violence directly here, we know that violence stems from the race, class, and gender conflicts that are encouraged in a racist, class-based, and sexist society. In addition, we think of family violence not only as that which occurs between family members but also as that which happens to families as they confront the institutional structures of race, class, and gender oppression.

MEDIA AND POPULAR CULTURE

Work, family, and other social institutions are patterned by race, class, and gender, yet many people do not see these patterns and even may claim that they do not exist. The effects of economic restructuring on people of color and women and the differences manifested as Americans try to meet the American family ideal can be difficult to understand. How can race, class, and gender exert such a powerful influence on our everyday lives, yet remain so invisible?

The products of cultural institutions, such as movies, books, television, advertisements in the mass media, the statistics and documents produced by government, and the textbooks, curriculum materials, and teaching methods of schools, explain in part why race, class, and gender relations are obscured. These and other cultural institutions reproduce ideas by identifying which ideas are valuable, which are not, and which should not be heard at all. In this way, the ideas of groups that are privileged within race, class, and gender relations are routinely heard, whereas the ideas of groups who are disadvantaged are silenced. Rebecca Brasfield's article "Re-Reading 'Sex in the City': Exposing the Hegemonic Feminist Narrative" points out that while the popular television show *Sex in the City* presents images of female empowerment, the show privileges the experiences of White middle-class women. It also silences and distorts the experiences of historically marginalized groups.

In part, invisibility and silencing result from deeply entrenched ideas about race, class, and gender in the social fabric of the United States. Because beliefs about race, class, and gender are so ordinary, they often go unnoticed. Robert B. Moore points this out in his essay "Racist Stereotyping in the English Language." Moore's analysis of the negative connotations associated with uses of the term *Black* and the almost uniformly positive connotations that accompany the term *White* reveal a pattern that, for most people, is hidden in plain sight. In everyday social interaction, people invoke these negative and positive meanings associated with "Black" and "White" without stopping to think about them at all. This everyday use obscures the historical origins of these meanings and makes the meanings of *Black* and *White* appear to be natural and timeless rather than socially constructed.

Mass media and similar cultural institutions have the ability to do harm to those stigmatized by them. The articles in this section remind us that the ideas produced by these cultural institutions are far from benign because they uphold social policies and affect individuals. Placing his analysis in a global human rights context, Ward Churchill ("Crimes against Humanity") discusses how American national identity has been constructed against the interests of Native Americans. He suggests that while Native American symbolism and imagery is ubiquitous in the United States, the derogatory use of Native American mascots and the like for sports teams stereotypes and stigmatizes contemporary Native Americans. Ironically, while Native Americans continue to experience what Churchill discusses as genocidal policies, distorted ideas about Native American culture continue to be celebrated at college and professional sporting events. Churchill feels so strongly about the damage done to Native Americans by pervasive negative symbolism that he views this practice as criminal.

Certain cultural institutions have gained greater importance in the context of global economic restructuring. With the emergence of a global mass media that circulates via television, music videos, cable, radio, and the Internet to all parts of the globe, racist and sexist ideas can travel quickly. Thus, those who control mass media are in a position to shape the basic ideas of society that come to be seen as truth. Gregory Mantsios's essay "Media Magic: Making Class Invisible" explores how mass media have been central in shaping ideas about social class. His work identifies the growing significance of mass media in framing how people think about race, class, and gender.

Mass media and other cultural institutions can help reproduce and mask social inequalities of race, class, and gender, but these same cultural institutions can be sites for challenging longstanding oppressions. In this sense, cultural institutions are rarely all good or all bad—more often, they are a little bit of both. The music industry generally and hip hop culture in particular illustrate these tendencies. Through their lyrics and their endorsement of liquor, gym shoes, and lavish lifestyles, many hip hop artists uphold prevailing ideas about race, class, and gender hierarchies. At the same time, hip hop has sparked a renewed activism in many African American and Latino youth. Imani Perry ("Who(se) Am I? The Identity and Image of Women in Hip-Hop") shows how female hip hop artists present contradictory messages about women's empowerment. Some Black female artists such as Eve create liberatory feminist messages that attend to issues of domestic violence and female empowerment. Yet the musical text and images in their videos silmultaneously support the objectification of Black women's bodies. Perry asserts that these contradictions between feminism and female subjugation go beyond the artist and instead reflect tensions between music as commodity, artistic creation, and the images made by record companies. Thus, mass media images are often the site of struggles between those who hold traditional views of race, class, and gender oppression as natural and inevitable and people of color, women, poor people, and other historically excluded groups who challenge these inequalities.

EDUCATION AND HEALTH

Intersecting power relations of race, class, and gender are fundamental in the construction of the American nation-state and its policies, just as the state is fundamental in shaping and defining these categories. Because the state is the system of legitimated power and authority in society, government policies determine people's rights. As the history of Jim Crow legislation shows, the U.S.

government has often upheld exclusionary practices. At the same time, as the Civil Rights Movement and subsequent civil rights legislation of the 1960s demonstrates, the state can be an important, albeit reluctant, ally in challenging unjust social policies.

Social policies can be developed on behalf of disadvantaged groups, and historically the state has had a significant role in developing such policies. The state can negotiate groups' rights and organize people's access to state and societal resources. Government can solve problems that individuals cannot solve on their own, although many people are now cynical about the state's ability to do so. Public confidence in government has eroded, with many observers believing that the government is incapable of solving social problems that they see as being caused by individual failure.

In this context, because ideally education prepares people to compete within the job market, the importance of education as a responsibility of the state has increased. Yet, American education has recently and frequently been described as an institution in crisis. People worry that children are not learning in school. High school dropout rates, especially among students from poor and working-class families of color, reveal deep problems in the system of education. In "Can Education Eliminate Race, Class, and Gender Inequality?" Roslyn Arlin Mickelson and Stephen Samuel Smith show how schools promote inequalities of race, class, and gender. Their analysis reveals the interplay between school policies that perpetuate existing hierarchies (such as credential inflation) and belief systems about race, class, and gender that are embedded in the so-called hidden curriculum. Students encounter these structures and belief systems while young, when their identities assume major importance. Schools thus become key sites where the inequalities of race, class, and gender are fostered and resisted.

Historically, social policies of racial and class segregation have been one important way that schools have fostered inequalities. Yet the elimination of formal segregation does not mean that desegregation as a policy yields educational equity. As Gary Orfield and Chungmei Lee point out in "Why Segregation Matters: Poverty and Educational Inequality," although actual desegregation policies were generated to attack schooling inequalities more than fifty years ago, racial isolation in schools persists, particularly among White students. In fact, Orfield and Lee assert that White students are more likely to attend racially isolated schools than minority students. The substantial concentration of Whites in certain schools provides White students with privileges not afforded to Black and Latino students who are more likely to attend schools that have high concentrations of poverty.

Schools also serve as sites of military recruitment. In "How a Scholarship Girl Becomes a Soldier: The Militarization of Latina/o Youth in Chicago

Public Schools," Gina Perez shows how working-class Puerto Rican and Latino youth in Chicago have become targets of who gets recruited into the military. The recruitment of racial-ethnic youth is not new, but Perez shows how the practices of military programs such as the Junior Reserve Officer Tranining Program (JROTC) illustrate intersections of race, class, gender, and sexuality. In the past, the army primarily recruited young Latino men who were considered "at risk" of falling into gangs and drug use. Today, the military also targets young Latinas, arguing that as women, they are "at risk" for unwed teenage pregnancy. Because the military seemingly offers more opportunities to escape poverty than college, Perez shows how the new push to recruit Latina youth reinforces structures of race, class, and gender inequalities. Yet these activities of the state remain largely invisible.

This connection between schooling and the military taps significant issues beyond providing employment opportunities for young working-class men and women. The military is also especially important in that it constitutes the legitimate use of force, typically in defense of American foreign policy. As the Perez article illustrates, race, class, and gender affect the composition of who joins the military and who goes to fight America's wars. Working-class kids from small towns and inner cities alike join the military, primarily in search of opportunities that elude them in their home communities. But if ideas about American citizenship are refracted through the lens of race, class, gender, ethnicity, and sexuality, one might ask who they defend and toward what ends? Working-class kids fight America's wars, yet who decides what those wars shall be? These are tough questions that get at the heart of the connections between schooling, employment, opportunity stuctures, and state policies. They are questions that are unlikely to be answered without attending to the complexities of race, class, and gender arguments.

The health care system also serves as an important area of struggle where intersections of race, class, and gender within American institutions are largely invisible. In "Life at the Top in America Isn't Just Better, It's Longer," Janny Scott shows how social class shapes understandings of illness, as well as the treatment and support people acquire in dealing with illness. Tracing an architect, utility worker, and a maid, who all expereinced heart attacks within the same time frame, Scott shows how social class greatly affected the life chances of each. Addressing conditions such as these requires a better understanding of how race, class, and gender create health disparities. In "Intersections of Race, Class, and Gender in Public Health Interventions," Amy J. Schulz, Nicholas Freudenberg, and Jessie Daniels present two case studies that showcase how scholars, activists, community members, and pratitioners employ an intersectional framework to lessen health disparities for Black men at risk for HIV/AIDS and Latino populations who face environmental degradation in their communities. Schulz, Freudenberg, and Daniels's analysis of health interventions provides us with a path to examine the potential for change when actors of the state and community join forces to end health disparities.

SOCIAL POLICIES, THE STATE, AND VIOLENCE

Because the state is the system of legitimated power and authority in society, government policies are important in shaping our understanding of important social issues. The violence permeating the United States is one such pressing social issue. School shootings, snipers, and street criminals dominate the news and project the idea that violence is rampant. Usually violence is depicted as the action of crazed individuals who are socially maladjusted, angry, or desperate. Although this may be the case for some, deeper questions about violence are raised when we think about violence in the context both of race, class, and gender relations and of state power. Who is perceived as violent shifts, as does who we see as victims of violence. The consequences of violence in society are numerous, not the least of which is the fear that images of violence create and the action of the state toward those perceived as violent. How can we apply our knowledge of race, class, and gender to understanding violence?

It is important to see violence not as acts of individual social deviance but as one outcome of how the state sanctions a politics of race, class, and gender that permeates a range of social institutions. C. Matthew Snipp's essay "The First Americans: American Indians" details the historical role of the state in forcing Native Americans from their homes and subordinating them to a new system of state control. Genocide, forced relocation, and regulation by state law have shaped Native American experience; state processes, whether in the form of war, legislation, or social policy, have contributed to the current place of Native Americans in U.S. society. At the same time, as Snipp notes, groups are resilient as they face adversity and often turn to the state for assistance. The history of civil rights legislation shows that the state can be an invaluable resource for addressing the wrongs of race, class, and gender injustice. In this sense, the state is both a source of oppression and an avenue for seeking justice.

Violent acts, the threat of violence, and more generalized policies based on the use of force find organizational homes in state institutions of social control—primarily the police, the military, and the criminal justice system. Intersections of race, class, and gender shape all of these institutions. In "Policing the National

Body: Sex, Race, and Criminalization," Jael Silliman demonstrates how intersections of race, class, and gender shape state power. Silliman links policies such as aggressive policing and the increasing incarceration of African American, Latino, and poor White men; the assault on reproductive rights of women of color; and the search of immigrants of color by the Immigration and Naturalization Service under the pretext of national security to an overarching system of policing the body politic.

Violence simultaneously permeates a range of other social institutions. Some social institutions are known for their use of force, while others are less peaceful than they appear. Men's violence against women and children (whether physical, emotional, or sexual) in families of all social classes and racial-ethnic compositions contradicts the dominant belief that the home is a place of tranquility and love. Sexual harassment on the job as a dimension of social control belies the belief that men and women encounter a similar work environment. In the media, depicting rape as pleasurable to women and portraying violence against Native Americans, Asian Americans, and African Americans in numerous movies as justified contributes to a generalized belief system condoning violence. Thus, violence is both hated and condoned in a society where the meaning, commission, and consequences of violence are deeply tied to race, class, and gender inequality.

When it comes to violence, it is important to remember that the police, the military, and the criminal justice system play an important part in punishing certain forms of violence while ignoring others. Through the policies themselves and differential enforcement of policies, state institutions shape what counts as violence. In this way, state policies link individual acts of violence to the more routine, systemic violence that reflects social inequalities of race, class, and gender. To illustrate this historically, lynchings of African American men can be seen as the random acts of unruly mobs; yet, the failure to arrest, prosecute, and convict those who did the lynching is a powerful testament to public endorsement of these seemingly individual acts of violence. Likewise, rape appears to be a private act, but it occurs as part of a generalized climate that condones violence against women. Yet race, class, and gender have shaped the very definition of *rape* and the treatment afforded rape survivors and their rapists. In "Rape, Racism, and the Law," Jennifer Wriggins examines how the legal system's treatment of rape has disproportionately targeted Black men for punishment and made Black women especially vulnerable by denying their sexual subordination. Although violence may be experienced individually, it occurs in specific organizational and institutional contexts shaped by race, class, and gender.

The state may also generate social policies that mask the kinds of violence encountered by its most vulnerable citizens. Through a discussion of eugenics, the authors of "Race, Poverty and Disability: Three Strikes and You're Out! Or Are You?" examine how common ways of thinking about disability intersect with similar thinking about race and poverty. Proponents of eugenics believed in "higher" and "lower" races and supported state-sponsored programs to control the population of the lower race. The article details how eugenics thinking may have disappeared from official public policy, yet its spirit persists in influencing contemporary public policy toward people of color living with disabilities in poverty.

Finally, one must ask, how can the United States build a national community that adheres to principles of fairness and justice within a global context? Thinking about state power and violence brings a new dimension to this complex question. After the events of 9/11, most in the United States felt fearful about violence—particularly terrorism and international war. For many in the United States who had been less touched in the past by threats of violence, these were new fears. But feeling unsafe is not new for many—those who, because of their class, their race, their sexual orientation, their gender, and/or their disability status have never been safe in the United States. For members of these groups, cries to defend freedom and democracy in the face of terrorist threats must be accompanied by the realization that basic social services, primarily, security, have long been denied to many living in the United States. The point is not to dismiss people's fears about new threats but to craft social policies that recongize bona fide threats both within the United States and in the world.

In this context, Rita Arditti's article "Women's Human Rights: It's About Time!" provides a provocative set of ideas for thinking about new directions. Arditti argues that shifting the focus from within U.S. borders to the global human rights context offers new possibilities to us all. She points out that much work in the global context recognizes the rights of all people, regardless of the categories of race, class, and gender detailed here. Moreover, she shows us how safety, security, and freedom from risk occurs not through denying the reality of categories of race, class, and gender but by embracing them.

REFERENCES

Childs, Erica Chito. 2005. *Navigating Interracial Borders: Black-White Couples and Their Social Worlds*. New Brunswick, NJ: Rutgers University Press.

Hondagneu-Sotelo, Pierrette. 2001. *Doméstica: Immigrant Workers Cleaning and Caring in the Shadows of Affluence*. Berkeley: University of California Press.

Thorne, Barrie, with Marilyn Yalom. 1992. *Rethinking the Family: Some Feminist Questions,* Rev. ed. Boston: Northeastern University Press.

U.S. Department of Labor. January 2005. *Employment and Earnings*. Washington, D.C.: U.S. Government Printing Office.

34

Race, Class, Gender, and Women's Works

TERESA AMOTT AND JULIE MATTHAEI

What social and economic factors determine and differentiate women's work lives? Why is it, for instance, that the work experiences of African American women are so different from those of European American women? Why have some women worked outside the home for pay, while others have provided for their families through unpaid work in the home? Why are most of the wealthy women in the United States of European descent, and why are so many women of color poor? . . .

Throughout U.S. history, economic differences among women (and men) have been constructed and organized along a number of social categories. In our analysis, we focus on the three categories which we see as most central—gender, race-ethnicity, and class with less discussion of others, such as age, sexual preference, and religion. We see these three social categories as interconnected, historical processes of domination and subordination. Thinking about gender, race-ethnicity, and class, then, necessitates thinking historically about power and economic exploitation. . . .

GENDER, RACE-ETHNICITY, AND CLASS
PROCESSES: HISTORICAL AND INTERCONNECTED

The concepts of gender, race-ethnicity, and class are neither trans-historical nor independent. Hence, it is artificial to discuss them outside of historical time and place, and separately from one another. At the same time, without such a set of concepts, it is impossible to make sense of women's disparate economic experiences.

Gender, race-ethnicity, and class are not natural or biological categories which are unchanging over time and across cultures. Rather, these categories are socially constructed: they arise and are transformed in history, and themselves transform history. Although societies rationalize them as natural or god-given,

SOURCE: From Teresa Amott and Julie Matthaei, *Race, Gender, and Work: A Multicultural Economic History of Women in the United States*, 2d ed. (New York: Monthly Review Press, 1996). Reprinted with permission of South End Press.

ideas of appropriate feminine and masculine behavior vary widely across history and culture. Concepts and practices of race-ethnicity, usually justified by religion or biology, also vary over time, reflecting the politics, economics, and ideology of a particular time and, in turn, reinforcing or transforming politics, economics, and ideology. For example, nineteenth-century European biologists Louis Agassiz and Count Arthur de Gobineau developed a taxonomy of race which divided humanity into separate and unequal racial species; this taxonomy was used to rationalize European colonization of Africa and Asia, and slavery in the United States. Class is perhaps the most historically specific category of all, clearly dependent upon the particular economic and social constellation of a society at a point in time. Still, notions of class as inherited or genetic continue to haunt us, harkening back to earlier eras in which lowly birth was thought to cause low intelligence and a predisposition to criminal activity.

Central to the historical transformation of gender, race-ethnicity, and class processes have been the struggles of subordinated groups to redefine or transcend them. For example, throughout the development of capitalism, workers' consciousness of themselves as workers and their struggles against class oppression have transformed capitalist-worker relationships, expanding workers' rights and powers. In the nineteenth century, educated white women escaped from the prevailing, domestic view of womanhood by arguing that homemaking included caring for the sick and the needy through volunteer work, social homemaking careers, and political organizing. In the 1960s, the transformation of racial-ethnic identity into a source of solidarity and pride was essential to movements of people of color, such as the Black Power and American Indian movements.

Race-ethnicity, gender, and class are interconnected, interdetermining historical processes, rather than separate systems. This is true in two senses, which we will explore in more detail below. First, it is often difficult to determine whether an economic practice constitutes class, race, or gender oppression: for example, slavery in the U.S. South was at the same time a system of class oppression (of slaves by owners) and of racial-ethnic oppression (of Africans by Europeans). Second, a person does not experience these different processes of domination and subordination independently of one another; in philosopher Elizabeth Spelman's metaphor, gender, race-ethnicity, and class are not separate "pop-beads" on a necklace of identity. Hence, there is no generic gender oppression which is experienced by all women regardless of their race-ethnicity or class. As Spelman puts it:

> . . . in the case of much feminist thought we may get the impression that a woman's identity consists of a sum of parts neatly divisible from one another, parts defined in terms of her race, gender, class, and so on. . . . On this view of personal identity (which might also be called pop-bead metaphysics), my being a woman means the same whether I am white or Black, rich or poor, French or Jamaican, Jewish or Muslim.[1]

The problems of "pop-bead metaphysics" also apply to historical analysis. In our reading of history, there is no common experience of gender across race-ethnicity and class, of race-ethnicity across class and gender lines, or of class across race-ethnicity and gender.

With these caveats in mind, let us examine the processes of gender, class, and race-ethnicity, their importance in the histories of women's works, and some of the ways in which these processes have been intertwined.

Gender

. . . Gender differences in the social lives of men and women are based on, but are not the same thing as, biological differences between the sexes. Gender is rooted in societies' beliefs that the sexes are naturally distinct and opposed social beings. These beliefs are turned into self-fulfilling prophecies through sex-role socialization: the biological sexes are assigned distinct and often unequal work and political positions, and turned into socially distinct genders.

Economists view the sexual division of labor as central to the gender differentiation of the sexes. By assigning the sexes to different and complementary tasks, the sexual division of labor turns them into different and complementary genders. . . . The work of males is at least partially, if not wholly, different from that of females, making "men" and "women" different economic and social beings. Sexual divisions of labor, not sexual difference alone, create difference and complementarity between "opposite" sexes. These differences, in turn, have been the basis for marriage in most societies. . . .

The concept of gender certainly helps us understand women's economic histories. . . . Each racial-ethnic group has had a sexual division of labor which has barred individuals from the activities of the opposite sex. Gender processes do differentiate women's lives in many ways from those of the men in their own racial-ethnic and class group. Further, gender relations in all groups tend to assign women to the intra-familial work of childrearing, as well as to place women in a subordinate position to the men of their class and racial-ethnic group.

But as soon as we have written these generalizations, exceptions pop into mind. Gender roles do not always correspond to sex. Some American Indian tribes allowed individuals to choose among gender roles: a female, for example, could choose a man's role, do men's work, and marry another female who lived out a woman's role. In the nineteenth century, some white females "passed" as men in order to escape the rigid mandates of gender roles. In many of these cases, women lived with and loved other women.

Even though childrearing is women's work in most societies, many women do not have children, and others do not perform their own child care or domestic work. Here, class is an especially important differentiating process. Upper-class women have been able to use their economic power to reassign some of the work of infant care—sometimes even breastfeeding—to lower-class women of their own or different racial-ethnic groups. These women, in turn, may have been forced to leave their infants alone or with relatives or friends. Finally, gender complementarity has not always led to social and economic inequality; for example, many American Indian women had real control over the home and benefited from a more egalitarian sharing of power between men and women.

Since the processes of sex-role socialization are historically distinct in different times and different cultures, they result in different conceptions of appropriate gender behavior. Both African American and Chicana girls, for instance, learn how to be women—but both must learn the specific gender roles which have developed within their racial-ethnic and class group and historical period. For example, for white middle-class homemakers in the 1950s, adherence to the concept of womanhood discouraged paid employment, while for poor Black women it meant employment as domestic servants for white middle-class women. Since racial-ethnic and class domination have differentiated the experiences of women, one cannot assume, as do many feminist theorists and activists, that all women have the same experience of gender oppression—or even that they will be on the same side of a struggle, not even when some women define that struggle as "feminist."

Not only is gender differentiation and oppression not a universal experience which creates a common "women's oppression," the sexual divisions of labor and family systems of people of color have been systematically disrupted by racial-ethnic and class processes. In the process of invasion and conquest, Europeans imposed their notions of male superiority on cultures with more egalitarian forms of gender relations, including many American Indian and African tribes. At the same time, European Americans were quick to abandon their notion of appropriate femininity when it conflicted with profits; for example, slave owners often assigned slave women to backbreaking labor in the fields.

Racial-ethnic and class oppression have also disrupted family life among people of color and the white working class. Europeans interfered with family relations within subordinated racial-ethnic communities through rape and forced cohabitation. Sometimes whites encouraged or forced reproduction, as when slaveowners forced slave women into sexual relations. On the other hand, whites have often used their power to curtail reproduction among peoples of color, and aggressive sterilization programs were practiced against Puerto Ricans and American Indians as late as the 1970s. Beginning in the late nineteenth century, white administrators took American Indian children from their parents to "civilize" them in boarding schools where they were forbidden to speak their own languages or wear their native dress. Slaveowners commonly split up slave families through sale to different and distant new owners. Nevertheless, African Americans were able to maintain strong family ties, and even augmented these with "fictive" or chosen kin. From the mid-nineteenth through the mid-twentieth centuries, many Asians were separated from their spouses or children by hiring policies and restrictions on immigration. Still, they maintained family life within these split households, and eventually succeeded in reuniting, sometimes after generations. Hence, for peoples of color, having children and maintaining families have been an essential part of the struggle against racist oppression. Not surprisingly, many women of color have rejected the white women's movement view of the family as the center of "women's oppression."

These examples reveal the limitations of gender as a single lens through which to view women's economic lives. Indeed, any attempt to understand women's experiences using gender alone cannot only cause misunderstanding, but can also interfere with the construction of broad-based movements against the oppressions experienced by women.

Race-Ethnicity

Like gender, race-ethnicity is based on a perceived physical difference, and rationalized as "natural" or "god-given." But whereas gender creates difference and inequality according to biological sex, race-ethnicity differentiates individuals according to skin color or other physical features.

In all of human history, individuals have lived in societies with distinct languages, cultures, and economic institutions; these ethnic differences have been perpetuated by intermarriage within, but rarely between, societies. However, ethnic differences can exist independently of a conception of race, and without a practice of racial-ethnic domination such as the Europeans practiced over the last three centuries. . . .

Does the concept of race-ethnicity help us understand the economic history of women in the United States? Certainly, for white racial-ethnic domination has been a central force in U.S. history. European colonization of North America entailed the displacement and murder of the continent's indigenous peoples, rationalized by the racist view that American Indians were savage heathens. The economy of the South was based on a racial-ethnic system, in which imported Africans were forced to work for white landowning families as slaves. U.S. military expansion in the nineteenth century brought more lands under U.S. control—the territories of northern Mexico (now the Southwest), the Philippines, and Puerto Rico—incorporating their peoples into the racial-ethnic hierarchy. And from the mid-nineteenth century onward, Asians were brought into Hawaii's plantation system to work for whites as semi-free laborers. In the twentieth century, racial-ethnic difference and inequality have been perpetuated by the segregation of people of color into different and inferior jobs, living conditions, schools, and political positions, and by the prohibition of intermarriage with whites in some states up until 1967.

Race-ethnicity is a key concept in understanding women's economic histories. But it is not without limitations. First, racial-ethnic processes have never operated independently of class and gender. In the previous section on gender, we saw how racial domination distorted gender and family relations among people of color. Racial domination has also been intricately linked to economic or class domination. As social scientists Michael Omi (Asian American) and Howard Winant (European American) explain, the early European arguments that people of color were without souls had direct economic meaning:

> At stake were not only the prospects for conversion, but the types of treatment to be accorded them. The expropriation of property, the denial of political rights, the introduction of slavery and other forms of coercive

labor, as well as outright extermination, all presupposed a worldview which distinguished Europeans—children of God, human beings, etc.—from "others." Such a worldview was needed to explain why some should be "free" and others enslaved, why some had rights to land and property while others did not.[2]

Indeed, many have argued that racial theories only developed after the economic process of colonization had started, as a justification for white domination of peoples of color.

The essentially economic nature of early racial-ethnic oppression in the United States makes it difficult to isolate whether peoples of color were subordinated in the emerging U.S. economy because of their race-ethnicity or their economic class. Whites displaced American Indians and Mexicans to obtain their land. Whites imported Africans to work as slaves and Asians to work as contract laborers. Puerto Ricans and Filipinas/os were victims of further U.S. expansionism. Race-ethnicity and class intertwined in the patterns of displacement for land, genocide, forced labor, and recruitment from the seventeenth through the twentieth centuries. While it is impossible, in our minds, to determine which came first in these instances—race-ethnicity or class—it is clear that they were intertwined and inseparable.

Privileging racial-ethnic analysis also leads one to deny the existence of class differences, both among whites and among people of color, which complicate and blur the racial-ethnic hierarchy. A racial-ethnic analysis implies that all whites are placed above all peoples of color. . . .

A minority of the dominated race is allowed some upward mobility and ranks economically above whites. At the same time, however, all whites have some people of color below them. For example, there are upper-class Black, Chicana, and Puerto Rican women who are more economically privileged than poor white women; however, there are always people of color who are less economically privileged than the poorest white woman. Finally, class oppression operates among women of the same racial-ethnic group.

A third problem with the analysis of racial domination is that such domination has not been a homogeneous process. Each subordinated racial-ethnic group has been oppressed and exploited differently by whites; for example, American Indians were killed and displaced, Africans were enslaved, and Filipinas/os and Puerto Ricans were colonized. Whites have also dominated whites; some European immigrant groups, particularly Southern and Eastern Europeans, were subjected to segregation and violence. In some cases, people of color have oppressed and exploited those in another group: Some American Indian tribes had African slaves; some Mexicans and Puerto Ricans displaced and murdered Indians and had African slaves. Because of these differences, racial oppression does not automatically bring unity of peoples of color across their own racial-ethnic differences, and feminists of color are not necessarily in solidarity with one another.

To sum up, we see that, as with gender, the concept of race-ethnicity is essential to our analysis of women's works. However, divorcing this concept from gender and class, again, leads to problems in both theory and practice.

Class

. . . We believe that the concepts of class and exploitation are crucial to understanding the work lives of women in early U.S. history, as well as in the modern, capitalist economy. Up through the nineteenth century, different class relations organized production in different regions of the United States. The South was dominated by slave agriculture; the Northeast by emerging industrial capitalism; the Southwest (then part of Mexico) by the *hacienda* system which carried over many elements of the feudal manor into large-scale production for the market; the rural Midwest by independent family farms that produced on a small scale for the market; and the American Indian West by a variety of tribal forms centered in hunting and gathering or agriculture, many characterized by cooperative, egalitarian economic relations. Living within these different labor systems, and in different class positions within them, women led very different economic lives. By the late nineteenth century, however, capitalism had become the dominant form of production, displacing artisans and other small producers along with slave plantations and tribal economies. Today, wage labor accounts for over 90 percent of employment; self employment, including family businesses, accounts for the remaining share. With the rise of capitalism, women were brought into the same labor system, and polarized according to the capitalist-wage laborer hierarchy.

At the same time as the wage labor form specific to capitalism became more prevalent, capitalist class relations became more complex and less transparent. Owners of wealth (stocks and bonds) now rarely direct the production process; instead, salaried managers, who may or may not own stock in the company, take on this function. While the capitalist class may be less identifiable, it still remains a small and dominant elite. . . .

Class can be a powerful concept in understanding women's economic lives, but there are limits to class analysis if it is kept separate from race-ethnicity and gender. First, as we saw in the race section above, the class relations which characterized the early U.S. economy were also racial-ethnic and gender formations. Slave owners were white and mostly male, slaves were Black. The displaced tribal economies were the societies of indigenous peoples. Independent family farmers were whites who farmed American Indian lands; they organized production in a patriarchal manner, with women and children's work defined by and subordinated to the male household head and property owner. After establishing their dominance in the pre-capitalist period, white men were able to perpetuate and institutionalize this dominance in the emerging capitalist system, particularly through the monopolization of managerial and other high-level jobs.

Second, the sexual division of labor within the family makes the determination of a woman's class complicated—determined not simply by her relationship to the production process, but also by that of her husband or father. For instance, if a woman is not in the labor force but her husband is a capitalist, then we might wish to categorize her as a member of the capitalist class. But what if that same woman worked as a personnel manager for a large corporation, or as a salesperson in an elegant boutique? Clearly, she derives upper-class status and access to income from her husband, but she is also, in her own right, a worker. Conversely, when women lose their husbands through divorce, widowhood, or desertion, they often change their class position in a downward direction. A second gender-related economic process overlooked by class analysis is the unpaid household labor performed by women for their fathers, husbands, and children— or by other women, for pay.

Third, while all workers are exploited by capitalists, they are not equally exploited, and gender and race-ethnicity play important roles in this differentiation. Men and women of the same racial-ethnic group have rarely performed the same jobs—this sex-typing and segregation is the labor market form of the sexual division of labor we studied above. Further, women of different racial-ethnic groups have rarely been employed at the same job, at least not within the same workplace or region. This racial-ethnic typing and segregation has both reflected and reinforced the racist economic practices upon which the U.S. economy was built.

Thus, jobs in the labor force hierarchy tend to be simultaneously race-typed and gender-typed. Picture in your mind a registered nurse. Most likely, you thought of a white woman. Picture a doctor. Again, you imagined a person of a particular gender (probably a man), and a race (probably a white person). If you think of a railroad porter, it is likely that a Black man comes to mind. Almost all jobs tend to be typed in such a way that stereotypes make it difficult for persons of the "wrong" race and/or gender to train for or obtain the job. Of course, there are regional and historical variations in the typing of jobs. On the West Coast, for example, Asian men performed much of the paid domestic work during the nineteenth century because women were in such short supply. In states where the African American population is very small, such as South Dakota or Vermont, domestic servants and hotel chambermaids are typically white. Nonetheless, the presence of variations in race-gender typing does not contradict the idea that jobs tend to take on racial-ethnic and gender characteristics with profound effects on the labor market opportunities of job-seekers. . . .

The racial-ethnic and gender processes operating in the labor market have transposed white and male domination from pre-capitalist structures into the labor market's class hierarchy. This hierarchy can be described by grouping jobs into different labor market sectors or segments: "primary," "secondary," and "underground." . . . The primary labor market—which has been monopolized by white men—offers high salaries, steady employment, and upward mobility. Its upper tier consists of white-collar salaried or self-employed workers with high status, autonomy, and, often, supervisory capacity. Wealth increases access to this sector, since it purchases elite education and provides helpful job connections. . . .

The lower tier of the primary sector, which still yields high earnings but involves less autonomy, contains many unionized blue-collar jobs. White working-class men have used union practices, mob violence, and intimidation to monopolize these jobs. By World War II, however, new ideologies of worker solidarity, embodied in the mass industrial unions, began to overcome the resistance of white male workers to the employment of people of color and white women in these jobs.

In contrast to both these primary tiers, the secondary sector offers low wages, few or no benefits, little opportunity for advancement, and unstable employment. People of color and most white women have been concentrated in these secondary sector jobs, where work is often part-time, temporary, or seasonal, and pay does not rise with increasing education or experience. Jobs in both tiers of the primary labor market have generally yielded family wages, earnings high enough to support a wife and children who are not in the labor force for pay. Men of color in the secondary sector have not been able to earn enough to support their families, and women of color have therefore participated in wage labor to a much higher degree than white women whose husbands held primary sector jobs.

Outside of the formal labor market is the underground sector, where the most marginalized labor force groups, including many people of color, earn their livings from illegal or quasi-legal work. This sector contains a great variety of jobs, including drug trafficking, crime, prostitution, work done by undocumented workers, and sweatshop work which violates labor standards such as minimum wages and job safety regulations.

NOTES

1.   Elizabeth V. Spelman, *Inessential Woman: Problems of Exclusion in Feminist Thought* (Boston: Beacon Press, 1988).

2.   Michael Omi and Howard Winant, *Racial Formation in the United States* (New York: Routledge & Kegan Paul, 1986), p. 58.

35

Racism in Toyland

CHRISTINE L. WILLIAMS

Not long ago I had to buy a present for a six-year-old. I had at least three choices for where to shop: The Toy Warehouse, a big-box superstore with a vast array of low-cost popular toys; Diamond Toys, a high-end chain store with

SOURCE: *Contexts,* Vol. 4, Issue 4, pp. 28–32, © 2005 by the American Sociological Association. Reprinted by permission of the University of California Press.

a more limited range of reputedly high-quality toys; or Tomatoes, a locally owned, neighborhood shop that sells a relatively small, offbeat assortment of traditional and politically correct toys. Can sociology offer any advice to consumers like me?

Unfortunately, many sociologists turn into utilitarian economists when it comes to analyzing shopping, assuming that customer behavior is determined only by price, convenience, and selection. But a number of social factors influence where we choose to shop, including the racial makeup of the store's workers and customers. In my book, *Inside Toyland: Working, Shopping, and Social Inequality* [University of California Press, 2006], I argue that racial inequality (and gender and class inequality as well) influence where we choose to shop, how we shop, and what we buy. The retail industry sustains such inequality through hiring policies that favor certain kinds of workers and advertising aimed at customers from specific racial or ethnic groups.

I noticed the connection between shopping and social inequality while working as a clerk at the Toy Warehouse and Diamond Toys for three months in 2001. These stores belonged to national chains, and both employed about 70 hourly employees. At the warehouse store, I was one of only three white women on the staff; most were African American, Hispanic, or second-generation Asian American. The "guests" (as we were required to call customers) were an amazing mix from every racial and ethnic group and social class. In contrast, only three African Americans worked at the upscale toy store; most clerks were white. Most of the customers were also white and middle to upper class. My experiences taught me to notice racial diversity (or its absence) wherever I shop.

"DON'T SHOP WHERE YOU CAN'T WORK!"

This slogan, popular during the Great Depression, rallied black protesters to demand equal access to jobs in stores, and many chains responded by hiring African Americans in predominately black neighborhoods: "We employ colored salesmen" signs appeared in Sears, Walgreens, and other stores eager for black customers.

Retail work is one of the most integrated occupations in the United States today. The proportions of whites, African Americans, Asian Americans, and Hispanics employed in retail jobs more or less match their representation in the labor force. But these statistics hide segregation at the store level. More than 15 percent of employees in shoe stores and variety stores, for instance, are black, but less than 5 percent of employees are black in stores that sell liquor, gardening equipment, or needlework supplies. Inside stores, there is further segregation by task; Whites usually have the top director and manager positions, and nonwhites have the lowest-paid, often invisible backroom jobs.

In the toy stores where I worked, the two most segregated jobs were the director positions (all white men) and the janitor jobs (all Latinas subcontracted from outside firms). The Toy Warehouse employed mostly African Americans in all the other positions, including cashiers. But over time I noticed that the managers preferred to assign younger and lighter-skinned women to this position. Older African-American women who wanted to work as cashiers had to struggle to get the assignment. Lazelle, for example, who was about 35, had been asking to work as a cashier for the two months she had been working there. She worked as a merchandiser, getting items from the storeroom, pricing items, and checking prices when bar codes were missing. Lazelle finally got her chance at the registers the same day that I started. We set up next to each other, and I noticed with a bit of envy how competent and confident on the register she was. (Later she told me she had worked registers at other stores, including fast-food restaurants.) I told her I had been hoping to get assigned as a merchandiser. I liked the idea of being free to walk around the store, talk with customers, and learn more about the toys. I had mentioned to the manager that I wanted that job, but she made it clear I was destined for cashiering and service desk (and later, to my horror, computer accounting). Lazelle looked at me like I was crazy. Most workers thought merchandising was the worst job in the store because it was so physically taxing. From her point of view, I had gotten the better job, no doubt because of my race, and it seemed to her that I wanted to throw that advantage away. (The manager may also have considered my background and educational credentials in assigning me to particular jobs.)

The preference for lighter-skinned women as cashiers reflects the importance of this job in the store's general operations. In discount stores, customers seldom talk with sales clerks. The cashier is the only person most customers deal with, giving her enormous symbolic—and economic—importance for the corporation. Transactions can break down if clerks do not treat customers as they expect. The preference for white and light-skinned women as cashiers should be understood in this light: In a racist and sexist society, managers generally believe that such women are the most friendly and solicitous, and thus most able to inspire trust and confidence in a commercial transaction.

At the upscale Diamond Toys, virtually all cashiers were white. Unlike the warehouse store, where cash registers were lined up at the front of the store, the upscale store had cash registers scattered throughout the different departments. The preference for white workers in these jobs (and throughout the store) seemed consistent with the marketing of the store's workers as "the ultimate toy experts." In retail work, professional expertise is typically associated with whiteness, much as it is in domestic service.

The purported expertise of salesclerks is one of the great deceptions of the retail industry. Here, where jobs pay little and turnover rates are high (estimated at more than 100 percent per year by the National Retail Federation), many clerks know almost nothing about the products they sell. I knew nothing about toys when I was put behind the cash register, and I received no training on the merchandise at either store. Any advice I gave I literally made up. But at the upscale store I was expected to help customers with their shopping decisions. They frequently asked questions like "What are going to be the hot toys for one-year-olds this Christmas?" or "What one item would you recommend for two sisters of different ages?" One mother asked me to help her pick out a $58 quartz watch for her seven-year-old son. A personal shopper phoned in and asked me to describe the three Britney Spears dolls we carried, help her pick out the "nicest" one, and then arrange to ship it to her employer's niece. Customers asked detailed questions about how the toys were meant to work, and they were especially curious about comparing the merits of the educational toys we offered (I was asked to compare the relative merits of the "Baby Mozart" and the "Baby Bach"). On my first day, I answered a phone call from a customer who asked me to pick out toys for a one-year-old girl and a boy who was two and a half, spend up to $100, and arrange to have the toys gift-wrapped and mailed to their recipients.

At Diamond Toys, most customers didn't mind waiting to talk with me. When the lines were long, they didn't make rude huffing noises or try to make eye contact with their fellow sufferers, as they often did at the Toy Warehouse. I couldn't help but think that the customers—mostly white—were more civil and polite at the upscale store because most of us workers were white. We were presumed to be professional, caring, and knowledgeable, even when we weren't. White customers seemed less respectful of minority service workers than white workers; they were willing to pay more and wait longer for the services of whites because they apparently assumed that whites were more refined and intelligent.

On several occasions at the warehouse store, I saw customers reveal racist attitudes toward my African-American coworkers. One night, after the store had closed, I saw Doris and Selma (fiftyish African Americans) escorting several white customers out. Getting straggling customers to leave the store after closing was always a big chore. Soon after, as I was being audited in the manager's office, Selma came in very upset because one of the women she and Doris escorted out had spit out her chewing gum at her. Doris and Selma were appalled. Doris said to them, "That is really disgusting; how could you do that?" And the woman said to Doris, "What's your name?" like she was going to report her. This got Selma so angry she said, "If you take her name, take mine too," and showed her name-tag. She told the woman that she was never welcome to come back to this store. Selma was very distressed. Talking back to customers was taboo, and she knew she could be fired for what she had said. The manager told her that some people are going to be gross and disgusting and what can you do? Clearly Selma and Doris would not get in trouble over this. But I sensed it was doubly humiliating to have to fear that she might be punished for talking back to a white woman who had spit at her.

Although I suffered from plenty of customer condescension at this store, at least putting up with racism was not part of my job. Once when Tanesha, a 23-year-old African American, was training me at the service desk, two white women elbowed up to the counter to complain to me about the long wait for service (they had waited about five minutes while we were serving another customer). I said something about being in training, and they thought I meant that I was training Tanesha. So they said, "Well, call up someone else to the register!" I said I'd have to ask Tanesha to do this. They demanded that I stop training her for a moment to call another person to the exchange desk. "No you don't understand," I said, "I'm the one who is in training; she knows what she is doing, and she is the only one who can call for backup, and she is in the middle of trying to accommodate this other customer." They seemed embarrassed at having assumed that the white woman was in charge. When they realized their mistake, they looked mortified and stepped back from the counter.

SHOPPING WHILE BLACK

During the civil rights era, equal access to stores was high on the list of demands for racial justice. Before Jim Crow laws were repealed, many stores restricted their facilities to whites only. Black customers often were not allowed to try on clothes, eat at lunch counters, or use public restroom facilities in stores.

Today the worst forms of racism have been eliminated. Gone are the "whites only" signs on restrooms and drinking fountains. But some stores build to exclude. The history of suburban malls is a history of intentional racial segregation. Even today, so-called desirable retail locations are characterized by limited access. In my city of Austin, Texas, local malls have opposed public bus service on the grounds that it would encourage undesirable (nonwhite) patrons.

But open access is not enough to ensure racial diversity. Diamond Toys was located in a racially diverse urban shopping district, next to subway and bus lines, yet nearly all its patrons were white. I didn't fully realize this until one day when Chandrika, an 18-year-old African-American gift wrapper, told me that she thought she saw one of her friends in the store. We weren't allowed to leave our section, so she asked the plainclothes security guard to look around and see if there was a black guy in the store. I asked her about this. Was it so unusual for an African-American teenager to be in the store that one black guy would be so apparent? After all, lots of young men came in to check out the new electronic toys and collectibles. Chandrika assured me that a young black man would definitely stand out.

One way that many stores show hostility to racial/ethnic minorities is through consumer racial profiling. Like racial profiling in police work, this involves detaining, searching, and harassing such people more often than is done for whites, usually because they are suspected of stealing. Some scholars have labeled this potential violation of people's rights "shopping while black." At the stores where I worked, clerks weren't allowed to pursue anyone suspected of stealing; that was the job of the plainclothes security workers. However, relations between customers and clerks sometimes broke down, and I saw double standards in the treatment of whites and minorities.

At the warehouse store, I was told to treat shoppers as if they were my mother (most shoppers at both stores were women). At the service desk, I was told to appease them by honoring all requests for returns, even if the merchandise had been used and worn out. The goal, my manager told me, was to make these shoppers so grateful that they would return to the store and spend $20,000 per child, the amount their marketers claimed was spent on an average child's toys.

In my experience, only middle-class white women could depend on this treatment. Nevertheless, I watched many white women throw fits, loudly complaining of shoddy service and merchandise with comments like "I will never shop in this store again!" Such arrogance no doubt came from being accustomed to having their demands met. On one occasion, when a white woman threw a tantrum because the bike she had ordered was not ready as scheduled, the manager offered her a $25 gift certificate for her troubles. She refused, demanded a refund, and left shouting that she would "never come to this store again." The manager then gathered the entire staff at the front of the store and chewed us out for being disorganized and incompetent.

Members of minority groups who wanted to return used merchandise or needed special consideration were rarely accommodated. The week before the bike incident, I was on a register that broke down in the middle of a credit card transaction. A middle-class black woman in her forties was buying inline skates for her ten-year-old daughter. The receipt came out of the register but not the slip for her to sign, so I had to call a manager, who came over and explained that she needed to go to another register and repeat the transaction. She refused, since it seemed to go through all right and she didn't want to be charged twice. She had to wait more than an hour to get this problem solved, and she wasn't offered any compensation. She didn't yell or make a scene; she waited stoically. I felt sorry for her and went to the service desk to tell a couple of my fellow workers what was happening while the managers tried to resolve it, and I said they should just give her the skates and let her go. My fellow workers thought that was the funniest thing they had ever heard. I said, "What about our policy of letting things go to make sure we keep loyal customers?" But they just laughed at me. Celeste said, "I want Christine to be the manager; she just lets the customers have whatever they want!"

Research has shown that African Americans suffer discrimination in public places, including stores; middle-class whites, on the other hand, are privileged. We do not recognize this precisely because it is so customary. Whites expect first-rate service; when they don't receive it, some feel victimized, even discriminated against, and some throw tantrums when they don't get what they want.

When African-American customers did shout or make a scene, the managers called or threatened to call the police. Each of these instances involved an African-American man complaining and demanding a refund. Once a young black man, denied a cash refund, threw a toy on the service desk, and accidentally hit the telephone, which flew off the desk and hit me on the side of the head, knocking me to the floor. Within minutes, three police officers arrived and asked me if I wanted to press assault charges. I did not. After all, angry people often threw merchandise on that desk, and what happened had been an accident. But at least I was appeased. At the end of my shift, the manager gave me a "toy buck" for "taking a hit" in the line of duty, which entitled me to a free Coke.

There wasn't much shouting or throwing at the upscale store. It protected itself from conflict by catering to an upper-class clientele, much as a gated community does. This is not to say that diversity always leads to conflict, but it did at the warehouse store because race, class, and gender differences existed under a layer of power differences within the store. Clerks and customers interacted in a context where these differences had been used to shape marketing agendas, hiring practices, and labor policies—all of which benefited specific groups (especially middle-class white men and women).

WHAT ABOUT TOMATOES?

There are alternatives to shopping at large chain stores, although their numbers are dwindling. The store I'm calling Tomatoes is a small, family-owned business in an upper-middle-class neighborhood on a busy street with lots of pedestrian traffic. It's been in the neighborhood for 25 years, owned by the same family. It sells an offbeat assortment of toys, including many traditional items like kites and wooden blocks, and a variety of toys I would call "politically correct." It didn't carry Barbie, for example, but it did have "Get Real Girls," female action figures that look like G.I. Joe's sisters. I laughed when I saw a pack of plastic "multicultural" food, including spaghetti, sushi, a taco, and a bagel (all marked "made in China").

Working conditions at the store seemed very relaxed compared to what I had experienced. The owner wore shorts and a Hawaiian shirt, and the workers dressed like punk college students, including weirdly dyed hair, piercings, and tattoos. They didn't wear uniforms (as we did at the other two stores). One clerk wore her T-shirt hiked up in a knot in the front and stuffed under her bra in the back. Clerks seemed to be on a friendly, first-name basis with several of the customers, who were mostly middle-class white women.

Although I didn't get hired at Tomatoes, after several visits I noticed social patterns in the store's organization. The owner and the manager were both white men, and all the clerks were young white women. The owner was the only one who was near my age (mid-40s). You can never be sure why you aren't hired, but my impression is that I wasn't young enough or hip enough to work there. Although Tomatoes allowed more autonomy and self-expression than the stores where I worked, race, class, and gender inequality were as much a part of the social organization there as in other retail stores.

CONCLUSION

I ended up buying my gift at Tomatoes—a children's book written and autographed by Marge Piercy. My decision reflects my identity and my social relationships. But what are the implications of my choice for social inequality? My purchase supported a store that was organized around racial exclusion, gender segregation, and class distinctions.

Everyone has to shop in our consumer society, yet the way shopping is organized bolsters social divisions. The racism of shopping is reflected in labor practices, store organization, and the guidelines, explicit or unspoken, for relations between clerks and customers. When deciding where and how to shop, consumers should be aware of what their choices imply with regard to racial justice and equality. Although an individual shopper can do little to change the overall social organization of shopping, raising awareness of the inequalities that our choices support must be a first step in imagining and then creating a better alternative.

36

The Indignities of Unemployment

KENNETH W. BROWN

I am a number, a statistic, a peak on a national bar graph. I am not physically or mentally disabled. But I am financially disabled; I am unemployed.

I am unemployed after nearly 14 years of being a loyal, hardworking employee. I bought into the company philosophy, followed the company credo and worked hard to help accomplish the corporate goal. As I was coddled, supported, and groomed, I grew to believe that I was an important member of the corporate family, only to be unceremoniously dumped after I had served my purpose and was no longer needed or welcomed, like a bastard son at a family reunion.

I am making every effort to find work. I have made countless phone inquiries, answered many classified ads and contacted people I have not spoken to in years, looking for job leads. I have attended job fairs in search of another chance to rejoin corporate America, struggling to maintain my sense of dignity as I walked about the crowded rooms—looking along with so many others for the next opportunity to reenter the economic mainstream.

As unemployment refuses to loosen its grasp, I find it increasingly difficult to remain hopeful, positive, and confident. Prolonged unemployment can do a number on your psyche, your soul, and your spirit. Unemployment can make you question your skills and abilities. And it can damage your self-esteem. I don't care how much confidence you have; unemployment can shake that confidence.

As I search for my next position, I do so with growing reluctance and trepidation. When I prepare for the next job interview, I practice my responses to the questions about myself, my strengths, my weaknesses, and my future goals.

I make sure I speak the King's English with perfect enunciation, ever mindful not to let street slang or my ethnicity slip into my conversation.

I often wonder whether my white counterparts are as conscious of their diction as I am of mine. I also wonder whether I am selling out or compromising myself too much. Do white men wonder if they are being too forceful, too eager to please, too assertive? Do they agonize as much as I do over how to appear skilled and capable without intimidating the people who have the power to hire? The thought of always keeping these ideas in mind is exhausting and sad. Is this the cross that African-American men and women must bear in order to be accepted into corporate America?

SOURCE: From *Essence* 26 (September 1995): 56.

Unemployment can make you question your faith and spiritual foundation. I pray for my unemployment to end. I pray for direction and guidance. I also pray for the day when I do not have to wonder whether my blackness is a barrier to reentry into the labor force. I am still waiting for a response. Unemployment makes me wonder if anyone is listening.

Being unemployed also makes me question my manhood. I was raised to be a responsible household member who would carry my share of the weight. But now that I am no longer holding up my end of the financial bargain, my wife and I exist on a single income: hers. While I am fortunate that she has been supportive throughout my long period of unemployment, I must fight hard against seeing myself as a failure or loser. I have never before considered myself a loser, and I refuse to be one. Every day I must remind myself that being without a job does not make me less of a man.

Some good things have resulted from my unemployment. I have gained a renewed awareness of life's blessings: the value of true friends and the love, support, encouragement and prayers of family. In many ways I have become a stronger person. I have resolved to survive unemployment. I resolve to become a statistic for working Americans, and not just another African-American male without a job.

37

"Soft" Skills and Race

PHILIP MOSS AND CHRIS TILLY

The growth of "soft" or social skills is a factor that has been neglected in research on the racial gap in labor market outcomes, despite the fact that employer surveys have repeatedly identified such skills as the most important hiring criterion for entry-level jobs (Capelli 1995). In this study, therefore, we examined the relationship between employer racial attitudes and their hiring practices, with special attention given to their use of soft skills. We defined soft skills as skills, abilities, and traits that pertain to personality, attitude, and behavior rather than to formal or technical knowledge. Based on 56 face-to-face interviews with employers, we sought to learn how employment gatekeepers conceived of soft skills in relation to hiring Black men for entry-level jobs.

According to our findings, employers reported an increasing need for soft skills—driven, they said, by heightened competitive pressure—and they rated Black men poorly in terms of such skills. Thus, the heightened competition

SOURCE: From *Work and Occupations* 23 (August 1996): 252–76 © 1996 Sage Publications, Inc. Reprinted by permission of Sage Publications, Inc.

that propels current business restructuring appears to contribute to increased labor market inequality by race. . . .

THE GROWING IMPORTANCE OF SOFT SKILLS

Again, we defined soft skills as skills, abilities, and traits that pertain to personality, attitude, and behavior rather than to formal or technical knowledge. Two clusters of soft skills were important to the employers surveyed. The first, *interaction,* involves ability to interact with customers, coworkers, and supervisors. This cluster includes friendliness, teamwork, ability to fit in, and appropriate affect, grooming, and attire. The interaction category is related to concepts of *emotional labor* (Hochschild 1983; Wharton 1993) and *nurturant social skills* (England 1992; Kilbourne, England, & Beron 1994). A second cluster, *motivation,* takes in characteristics such as enthusiasm, positive work attitude, commitment, dependability, and willingness to learn. We distinguish both from hard skills, including skills in math, reading, and writing; knowledge of particular job procedures; "brightness;" ability to learn; educational attainment; and physical strength.

Interaction and motivation skills differ from one another, and in much of our analysis, we distinguish between the two. However, we grouped them together under the rubric of soft skills because employers often subsumed both in terms like *attitude* and because many employers viewed both as more immutable than hard skills. Of course, soft skills are in part culturally defined, and therefore employer assessments of soft skills will be confounded by differences in culture and by racial stereotyping. Indeed, the word *skills* is to some extent a misnomer, although employers most definitely conceptualize these attributes as contributing to individual productivity differences.

We asked almost all interviewees to identify the most important qualities they looked for when hiring entry-level workers. As Table 1 shows, interaction skills were by far the most important qualification in retail. Motivation and hard skills received roughly equal emphasis in auto parts and insurance. Only in the public sector were hard skills mentioned most often. Overall, 86% of respondents

T A B L E   1   **Most Important Qualities Looked for in Entry-Level Employees**

Industry	Hard Skills	Interaction	Motivation
Auto parts manufacturing	58	32	63
Retail clothing	22	78	39
Insurance	67	67	78
Public sector	100	60	60

Frequency With Which Each Category Was Mentioned (Percent)

included soft skills in their list of the most important hiring criteria, and almost half mentioned soft skills first in that list. . . .

Respondents in all four industries stated that competitive pressures have led to growing soft-skill needs (Moss & Tilly 1996). For instance, auto makers are pushing their suppliers to cut costs and increase quality. In response, many of the parts manufacturers are escalating basic and technical skill requirements. Some, however, are demanding more soft skills as well. For instance, a human resource manager at an alloy casting plant told us:

> Hiring used to be based on 90% experience, 10% attitude or work ethic. I find that changing . . . due to the [emphasis on] team work and total quality. Attitudes and people getting along with one another . . .—this is a big part now. I would almost say it's 50% to 60% being experience and the other 40% being attitude, work ethic, teamwork.

In retail clothing, heightened emphasis on customer service—again, spurred by intensified competition—has led companies to screen more carefully for soft skills when hiring sales clerks. One discount clothing chain has adopted the slogan "fast, fun, and friendly." According to a regional personnel representative for the chain, this means that now

> I tell my . . . personnel managers, "If they don't smile, don't hire them." I don't care how well-educated they are, how well-versed they are in retail, if they can't smile, they're not going to make a customer feel welcome. And we don't want them in our store.

On the other hand, respondents reported declining hard skill needs among sales clerks, due to optical scanning equipment and computerized cash registers.

Insurance companies increasingly demand computer literacy among their clerical workers. In addition, financial deregulation has heightened competition, leading insurers to adopt several strategies: downsizing, reorganization of work, and greater stress on customer service. As one human resource manager put it, "on a scale of 1 to 10 it [customer relations] is a 9.999. . . . There is much more emphasis now being placed on it."

Finally, in the public sector, budget cuts have combined with political demands for greater productivity and quality of service. In our public sector interviews, two thirds of respondents reported that they were looking for more skilled people. Although some of these respondents noted a greater need for some basic or technical skill, all mentioned a need for greater customer skills—in jobs ranging from clerk-typist to hospital housekeeper.

SOFT SKILLS AND RACE IN THE
EYES OF EMPLOYERS

The emphasis employers place on soft skills disadvantages Black male job applicants. This is because many employers see Black men as lacking in precisely the skills they consider increasingly important. Indeed, in our sample, the employers placing the greatest emphasis on soft skills were those most likely to have negative views of Black men as workers. The views employers hold of Black men in this regard were partly stereotype, partly cultural gap, and partly an accurate perception of the skills that many less educated Black men bring to the labor market. . . .

Interaction Skills

Employers voiced two main sets of concerns about the ability of Black men to interact effectively with customers and coworkers. First, a substantial minority of respondents—32%—described Black men as defensive, hostile, or having a difficult "attitude." The content of these comments ranged widely. A Latino store manager in a Black area of Los Angeles, who hires mostly Latinos, flatly stated, "You know, a lot of people are afraid, they [Black men] project a certain image that makes you back off. . . . They're really scary." When asked, "How much of that do you think is perception, how much do you think is actually reality?" he responded, "I think 80% is reality. 80% of it's factual."

Other respondents stated that managers see Black men as difficult to control. For example, the Black female personnel manager of a Detroit retail store commented,

> Employers are sometimes intimidated by an uneducated Black male to come in. Their appearance really isn't up to par, their language, how they go about an interview. Whereas females Black or White, most people do feel, "I could control this person." . . . A lot of times people are physically intimidated by Black men. . . . The majority of our employers are not Black. And if you think that person may be a problem, [that] young Black men normally are bad, or [that] the ones in this area [are], you say, "I'm not going to hire that person, because I don't want trouble."

A White female personnel official from a Los Angeles public sector department offered a related perspective, laying part of the blame with White supervisors:

> . . . And I also think that part of the problem is that the supervisors and managers of these people have their own sets of expectations and their own sets of goals that don't address the diversity of these people, and it's kind of like, well, hell, if they're going to come work for me, they're going to damn well do it my way . . . . And my own personal feeling is that a lot of these young Black men who are being tough scare some of their supervisors. And so rather than address their behavior problems and deal with the issues, they will back away until they can find a way to get rid of them. We have a tendency to fear what we're not real familiar with.

Although a few respondents provided this level of detail, most of the negative responses were far briefer: "a lot of Black males have a chip on their shoulder," or "I get a strong sense of, with Black males at times of a hostility, of 'I deserve so and so, this belongs to me.'"

Our questions probed primarily for generalizations, but some respondents noted variation within race and gender categories. Even the store manager who described Black men as "really scary" added, "You know, there's a lot of Black males that are nice; they usually project a different image. They don't want to project the same image as gangsters."

In addition to negative views of the demeanor of Black men, we found that many retail employers saw the racial composition of employment itself as an issue for customers. The Black male personnel manager of a large retail store located in a Detroit-area suburban mall stated that because the labor market area is 90% Black, "we are forced to have an Affirmative Action program for nonminorities in this particular store." In fact, the store has shifted away from walk-in applications to in-store or mail recruiting from the store's customer base. Given that the mall sits in an integrated suburb of Detroit, his statement implies a fear that an all-Black workforce would erode the White suburban customer base.

In subsequent interviews, we asked retail informants explicitly about attempts to keep the racial mix of store employees similar to that of customers. Seven of the 10 retail informants whom we asked, responded that this was indeed a management concern. Not all of them approved of the customer attitudes to which they were responding. For example, a White female personnel manager at a Los Angeles store said,

> At [a store she was posted at previously] we had a lot of customer complaints because it's primarily White, and we were always getting complaints that there were all Black employees and it's because they were Black. That would be the first thing the customer would bring up was "Black." It was because they were Black that they didn't do their job right.

Nonetheless, this informant and others—Black, White, and Latino alike—viewed the goal of race-matching staff with customers as a legitimate management objective.

Motivation

Forty percent of respondents voiced perceptions of Black men as unmotivated employees. Once again, comments varied widely in substance. A Latino female personnel officer of a Los Angeles retail distribution warehouse, whose workforce is 72% Latino and only 6% Black, stated, "Black men are lazy. . . . Who is going to turn over? The uneducated Black." The White male owner of a small Detroit area plastic parts plant (46% Black, 54% White) said that in his experience, Black men "just don't care—like everybody owes them." "Black kids don't want to work," was the opinion of a White male owner of a small auto parts rebuilding shop in Los Angeles, whose workforce was entirely Hispanic female. "Black men are not responsible," added a Latino female personnel supervisor for a Los Angeles auto parts manufacturer located next to a major Black area but with a workforce that was 85% Latino and less than 1% Black.

This is still a minority viewpoint. The majority of respondents stated that they saw no differences in work ethic by race, although surely, in some cases, they were simply proffering the socially approved response. As in the case of interaction, some respondents discussed variation within race/gender categories. A number of employers invoked such variation to dismiss racial differences:

> We have problem employees, who choose not to come to work as often as others, but that cuts across all racial lines, you know, so no, I wouldn't say that's the case as a blanket statement. We have problem people, but they're just problem people. Some of them happen to be Black and some of them happen to be White, but I wouldn't say it's any better or worse in any one group or the other.

Other managers attributed apparent racial differences to class or neighborhood effects. Yet others noted distinctions within racial groups but still rated Black employees lower on average. For example, in a Los Angeles area discount store where the main workforce was Latino, the manager opined:

> I think the Hispanic people have a very serious work ethic. I have a lot of respect for them. They take pride in what they do. Some of the Black folks that I've worked with do, but I'd say a majority of them are just there putting in the time and kind of playing around.

In fact, although only a minority of respondents questioned the work ethic of Blacks, a substantial majority agreed with the idea that immigrants have a stronger work ethic than native-born workers—81% of Detroit respondents who ventured an opinion on this agreed, and 88% of Los Angeles respondents. This bodes ill for less-skilled Black workers, particularly in Los Angeles, because they increasingly compete with immigrant workers for jobs. . . .

How and Why Employers Form These Perceptions

Employers indicated that they based their perceptions of Black men on experiences with past and present employees, on impressions of applicants, and on more general impressions from the media and from experiences outside work. About half of respondents referred to their own employees, sometimes arguing that the immediacy of these observations rendered them objective. Stated a store manager,

> I think [Black men] feel things should be given to them and not earned. And because of that, they don't earn the right to keep jobs. Now that may be, you know, someone would say I may have an attitude problem, but that, I just look at pure facts. I mean with the people that I've had work for me. . . .

In a number of cases, informants pleaded ignorance due to a lack of Black applicants: "It's hard for us to say [why Black men do poorly in the labor market], because we don't have a large enough community, so we don't interview very many Blacks, and we don't reject many Blacks."

Other employers, especially those in White or Hispanic areas, referred to contact with Blacks outside the workplace. A White male insurance manager outside Detroit reported:

> I am involved and attend an urban church, and we attend there working with the homeless and the retarded and people of that nature, [and] unless something is done to help these Black males, it is just a sorry situation.

Although no respondent specifically cited the media as a source for their impressions of Black men, the impact of the media was evident, as in these two comments by White Detroit area manufacturing managers:

MANAGER 1: We have a lot guys out there with cocaine in their pocket and Uzis in the trunk.

MANAGER 2: I think a lot of [the difficulty Black men face in the labor market] is based on their inability to complete schooling early on, for whatever reason. I don't know a lot of the statistics of the Black race but I do see that. I think that is a good fact . . . . It's going to take unfortunately a heck of a long time to fully eliminate discrimination. We hear about it in the news still, and it's a shame.

. . . Even more important than how employers form their perceptions of Black men is why they hold these views. We argue that the negative views are a complicated combination of stereotypes, realities, and conflicting cultures. The evidence that stereotypes are involved is straightforward. On the one hand, some respondents voiced clearly stereotypical attitudes: "Black men are lazy. . . . They've got no respect for anyone"; and "I'm no doctor . . . but I'm convinced, having dealt with grievance and unrest . . . that Black men, and to some extent Black women, do not deal with stress physically as well as some other races."

On the other hand, certain managers charged that other managers harbored stereotypical views:

> People have a tendency, . . . even if they don't voice it, they deal with their stereotypes. It's easy to identify someone as a female or identify someone as [a] Black male or a White male. And so whatever it is that they have of their expectations of those people, they will project that.

In particular, some managers claimed that their peers engaged in statistical discrimination or generalized from a visible but unrepresentative subset of Black men:

> I think that many employers may feel that because of the large numbers of Black males who are in prison and have problems, that there is a tendency for those who are out and in the workforce to do mischievous things. That's unfortunate.

And one informant, a Black female manager at a Detroit area utility, contended that employers hold Blacks to a different standard than Whites:

> When Blacks and Hispanics and whomever come into the work group, and they are not part of the majority, then one thing they need to know is that they cannot always do what they see others doing, and that is key. That is a lesson that needs to be taught. I think the rules aren't always the same, and it may not always be intentional. A lot is institutional and lot of it is people just not understanding others.

We believe that the perceptions stated by employers contain an element of reality as well. We have no independent way of verifying the statements employers made. However, some comments offered substantial detail and specificity, and some referred to "objective" measures such as absenteeism. . . .

Wilson (1987) argued that soft-skill problems are the product of poor neighborhoods; a number of our informants argued instead that such skills—and particularly motivation—are endogenous to the workplace and labor market. As a Black human resource official of a Detroit-area insurer expressed it,

> I think business drives the work ethic. . . . If business is lax . . . then people have casual attitudes about their jobs. . . . You are one thing up to the point of entering the business world, but then you are something else. I'm not the same person I was 15 years ago. I had to take on certain thoughts and attitudes whether I liked it or not.

Several others agreed that motivation is more a function of management than of the workforce. When asked about racial differences in the work ethic, a White male manager of contracted public-sector workers mused,

> I think it's how you motivate each group. Two or three years ago, I would have probably said, well, the Black race isn't as motivated as the Oriental or the Hispanic. But I've seen that if you motivate, that you have to motivate each group differently.

A White male public-sector human resource official added that work ethic may vary by job:

> If I take security, or I take the basic labor jobs, I'm not so sure that when they were Caucasian-dominated 20 years ago, that people weren't leaning on a shovel and goldbricking. Many times, the classifications we normally associate with being more lazy or finding ways to avoid work, are the entry-level, lower skilled ones. And now those happen to be dominated by Blacks and to a lesser extent Hispanics right now.

A small number of informants also argued that workers can be trained to relate well to customers. Even a store manager who commented that "it does take a certain kind of person" to be "fast, fun, and friendly," added, "but if you work with a person, I think that you could pretty much [get them to] be fast, fun, and friendly." . . .

In addition to stereotype and reality, interviewees spoke of cultural gaps between young Black men and their supervisors, coworkers, and customers—especially as an explanation of difficulties in interaction. A Black human resource manager at an insurance company described the problems of cultural translation:

> I think that, as I attend executive meetings, and in many cases, I'm the only Black man there, the cultural diversity and the strangeness that different people bring to one another—oftentimes people aren't prepared to receive what another person may be prepared to offer. And I think that through that lack of communication, a lot of times things are misunderstood. When problems occur, if I work for you and you had a problem with me, you may not know how to approach me and vice versa, I may not know how to approach you.

White respondents also referred to "a difference in understanding," but were more likely to pose it as a failure of the Black men themselves: "[Young Black men] don't present themselves well to the employer, just because they don't know, they don't realize how they're communicating, or not communicating." This conforms with the view, expressed in focus groups by young, inner-city Black and Latino men, that code switching—being able to present oneself and communicate in ways acceptable to majority White culture—is the most important skill needed to find and keep a job (Jobs for the Future 1995). . . .

CONCLUSIONS

We find that due to competitive pressure, employers are demanding more soft skills, even in low-skill jobs. Soft skills include interaction and motivation, and employers are valuing both more highly. However, many managers perceive Black men as possessing fewer soft skills, along both dimensions. Thus, the same increases in competitive pressure that drive corporate downsizing and restructuring are contributing to widening racial inequality in labor market outcomes.

Employers base their perceptions of Black men on assessments of current or past employees and applicants, as well as interaction outside the workplace, and media images of Blacks. Three factors underlie negative evaluations of Black men as workers: racial stereotypes, cultural differences between employers and young Black men, and actual skill differences. The actual skill differences themselves are in part endogenous to the work situation. Moreover, in a work world characterized by increasing levels of interaction, racially biased attitudes held by customers or coworkers of other racial groups can themselves lead to lower measured productivity—that is, productivity differences can be the direct result of discrimination.

REFERENCES

Capelli, P. (1995). Is the "skills gap" really about attitudes? *California Management Review*, 37(4), 108–124.

England, P. (1992). *Comparable worth: Theories and evidence.* New York: Aldine de Gruyter.

Hochschild, A. R. (1983). *The managed heart: Commercialization of human feeling.* Berkeley: University of California Press.

Jobs for the Future. (1995). Information from focus groups of young, inner city men of color, for work in progress. Unpublished manuscripts. Boston, MA: Author, for the Annie E. Casey Foundation.

Kilbourne, B. S., England, P., & Beron, K. (1994). Effects of changing individual, occupational, and industrial characteristics on changes in earnings: Intersections of race and gender. *Social Forces, 72,* 1149–1176.

Moss, P., & Tilly, C. (1996). *Growing demand for "soft" skills in four industries: Evidence from in-depth employer interviews* (Working Paper No. 93). New York: Russell Sage Foundation.

Wharton, A. (1993). The affective consequences of service work: Managing emotions on the job. *Work and Occupations, 20*(2), 205–232.

Wilson, W. J. (1987). *The truly disadvantaged.* Chicago: University of Chicago Press.

38

The Invisible Poor

KATHERINE S. NEWMAN

Forty years ago, when it first began to dawn on this prosperous nation that the end of the Great Depression had not cured widespread poverty, popular representations of the poor focused on Kentucky hollows where Appalachian children sat on wooden stoops, their bellies swollen and faces blank. Miners whose livelihoods had disappeared with the end of the coal boom were left behind like refugees from another era, their families consigned to a threadbare existence. We understood that these were working folk who had fallen on hard times. The whole country reeled at the thought that such desperate conditions existed in the heartland—among good (white) people—and the war on poverty was born in order to lift up the downtrodden.

In the intervening years, however, as many writers have noted, poverty has been racialized. Even though the majority of the poor are still white and working—as they were in the 1930s and thereafter—the public impression is quite clearly the reverse: poverty wears a black face and is presumed to follow from an unwillingness to enter the labor force. As Herbert Gans explains in his compelling

SOURCE: From Katherine S. Newman, *No Shame in My Game: The Working Poor in the Inner City,* copyright © 1999 by Russell Sage Foundation. Used by permission of Alfred A. Knopf, a division of Random House, Inc.

book *The War Against the Poor*, the tendency to racialize the undeserving, the objects of society's contempt, is of long standing. The English poor laws of the mid-nineteenth century distinguished sharply between the shiftless and the working poor, associating the former with the Irish, who were regarded as an undesirable racial caste. In the United States, successive waves of immigrants have taken their place in the unholy category of the undeserving poor, including the Italians, the Slavs, and the Eastern European Jews who filled the boweries of the eastern seaboard in the early twentieth century.

But the Great Migration of African-Americans out of the rural South to the industrialized cities of the North in the twentieth century altered forever the racial dynamics of urban poverty. And urban poverty has long been far more visible, to the media and hence to the public at large, than its rural counterpart, however devastating the latter may be. Initially pulled by war-related job opportunities and pushed by agricultural mechanization that put a damper on southern demands for labor, blacks flooded into cities like Chicago, Milwaukee, Akron, New York, Pittsburgh, and elsewhere, only to be met by intense white hostility and overt policies of containment. They were confined in segregated communities that rapidly became overcrowded. . . .

As the jobs that initially sustained the northern migrants began to dry up, the consequences of extreme segregation became all too evident: places like the Cabrini-Green housing project were devastated and volatile. The underground economy and the welfare system were all that remained as forms of subsistence in some projects, and the association of African-Americans who were not gainfully employed with the damning label of "undeserving poor" took root. It was an easy linkage to make in a country obsessed by race.

It is also misleading because in fact the largest group of poor people in the United States are not those on welfare. They are the working poor whose earnings are so meager that despite their best efforts, they cannot afford decent housing, diets, health care, or child care. The debilitating conditions that impinge upon the working poor—substandard housing, crumbling schools, inaccessible health care—are hardly different from those that surround their nonworking counterparts. For many, indeed, these difficulties are measurably worse because too many of the working poor lack access to government supports that cushion those out of the labor force: subsidized housing, medical care, and food stamps.

DEFINING WORKING POVERTY

Debates over how to define poverty have raged for many years now, and not just because scholars love to argue: federal dollars ebb and flow depending on the definition. The technicalities of these debates need not concern us, but some of the principles that underlie them are relevant because they determine whose lives we are talking about when we use the term "working poor." . . .

We might want to think of the working poor as those employees who receive the minimum wage. This is hardly a perfect measure, for many of these minimum-wage earners are young people who live with their middle-class families. While their earnings are low, they do not live in poor households. Still, this is not universally true: only 36 percent of minimum-wage workers are teenagers; 42 percent are adults twenty-five years old and older. Some scholars dismiss the problems of minimum-wage workers because so many of them are women who work part-time (and are therefore presumed to be bringing in "pin money" rather than the necessities of life). While 65 percent of the nation's low-wage workers are women, the gender picture changes dramatically when we examine only *full-time,* year-round minimum wage workers: about 60 percent of them are men. . . .

When readers picture this low-wage workforce, they are likely to imagine hamburger flippers. Service workers of this kind are indeed among the working poor. Yet a surprisingly varied group of American employees take home wages low enough to pull their households below the poverty line. Household service workers (housekeepers, child care workers, and cooks) had a poverty rate of 21.8 percent during 1996. The rate for dental assistants, bartenders, hairdressers, and waitresses was 13.2 percent, and that for operators, fabricators, and laborers was 8 percent.

AGE, FAMILY STATUS, EDUCATION, AND RACE

With these definitions of the working poor in mind, we might ask who they are in a sociological sense: what kinds of people find themselves in this category? The short answer is that the nation's young, its single parents, the poorly educated, and minorities are more likely than other workers to be poor.

As America's youth enter an increasingly inhospitable labor market, they have found themselves at a disadvantage over those who preceded them in better years. Poverty rates for young workers have nearly doubled in the last twenty years. Family status has a powerful impact on the working poor as well. Not surprisingly, families in which both husband and wife work are the least likely to be poor. Working single mothers, on the other hand, are the most likely to be poor—the poverty rate of families supported by single mothers is almost four times that of married-couple families with at least one worker. Single-parent families with mothers at the helm are almost twice as likely to be poor as families maintained solely by men, a reflection of the weak position of women in the labor market. But the predicament for single men should not be understated: 10 percent of America's working men who maintain families without a wife suffer poverty as well. Clearly, lacking a second earner—male or female—a low-wage worker is at great risk for falling below the poverty line.

We know that our changing economy favors the well-educated. Their incomes have risen while high school dropouts have seen a precipitous loss in their wages and their employment rates and a concomitant increase in their poverty. . . .

Race complicates this picture. In an all-too-familiar story of America's racial divide, we discover that the white population has a significantly lower proportion of working poor (5.4 percent using the BLS[*] definition) than either

---

[*]Editors' note: Bureau of Labor Statistics.

African-Americans (13.3 percent) or Hispanics (14.6 percent). This is not just a matter of educational disadvantage concentrated among minorities: African-Americans in the labor force are more likely to be poor than their white counterparts at all levels of the educational continuum. . . .

The clamor over welfare reform that crested in 1996 further tightened the linkage in the public mind between the indigent poor and the whole concept of poverty. The . . . data presented here suggest that this is a relatively minor part of the problem. More important by far are the trends that have increased the size of the *working* population that cannot make ends meet and the millions of children growing up in their households.

THE WORKING POOR IN HARLEM

Central Harlem is one of the poorest parts of New York City. Nearly 30 percent of its households were on public assistance in 1990. Poverty stands at 40 percent. Harlem's walls are laden with graffiti, windows are frequently shattered, and housing projects—magnets for troublemakers—rise above the skyline and dominate the once-elegant brownstones that are now boarded up, their stoops crumbling in decay. Running through the middle of this enclave is a bustling commercial strip, 125th Street (Martin Luther King, Jr. Boulevard), that has a long and glorious history as the crucible of the Harlem Renaissance. The Apollo Theatre, still the grand center of Harlem culture, is on 125th Street, as is a large state office building, named after one of the most famous of the community's native sons, Adam Clayton Powell, Jr.

Harlem residents do much of their shopping along 125th Street, for it is there that the most accessible drugstores, clothing shops, furniture marts, and fast food restaurants are to be found. To buy anything at a fancier store requires a subway trip or bus ride to another neighborhood. Shopping is therefore a community ritual, a constant public parade through the heart of America's best-known African-American enclave.

The main thoroughfare is also the place many Harlem residents turn to when they try to find work. Hence it is also the place where I began to look for the working-poor families. . . . I found them by way of four fast food restaurants that cater to the African-American and Latino clientele in the center of Harlem and on its periphery, where immigrant families from the Dominican Republic and Puerto Rico cluster. Beginning with fast food eateries was not an accident, for the stereotype of hamburger-flipping as the emblematic low-wage job turns out to be quite accurate. *More than one in every fifteen Americans has worked in the fast food industry.* Minority youth are a big part of this labor force: some one in eight of them has worked in the business. Hence the public image of the "McJob" accords fairly well with reality, for the fast food industry is a critical gateway into the labor market for thousands of people who live in inner-city neighborhoods. It is from this starting point that they hope to launch careers that will take them out of the minimum-wage bracket into a job that pays enough to sustain a family.

As is true in any study of a particular occupational group, fast food workers do not necessarily represent the entire universe of low-wage workers. One could just as usefully study people who scan in groceries in supermarkets or clerks in low-priced clothing shops. The question naturally arises, then, how far we can generalize from the experience of Harlem's fast food workers to the working poor elsewhere in the country. . . .

Race/Ethnicity

Nearly all of the residents of central Harlem are African-American. But the workforce we encountered in Harlem's fast food establishments was far more diverse: only slightly over half of its workers are black. A large group, nearly one-quarter, are immigrants who were either born in the Dominican Republic and immigrated to New York as children or first-generation Americans of Dominican parentage. They are members of one of the fastest growing groups of newcomers to New York. . . .

Sons and daughters of Dominican immigrants tend to live in enclaves dominated by friends, family members, and other co-ethnics, communities where most children begin life as monolingual speakers of Spanish. Eventually, of course, they learn the English language and American ways as they move through New York's school system and rub shoulders with outsiders who do not share their culture. Nonetheless, the identification they feel with their fellow immigrants—an identity reinforced by frequent visits and long stays in their "homeland"—is so strong that they think of themselves as Dominican even though many were born in New York.

Sixteen percent of the people whom we encountered in the restaurant workforce were Latinos from elsewhere in the Caribbean, principally Puerto Rico, which has for decades sent workers to the U.S. mainland for jobs, and pulled them back with family ties that run deep. Return migration, or, rather, circular patterns of migration, are so common in this community that a hybrid culture has developed: the Nuyoricans. Puerto Ricans are the largest Hispanic group in New York; one-half of all the Latinos in the city are of Puerto Rican origin. In fact, they are among the largest ethnic groups of any kind, numbering nearly a million and increasing at a rapid rate. . . .

Puerto Ricans and Dominicans represent the two largest groups of Latino immigrants in New York City. They are also among the poorest of the city's residents: Hispanics have median household incomes that are two-thirds of the median for whites or less, and their poverty rates are at least one and a half times greater than those for whites. Nuyoricans and Dominicans, who are important entrants to Harlem's low-wage economy, are also among the poorest of the city's Hispanics, falling well behind the smaller and more prosperous groups of Argentines, Panamanians, and Peruvians. . . .

Although the two groups of Latinos are comparable in many respects (age, fertility, family structure), Puerto Ricans tend to have higher earnings.

Dominicans in New York—and in my study—tend to make up for this disadvantage by increasing the number of people per household who are in the labor market. Both groups, however, are at great risk for joining the ranks of the working poor. In this they resemble the native-born African-Americans in Harlem, whose options are similarly limited by the pressures of racial discrimination, an imploding labor market, and a weak educational system that does little to position New York's poverty population for the good jobs that remain.

Age

We usually think of a fast food job as a youngster's first stepping-stone into the labor market, a part-time foray for pocket money and a foretaste of an adult lifetime of full-time employment. In most of the nation's suburbs, and in some of its more affluent cities, that is exactly who occupies this employment niche. Not so in central Harlem. . . . While a significant number of high-school-age kids are working in these restaurants, most of the employees are considerably older. Seventy percent of the workers we interviewed are over nineteen. Thirty-five percent are well into adulthood, twenty-three years old and older. This stands in marked contrast to national figures on the fast food workforce. Nationwide, nearly three-quarters of the workers in this industry are twenty years old or younger, with more than one-quarter in the high-school-age group. No doubt the differences reflect the relative paucity of opportunities in a depressed economy like that of central Harlem. Older people have fewer better-paying prospects and are therefore "pushing down" into jobs that would qualify as entry-level way stations elsewhere.

When we look at the local age figures for African-Americans, the contrast with the national average-aged worker grows sharper. . . . One-half of the African-Americans in this workforce are twenty-three years old and older, reflecting once again the more restricted nature of the labor market opportunities open to them. Dominicans, by contrast, look more like the stereotypical young worker in a hamburger flipping job: half of them are in the youngest age group and only 16 percent are as old as the majority of black fast food workers.

Education

Fast food jobs are often thought to act as magnets, pulling young people away from school, distracting their attention from the kind of "human capital" investment that will pay off in the long run. . . . Contrary to prevailing stereotypes of fast food workers as high school dropouts—these inner-city workers are both quite strongly attached to the educational enterprise and better educated as a group than we might have expected. About 70 percent of the youngest workers (fifteen-to-eighteen-year-olds) are in a regular high school program, and another 10 percent are enrolled in an alternative high school. Eight percent are going to college. Less than 10 percent, most of whom are immigrant workers who completed their schooling in Latin America, are completely detached from the world of education. . . .

For many people in these age groups, education is an ongoing process that will continue for many years into adulthood. It is instructive, then, to consider what they hope to accomplish over the long run—what the *aspirations* of these low-wage workers look like. Perhaps in part because of their encounter with the raw world of low-wage employment and their desire to find something better over the long run, Burger Barn employees have very high expectations for their educational futures. Over half of the youngest workers expect to go on to college. African-Americans have higher educational goals than do their Latino counterparts, and part-time workers (many of whom are taking courses while working) expect to do better than those who work full-time.

Whether these aspirations will be realized or not is hard to forecast, but the data on the actual enrollment patterns of these fast food workers are not particularly encouraging. The older the workers, the less likely they are to be enrolled in courses of any kind, and most of those who are in school are taking vocational, technical, or para-professional courses, rather than liberal arts courses of the kind that is typical for the four-year institutions they aspire to attend. Older workers with adult responsibilities may find higher education out of reach.

Still, we see an extraordinary commitment to higher education in aspiration if not in practice, even among those fast food employees who are already quite a bit older than the normal middle-class college student. The numbers reflect what we found in the course of many months of conversation with Harlem's low-wage workers: an enduring belief in education as an essential credential for mobility in the labor market. This portrait could not possibly be farther from the stereotypical picture of the inner-city minority (usually described as a high school dropout with no appreciation of the value of education).

Gender

We have known for some time that low-wage work is often women's work. Nationally, women are concentrated in the jobs that cluster at the bottom of the income distribution. Fast food jobs, prototypical examples of low-wage work, are overwhelmingly held by females in the United States: in 1984, two-thirds of this labor force was female. Not so in Harlem. Nearly half of the workers in these ghetto restaurants were men. This suggests that men face a steeper uphill battle in finding better jobs in communities like Harlem, ultimately "slipping down" into sectors that, elsewhere, would be largely women's employment preserves.

Family/Household Structure

There are many ways to describe . . . Harlem's fast food labor force, and we have already encountered some—race, gender, and age being among the most common demographic descriptors. One other needs to be added, however, if we are to understand who these people are: family structure. In suburban America, we would expect to find that the fast food labor force was composed mainly of young people who are living at home with their parents and passing through this work experience as a way station to a more remunerative future. We already know that Harlem's restaurant workers are older. But what kinds of families do these low-wage workers come from, or more particularly, what do the households they live in look like?

Almost one-third of these Harlem workers live with a single parent (overwhelmingly a mother). Only 13 percent of these workers are children living in a nuclear household (with both parents). Over half of the workers we studied are over the age of eighteen, but still live with their parents or other relatives. These are young adults who cannot earn enough money to launch independent lives and are therefore forced to remain at home, where resources can be pooled and poverty managed collaboratively in the family.

More than one-third of these Harlem workers were parents trying to support families on the strength of these minimum-wage positions. Very few had children under the age of six, most likely because it is so hard to pay for child care out of wages this low. Those who did have young children were older (twenty-five to thirty-four) and worked full-time; one-third were trying to make ends meet alone, as single parents.

The literature on the urban underclass posits a physical separation of welfare recipients and working people, a separation that supposedly underwrites divergent cultures. Aid to Families with Dependent Children, we are told, spawned negative socialization patterns, most especially an unfamiliarity with (or lack of appreciation for) the world of work. When we look at real families, though, we find that what look like separate worlds to the Bureau of Labor Statistics are whole family units that combine work and welfare, and always have.

Twenty-nine percent of central Harlem's households received support from the welfare system in 1995. However, many of the same families also contained members who were in the labor force. Indeed, among restaurant workers, about one-quarter were living in households where someone was receiving AFDC income, though wages remain the overwhelming source of family support for virtually all of these households. Two in five of the Burger Barn workers are the only formally employed person in their household. This is more than designer sneakers and gold chains: the income these young workers receive is a critical source of support in poor households in Harlem.

Rather than paint welfare and work as different worlds, it makes far more sense to describe them as two halves of a single coin, as an integrated economic system at the very bottom of our social structure. Kyesha's family is a clear example of this fusion. Her mother needs the income her working child brings into the house; Kyesha needs the subsidies (housing, medical care, etc.) that state aid provides to her mother. Only because the two domains are linked can this family manage to make ends meet, and then just barely. Of course, many of these restaurant workers are in families that have no contact with the state welfare system at all. Instead, they are wage workers whose family members are also working for a living. Our understanding of their parents' occupational situations is complicated by the fractured family structure that so many are embedded in: 27 percent of their fathers are deceased, and 14 percent of their mothers as well. Since very few of the people we studied were over the age of forty, this suggests a pattern of early mortality among their parents, with all the difficulties this brings about in the lives of young people. Divorce and the incidence of never-married mothers produce enough distance between children and their fathers that many know only that their fathers are alive, and are not aware of what they do for a living.

Still, we do know something about the employment patterns of the parents of our restaurant workers. Over half of their mothers are working, with medical services—hospitals, home care agencies, and nursing homes—providing by far the largest source of employment. The health care industry was one of the few sectors that continued to grow in New York City throughout the relentless recession of the early 1990s; it absorbed many of the inner city's working poor.

What we know about the fathers of these fast food workers is probably less reliable than the information about mothers, simply because the ties between children and their fathers are more tenuous. Nonetheless, 46 percent of the fathers are reported by their sons and daughters to be employed. For many of them, we have no information about their occupation at all because their children do not know where they work. Those for whom we have information are employed as skilled craftsmen and transportation workers. The rest are scattered in janitorial services, factory labor, and retail trades. A number of them are retired and 10 percent are presently unemployed but once worked. . . .

Ongoing changes in the American economy have pulled the rug out from under the low-wage labor market. The continued fiscal instability of cities like New York is spurring cuts in public employment and critical services. Federal government retrenchment has reduced funding for everything from housing subsidies to Medicaid. These trends conspire to make the problems of the working poor more severe than they used to be. That's the bad news. The good news is that despite all of these difficulties, the nation's working poor continue to seek their salvation in the labor market. That such a commitment persists when the economic rewards are so minimal is testimony to the durability of the work ethic, to the powerful reach of mainstream American culture, which has always placed work at the center of our collective moral existence.

REFERENCES

Gans, Herbert. 1995. *The War Against the Poor.* New York: Basic Books.

39

Our Mothers' Grief

Racial-Ethnic Women and the Maintenance of Families

BONNIE THORNTON DILL

REPRODUCTIVE LABOR[1] FOR WHITE WOMEN
IN EARLY AMERICA

In eighteenth- and nineteenth-century America, the lives of white[2] women in the United States were circumscribed within a legal and social system based on patriarchal authority. This authority took two forms: public and private. The social, legal, and economic position of women in this society was controlled through the private aspects of patriarchy and defined in terms of their relationship to families headed by men. The society was structured to confine white wives to reproductive labor within the domestic sphere. At the same time, the formation, preservation, and protection of families among white settlers was seen as crucial to the growth and development of American society. Building, maintaining, and supporting families was a concern of the State and of those organizations that prefigured the State. Thus, while white women had few legal rights as women, they were protected through public forms of patriarchy that acknowledged and supported their family roles of wives, mothers, and daughters because they were vital instruments for building American society. . . .

In colonial America, white women were seen as vital contributors to the stabilization and growth of society. They were therefore accorded some legal and economic recognition through a patriarchal family structure. . . .

Throughout the colonial period, women's reproductive labor in the family was an integral part of the daily operation of small-scale family farms or artisan's shops. According to Kessler-Harris (1981), a gender-based division of labor was common, but not rigid. The participation of women in work that was essential to family survival reinforced the importance of their contributions to both the protection of the family and the growth of society.

Between the end of the eighteenth and mid-nineteenth century, what is labeled the "modern American family" developed. The growth of industrialization and an urban middle class, along with the accumulation of agrarian wealth

SOURCE: From *Journal of Family History* 13 (1988): 415–31. Reprinted by permission.

among Southern planters, had two results that are particularly pertinent to this discussion. First, class differentiation increased and sharpened, and with it, distinctions in the content and nature of women's family lives. Second, the organization of industrial labor resulted in the separation of home and family and the assignment to women of a separate sphere of activity focused on childcare and home maintenance. Whereas men's activities became increasingly focused upon the industrial competitive sphere of work, "women's activities were increasingly confined to the care of children, the nurturing of the husband, and the physical maintenance of the home" (Degler 1980, p. 26).

This separate sphere of domesticity and piety became both an ideal for all white women as well as a source of important distinctions between them. As Matthaei (1982) points out, tied to the notion of wife as homemaker is a definition of masculinity in which the husband's successful role performance was measured by his ability to keep his wife in the homemaker role. The entry of white women into the labor force came to be linked with the husband's assumed inability to fulfill his provider role.

For wealthy and middle-class women, the growth of the domestic sphere offered a potential for creative development as homemakers and mothers. Given ample financial support from their husband's earnings, some of these women were able to concentrate their energies on the development and elaboration of the more intangible elements of this separate sphere. They were also able to hire other women to perform the daily tasks such as cleaning, laundry, cooking, and ironing. Kessler-Harris cautions, however, that the separation of productive labor from the home did not seriously diminish the amount of physical drudgery associated with housework, even for middle-class women. . . . In effect, household labor was transformed from economic productivity done by members of the family group to home maintenance; childcare and moral uplift done by an isolated woman who perhaps supervised some servants.

Working-class white women experienced this same transformation but their families' acceptance of the domestic code meant that their labor in the home intensified. Given the meager earnings of working-class men, working-class families had to develop alternative strategies to both survive and keep the wives at home. The result was that working-class women's reproductive labor increased to fill the gap between family need and family income. Women increased their own production of household goods through things such as canning and sewing; and by developing other sources of income, including boarders and homework. A final and very important source of other income was wages earned by the participation of sons and daughters in the labor force. In fact, Matthaei argues that "the domestic homemaking of married women was supported by the labors of their daughters" (1982, p. 130). . . .

Another way in which white women's family roles were socially acknowledged and protected was through the existence of a separate sphere for women. The code of domesticity, attainable for affluent women, became an ideal toward which nonaffluent women aspired. Notwithstanding the personal constraints placed on women's development, the notion of separate spheres promoted the growth and stability of family life among the white middle class and became the basis for working-class men's efforts to achieve a family wage, so that they could keep their wives at home. Also, women gained a distinct sphere of authority and expertise that yielded them special recognition.

During the eighteenth and nineteenth centuries, American society accorded considerable importance to the development and sustenance of European immigrant families. As primary laborers in the reproduction and maintenance of family life, women were acknowledged and accorded the privileges and protections deemed socially appropriate to their family roles. This argument acknowledges the fact that the family structure denied these women many rights and privileges and seriously constrained their individual growth and development. Because women gained social recognition primarily through their membership in families, their personal rights were few and privileges were subject to the will of the male head of the household. Nevertheless, the recognition of women's reproductive labor as an essential building block of the family, combined with a view of the family as the cornerstone of the nation, distinguished the experiences of the white, dominant culture from those of racial ethnics.

Thus, in its founding, American society initiated legal, economic, and social practices designed to promote the growth of family life among European colonists. The reception colonial families found in the United States contrasts sharply with the lack of attention given to the families of racial-ethnics. Although the presence of racial-ethnics was equally as important for the growth of the nation, their political, economic, legal, and social status was quite different.

REPRODUCTIVE LABOR AMONG RACIAL-ETHNICS IN EARLY AMERICA

Unlike white women, racial-ethnic women experienced the oppressions of a patriarchal society but were denied the protections and buffering of a patriarchal family. Their families suffered as a direct result of the organization of the labor systems in which they participated.

Racial-ethnics were brought to this country to meet the need for a cheap and exploitable labor force. Little attention was given to their family and community life except as it related to their economic productivity. Labor, and not the existence or maintenance of families, was the critical aspect of their role in building the nation. Thus they were denied the social structural supports necessary to make *their* families a vital element in the social order. Family membership was not a key means of access to participation in the wider society. The lack of social, legal, and economic support for racial-ethnic families intensified and extended women's reproductive labor, created tensions and strains in family relationships, and set the stage for a variety of creative and adaptive forms of resistance.

AFRICAN-AMERICAN SLAVES

Among students of slavery, there has been considerable debate over the relative "harshness" of American slavery, and the degree to which slaves were permitted or encouraged to form families. It is generally acknowledged that many slave-owners found it economically advantageous to encourage family formation as a way of reproducing and perpetuating the slave labor force. This became increasingly true after 1807 when the importation of African slaves was explicitly prohibited. The existence of these families and many aspects of their functioning, however, were directly controlled by the master. In other words, slaves married and formed families but these groupings were completely subject to the master's decision to let them remain intact. One study has estimated that about 32 percent of all recorded slave marriages were disrupted by sale, about 45 percent by death of a spouse, about 10 percent by choice, with the remaining 13 percent not disrupted at all (Blassingame 1972, pp. 90–92). African slaves thus quickly learned that they had a limited degree of control over the formation and maintenance of their marriages and could not be assured of keeping their children with them. The threat of disruption was perhaps the most direct and pervasive cultural assault[3] on families that slaves encountered. Yet there were a number of other aspects of the slave system which reinforced the precariousness of slave family life.

In contrast to some African traditions and the Euro-American patterns of the period, slave men were not the main provider or authority figure in the family. The mother-child tie was basic and of greatest interest to the slaveowner because it was critical in the reproduction of the labor force.

In addition to the lack of authority and economic autonomy experienced by the husband-father in the slave family, use of the rape of women slaves as a weapon of terror and control further undermined the integrity of the slave family. . . . The slave family, therefore, was at the heart of a peculiar tension in the master–slave relationship. On the one hand, slaveowners sought to encourage familial ties among slaves because, as Matthaei (1982) states: ". . . these provided the basis of the development of the slave into a self-conscious socialized human being" (p. 81). They also hoped and believed that this socialization process would help children learn to accept their place in society as slaves. Yet the master's need to control and intervene in the familial life of the slaves is indicative of the other side of this tension. Family ties had the potential for becoming a competing and more potent source of allegiance than the slavemaster himself. Also, kin were as likely to socialize children in forms of resistance as in acts of compliance.

It was within this context of surveillance, assault, and ambivalence that slave women's reproductive labor took place. She and her menfolk had the task of preserving the human and family ties that could ultimately give them a reason for living. They had to socialize their children to believe in the possibility of a life in which they were not enslaved. The slave woman's labor on behalf of the family was, as Davis (1971) has pointed out, the only labor the slave engaged in that could not be directly appropriated by the slaveowner for his own profit.

Yet, its indirect appropriation, as labor crucial to the reproduction of the slave-owner's labor force, was the source of strong ambivalence for many slave women. Whereas some mothers murdered their babies to keep them from being slaves, many sought within the family sphere a degree of autonomy and creativity denied them in other realms of the society. The maintenance of a distinct African-American culture is testimony to the ways in which slaves maintained a degree of cultural autonomy and resisted the creation of a slave family that only served the needs of the master.

Gutman (1976) provides evidence of the ways in which slaves expressed a unique Afro-American culture through their family practices. He provides data on naming patterns and kinship ties among slaves that flies in the face of the dominant ideology of the period. That ideology argued that slaves were immoral and had little concern for or appreciation of family life.

Yet Gutman demonstrated that within a system which denied the father authority over his family, slave boys were frequently named after their fathers, and many children were named after blood relatives as a way of maintaining family ties. Gutman also suggested that after emancipation a number of slaves took the names of former owners in order to reestablish family ties that had been disrupted earlier. On plantation after plantation, Gutman found considerable evidence of the building and maintenance of extensive kinship ties among slaves. In instances where slave families had been disrupted, slaves in new communities reconstituted the kinds of family and kin ties that came to characterize Black family life throughout the South. These patterns included, but were not limited to, a belief in the importance of marriage as a long-term commitment, rules of exogamy that included marriage between first cousins, and acceptance of women who had children outside of marriage. Kinship networks were an important source of resistance to the organization of labor that treated the individual slave, and not the family, as the unit of labor (Caulfield 1974).

Another interesting indicator of the slaves' maintenance of some degree of cultural autonomy has been pointed out by Wright (1981) in her discussion of slave housing. Until the early 1800s, slaves were often permitted to build their housing according to their own design and taste. During that period, housing built in an African style was quite common in the slave quarters. By 1830, however, slaveowners had begun to control the design and arrangement of slave housing and had introduced a degree of conformity and regularity to it that left little room for the slave's personalization of the home. Nevertheless, slaves did use some of their own techniques in construction and often hid it from their masters. . . .

Housing is important in discussions of family because its design reflects sociocultural attitudes about family life. The housing that slaveowners provided for their slaves reflected a view of Black family life consistent with the stereotypes of the period. While the existence of slave families was acknowledged, it certainly was not nurtured. Thus, cabins were crowded, often containing more than one family, and there were no provisions for privacy. Slaves had to create their own. . . .

Perhaps most critical in developing an understanding of slave women's reproductive labor is the gender-based division of labor in the domestic sphere. The organization of slave labor enforced considerable equality among men and women. The ways in which equality in the labor force was translated into the family sphere is somewhat speculative. . . .

We know, for example, that slave women experienced what has recently been called the "double day" before most other women in this society. Slave narratives (Jones 1985; White 1985; Blassingame 1977) reveal that women had primary responsibility for their family's domestic chores. They cooked (although on some plantations meals were prepared for all of the slaves), sewed, cared for their children, and cleaned house, all after completing a full day of labor for the master. Blassingame (1972) and others have pointed out that slave men engaged in hunting, trapping, perhaps some gardening, and furniture making as ways of contributing to the maintenance of their families. Clearly, a gender-based division of labor did exist within the family and it appears that women bore the larger share of the burden for housekeeping and child care. . . .

Black men were denied the male resources of a patriarchal society and therefore were unable to turn gender distinctions into female subordination, even if that had been their desire. Black women, on the other hand, were denied support and protection for their roles as mothers and wives and thus had to modify and structure those roles around the demands of their labor. Thus, reproductive labor for slave women was intensified in several ways: by the demands of slave labor that forced them into the double-day of work; by the desire and need to maintain family ties in the face of a system that gave them only limited recognition; by the stresses of building a family with men who were denied the standard social privileges of manhood; and by the struggle to raise children who could survive in a hostile environment.

This intensification of reproductive labor made networks of kin and quasi-kin important instruments in carrying out the reproductive tasks of the slave community. Given an African cultural heritage where kinship ties formed the basis of social relations, it is not at all surprising that African American slaves developed an extensive system of kinship ties and obligations (Gutman 1976; Sudarkasa 1981). Research on Black families in slavery provides considerable documentation of participation of extended kin in child rearing, childbirth, and other domestic, social, and economic activities (Gutman 1976; Blassingame 1972; Genovese and Miller 1974). . . .

With individual households, the gender-based division of labor experienced some important shifts during emancipation. In their first real opportunity to establish family life beyond the controls and constraints imposed by a slavemaster, family life among Black sharecroppers changed radically. Most women, at least those who were wives and daughters of able-bodied men, withdrew from field labor and concentrated on their domestic duties in the home. Husbands took primary responsibility for the fieldwork and for relations with the owners, such as signing contracts on behalf of the family. Black women were severely criticized by whites for removing themselves from field labor because they were seen to be aspiring to a model of womanhood that was considered inappropriate for them. This reorganization of female labor, however, represented an attempt on the part of Blacks to protect women from some of the abuses of the slave system and to thus secure their family life. It was more likely a response to the particular set of circumstances that the newly freed slaves faced than a reaction to the lives of their former masters. Jones (1985) argues that these patterns were "particularly significant" because at a time when industrial development was introducing a labor system that divided male and female labor, the freed Black family was establishing a pattern of joint work and complementary tasks between males and females that was reminiscent of the preindustrial American families. Unfortunately, these former slaves had to do this without the institutional supports that white farm families had in the midst of a sharecropping system that deprived them of economic independence.

CHINESE SOJOURNERS

An increase in the African slave population was a desired goal. Therefore, Africans were permitted and even encouraged at times to form families subject to the authority and whim of the master. By sharp contrast, Chinese people were explicitly denied the right to form families in the United States through both law and social practice. Although male laborers began coming to the United States in sizable numbers in the middle of the nineteenth century, it was more than a century before an appreciable number of children of Chinese parents were born in America. Tom, a respondent in Nee and Nee's (1973) book, *Longtime Californ'*, says: "One thing about Chinese men in America was you had to be either a merchant or a big gambler, have lot of side money to have a family here. A working man, an ordinary man, just can't!" (p. 80).

Working in the United States was a means of gaining support for one's family with an end of obtaining sufficient capital to return to China and purchase land. The practice of sojourning was reinforced by laws preventing Chinese laborers from becoming citizens, and by restrictions on their entry into this country. Chinese laborers who arrived before 1882 could not bring their wives and were prevented by law from marrying whites. Thus, it is likely that the number of Chinese-American families might have been negligible had it not been for two things: the San Francisco earthquake and fire in 1906, which destroyed all municipal records; and the ingenuity and persistence of the Chinese people who used the opportunity created by the earthquake to increase their numbers in the United States. Since relatives of citizens were permitted entry, American-born Chinese (real and claimed) would visit China, report the birth of a son, and thus create an entry slot. Years later the slot could be used by a relative or purchased. The purchasers were called "paper sons." Paper sons became a major mechanism for increasing the Chinese population, but it was a slow process and the sojourner community remained predominantly male for decades.

The high concentration of males in the Chinese community before 1920 resulted in a split-household form of family. . . .

The women who were in the United States during this period consisted of a small number who were wives and daughters of merchants and a larger percentage who were prostitutes. Hirata (1979) has suggested that Chinese prostitution was an important element in helping to maintain the split-household family. In conjunction with laws prohibiting intermarriage, Chinese prostitution helped men avoid long-term relationships with women in the United States and ensured that the bulk of their meager earnings would continue to support the family at home.

The reproductive labor of Chinese women, therefore, took on two dimensions primarily because of the split-household family form. Wives who remained in China were forced to raise children and care for in-laws on the meager remittances of their sojourning husband. Although we know few details about their lives, it is clear that the everyday work of bearing and maintaining children and a household fell entirely on their shoulders. Those women who immigrated and worked as prostitutes performed the more nurturant aspects of reproductive labor, that is, providing emotional and sexual companionship for men who were far from home. Yet their role as prostitute was more likely a means of supporting their families at home in China than a chosen vocation.

The Chinese family system during the nineteenth century was a patriarchal one wherein girls had little value. In fact, they were considered only temporary members of their father's family because when they married, they became members of their husband's families. They also had little social value: girls were sold by some poor parents to work as prostitutes, concubines, or servants. This saved the family the expense of raising them, and their earnings also became a source of family income. For most girls, however, marriages were arranged and families sought useful connections through this process.

With the development of a sojourning pattern in the United States, some Chinese women in those regions of China where this pattern was more prevalent would be sold to become prostitutes in the United States. Most, however, were married off to men whom they saw only once or twice in the 20- or 30-year period during which he was sojourning in the United States. Her status as wife ensured that a portion of the meager wages he earned would be returned to his family in China. This arrangement required considerable sacrifice and adjustment on the part of wives who remained in China and those who joined their husbands after a long separation. . . .

Despite these handicaps, Chinese people collaborated to establish the opportunity to form families and settle in the United States. In some cases it took as long as three generations for a child to be born on United States soil. . . .

CHICANOS

Africans were uprooted from their native lands and encouraged to have families in order to increase the slave labor force. Chinese people were immigrant laborers whose "permanent" presence in the country was denied. By contrast,

Mexican-Americans were colonized and their traditional family life was disrupted by war and the imposition of a new set of laws and conditions of labor. The hardships faced by Chicano families, therefore, were the result of the United States colonization of the indigenous Mexican population, accompanied by the beginnings of industrial development in the region. The treaty of Guadalupe Hidalgo, signed in 1848, granted American citizenship to Mexicans living in what is now called the Southwest. The American takeover, however, resulted in the gradual displacement of Mexicans from the land and their incorporation into a colonial labor force (Barrera 1979). In addition, Mexicans who immigrated into the United States after 1848 were also absorbed into the labor force.

Whether natives of Northern Mexico (which became the United States after 1848) or immigrants from Southern Mexico, Chicanos were a largely peasant population whose lives were defined by a feudal economy and a daily struggle on the land for economic survival. Patriarchal families were important instruments of community life and nuclear family units were linked together through an elaborate system of kinship and godparenting. Traditional life was characterized by hard work and a fairly distinct pattern of sex-role segregation. . . .

As the primary caretakers of hearth and home in a rural environment, *Las Chicanas* labor made a vital and important contribution to family survival. . . .

Although some scholars have argued that family rituals and community life showed little change before World War I (Saragoza 1983), the American conquest of Mexican lands, the introduction of a new system of labor, the loss of Mexican-owned land through the inability to document ownership, plus the transient nature of most of the jobs in which Chicanos were employed, resulted in the gradual erosion of this pastoral way of life. Families were uprooted as the economic basis for family life changed. Some immigrated from Mexico in search of a better standard of living and worked in the mines and railroads. Others who were native to the Southwest faced a job market that no longer required their skills and moved into mining, railroad, and agricultural labor in search of a means of earning a living. According to Camarillo (1979), the influx of Anglo[4] capital into the pastoral economy of Santa Barbara rendered obsolete the skills of many Chicano males who had worked as ranchhands and farmers prior to the urbanization of that economy. While some women and children accompanied their husbands to the railroad and mine camps, they often did so despite prohibitions against it. Initially many of these camps discouraged or prohibited family settlement.

The American period (post-1848) was characterized by considerable transiency for the Chicano population. Its impact on families is seen in the growth of female-headed households, which was reflected in the data as early as 1860. Griswold del Castillo (1979) found a sharp increase in female-headed households in Los Angeles, from a low of 13 percent in 1844 to 31 percent in 1880. Camarillo (1979, p. 120) documents a similar increase in Santa Barbara from 15 percent in 1844 to 30 percent by 1880. These increases appear to be due not so much to divorce, which was infrequent in this Catholic population, but to widowhood and temporary abandonment in search of work. Given the hazardous nature of work in the mines and railroad camps, the death of a husband, father or son who was laboring in these sites was not uncommon. Griswold del Castillo (1979) reports a higher death rate among men than women in Los Angeles. The rise in female-headed households, therefore, reflects the instabilities and insecurities introduced into women's lives as a result of the changing social organization of work.

One outcome, the increasing participation of women and children in the labor force was primarily a response to economic factors that required the modification of traditional values. . . .

Slowly, entire families were encouraged to go to railroad workcamps and were eventually incorporated into the agricultural labor market. This was a response both to the extremely low wages paid to Chicano laborers and to the preferences of employers who saw family labor as a way of stabilizing the workforce. For Chicanos, engaging all family members in agricultural work was a means of increasing their earnings to a level close to subsistence for the entire group and of keeping the family unit together. . . .

While the extended family has remained an important element of Chicano life, it was eroded in the American period in several ways. Griswold del Castillo (1979), for example, points out that in 1845 about 71 percent of Angelenos lived in extended families and that by 1880, fewer than half did. This decrease in extended families appears to be a response to the changed economic conditions and to the instabilities generated by the new sociopolitical structure. Additionally, the imposition of American law and custom ignored and ultimately undermined some aspects of the extended family. The extended family in traditional Mexican life consisted of an important set of familial, religious, and community obligations. Women, while valued primarily for their domesticity, had certain legal and property rights that acknowledged the importance of their work, their families of origin and their children. . . .

In the face of the legal, social, and economic changes that occurred during the American period, Chicanas were forced to cope with a series of dislocations in traditional life. They were caught between conflicting pressures to maintain traditional women's roles and family customs and the need to participate in the economic support of their families by working outside the home. During this period the preservation of some traditional customs became an important force for resisting complete disarray. . . .

Of vital importance to the integrity of traditional culture was the perpetuation of the Spanish language. Factors that aided in the maintenance of other aspects of Mexican culture also helped in sustaining the language. However, entry into English-language public schools introduced the children and their families to systematic efforts to erase their native tongue. . . .

Another key factor in conserving Chicano culture was the extended family network, particularly the system of *compadrazgo* or godparenting. Although the full extent of the impact of the American period on the Chicano extended family is not known, it is generally acknowledged that this family system, though lacking many legal and social sanctions, played an important role in the preservation of the Mexican community (Camarillo 1979, p. 13). In Mexican society, godparents were an important way of linking family and community through respected friends or authorities. Named at the important rites of passage in a child's life, such as birth, confirmation, first communion, and marriage, *compadrazgo* created a moral obligation for godparents to act as guardians, to provide financial assistance in times of need, and to substitute in case of the death of a parent. Camarillo (1979) points out that in traditional society these bonds cut across class and racial lines. . . .

The extended family network—which included godparents—expanded the support groups for women who were widowed or temporarily abandoned and for those who were in seasonal, part-, or full-time work. It suggests, therefore, the potential for an exchange of services among poor people whose income did not provide the basis for family subsistence. . . . This family form is important to the continued cultural autonomy of the Chicano community.

CONCLUSION: OUR MOTHERS' GRIEF

Reproductive labor for Afro-American, Chinese-American, and Mexican-American women in the nineteenth century centered on the struggle to maintain family units in the face of a variety of cultural assaults. Treated primarily as individual units of labor rather than as members of family groups, these women labored to maintain, sustain, stabilize, and reproduce their families while working in both the public (productive) and private (reproductive) spheres. Thus, the concept of reproductive labor, when applied to women of color, must be modified to account for the fact that labor in the productive sphere was required to achieve even minimal levels of family subsistence. Long after industrialization had begun to reshape family roles among middle-class white families, driving white women into a cult of domesticity, women of color were coping with an extended day. This day included subsistence labor outside the family and domestic labor within the family. For slaves, domestics, migrant farm laborers, seasonal factory-workers, and prostitutes, the distinctions between labor that reproduced family life and which economically sustained it were minimized. The expanded workday was one of the primary ways in which reproductive labor increased.

Racial-ethnic families were sustained and maintained in the face of various forms of disruption. Yet racial-ethnic women and their families paid a high price in the process. High rates of infant mortality, a shortened life span, the early onset of crippling and debilitating disease provided some insight into the costs of survival.

The poor quality of housing and the neglect of communities further increased reproductive labor. Not only did racial-ethnic women work hard outside the home for a mere subsistence, they worked very hard inside the home to achieve even minimal standards of privacy and cleanliness. They were continually faced with disease and illness that directly resulted from the absence of basic sanitation. The fact that some African women murdered their children to prevent them from becoming slaves is an indication of the emotional strain associated with bearing and raising children while participating in the colonial labor system.

We have uncovered little information about the use of birth control, the prevalence of infanticide, or the motivations that may have generated these or other behaviors. We can surmise, however, that no matter how much children were accepted, loved, or valued among any of these groups of people, their futures in a colonial labor system were a source of grief for their mothers. For those children who were born, the task of keeping them alive, of helping them to understand and participate in a system that exploited them, and the challenge of maintaining a measure—no matter how small—of cultural integrity, intensified reproductive labor.

Being a racial-ethnic woman in nineteenth-century American society meant having extra work both inside and outside the home. It meant having a contradictory relationship to the norms and values about women that were being generated in the dominant white culture. As pointed out earlier, the notion of separate spheres of male and female labor had contradictory outcomes for the nineteenth-century whites. It was the basis for the confinement of women to the household and for much of the protective legislation that subsequently developed. At the same time, it sustained white families by providing social acknowledgment and support to women in the performance of their family roles. For racial-ethnic women, however, the notion of separate spheres served to reinforce their subordinate status and became, in effect, another assault. As they increased their work outside the home, they were forced into a productive labor sphere that was organized for men and "desperate" women who were so unfortunate or immoral that they could not confine their work to the domestic sphere. In the productive sphere, racial-ethnic women faced exploitative jobs and depressed wages. In the reproductive sphere, however, they were denied the opportunity to embrace the dominant ideological definition of "good" wife or mother. In essence, they were faced with a double-bind situation, one that required their participation in the labor force to sustain family life but damned them as women, wives, and mothers because they did not confine their labor to the home. Thus, the conflict between ideology and reality in the lives of racial-ethnic women during the nineteenth century sets the stage for stereotypes, issues of self-esteem, and conflicts around gender-role prescriptions that surface more fully in the twentieth century. Further, the tensions and conflicts that characterized their lives during this period provided the impulse for community activism to jointly address the inequities, which they and their children and families faced.

NOTES

1. The term *reproductive labor* is used to refer to all of the work of women in the home. This includes but is not limited to: the buying and preparation of food and clothing, provision of emotional support and nurturance for all family members, bearing children, and planning, organizing, and carrying out a wide variety of tasks associated with their socialization. All of these activities are necessary for the growth of

patriarchal capitalism because they maintain, sustain, stabilize, and *reproduce* (both biologically and socially) the labor force.

2. The term *white* is a global construct used to characterize peoples of European descent who migrated to and helped colonize America. In the seventeenth century, most of these immigrants were from the British Isles. However, during the time period covered by this article, European immigrants became increasingly diverse. It is a limitation of this article that time and space does not permit a fuller discussion of the variations in the white European immigrant experience. For the purposes of the argument made herein and of the contrast it seeks to draw between the experiences of mainstream (European) cultural groups and that of racial/ethnic minorities, the differences among European settlers are joined and the broad similarities emphasized.

3. Cultural assaults, according to Caulfield (1974), are benign and systematic attacks on the institutions and forms of social organization that are fundamental to the maintenance and flourishing of a group's culture.

4. This term is used to refer to white Americans of European ancestry.

REFERENCES

Barrera, Mario. 1979. *Race and Class in the Southwest*. South Bend, IN: Notre Dame University Press.

Blassingame, John. 1972. *The Slave Community: Plantation Life in the Antebellum South*. New York: Oxford University Press.

Blassingame, John. 1977. *Slave Testimony: Two Centuries of Letters, Speeches, Interviews, and Autobiographies*. Baton Rouge, LA: Louisiana State University Press.

Camarillo, Albert. 1979. *Chicanos in a Changing Society*. Cambridge, MA: Harvard University Press.

Caulfield, Mina Davis. 1974. "Imperialism, the Family, and Cultures of Resistance." *Socialist Review* 4(2)(October): 67–85.

Davis, Angela. 1971. "The Black Woman's Role in the Community of Slaves." *Black Scholar* 3(4)(December): 2–15.

Degler, Carl. 1980. *At Odds*. New York: Oxford University Press.

Genovese, Eugene D., and Elinor Miller, eds. 1974. *Plantation, Town, and County: Essays on the Local History of American Slave Society*. Urbana: University of Illinois Press.

Griswold del Castillo, Richard. 1979. *The Los Angeles Barrio: 1850–1890*. Los Angeles: The University of California Press.

Gutman, Herbert. 1976. *The Black Family in Slavery and Freedom: 1750–1925*. New York: Pantheon.

Hirata, Lucie Cheng. 1979. "Free, Indentured, Enslaved: Chinese Prostitutes in Nineteenth-Century America." *Signs 5* (Autumn): 3–29.

Jones, Jacqueline. 1985. *Labor of Love, Labor of Sorrow*. New York: Basic Books.

Kessler-Harris, Alice. 1981. *Women Have Always Worked*. Old Westbury: The Feminist Press.

Matthaei, Julie. 1982. *An Economic History of Women in America*. New York: Schocken Books.

Nee, Victor G., and Brett de Bary Nee. 1973. *Longtime Californ'*. New York: Pantheon Books.

Saragoza, Alex M. 1983. "The Conceptualization of the History of the Chicano Family: Work, Family, and Migration in Chicanos." Research Proceedings of the Symposium on Chicano Research and Public Policy. Stanford, CA: Stanford University, Center for Chicano Research.

Sudarkasa, Niara. 1981. "Interpreting the African Heritage in Afro-American Family Organization." Pp. 37–53 in *Black Families,* edited by Harriette Pipes McAdoo. Beverly Hills, CA: Sage Publications.

Wright, Deborah Gray. 1985. *Ar'n't I a Woman?: Female Slaves in the Plantation South.* New York: W. W. Norton.

Wright, Gwendolyn. 1981. *Building the Dream: A Social History of Housing in America.* New York: Pantheon Books.

40

Navigating Interracial Borders

Black-White Couples and Their Social Worlds

ERICA CHITO CHILDS

The 1967 Academy Award–winning movie *Guess Who's Coming to Dinner* concluded with a warning from a white father to his daughter and her "Negro" fiancé. That same year, the Supreme Court overturned any laws against interracial marriage as unconstitutional. Yet how does the contemporary U.S. racial landscape compare? In this ever-changing world of race and color, where do black-white couples fit, and has this unimaginable opposition disappeared?

While significant changes have occurred in the realm of race relations largely from the civil rights struggle of the 1960s, U.S. society still has racial borders. Most citizens live, work, and socialize with others of the same race—as if living within borders, so to speak—even though there are no longer legal barriers such as separate facilities or laws against intermarriage. Yet if these largely separate racial worlds exist, what social world(s) do black-white couples live in and how do they navigate these racial borders? Even more important, how do white communities and black communities view and respond to black-white couples? In other words, do they navigate the racial borders by enforcing, ignoring, or actively trying to

SOURCE: From Erica Chito Childs, *Navigating Interracial Borders: Black-White Couples and Their Social World* (New Brunswick, NJ: Rutgers University Press, © 2005.)

dismantle them? My goal is to explore these issues to better understand the contemporary beliefs and practices surrounding black-white couples. . . . My data comes from varied sources, including Web sites, black-white couples, Hollywood films, white communities, and black communities. . . .

My own story also brought me to this research. The social world of black-white couples is the world I navigate. From my own experiences, I have seen the ways most whites respond to an interracial relationship. Growing up white, second-generation Portuguese in a predominantly white Rhode Island suburb, race never was an issue, or at least not one I heard about. After moving to Los Angeles during high school and beginning college, I entered into a relationship with an African American man (who I eventually married), never imagining what it would bring. My family did not disown me or hurl racial slurs. Still, in many ways I learned what it meant to be an "interracial couple" and how this was not what my family, community, or countless unknown individuals had scripted for me. Not many whites ever said outright that they were opposed to the relationship, yet their words and actions signaled otherwise.

One of the most telling examples occurred a few years into our relationship. An issue arose when my oldest sister's daughter wanted to attend her prom with an African American schoolmate she was dating. My sister and her husband refused to let him in the house the night of the prom or any other time because, they said, he was "not right" for her. It was clear to everyone, however, that skin color was the problem. To this day, my niece will tell you that her parents would never have accepted her with a black man. Yet my sister and her family never expressed any opposition to my relationship and even seemed supportive, in terms of inviting us over to their house, giving wedding and holiday gifts, and so forth. Although my sister never openly objected to my relationship, she drew the line with her daughter—quite literally enforcing a racial boundary to protect her daughter and family from blackness. For me, this personal story and the countless stories of other interracial couples point to the necessity of examining societal attitudes, beliefs, images, and practices regarding race and, more specifically, black-white relations. Interracial couples—because of their location on the line between white and black—often witness or bring forth racialized responses from both whites and blacks. As with my sister, opposition may exist yet is not visible until a close family member or friend becomes involved or wants to become involved interracially. . . .

INTERRACIAL RELATIONSHIPS AS A
MINER'S CANARY

It is these community and societal responses, as well as the images and beliefs produced and reproduced about these unions that provide the framework within which to understand the issue of interracial couplings. Underlying these responses and images is a racial ideology, or, in other words, a dominant discourse, that posits interracial couples and relationships as deviant. Still, the significance of these discourses and what exactly they reveal about race in society can be hard to see. For some of us the effects of race are all too clear, while for others race—and the accompanying advantages and disadvantages— remain invisible. As a white woman, it was only through my relationship, and raising my two children, that I came to see how race permeates everything in society. Being white yet now part of a multiracial family, I experienced, heard, and even thought things much differently than before, primarily because whites and blacks responded to me differently than before. I think of the metaphor of the "miner's canary"—the canaries miners use to alert them to a poisonous atmosphere. In *The Miner's Canary: Enlisting Race, Resisting Power, Transforming Power*, Lani Guinier and Gerald Torres argue that the experiences of racial minorities, like the miner's canary, can expose the underlying problems in society that ultimately affect everyone, not just minorities. In many ways, the experiences of black-white couples are a miner's canary, revealing problems of race that otherwise can remain hidden, especially to whites. The issues surrounding interracial couples—racialized/sexualized stereotypes, perceptions of difference, familial opposition, lack of community acceptance—should not be looked at as individual problems, but rather as a reflection of the larger racial issues that divide the races. Since interracial couples exist on the color-line in society—a "borderland" between white and black—their experiences and the ways communities respond to these relationships can be used as a lens through which we can understand contemporary race relations. . . .

LIFE ON THE BORDER: NARRATIVES OF
BLACK-WHITE COUPLES

From 1999 to 2001, I interviewed fifteen black-white heterosexual couples who were referred to me through personal and professional contacts, and some of whom I encountered randomly in public. They ranged in age from twenty to sixty-nine and all were in committed relationships of two to twenty-five years. Nine were married. The couples' education levels varied. All respondents had finished high school or its equivalent; twenty-one respondents had attended some college and/or had received a bachelor's degree; and four respondents had advanced degrees. The socioeconomic status of the couples ranged from working class to upper middle class. The respondents included a college student, waitress, manager, factory worker, university professor, social worker, salesperson, and postal worker. All couples lived in the northeastern United States, from Maine to Pennsylvania, yet many of the couples had traveled extensively and had lived in other parts of the country, including California, Florida, and the South.

I interviewed the couples together, since I was interested in their experiences, accounts, narratives, and the ways they construct their lives and create

their "selves" and their identities as "interracial couples." The interviews lasted for two to three hours, and I ended up with more than forty hours of interview data. . . . These accounts are seen not only as "descriptions, opinions, images, or attitudes about race relations but also as 'systems of knowledge' and 'systems of values' in their own right, used for the discovery and organization of reality."

THE SEPARATE WORLDS OF WHITES AND BLACKS

To explore the larger cultural and sociopolitical meanings that black-white couplings have for both the white and black communities in which they occur, a significant portion of this work is based on original qualitative research in white communities and black communities about their ideas, beliefs, and views on interracial sexuality and marriage. Community research was conducted to further explore the responses to interracial couples that are found in social groups and communities—family, friends, neighbors, religious groups, schools, etc. The ways that these couples provide the occasion for groups to express and play out their ideas and prejudices about race and sex are integral to understanding the social construction of interracial couples. . . .

A black person and a white person coming together has been given many names—miscegenation, amalgamation, race mixing, and jungle fever—conjuring up multiple images of sex, race, and taboo. Black-white relationships and marriages have long been viewed as a sign of improving race relations and assimilation, yet these unions have also been met with opposition from both white and black communities. Overall, there is an inherent assumption that interracial couples are somehow different from same-race couples. Within the United States, the responses to black-white couplings have ranged from disgust to curiosity to endorsement, with the couples being portrayed as many things—among them, deviant, unnatural, pathological, exotic, but always sexual. Even the way that couples are labeled or defined as "interracial" tells us something about societal expectations. We name what is different. For example, a male couple is more likely to be called a "gay couple" than a gender-mixed couple is to be called a "heterosexual couple."

Encompassed by the history of race relations and existing interracial images, how do black-white couples view themselves, their relationships, and the responses of their families and communities? And how do they interpret these familial and community responses? Black-white couples, like all of us, make meaning out of their experiences in the available interpretive frameworks and often inescapable rules of race relations in this country. . . .

Black-white couples come together across the boundaries of race and perceived racial difference seemingly against the opposition of their communities. This is not to say, however, that the couples are free from racialized thinking, whether it be in their use of color-blind discourse or their own racial preferences, such as to date only interracially or to live in all-white neighborhoods.

Nonetheless, these couples create multiracial families, not only creating multiracial families of their own but also changing the racial dynamic of the families from which they come. What significance does this have for the institution of family, and how does this play out for the white and black families to whom it occurs?

It might be expected that the family is the source of the greatest hostility toward interracial relationships. It is in families that the meanings and attachments to racial categories are constructed and learned; one's family is often "the most critical site for the generation and reproduction of racial formations." This includes who is and is not an acceptable marriage partner. In white and black families, certain discourses are used when discussing black-white relationships that reproduce the image of these unions as different, deviant, even dangerous. Interracial relationships and marriage often bring forth certain racialized attitudes and beliefs about family and identity that otherwise may have remained hidden. The ways that white and black families understand and respond to black-white interracial couples and the racialized discourses they use are inextricably tied to ideas of family, community, and identity. White and black families' (and communities') interpretations and responses to interracial couples are part of these available discourses on race and race relations in our society. Many times, black-white couples provide the occasion for families to express and play out their ideas and prejudices about race and sex, which is integral to understanding the social construction of "interracial couples" within America today.

ALL IN THE FAMILY: WHITE FAMILIES

Among whites, the issue of interracial marriage is often a controversial topic, and even more so when they are asked to discuss their own views of their family's views. The white community respondents in this study were hesitant to discuss their personal views on family members becoming involved interracially. One strategy used by the respondents was to discuss other families they knew rather than their own views. For example, during the white focus group interviews, the first responses came from two individuals who had some experience or knowledge of interracial couples or families. Sara discussed a friend who adopted two "very dark" black children and how white people would stare at her and the children in public. Anne mentioned her niece who married a black man and said "the family is definitely against it," which causes them problems.

In group interviews with white college students, a number of the students also used stories about other interracial families they knew to explain why their family would prefer they marry someone of the same race. One college student said her family would have a problem with her marrying interracially, explaining that their opinion is based on their experiences with an interracial family in their neighborhood. She had babysat for this family and, according to her, they had "social issues because the dad was *real dark* and the mom was white, and the kids just had major issues." Her choice of words reveals the importance of color and the use of a discursive strategy such as referring to the children as a problem and not the relationship itself. . . .

White "Concern" and Preference

For the white college students interviewed, the role of family is key and certainly influences their decisions not to date interracially. The majority of white students expressed verbally or by raising their hand that their parents, white parents in general, would have a difficult time with an interracial partner for a number of reasons. Yet their parents' opposition was described in nonracial terms, much like their own views. For example, one white male student said, "All parents find something wrong [with the person their child chooses] if they're not exactly as they imagined. My parents would be *surprised* because [an interracial relationship] is not what they are expecting, so it would be difficult in that sense." Many students simply stated something to the effect of "my parents aren't prejudiced; they just wouldn't want me to marry a black guy."

Other students described their families' views against interracial dating as based on the meanings of family and marriage in general. For example, one male student stated that his parents would have a problem with an interracial relationship, because "they brought me up to date within [my] race. They're not like racist, but [they say] just keep with your own culture, be proud of who you are and carry that on to your kids." Some students cited the difference between dating and marriage. As one female college student said, "It becomes more of an issue when you get to be juniors or seniors, because you start thinking long-term, about what your parents will say, and dating a black person just isn't an option." Another student described dating interracially in similar terms: "Dating's not an issue, they always encouraged [me] to interact with all people. For the future and who you're gonna spend the rest of your life with, there's a difference between being with someone of the same race and someone of a different race, [interracial marriage] is not like it's wrong but . . . just that it would be too difficult."

In one of the discussions of family with the white college students, an interesting incident occurred. One of the male students, who was vocal throughout the focus group, sat listening to the other students and looking around the room. Finally he interrupted the discussion: "Are you kidding me? . . . parents would shit, they'd have a freaking heart attack. [*He dramatically grabs the front of his shirt and imitates a growling father's voice*]. "Uhh, son, how could you do this to us, the family?" Everyone laughed at his performance, but this student's use of humor to depict his father (or a white father in general) having "a heart attack" is an interesting discursive strategy. Although he presented it in a comical way, his statement does not seem unrealistic to the group. Often, jocular speech is used to convey a serious message in order to avoid being labeled racist or prejudiced.

When asked why their families would respond in these ways to an interracial relationship, the group largely cited the "opposition of the larger society" as the reason why they and/or their family personally would prefer that their family not become involved interracially. Among the white college students, parents were described as "concerned about how difficult it would be." . . .

IT'S A FAMILY AFFAIR: BLACK FAMILIES

Black families, like white families, can operate as a deterrent to interracial relationships. A family member becoming involved with a white individual is seen as problematic on a number of levels, yet the black families often raised different issues than the white families, such as the importance of "marrying black" and the negative meanings attached to becoming involved interracially. These issues figured prominently among the black partners, the black community respondents, and even black popular culture.

Among the black college students and community respondents, a main issue was the emphasis on marrying within their race, explicitly identifying race as the issue, unlike the white communities. Most students discussed how their parents would have a problem. One black female student stated, "My family would outright disown me" (which received a number of affirmations from the group). Another college student commented on the beliefs her family instilled in her: "My family raised me to be very proud of who I am, a black woman, and they instilled in me the belief that I would never want to be with anyone but a strong black man." Other students described incidents or comments they had heard that let them know they were expected to marry black. Similarly, Leslie, a black community respondent, stated she was raised by her parents to date anyone, but they were "adamant about me marrying black." She added that they told her not to "even think about marrying interracially." Allen also stated that his parents told him that "high school dating is fine but not marriage, ohhh nooo!" Couples like Gwen and Bill, Chris and Victoria, and others also recounted how the black partner's family had more difficulty accepting the relationship when they realized that the couple was getting married as opposed to just dating.

A significant piece of black familial opposition involves the perceived racism of whites in the larger society. Black families were described as having a hard time accepting a family member getting involved with a white person because of lingering racism and a distrust in whites in general. For example, one black college student stated that her family would have a problem if she brought home a white man, "because they would always be wondering what his family was saying, you know, do they talk about me behind my back?" Other students' responses echoed these views, such as one student who remarked, "My mom would have a problem with it. She just doesn't trust white people."

Also, black community members such as Alice and Jean argued that their opposition to a family member getting involved with a white person was based on the belief that the white individual (or their white family/neighborhood) would mistreat their family member. All but two of the black community members expressed the concern that since white society is racist there is no reason to become involved with a white person. . . .

The black families related to this study objected to having a white person in their family and intimate social circles because they viewed whites as the "enemy" and their presence as a sign that the black partner is not committed to his or her community or family. Not surprisingly, black individuals opposed interracial marriage much more than interracial dating, primarily because marriage represents a legitimation of the union and formally brings the white partner into the family.

Despite such opposition, the black college students argued that black families are still more accepting than white families. . . . While white families discourage their family members from engaging in interracial relationships to maintain white privilege, black families discourage these unions to maintain the strength and solidarity of black communities. Black families view interracial relationships as a loss in many ways—the loss of individuals to white society, the weakening of families and communities, and the devaluing of blackness. . . .

FAMILY MATTERS

. . . The familial responses discussed throughout the research clearly demonstrate how images of oneself and one's family is linked to the concept of race and otherness, especially within the construction of families, both white and black. White families often objected to the idea of a member dating interracially, not because they met the black individual and were confronted with overwhelming "racial" differences, but because they were merely responding based on their ideas and beliefs about interracial relationships and blacks in general. . . .

Being in an interracial relationship often brings forth problems within families, since white and black families overwhelmingly want to remain monoracial . . . .

This . . . highlights how difficult it is to negotiate issues of race and family. Discussing race and, more important, views on interracial relationships in nonracial terms often makes it even more difficult for individuals to challenge and confront their family's views. By using phrases such as "It's not my personal preference" or "I just worry about the problems you will face in society," families and individuals are able to oppose interracial relationships without appearing prejudiced or racist. A family member's opposition, however, is also tied to the issue of identity—the identity of the family and the individuals involved. Families express concern over the identities of the biracial children who will be produced, and in many ways there is a tendency for white families to worry that the children will be "too black," and the black families worry about the children being "more white." The white individuals who enter into a relationship with a black person are seen as "less white" and as tainting the white family, while the black individuals who get involved are seen as "not black enough" and as leaving their blackness behind. All of these fears and beliefs demonstrate the centrality of race to the constructions of families and identities and, more important, the socially constructed nature of race, if one's relationship can change one's "race" in this society still divided by racial boundaries. Ultimately, black–white couples and biracial children are forced to exist somewhere in between, with or without their families.

41

Straight Is to Gay as Family Is to No Family

KATH WESTON

IS "STRAIGHT" TO "GAY" AS "FAMILY"
IS TO "NO FAMILY"?

For years, and in an amazing variety of contexts, claiming a lesbian or gay identity has been portrayed as a rejection of "the family" and a departure from kinship. In media portrayals of AIDS, Simon Watney . . . observes that "we are invited to imagine some absolute divide between the two domains of 'gay life' and 'the family,' as if gay men grew up, were educated, worked and lived our lives in total isolation from the rest of society." Two presuppositions lend a dubious credence to such imagery: the belief that gay men and lesbians do not have children or establish lasting relationships, and the belief that they invariably alienate adoptive and blood kin once their sexual identities become known. By presenting "the family" as a unitary object, these depictions also imply that everyone participates in identical sorts of kinship relations and subscribes to one universally agreed–upon definition of family.

Representations that exclude lesbians and gay men from "the family" invoke what Blanche Wiesen Cook . . . has called "the assumption that gay people do not love and do not work," the reduction of lesbians and gay men to sexual identity, and sexual identity to sex alone. In the United States, sex apart from

SOURCE: From Kath Weston, *Families We Choose: Lesbians, Gays, Kinship* (New York: Columbia University Press, 1991), pp. 22–29. Reprinted by permission of Columbia University Press.

heterosexual marriage tends to introduce a wild card into social relations, signifying unbridled lust and the limits of individualism. If heterosexual intercourse can bring people into enduring association via the creation of kinship ties, lesbian and gay sexuality in these depictions isolates individuals from one another rather than weaving them into a social fabric. To assert that straight people "naturally" have access to family, while gay people are destined to move toward a future of solitude and loneliness, is not only to tie kinship closely to procreation, but also to treat gay men and lesbians as members of a nonprocreative species set apart from the rest of humanity. . . .

It is but a short step from positioning lesbians and gay men somewhere beyond "the family"—unencumbered by relations of kinship, responsibility, or affection—to portraying them as a menace to family and society. A person or group must first be outside and other in order to invade, endanger, and threaten. My own impression from fieldwork corroborates Frances Fitzgerald's . . . observation that many heterosexuals believe not only that gay people have gained considerable political power, but also that the absolute number of lesbians and gay men (rather than their visibility) has increased in recent years. Inflammatory rhetoric that plays on fears about the "spread" of gay identity and of AIDS finds a disturbing parallel in the imagery used by fascists to describe syphilis at mid-century, when "the healthy" confronted "the degenerate" while the fate of civilization hung in the balance. . . .

A long sociological tradition in the United States of studying "the family" under siege or in various states of dissolution lent credibility to charges that this institution required protection from "the homosexual threat." . . .

. . . By shifting without signal between reproduction's meaning of physical procreation and its sense as the perpetuation of society as a whole, the characterization of lesbians and gay men as nonreproductive beings links their supposed attacks on "the family" to attacks on society in the broadest sense. Speaking of parents who had refused to accept her lesbian identity, a Jewish woman explained, "They feel like I'm finishing off Hitler's job." The plausibility of the contention that gay people pose a threat to "the family" (and, through the family, to ethnicity) depends upon a view of family grounded in heterosexual relations, combined with the conviction that gay men and lesbians are incapable of procreation, parenting, and establishing kinship ties.

Some lesbians and gay men . . . had embraced the popular equation of their sexual identities with the renunciation of access to kinship, particularly when first coming out. "My image of gay life was very lonely, very weird, no family," Rafael Ortiz recollected. "I assumed that my family was gone now—that's it." After Bob Korkowski began to call himself gay, he wrote a series of poems in which an orphan was the central character. Bob said the poetry expressed his fear of "having to give up my family because I was queer." When I spoke with Rona Bren after she had been home with the flu, she told me that whenever she was sick, she relived old fears. That day she had remembered her mother's grim prediction: "You'll be a lesbian and you'll be alone the rest of your life. Even a dog shouldn't be alone."

Looking backward and forward across the life cycle, people who equated their adoption of a lesbian or gay identity with a renunciation of family did so in the double-sided sense of fearing rejection by the families in which they had grown up, and not expecting to marry or have children as adults. Although few in numbers, there were still those who had considered "going straight" or getting married specifically in order to "have a family." Vic Kochifos thought he understood why:

It's a whole lot easier being straight in the world than it is being gay. . . . You have built-in loved ones: wife, husband, kids, extended family. It just works easier. And when you want to do something that requires children, and you want to have a feeling of knowing that there's gonna be someone around who cares about you when you're 85 years old, there are thoughts that go through your head, sure. There must be. There's a way of doing it gay, but it's a whole lot harder, and it's less secure.

Bernie Margolis had been sexually involved with men since he was in his teens, but for years had been married to a woman with whom he had several children. At age 67 he regretted having grown to adulthood before the current discussion of gay families, with its focus on redefining kinship and constructing new sorts of parenting arrangements.

I didn't want to give up the possibility of becoming a family person. Of having kids of my own to carry on whatever I built up. . . . My mother was always talking about [how] she's looking forward to the day when she would bring her children under the canopy to get married. It never occurred to her that I wouldn't be married. It probably never occurred to me either.

The very categories "good family person" and "good family man" had seemed to Bernie intrinsically opposed to a gay identity. In his fifties at the time I interviewed him, Stephen Richter attributed never having become a father to "not having the relationship with the woman." Because he had envisioned parenting and procreation only in the context of a heterosexual relationship, regarding the two as completely bound up with one another, Stephen had never considered children an option.

Older gay men and lesbians were not the only ones whose adult lives had been shaped by ideologies that banish gay people from the domain of kinship. Explaining why he felt uncomfortable participating in "family occasions," a young man who had no particular interest in raising a child commented, "When families get together, what do they talk about? Who's getting married, who's having children. And who's not, okay? Well, look who's not." Very few of the lesbians and gay men I met believed that claiming a gay identity automatically requires leaving kinship behind. In some cases people described this equation as an outmoded view that contrasted sharply with revised notions of what constitutes a family.

Well-meaning defenders of lesbian and gay identity sometimes assert that gays are not inherently "anti-family," in ways that perpetuate the association of heterosexual identity with exclusive access to kinship. Charles Silverstein . . . , for instance, contends that lesbians and gay men may place more importance on maintaining family ties than heterosexuals do because gay people do not marry and raise children. Here the affirmation that gays and lesbians are capable of fostering enduring kinship ties ends up reinforcing the implication that they cannot establish "families of their own," presumably because the author regards kinship as unshakably rooted in heterosexual alliance and procreation. In contrast, discourse on gay families cuts across the politically loaded couplet of "pro-family" and "anti-family" that places gay men and lesbians in an inherently antagonistic relation to kinship solely on the basis of their nonprocreative sexualities. "Homosexuality is not what is breaking up the Black family," declared Barbara Smith . . . , a black lesbian writer, activist, and speaker at the 1987 Gay and Lesbian March on Washington. "Homophobia is. My Black gay brothers and my Black lesbian sisters are members of Black families, both the ones we were born into and the ones we create."

At the height of gay liberation, activists had attempted to develop alternatives to "the family," whereas by the 1980s many lesbians and gay men were struggling to legitimate gay families as a form of kinship. . . . Gay or chosen families might incorporate friends, lovers, or children, in any combination. Organized through ideologies of love, choice, and creation, gay families have been defined through a contrast with what many gay men and lesbians . . . called "straight," "biological," or "blood" family. If families we choose were the families lesbians and gay men created for themselves, straight family represented the families in which most had grown to adulthood.

What does it mean to say that these two categories of family have been defined through contrast? One thing it emphatically does *not* mean is that heterosexuals share a single coherent form of family (although some of the lesbians and gay men doing the defining believed this to be the case). I am not arguing here for the existence of some central, unified kinship system vis-à-vis which gay people have distinguished their own practice and understanding of family. In the United States, race, class, gender, ethnicity, regional origin, and context all inform differences in household organization, as well as differences in notions of family and what it means to call someone kin.

In any relational definition, the juxtaposition of two terms gives meaning to both. Just as light would not be meaningful without some notion of darkness, so gay or chosen families cannot be understood apart from the families lesbians and gay men call "biological," "blood," or "straight." Like others in their society, most gay people . . . considered biology a matter of "natural fact." When they applied the terms "blood" and "biology" to kinship, however, they tended to depict families more consistently organized by procreation, more rigidly grounded in genealogy, and more uniform in their conceptualization than anthropologists know most families to be. For many lesbians and gay men, blood family represented not some naturally given unit that provided a base for all forms of kinship, but rather a procreative principle that organized only one possible *type* of kinship. In their descriptions they situated gay families at the opposite end of a spectrum of determination, subject to no constraints beyond a logic of "free" choice that ordered membership. To the extent that gay men and lesbians mapped "biology" and "choice" onto identities already opposed to one another (straight and gay, respectively), they polarized these two types of family along an axis of sexual identity.

The following chart recapitulates the ideological transformation generated as lesbians and gay men began to inscribe themselves within the domain of kinship.

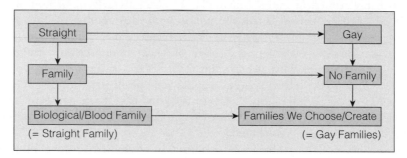

What this chart presents is not some static substitution set, but a historically motivated succession. To move across or down the chart is to move through time. Following along from left to right, time appears as process, periodized with reference to the experience of coming out. In the first opposition, coming out defines the transition from a straight to a gay identity. For the person who maintains an exclusively biogenetic notion of kinship, coming out can mark the renunciation of kinship, the shift from "family" to "no family" portrayed in the second opposition. In the third line, individuals who accepted the possibility of gay families after coming out could experience themselves making a transition from the biological or blood families in which they had grown up to the establishment of their own chosen families.

Moving from top to bottom, the chart depicts the historical time that inaugurated contemporary discourse on gay kinship. "Straight" changes from a category with an exclusive claim on kinship to an identity allied with a specific kind of family symbolized by biology or blood. Lesbians and gay men, originally relegated to the status of people without family, later lay claim to a distinctive type of family characterized as families we choose or create. While dominant cultural representations have asserted that straight is to gay as family is to no family (lines 1 and 2), at a certain point in history gay people began to contend that straight is to gay as blood family is to chosen families (lines 1 and 3).

What provided the impetus for this ideological shift? Transformations in the relation of lesbians and gay men to kinship are inseparable from sociohistorical developments: changes in the context for disclosing a lesbian or gay identity to others, attempts to build urban gay "community," cultural inferences about relationships between "same-gender" partners, and the lesbian baby boom associated with alternative (artificial) insemination. . . . If . . . kinship is something people use to act as well as to think, then its transformations should have unfolded not only on the "big screen" of history, but also on the more modest stage of day-to-day life, where individuals have actively engaged novel ideological distinctions and contested representations that would exclude them from kinship.

42

Unequal Childhoods

ANNETTE LAREAU

Laughing and yelling, a white fourth-grader named Garrett Tallinger splashes around in the swimming pool in the backyard of his four-bedroom home in the suburbs on a late spring afternoon. As on most evenings, after a quick dinner his father drives him to soccer practice. This is only one of Garrett's many activities. His brother has a baseball game at a different location. There are evenings when the boys' parents relax, sipping a glass of wine. Tonight is not one of them. As they rush to change out of their work clothes and get the children ready for practice, Mr. and Mrs. Tallinger are harried.

Only ten minutes away, a Black fourth-grader, Alexander Williams, is riding home from a school open house. His mother is driving their beige, leather-upholstered Lexus. It is 9:00 P.M. on a Wednesday evening. Ms. Williams is tired from work and has a long Thursday ahead of her. She will get up at 4:45 A.M. to go out of town on business and will not return before 9:00 P.M. On Saturday morning, she will chauffeur Alexander to a private piano lesson at 8:15 A.M., which will be followed by a choir rehearsal and then a soccer game. As they ride in the dark, Alexander's mother, in a quiet voice, talks with her son, asking him questions and eliciting his opinions.

Discussions between parents and children are a hallmark of middle-class child rearing. Like many middle-class parents, Ms. Williams and her husband see themselves as "developing" Alexander to cultivate his talents in a concerted fashion. Organized activities, established and controlled by mothers and fathers, dominate the lives of middle-class children such as Garrett and Alexander. By making certain their children have these and other experiences, middle-class parents engage in a process of *concerted cultivation*. From this, a robust sense of entitlement takes root in the children. This sense of entitlement plays an especially important role in institutional settings, where middle-class children learn to question adults and address them as relative equals.

Only twenty minutes away, in blue-collar neighborhoods, and slightly farther away, in public housing projects, childhood looks different. Mr. Yanelli, a white working-class father, picks up his son Little Billy, a fourth-grader, from an after-school program. They come home and Mr. Yanelli drinks a beer while Little Billy first watches television, then rides his bike and plays in the street. Other nights, he and his Dad sit on the sidewalk outside their house and play

SOURCE: From Annette Lareau, *Unequal Childhoods: Class, Race, and Family Life*. (Los Angeles: University of California Press, © 2003.) Reprinted by permission.

cards. At about 5:30 P.M. Billy's mother gets home from her job as a house cleaner. She fixes dinner and the entire family sits down to eat together. Extended family are a prominent part of their lives. Ms. Yanelli touches base with her "entire family every day" by phone. Many nights Little Billy's uncle stops by, sometimes bringing Little Billy's youngest cousin. In the spring, Little Billy plays baseball on a local team. Unlike for Garrett and Alexander, who have at least four activities a week, for Little Billy, baseball is his only organized activity outside of school during the entire year. Down the road, a white working-class girl, Wendy Driver, also spends the evening with her girl cousins, as they watch a video and eat popcorn, crowded together on the living room floor.

Farther away, a Black fourth-grade boy, Harold McAllister, plays outside on a summer evening in the public housing project in which he lives. His two male cousins are there that night, as they often are. After an afternoon spent unsuccessfully searching for a ball so they could play basketball, the boys had resorted to watching sports on television. Now they head outdoors for a twilight water balloon fight. Harold tries to get his neighbor, Miss Latifa, wet. People sit in white plastic lawn chairs outside the row of apartments. Music and television sounds waft through the open windows and doors.

The adults in the lives of Billy, Wendy, and Harold want the best for them. Formidable economic constraints make it a major life task for these parents to put food on the table, arrange for housing, negotiate unsafe neighborhoods, take children to the doctor (often waiting for city buses that do not come), clean children's clothes, and get children to bed and have them ready for school the next morning. But unlike middle-class parents, these adults do not consider the concerted development of children, particularly through organized leisure activities, an essential aspect of good parenting. Unlike the Tallingers and Williamses, these mothers and fathers do not focus on concerted cultivation. For them, the crucial responsibilities of parenthood do not lie in eliciting their children's feelings, opinions, and thoughts. Rather, they see a clear boundary between adults and children. Parents tend to use directives: they tell their children what to do rather than persuading them with reasoning. Unlike their middle-class counterparts, who have a steady diet of adult organized activities, the working-class and poor children have more control over the character of their leisure activities. Most children are free to go out and play with friends and relatives who typically live close by. Their parents and guardians facilitate the *accomplishment of natural growth*. Yet these children and their parents interact with central institutions in the society, such as schools, which firmly and decisively promote strategies of concerted cultivation in child rearing. For working-class and poor families, the cultural logic of child rearing at home is out of synch with the standards of institutions. As a result, while children whose parents adopt strategies of concerted cultivation appear to gain a sense of entitlement, children such as Billy Yanelli, Wendy Driver, and Harold McAllister appear to gain an emerging sense of distance, distrust, and constraint in their institutional experiences.

America may be the land of opportunity, but it is also a land of inequality. This research identifies the largely invisible but powerful ways that parents' social class impacts chilren's life experiences. It shows, using indepth observations and interviews with middle-class (including members of the upper- middle-class), working-class, and poor families, that inequality permeates the fabric of the culture. . . . I report the results of intensive observational research for a total of twelve families when their children were nine and ten years old. I argue that key elements of family life cohere to form a cultural logic of child rearing. In other words, the differences among families seem to cluster together in meaningful patterns. In this historical moment, middle-class parents tend to adopt a cultural logic of child rearing that stresses the concerted cultivation of children. Working-class and poor parents, by contrast, tend to undertake the accomplishment of natural growth. In the accomplishment of natural growth, children experience long stretches of leisure time, child-initiated play, clear boundaries between adults and children, and daily interactions with kin. Working-class and poor children, despite tremendous economic strain, often have more "childlike" lives, with autonomy from adults and control over their extended leisure time. Although middle-class children miss out on kin relationships and leisure time, they appear to (at least potentially) gain important institutional advantages. From the experience of concerted cultivation, they acquire skills that could be valuable in the future when they enter the world of work. Middle-class white and Black children in my study did exhibit some key differences; yet the biggest gaps were not within social classes but, as I show, across them. It is these class differences and how they are enacted in family life and child rearing that shape the ways children view themselves in relation to the rest of the world.

CULTURAL REPERTOIRES

Professionals who work with children, such as teachers, doctors, and counselors, generally agree about how children should be raised. Of course, from time to time they may disagree on the ways standards should be enacted for an individual child or family. For example, teachers may disagree about whether or not parents should stop and correct a child who mispronounces a word while reading. Counselors may disagree over whether a mother is being too protective of her child. Still, there is little dispute among professionals on the broad principles for promoting educational development in children through proper parenting. These standards include the importance of talking with children, developing their educational interests, and playing an active role in their schooling. Similarly, parenting guidelines typically stress the importance of reasoning with children and teaching them to solve problems through negotiation rather than with physical force. Because these guidelines are so generally accepted, and because they focus on a set of practices concerning how parents should raise children, they form a *dominant set of cultural repertoires* about how children should be raised. This widespread agreement among professionals about the broad principles for child rearing permeates our society. A small number of experts thus potentially shape the behavior of a large number of parents.

Professionals' advice regarding the best way to raise children has changed regularly over the last two centuries. From strong opinions about the merits of bottle feeding, being stern with children, and utilizing physical punishment (with dire warnings of problematic outcomes should parents indulge children), there have been shifts to equally strongly worded recommendations about the benefits of breast feeding, displaying emotional warmth toward children, and using reasoning and negotiation as mechanisms of parental control. Middle-class parents appear to shift their behaviors in a variety of spheres more rapidly and more thoroughly than do working-class or poor parents. As professionals have shifted their recommendations from bottle feeding to breast feeding, from stern approaches to warmth and empathy, and from spanking to time-outs, it is middle-class parents who have responded most promptly. Moreover, in recent decades, middle-class children in the United States have had to face the prospect of "declining fortunes." Worried about how their children will get ahead, middle-class parents are increasingly determined to make sure that their children are not excluded from any opportunity that might eventually contribute to their advancement.

Middle-class parents who comply with current professional standards and engage in a pattern of concerted cultivation deliberately try to stimulate their children's development and foster their cognitive and social skills. The commitment among working-class and poor families to provide comfort, food, shelter, and other basic support requires ongoing effort, given economic challenges and the formidable demands of child rearing. But it stops short of the deliberate cultivation of children and their leisure activities that occurs in middle-class families. For working-class and poor families, sustaining children's natural growth is viewed as an accomplishment.

What is the outcome of these different philosophies and approaches to child rearing? Quite simply, they appear to lead to the *transmission of differential advantages* to children. In this study, there was quite a bit more talking in middle-class homes than in working-class and poor homes, leading to the development of greater verbal agility, larger vocabularies, more comfort with authority figures, and more familiarity with abstract concepts. Importantly, children also developed skill differences in interacting with authority figures in institutions and at home. Middle-class children such as Garrett Tallinger and Alexander Williams learn, as young boys, to shake the hands of adults and look them in the eye. In studies of job interviews, investigators have found that potential employees have less than one minute to make a good impression. Researchers stress the importance of eye contact, firm handshakes, and displaying comfort with bosses during the interview. In poor families like Harold McAllister's, however, family members usually do not look each other in the eye when conversing. In addition, as Elijah Anderson points out, they live in neighborhoods where it can be dangerous to look people in the eye too long. The types of social competence transmitted in the McAllister family are valuable, but they are potentially less valuable (in employment interviews, for example) than those learned by Garrett Tallinger and Alexander Williams.

The white and Black middle-class children in this study also exhibited an emergent version of the *sense of entitlement* characteristic of the middle-class. They acted as though they had a right to pursue their own individual preferences and to actively manage interactions in institutional settings. They appeared comfortable in these settings; they were open to sharing information and asking for attention. Although some children were more outgoing than others, it was common practice among middle-class children to shift interactions to suit *their* preferences. Alexander Williams knew how to get the doctor to listen to his concerns (about the bumps under his arm from his new deodorant). His mother explicitly trained and encouraged him to speak up with the doctor. Similarly, a Black middle-class girl, Stacey Marshall, was taught by her mother to expect the gymnastics teacher to accommodate her individual learning style. Thus, middle-class children were trained in "the rules of the game" that govern interactions with institutional representatives. They were not conversant in other important social skills, however, such as organizing their time for hours on end during weekends and summers, spending long periods of time away from adults, or hanging out with adults in a nonobtrusive, subordinate fashion. Middle-class children also learned (by imitation and by direct training) how to make the rules work in their favor. Here, the enormous stress on reasoning and negotiation in the home also has a potential advantage for future institutional negotiations. Additionally, those in authority responded positively to such interactions. Even in fourth grade, middle-class children appeared to be acting on their own behalf to gain advantages. They made special requests of teachers and doctors to adjust procedures to accommodate their desires.

The working-class and poor children, by contrast, showed an emerging *sense of constraint* in their interactions in institutional settings. They were less likely to try to customize interactions to suit their own preferences. Like their parents, the children accepted the actions of persons in authority (although at times they also covertly resisted them). Working-class and poor parents sometimes were not as aware of their children's school situation (as when their children were not doing homework). Other times, they dismissed the school rules as unreasonable. For example, Wendy Driver's mother told her to "punch" a boy who was pestering her in class; Billy Yanelli's parents were proud of him when he "beat up" another boy on the playground, even though Billy was then suspended from school. Parents also had trouble getting "the school" to respond to their concerns. When Ms. Yanelli complained that she "hates" the school, she gave her son a lesson in powerlessness and frustration in the face of an important institution. Middle-class children such as Stacey Marshall learned to make demands on professionals, and when they succeeded in making the rules work in their favor they augmented their "cultural capital" (i.e., skills individuals inherit that can then be translated into different forms of value as they move through various institutions) for the future. When working-class and poor children confronted institutions, however, they generally were unable to make the rules work in their favor nor did they obtain capital for adulthood. Because of these patterns of legitimization, children raised according to the logic of concerted cultivation can gain advantages, in the form of an emerging sense of entitlement, while children raised according to the logic of natural growth tend to develop an emerging sense of constraint.

SOCIAL STRATIFICATION AND INDIVIDUALISM

Public discourse in America typically presents the life accomplishments of a person as the result of her or his individual qualities. Songs like "I Did It My Way," memoirs, television shows, and magazine articles, celebrate the individual. Typically, individual outcomes are connected to individual effort and talent, such as being a "type A" personality, being a hard worker, or showing leadership. These cultural beliefs provide a framework for Americans' views of inequality.

Indeed, Americans are much more comfortable recognizing the power of individual initiative than recognizing the power of social class. Studies show that Americans generally believe that responsibility for their accomplishments rests on their individual efforts. Less than one-fifth see "race, gender, religion, or class as very important for 'getting ahead in life.'" Compared to Europeans, individuals in the United States are much more likely to believe they can improve their standard of living. Put differently, Americans believe in the American dream: "The American dream that we were all raised on is a simple but powerful one—if you work hard and play by the rules, you should be given a chance to go as far as your God-given ability will take you." This American ideology that each individual is responsible for his or her life outcomes is the expressed belief of the vast majority of Americans, rich and poor.

Yet there is no question that society is stratified. . . . Highly valued resources such as the possession of wealth; having an interesting, well-paying, and complex job; having a good education; and owning a home, are not evenly distributed throughout the society. Moreover, these resources are transferred across generations: One of the best predictors of whether a child will one day graduate from college is whether his or her parents are college graduates. Of course, relations of this sort are not absolute: Perhaps two-thirds of the members of society ultimately reproduce their parents' level of educational attainment, while about one-third take a different path. Still, there is no question that we live in a society characterized by considerable gaps in resources or, put differently, by substantial *inequality*. . . . I demonstrate the existence of a cultural logic of child rearing that tends to differ according to families' social class positions. I see these interweaving practices as coming together in a messy but still recognizable way. In contrast to many, I suggest that social class does have a powerful impact in shaping the daily rhythms of family life. . . .

THE POWER OF SOCIAL CLASS

In the United States, people disagree about the importance of social class in daily life. Many Americans believe that this country is fundamentally *open*. They assume the society is best understood as a collection of individuals. They believe that people who demonstrate hard work, effort, and talent are likely to achieve upward mobility. Put differently, many Americans believe in the American Dream. In this view, children should have roughly equal life chances. The extent to which life chances vary can be traced to differences in aspirations, talent, and hard work on the part of individuals. This perspective rejects the notion that parents' social location systematically shapes children's life experiences and outcomes. Instead, outcomes are seen as resting more in the hands of individuals. . . .

THE LIMITS OF SOCIAL CLASS

Among the families we observed, some aspects of daily life did not vary systematically by social class. There were episodes of laughter, emotional connection, and happiness as well as quiet comfort in every family. Harold McAllister and his mother laughed together as he almost dropped his hot dog but then, in an awkward grab, caught it. After a baseball game, Mr. Williams rubbed Alexander's head affectionately and called him "handsome." Ms. Handlon gave her daughter a big squeeze around her shoulders after the Christmas Eve pageant, and Melanie beamed. One summer afternoon, Mr. Yanelli and Billy played cards together, sitting cross-legged on the sidewalk. These moments of connection seemed deeply meaningful to both children and parents in all social classes, even as they take different shape by social class, in terms of language, activity, and character.

All the families we observed also had rituals: favorite meals they often ate, television programs they watched, toys or games that were very important, family outings they looked forward to, and other common experiences. The content of their rituals varied (especially by social class); what did not vary was that the children enjoyed these experiences and they provided a sense of membership in a family. Also, in all social classes, a substantial part of the children's days was spent in repetitive rituals: getting up, making the bed, taking a shower, getting dressed, brushing hair and teeth, eating breakfast, finding school books and papers, and waiting for adults to get ready. These moments were interspersed with hours, days, and weeks of household work, tedious demands, mundane tasks, and tension. This was true for all families, regardless of social class. Nor were any families immune to life tragedies: across all social classes there were premature deaths due to car accidents or suicides. Across all social classes children and parents had different temperaments: some were shy and quiet; some were outgoing and talkative. Some had a sense of humor and some did not. The degree of organization and orderliness in daily life also did not vary systematically by social class. Some houses were clean and some were a disaster. Some of the messiest ones were middle-class homes in which the entryway was a paragon of order but the living spaces, particularly the upstairs, were in a tumble. Despite the formidable differences among the families detailed in each home, after a few visits, the research assistants and I found that the surroundings felt normal, comfortable, and safe. Put differently, they all felt like home.

CONCERTED CULTIVATION AND THE
ACCOMPLISHMENT OF NATURAL GROWTH

Despite these important areas of shared practices, social class made a significant difference in the routines of children's daily lives. The white and Black middle-class parents engaged in practices of *concerted cultivation*. In these families, parents actively fostered and assessed their children's talents, opinions, and skills. They scheduled their children for activities. They reasoned with them. They hovered over them and outside the home they did not hesitate to intervene on the children's behalf. They made a deliberate and sustained effort to stimulate children's development and to cultivate their cognitive and social skills. The working-class and poor parents viewed children's development as unfolding spontaneously, as long as they were provided with comfort, food, shelter, and other basic support. I have called this cultural logic of child rearing the *accomplishment of natural growth*. As with concerted cultivation, this commitment, too, required ongoing effort; sustaining children's natural growth despite formidable life challenges is properly viewed as accomplishment. Parents who relied on natural growth generally organized their children's lives so they spent time in and around home, in informal play with peers, siblings, and cousins. As a result, the children had more autonomy regarding leisure time and more opportunities for child-initiated play. They also were more responsible for their lives outside the home. Unlike in middle-class families, adult-organized activities were uncommon. Instead of the relentless focus on reasoning and negotiation that took place in middle-class families, there was less speech (including less whining and badgering) in working-class and poor homes. Boundaries between adults and children were clearly marked; parents generally used language not as an aim in itself but more as a conduit for social life. Directives were common. In their institutional encounters, working-class and poor parents turned over responsibility to professionals; when parents did try to intervene, they felt that they were less capable and less efficacious than they would have liked. While working-class and poor children differed in important ways, particularly in the stability of their lives, surprisingly there was not a major difference between them in their cultural logic of child rearing. Instead, in this study the cultural divide appeared to be between the middle class and everyone else. . . .

HOW DOES IT MATTER?

Both concerted cultivation and the accomplishment of natural growth offer intrinsic benefits (and burdens) for parents and their children. Nevertheless, these practices are accorded different social values by important social institutions. There are signs that some family cultural practices, notably those associated with concerted cultivation, give children advantages that other cultural practices do not.

In terms of the rhythms of daily life, both concerted cultivation and the accomplishment of natural growth have advantages and disadvantages. Middle-class children learn to develop and value an individualized sense of self. Middle-class children are allowed to participate in a variety of coveted activities: gymnastics, soccer, summer camps, and so on. These activities improve their skills and teach them, as Mr. Tallinger noted, to be better athletes than their parents were at comparable ages. They learn to handle moments of humiliation on the field as well as moments of glory. Middle-class children learn, as Mr. Williams noted, the difference between baroque and classical music. They learn to perform. They learn to present themselves. But this cultivation has a cost. Family schedules are disrupted. Dinner hours are very hard to arrange. Siblings such as Spencer and Sam Tallinger spend dreary hours waiting at athletic fields and riding in the car going from one event to another. Family life, despite quiet interludes, is frequently frenetic. Parents, especially mothers, must reconcile conflicting priorities, juggling events whose deadlines are much tighter than the deadlines connected to serving meals or getting children ready for bed. The domination of children's activities can take a toll on families. At times, everyone in the middle-class families—including ten-year-old children—seemed exhausted. Thus, there are formidable costs, as well as benefits to this child-rearing approach.

Working-class and poor children also had advantages, as well as costs, from the cultural logic of child rearing they experienced. Working-class and poor children learned to entertain themselves. They played outside, creating their own games, as Tyrec Taylor did with his friends. They did not complain of being bored. Working-class and poor children also appeared to have boundless energy. They did not have the exhaustion that we saw in middle-class children the same age. Some working-class and poor children longed to be in organized activities—Katie Brindle wanted to take ballet and Harold McAllister wanted to play football. When finances, a lack of transportation, and limited availability of programs conspired to prevent or limit their participation, they were disappointed. Many were also deeply aware of the economic constraints and the limited consumption permitted by their family's budget. Living spaces were small, and often there was not much privacy. The television was almost always on and, like many middle-class children growing up in the 1950s, working-class and poor children watched unrestricted amounts of television. As a result, family members spent more time together in shared space than occurred in middle-class homes. Indeed, family ties were very strong, particularly among siblings. Working-class and poor children also developed very close ties with their cousins and other extended family members.

Within the home, these two approaches to child rearing each have identifiable strengths and weaknesses. When we turn to examining institutional dynamics outside the home, however, the unequal benefits of middle-class children's lives compared to working-class and poor children's lives become clearer. In crucial ways, middle-class family members appeared reasonably comfortable and entitled, while working-class and poor family members appeared uncomfortable and constrained. . . .

This pattern occurred in school interactions, as well. Some working-class and poor parents had warm and friendly relations with educators. Overall, however, working-class and poor parents in this study had much more distance or separation from the school than did middle-class mothers.

Other working-class and poor parents also appeared baffled, intimidated, and subdued in parent-teacher conferences. Ms. Driver, frantically worried because Wendy, a fourth-grader, was not yet able to read, resisted intervening, saying, "I don't want to jump into anything and find it is the wrong thing." When working-class and poor parents did try to intervene in their children's educational experiences, they often felt ineffectual. . . .

Overall, the routine rituals of family life are not equally legitimized in the broader society. Parents' efforts to reason with children (even two-year-olds) are seen as more educationally valuable than parents' use of directives. Spending time playing soccer or baseball is deemed by professionals as more valuable than time spent watching television. Moreover, differences in the cultural logic of child rearing are attached to unequal currency in the broader society. The middle-class strategy of concerted cultivation appears to have greater promise of being capitalized into social profits than does the strategy of the accomplishment of natural growth found in working-class and poor homes.

43

Doméstica

PIERRETTE HONDAGNEU-SOTELO

Particular regional formations have historically characterized the racialization of paid domestic work in the United States. Relationships between domestic employees and employers have always been imbued with racial meanings: white "masters and mistresses" have been cast as pure and superior, and "maids and servants," drawn from specific racial-ethnic groups (varying by region), have been cast as dirty and socially inferior. The occupational racialization we see now in Los Angeles or New York City continues this American legacy, but it also draws to a much greater extent on globalization and immigration.

In the United States today, immigrant women from a few non-European nations are established as paid domestic workers. These women—who hail primarily from Mexico, Central America, and the Caribbean and who are perceived as "nonwhite" in Anglo-American contexts—hold various legal statuses. Some are legal permanent residents or naturalized U.S. citizens, many as beneficiaries of the 1986 Immigration Reform and Control Act's amnesty-legalization program. Central American women, most of whom entered the United States after the 1982 cutoff date for amnesty, did not qualify for legalization, so in the 1990s they generally either remained undocumented or held a series of temporary

SOURCE: From Pierrette Hondagneu-Sotelo, *Doméstica: Immigrant Workers Caring and Cleaning in America* (Berkeley: University of California Press, 2001) pp. 13–22. Reprinted by permission of the University of California Press.

work permits, granted to delay their return to war-ravaged countries. Domestic workers who are working without papers clearly face extra burdens and risks: criminalization of employment, denial of social entitlements, and status as outlaws anywhere in the nation. If they complain about their jobs, they may be threatened with deportation. Undocumented immigrant workers, however, are not the only vulnerable ones. In the 1990s, even legal permanent residents and naturalized citizens saw their rights and privileges diminish, as campaigns against illegal immigration metastasized into more generalized xenophobic attacks on all immigrants, including those here with legal authorization. Immigration status has clearly become an important axis of inequality, one interwoven with relations of race, class, and gender, and it facilitates the exploitation of immigrant domestic workers.

Yet race and immigration are interacting in an important new way, which Latina immigrant domestic workers exemplify: their position as "foreigners" and "immigrants" allows employers, and the society at large, to perceive them as outsiders and thereby overlook the contemporary racialization of the occupation. Immigration does not trump race but, combined with the dominant ideology of a "color-blind" society, manages to shroud it.

With few exceptions, domestic work has always been reserved for poor women, for immigrant women, and for women of color; but over the last century, paid domestic workers have become more homogenous, reflecting the subordinations of both race and nationality/immigration status. In the late nineteenth century, this occupation was the most likely source of employment for U.S.-born women. In 1870, according to the historian David M. Katzman, two-thirds of all nonagricultural female wage earners worked as domestics in private homes. The proportion steadily declined to a little over one-third by 1900, and to one-fifth by 1930. Alternative employment opportunities for women expanded in the mid- and late twentieth century, so by 1990, fewer than 1 percent of employed American women were engaged in domestic work. Census figures, of course, are notoriously unreliable in documenting this increasingly undocumentable, "under-the-table" occupation, but the trend is clear: paid domestic work has gone from being *either* an immigrant woman's job *or* a minority woman's job to one that is now filled by women who, as Latina and Caribbean immigrants, embody subordinate status both racially and as immigrants.

Regional racializations of the occupation were already deeply marked in the late nineteenth and early twentieth centuries, as the occupation recruited women from subordinate racial-ethnic groups. In northeastern and midwestern cities of the late nineteenth century, single young Irish, German, and Scandinavian immigrants and women who had migrated from the country to the city typically worked as live-in "domestic help," often leaving the occupation when they married. During this period, the Irish were the main target of xenophobic vilification. With the onset of World War I, European immigration declined and job opportunities in manufacturing opened up for whites, and black migration from the South enabled white employers to recruit black women for domestic jobs in the Northeast. Black women had always predominated as a servant caste in the

South, whether in slavery or after, and by 1920 they constituted the single largest group in paid domestic work in both the South and the Northeast. Unlike European immigrant women, black women experienced neither individual nor intergenerational mobility out of the occupation, but they succeeded in transforming the occupation from one characterized by live-in arrangements, with no separation between work and social life, to live-out "day work"—a transformation aided by urbanization, new interurban transportation systems, and smaller urban residences.

In the Southwest and the West of the late nineteenth and early twentieth centuries, the occupation was filled with Mexican American and Mexican immigrant women, as well as Asian, African American, and Native American women and, briefly, Asian men. Asian immigrant men were among the first recruits for domestic work in the West. California exceptionalism—its Anglo-American conquest from Mexico in 1848, its ensuing rapid development and overnight influx of Anglo settlers and miners, and its scarcity of women—initially created many domestic jobs in the northern part of the territory for Chinese "house-boys," laundrymen, and cooks, and later for Japanese men, followed by Japanese immigrant women and their U.S.-born daughters, the nisei, who remained in domestic work until World War II. Asian American women's experiences . . . provide an intermediate case of intergenerational mobility out of domestic work between that of black and Chicana women who found themselves, generation after generation, stuck in the occupational ghetto of domestic work and that of European immigrant women of the early twentieth century who quickly moved up the mobility ladder.

For Mexican American women and their daughters, domestic work became a dead-end job. From the 1880s until World War II, it provided the largest source of nonagricultural employment for Mexican and Chicana women throughout the Southwest. During this period, domestic vocational training schools, teaching manuals, and Americanization efforts deliberately channeled them into domestic jobs. Continuing well into the 1970s throughout the Southwest, and up to the present in particular regions, U.S.-born Mexican American women have worked as domestics. Over that time, the job has changed. Much as black women helped transform the domestic occupation from live-in to live-out work in the early twentieth century, Chicanas in the Southwest increasingly preferred contractual housecleaning work . . . to live-in or daily live-out domestic work.

While black women dominated the occupation throughout the nation during the 1950s and 1960s, there is strong evidence that many left it during the late 1960s. The 1970 census marked the first time that domestic work did not account for the largest segment of employed black women; and the proportion of black women in domestic work continued to drop dramatically in the 1970s and 1980s, falling from 16.4 percent in 1972 to 7.4 percent in 1980, then to 3.5 percent by the end of the 1980s. By opening up public-sector jobs to black women, the Civil Rights Act of 1964 made it possible for them to leave private domestic service. Consequently, both African American and Mexican American women moved into jobs from which they had been previously barred, as secretaries, sales clerks, and public-sector employees, and into the expanding number of relatively low-paid service jobs in convalescent homes, hospitals, cafeterias, and hotels.

These occupational adjustments and opportunities did not go unnoticed. In a 1973 *Los Angeles Times* article, a manager with thirty years of experience in domestic employment agencies reported, "Our Mexican girls are nice, but the blacks are hostile." Speaking very candidly about her contrasting perceptions of Latina immigrant and African American women domestic workers, she said of black women, "you can feel their anger. They would rather work at Grant's for $1.65 an hour than do housework. To them it denotes a lowering of self." By the 1970s black women in the occupation were growing older, and their daughters were refusing to take jobs imbued with servitude and racial subordination. Domestic work, with its historical legacy in slavery, was roundly rejected. Not only expanding job opportunities but also the black power movement, with its emphasis on self-determination and pride, dissuaded younger generations of African American women from entering domestic work.

It was at this moment that newspaper reports, census data, and anecdotal accounts first register the occupation's demographic shift toward Latina immigrants, a change especially pronounced in areas with high levels of Latino immigration. In Los Angeles, for example, the percentage of African American women working as domestics in private households fell from 35 percent to 4 percent from 1970 to 1990, while foreign-born Latinas increased their representation from 9 percent to 68 percent. Again, since census counts routinely underestimate the poor and those who speak limited or no English, the women in this group may represent an even larger proportion of private domestic workers.

Ethnographic case studies conducted not only in Los Angeles but also in Washington, D.C., San Francisco, San Diego, Houston, El Paso, suburban areas of Long Island, and New York City provide many details about the experiences of Mexican, Caribbean, and Central American women who now predominate in these metropolitan centers as nanny/housekeepers and housecleaners. Like the black women who migrated from the rural South to northern cities in the early twentieth century, Latina immigrant women newly arrived in U.S. cities and suburbs in the 1970s, 1980s, and 1990s often started as live-ins, sometimes first performing unpaid household work for kin before taking on very low paying live-in jobs for other families. Live-in jobs, shunned by better-established immigrant women, appeal to new arrivals who want to minimize their living costs and begin sending their earnings home. Vibrant social networks channel Latina immigrants into these jobs, where the long hours and the social isolation can be overwhelming. As time passes, many of the women seek live-out domestic jobs. Despite the decline in live-in employment arrangements at the century's midpoint, the twentieth century ended in the United States much as it began, with a resurgence of live-in jobs filled by women of color—now Latina immigrants.

Two factors of the late twentieth century were especially important in creating this scenario. First, . . . globalization has promoted higher rates of immigration. The expansion of U.S. private investment and trade; the opening of U.S.

multinational assembly plants (employing mostly women) along the U.S.-Mexico border and in Caribbean and Central American nations, facilitated by government legislative efforts such as the Border Industrialization Program, the North American Free Trade Agreement, and the Caribbean Basin Initiative; the spreading influence of U.S. mass media; and U.S. military aid in Central America have all helped rearrange local economies and stimulate U.S.-bound migration from the Caribbean, Mexico, and Central America. Women from these countries have entered the United States at a propitious time for families looking to employ housecleaners and nannies.

Second, increased immigration led to the racialized xenophobia of the 1990s. The rhetoric of these campaigns shifted focus, from attacking immigrants for lowering wages and competing for jobs to seeking to bar immigrants' access to social entitlements and welfare. In the 1990s, legislation codified this racialized nativism, in large part taking aim at women and children. In 1994 California's Proposition 187, targeting Latina immigrants and their children, won at the polls; and although its denial of all public education and of publicly funded health care was ruled unconstitutional by the courts, the vote helped usher in new federal legislation. In 1996 federal welfare reform, particularly the Immigration Reform Act and Individual Responsibility Act (IRAIRA), codified the legal and social disenfranchisement of legal permanent residents and undocumented immigrants. At the same time, language—and in particular the Spanish language—was becoming racialized; virulent "English Only" and anti-bilingual education campaigns and ballot initiatives spread.

Because Latina immigrants are disenfranchised as immigrants and foreigners, Americans can overlook the current racialization of the job. On the one hand, racial hostilities and fears may be lessened as increasing numbers of Latina and Caribbean nannies care for tow-headed children. As Sau-ling C. Wong suggests in an analysis of recent films, "in a society undergoing radical demographic and economic changes, the figure of the person of color patiently mothering white folks serves to allay racial anxieties."[1] Stereotypical images of Latinas as innately warm, loving, and caring certainly round out this picture. Yet on the other hand, the status of these Latinas as immigrants today serves to legitimize their social, economic, and political subordination and their disproportionate concentration in paid domestic work.

Such legitimation makes it possible to ignore American racism and discrimination. Thus the abuses that Latina domestic workers suffer in domestic jobs can be explained away because the women themselves are foreign and unassimilable. If they fail to realize the American Dream, according to this distorted narrative, it is because they are lazy and unmotivated or simply because they are "illegal" and do not merit equal opportunities with U.S.-born American citizens. Contemporary paid domestic work in the United States remains a job performed by women of color, by black and brown women from the Caribbean, Central America, and Mexico. This racialization of domestic work is masked by the ideology of "a color-blind society" and by the focus on immigrant "foreignness."

GLOBAL TRENDS IN PAID DOMESTIC WORK

Just as paid domestic work has expanded in the United States, so too it appears to have grown in many other postindustrial societies—in Canada and in parts of Europe—in the "newly industrialized countries" (NICs) of Asia, and in the oil-rich nations of the Middle East. Around the globe Caribbean, Mexican, Central American, Peruvian, Sri Lankan, Indonesian, Eastern European, and Filipina women—the latter in disproportionately great numbers—predominate in these jobs. World wide, paid domestic work continues its long legacy as a racialized and gendered occupation, but today divisions of nation and citizenship are increasingly salient. Rhacel Parreñas, who has studied Filipina domestic workers, refers to this development as the "international division of reproductive labor," and Anthony Richmond has called it part of a broad, new "global apartheid."[2]

. . . We must remember that the inequality of nations is a key factor in the globalization of contemporary paid domestic work. This inequality has had three results. First, around the globe, paid domestic work is increasingly performed by women who leave their own nations, their communities, and often their families of origin to do it. Second, the occupation draws not only women from the poor socioeconomic classes but also women of relatively high status in their own countries—countries that colonialism made much poorer than those countries where they go to do domestic work. Thus it is not unusual to find middle-class, college-educated women working in other nations as private domestic workers. Third, the development of service-based economies in postindustrial nations favors the international migration of women laborers. Unlike in earlier industrial eras, today the demand for gendered labor favors migrant women's services.

Nations use vastly different methods to "import" domestic workers from other countries. Some countries have developed highly regulated, government-operated, contract labor programs that have institutionalized both the recruitment and working conditions of migrant domestic workers. Canada and Hong Kong exemplify this approach. Since 1981 the Canadian federal government has formally recruited thousands of women to work as live-in nanny/housekeepers for Canadian families. Most come from third world countries, the majority in the 1980s from the Caribbean and in the 1990s from the Philippines; and once in Canada, they must remain in live-in domestic service for two years, until they obtain their landed immigrant status, the equivalent of the U.S. "green card." During this period, they must work in conditions reminiscent of formal indentured servitude and they may not quit their jobs or collectively organize to improve job conditions.

Similarly, since 1973 Hong Kong has relied on the formal recruitment of domestic workers, mostly Filipinas, to work on a full-time, live-in basis for Chinese families. Of the 150,000 foreign domestic workers in Hong Kong in 1995, 130,000 hailed from the Philippines, with smaller numbers drawn from Thailand, Indonesia, India, Sri Lanka, and Nepal. Just as it is now rare to find African American women employed in private domestic work in Los Angeles, so too have Chinese women vanished from the occupation in Hong Kong. . . .

In the larger global context, the United States remains distinctive, as it follows a more laissez-faire approach to incorporating immigrant women into paid domestic work. Unlike in Hong Kong and Canada, here there is no formal government system or policy to legally contract with foreign domestic workers. In the past, private employers in the United States were able to "sponsor" individual immigrant women working as domestics for their green cards, sometimes personally recruiting them while they were vacationing or working in foreign countries, but this route is unusual in Los Angeles today. For such labor certification, the sponsor must document that there is a shortage of labor able to perform a particular, specialized job—and in Los Angeles and many other parts of the country, demonstrating a shortage of domestic workers has become increasingly difficult. And it is apparently unnecessary, as the significant demand for domestic workers in the United States is largely filled not through formal channels but through informal recruitment from the growing number of Caribbean and Latina immigrant women who are already living (legally or illegally) in the United States. The Immigration and Naturalization Service, the federal agency charged with stopping illegal migration, has historically served the interest of domestic employers and winked at the hiring of undocumented immigrant women in private homes.

As we compare the hyperregulated employment systems in Hong Kong and Canada with the U.S. approach to domestic work, we must distinguish between the regulation of labor and the regulation of foreign domestic workers. . . . Here, the United States is again an exception. U.S. labor regulations *do* cover private domestic work—but no one knows about them. . . . Domestic workers' wages and hours are governed by state and federal law, and special regulations cover such details as limits on permissible deductions for breakage and for boarding costs of live-in workers. These regulations did not fall from the sky: they are the result of several important, historic campaigns organized by and for paid domestic workers. Most U.S. employers now know . . . about their obligations for employment taxes—though these obligations are still widely ignored—but few employers and perhaps fewer employees know about the labor laws pertaining to private domestic work. It's almost as though these regulations did not exist. At the same time, the United States does not maintain separate immigration policies for domestic workers, of the sort that mandate live-in employment or decree instant deportation if workers quit their jobs.

This duality has two consequences. On the one hand, both the absence of hyperregulation of domestic workers and the ignorance about existing labor laws further reinforce the belief that paid domestic work is not a real job. Domestic work remains an arrangement that is thought of as private: it remains informal, "in the shadows," and outside the purview of the state and other regulating agencies. On the other hand, the absence of state monitoring of domestic job contracts and of domestic workers' personal movement, privacy, and bodily adornment suggests an opening to upgrade domestic jobs in the United States. Unlike in Hong Kong and Canada, for example, where state regulations prevent Filipina domestic workers from quitting jobs that they find unsatisfactory or abusive, in Los Angeles, Latina immigrant domestic workers can . . . quit their jobs.

Certainly they face limited options when they seek jobs outside of private homes, but it is important to note that they are not yoked by law to the same boss and the same job.

The absence of a neocolonialist, state-operated, contractual system for domestic work thus represents an opportunity to seek better job conditions. The chance of success might be improved if existing labor regulations were strengthened, if domestic workers were to work at collective organizing, and if informational and educational outreach to the domestic workers were undertaken. But to be effective, these efforts must occur in tandem with a new recognition that the relationships in paid domestic work are relations of employment.

NOTES

1.   Sau-ling C. Wong. 1994. "Diverted Mothering: Representations of Caregivers of Color in the Age of Multiculturalism." pp. 67–91 in *Mothering: Ideology, Experience, and Agency*, edited by Evelyn Nakano Glenn, Grace Chang, and Linda Pennie Forcey. New York: Routledge, p. 69.

2.   Rhacel Salazar Parreñas. 2000. "Migrant Filipina Domestic Workers and the International Division of Reproductive Labor." *Gender & Society* 14:560–580; Anthony Richmond. 1994. *Global Apartheid: Refugees, Racism, and the New World Order*. Toronto: Oxford University Press.

44

Rereading *Sex and the City*

Exposing the Hegemonic Feminist Narrative

REBECCA BRASFIELD

*S*ex and the City is an Emmy award–winning cable television program. The
. . . show originally aired on HBO for six seasons from 1998–2004. . . . Based
in New York City, the show is about four single women in their thirties and
forties, navigating the often complicated and chaotic dating scene. Carrie
Bradshaw, the protagonist, narrates each episode as she seeks insight and answers
to relationship dilemmas. Employed as a columnist, Carrie writes a weekly article
titled "Sex and the City" for the *New York Star* (a fictional newspaper). She is
known for her designer shoe obsession and unique and glamorous fashion statements. Carrie's on-again, off-again relationship with Mr. Big anchors *Sex and the
City*'s primary story line.

Although Carrie provides the major plotlines, *Sex and the City* also follows the
stories of her three friends: Miranda Hobbs, Samantha Jones, and Charlotte York.
Miranda is a career-oriented Harvard Law School graduate who eventually becomes
partner at her law firm. Miranda is the first of her friends to have a baby and is a
single mother for the first months of her child's life. Miranda's cynicism toward relationships is the essence of her character. She has a sarcastic sense of humor and usually counterbalances the views of her friends by providing what might be viewed as
a voice of reason. The only character to don a short hairstyle (majority of seasons),
Miranda appears to cut straight to the point in her analysis of dating. . . .

Often characterized by portraying a man's view on sex and relationships,
Samantha is known for her numerous sexual encounters. Confident and secure
with her sexuality, Samantha could be described as promiscuous. Her disinterest
in conventional relationships separates her from the other women. Employed as
a successful public relations executive, Samantha is in her forties, making her the
oldest of the four women. . . .

Frequently offended by Samantha's lustful views on sex, Charlotte is, by far,
the most sexually conservative of the group. Charlotte works as an art dealer until
deciding to end her career to concentrate on raising a family. Charlotte's views on
relationships are traditional, making her the voice of romantic love. . . .

SOURCE: From *Journal of Popular Film and Television*, Fall 2006, p. 130–138. Reprinted
by permission of the Helen Dwight Reid Educational Foundation. Published by Heldref
Publications, 1319 Eighteenth Street. NW, Washington, D.C. 20036-1802. Copyright
© 2006.

Carrie, Miranda, Samantha, and Charlotte represent a continuum of women's views and dilemmas when it comes to sex, love, and dating. The range of perspectives may be one of the reasons why *Sex and the City* sparks so much interest, enthusiasm, and criticism. . . .

HEGEMONIC FEMINIST PRACTICES
AND *SEX AND THE CITY*

. . . *Sex and the City* provides an excellent example of how hegemonic feminism looks, how it thinks, and what it does. . . .

Sex and the City, as a medium for social analysis, reflects almost exclusively the perspectives and values of white, middle-class, heterosexual women who define themselves primarily as oppressed victims of patriarchy. Carrie, Miranda, Samantha, and Charlotte are protagonists and subjects whose voices are heard. The telling of their stories centers their perspective. Their voices and narration dominate the discourse and as viewers we comprehend their experiences through their thoughts, feelings, and behaviors. Are the views presented by *Sex and the City* representative of a hegemonic discourse or do these views represent socially constructed, apolitical perspectives? White, middle-class, heterosexual women are centered subjects, and their values and attitudes comprise the program's underlying master narrative.

Sex and the City's master narrative is that the women's aim is to gain equal power to white, heterosexual, middle-class men within the existing hegemonic social structure. This reform narrative solely addresses the centered subjects. By developing the subjectivity of centered subjects, while simultaneously exploiting marginalized groups, *Sex and the City* sustains a hegemonic feminist discourse.

RACISM, ETHNOCENTRISM, AND *SEX*
AND THE CITY

In her assessment of the contemporary women's movement, Audre Lorde points out that, "White women focus upon their oppression as women and ignore differences of race, sexual preference, class and age. There is a pretense to homogeneity of experience covered by the word sisterhood that does not in fact exist" (Lorde 289). Throughout six seasons of *Sex and the City*, viewers are introduced to tokenized racism and ethnocentrism dominant storylines. These episodes explore the women's experiences with nonwhite and non-American-born characters whose race or ethnicity serve as the focus of their interaction.

African American characters, like other marginalized groups, are mostly absent from the hit series. When they do appear, they are cast in unimaginative, stereotypical roles. In episode 35, titled "No Ifs, Ands or Butts," an African American brother and sister are featured. Samantha, known for her promiscuity, dates Chivon. Chivon's sister Adeena confronts Samantha by telling her that she does not want her brother dating a white woman. Carrie, Miranda, and Charlotte dance around the stereotypical Mandingo representations of black men, while Samantha rejects their overt racism in favor of the covert type. Samantha declares, "I don't see color, I see conquests." Conveniently adopting the color-blind standpoint, Samantha's character avoids appearing racist by erasing the racial dimension of his identity.

On one hand, Samantha is viewed as the liberal white woman who dates interracially because she has moved beyond superficial color politics, while on the other, Adeena reads as the angry black woman who hysterically sees color and practices separatism. The power struggle or conflict is located between the white woman (Samantha) and the black woman (Adeena). The women are friendly, until Chivon, the black man, is positioned between them. When Samantha cannot "conquer" Chivon, she emasculates him by calling him a "pussy." The passive black man and his angry sister could not be conquered, so they had to be dismissed and subjugated.

Sex and the City castrates an African American man who allies himself with an African American woman. That Adeena is Chivon's sister, rather than an attractive single black woman, is no mere accident and further demonstrates the show's allegiance to hegemonic feminism. It is too far-fetched to consider that a black man is more sexually attracted to a black woman than to a white woman. Family ties and the fear of a crazy black woman make more sense and are more believable to the viewer. *Sex and the City* is careful to employ plotlines promoting limited and racist depictions of persons of color. . . .

African Americans are not the only racial-ethnic group exploited by *Sex and the City*. When the woman giggles while taking Miranda's take-out order, Miranda misinterprets this stereotypical laughter for negative judgments ("Cock-a-Doodle-Doo," episode 48). The truth is, the Chinese restaurant employee simply giggles and smiles in response to any order. Miranda is relieved to learn that this Asian stereotype is actually a reality. Her self-esteem is restored through the recognition of this stereotype. Charlotte seeks out racial minorities to meet her needs when all else fails. She has a sexual affair with the tanned-skinned gardener after many unsuccessful attempts at intercourse with her husband Trey ("What Goes Around Comes Around," episode 47). Charlotte pursues overseas adoption for a Mandarin baby as her last resort in raising a family. In her second marriage, to Harry, Charlotte sees a Chinese fertility doctor, prior to becoming pregnant. Again and again, nonwhite characters are tokenized for stereotypical story line purposes. . . .

SEXISM, PATRIARCHY, AND *SEX AND THE CITY*

Gender and sex are important themes in rereading *Sex and the City*, because so much of the women's identities is determined by their views of masculinity and its dominance over their lives. Whether one of the women is being exploited by a man or has internalized patriarchal thinking, sex and gender issues are always present. . . .

Sex and the City's centered subjects are portrayed as biological women. It serves as no surprise that *Sex and the City* positions biological women higher in the social hierarchy than transgendered women. In a patriarchal society, men becoming women will assume an oppressed position to both biological men and biological women. Men and women, as gendered constructs can be viewed as relational or even relative identities. . . .

Male-to-female transitioning contributes to our understanding of sex and gender. Marcia Yudkin offers a critical perspective in "Transsexualism and Women." In her investigation of the concept of woman, she proposes three levels of identity: biological, social, and psychological. Framing the concept of woman with levels suggests there is a hierarchy of womanness or womanhood. The biological level refers to sex organs, the social level describes the sex roles enacted or sustained, and the psychological level is the subjective identification with a gender identity. Yudkin specifies that transsexuals identify with the "opposite sex role" (101), as opposed to the opposite sex. . . .

The trilevel framework for understanding sex and gender reflects feminist discourses rampant within *Sex and the City* story lines. Episode 48 of *Sex and the City* ("Cock-a-Doodle Doo") features the "pre-op [operation] up my ass crew." The preoperative transsexual males to females work the streets as prostitutes. They are loud, dirty-talking "working girls" whose late night noise disrupts Samantha's sleep. All crew members are cast as persons of color. Following a series of back and forth battles of the divas, the transgendered prostitutes of color are eventually defeated by the embodiment of woman and assume their place in the hierarchy.

While there are intersections clearly visible within this hierarchy, it is *Sex and the City*'s view of sex and gender that is formative. Socially and psychologically, the pre-ops identify themselves as women. They demonstrate stereotypically feminine characteristics such as wearing long hairstyles, makeup, and women's clothing. Yet *Sex and the City* writers are careful to point out that these "women" are preoperative and have not yet become biological women. The pre-op "up my ass crew" are transsexuals whose sex transition is incomplete, thus making them merely transgendered instead of women.

Further characterizing this hierarchical view of "woman" is reflected by depicting preoperative transsexuals as prostitutes. Sex is situated as a commodity that can be bought and sold. Therefore, the centered subjects of *Sex and the City* are inherently privileged with female biological sex organs. The battle of Samantha versus the "pre-ops" is not only about the dominance of biological women over nonbiological women, but it can also be conceptualized as biological women's triumph over biological men. The absence of female biological sex organs renders "The Pre-ops" mere female impersonators, performing gender. They are no match for "real women."

Confusion regarding sex and gender provide challenges for the *Sex and the City* women. "Evolution" (episode 23) tells the story of Stephen, an effeminate man who dates Charlotte. Stephen resides in Chelsea and is employed as a pastry chef. Charlotte is "so confused. Is he gay or is he straight?" Stephen is effeminate in mannerism and has an acute awareness of fashion and cooking. With this interest in women-associated domains, Stephen fits Charlotte's schema of a gay man. Even after a goodnight kiss and a second date, "Charlotte wanted to be open-minded, but Stephen seemed to be making it as hard as possible." Unable to resolve her inner conflict and inability to understand Stephen, Charlotte recruits a team of experts (Carrie and Stanford) to help her determine Stephen's sexual orientation.

The purpose of discussing this episode in regard to sex and gender issues is to explicate the master narrative of *Sex and the City*. While on the surface, gender and sexual orientation are clearly intersections within this story line, a deeper understanding of this presentation shows us that patriarchal thinking is internalized within these women. . . . Writers of this episode make clear that Charlotte's quest is to discern Stephen's sexual orientation in the face of questionable masculinity. We do not read the story as Charlotte's inability to accept that Stephen is heterosexual, or not gay. Viewers are also discouraged from interpreting "Evolution" as Charlotte's exploration of Stephen's gender identity, which is what seems to be the purpose of the story line.

Instead, *Sex and the City* opts for a mundane discussion of an evolutionary phenomenon: "The gay straight man was a new string of heterosexuals spawned in Manhattan as a result of overexposure to fashion, exotic cuisine, musical theater and antique furniture." This episode is powerful in that it shows us how difficult it is to move beyond our narrow frameworks for understanding gender. In fact, Charlotte enjoys Stephen's company and a two-orgasm sexual encounter, but in the end cannot accept his feminine characteristics. "Her masculine side wasn't evolved enough for a man whose feminine side was as highly evolved as Stephen's." The *Sex and the City* women are not attracted to men who have evolved into homosexuals, and they reject notions that they can have romantic relationships with men who display their feminine sides. . . .

HETEROSEXISM, HOMOPHOBIA, AND *SEX AND THE CITY*

Sexual orientation is a regular theme for *Sex and the City*. Stanford Blache is a white, gay character on the show. During season four, Samantha has a "lesbian relationship" with a Brazilian artist named Maria. *Sex and the City* exploits non-heterosexual orientations. Bisexual, lesbian, and gay male characters are all presented in ways that marginalize their existence and reify the dominance of heterosexuality.

Bisexuality is a deeply misunderstood sexual orientation that receives little research attention. Currently, there is no specific model of bisexual identity development. . . . Furthermore, . . . many persons engaging in bisexual behavior do not label themselves as such. The lack of research, a clear definition of the term, and invisibility contribute to our society's adoption of myths. These misconceptions marginalize bisexuality and uphold binary systems of sexual orientation. . . .

[I]n *Sex and the City*'s "Boy, Girl, Boy, Girl" (episode 34), Carrie dates Sean, a younger man, who she casually learns has dated both men and women. Carrie becomes preoccupied by trying to figure out whom he is more attracted to, men or women. She wants to understand how bisexuality works. Eventually, the focus moves away from the development of their relationship to Carrie's persistent confusion as she tries to comprehend and fit in with Sean's bisexual lifestyle. Throughout the episode, viewers are treated to a review of the myths of bisexuality.

Joy Morgenstern outlines myths of bisexuality. One is that bisexuals are sexually promiscuous nymphomaniacs. Episode 34 ("Boy, Girl") ends with a gathering of Sean's bisexual friends. They are introduced to Carrie by detailing their previous romantic partnerships with each other. These bisexuals, you see, have all dated each other. A second myth of bisexuality is that bisexuals are gay people who are still in the closet. Bisexuality is then viewed as a transitional phase that will end in homosexuality or heterosexuality. . . . *Sex and the City* builds on this myth by casting Sean's character as younger than Carrie. Being in his twenties, Sean is still developing and transitioning. According to the myth, one day he will self-actualize and end his exploration.

And a third myth of bisexuality that we see in this *Sex and the City* episode is that bisexuals are indecisive neurotics who will never be sexually satisfied. Sean and his friends decide to play a game of spin the bottle at the party. Not only does this build on the characterization of bisexuality as a developmental phase, but it also lends itself to this indecisive myth. Just spin a bottle and have a sexual experience with whomever and whatever. Bisexuality is further marginalized by being cast as a game for which Carrie is "too old." Bisexuality is regarded as the problem, rather than Carrie's stereotypical and hegemonic views of it. As a centered subject, she chooses to relegate this sexual orientation to a status lower than that of her own. . . .

CLASS EXPLOITATION AND *SEX AND THE CITY*

. . . The women of *Sex and the City* enjoy economic privilege. As professional women, we learn that they are formally educated and able to independently support themselves. Yet their economic privilege does not exist within a vacuum. This class privilege is supported by a hierarchy that remains intact and a capitalist system that demands it. "The Caste System" (episode 22) reveals how the system—which has provided upward mobility for the *Sex and the City* women—continues to exploit working-class men and women.

Foreign domestic servants' labor provides the economically privileged increased freedom, at a low cost. In "The Caste System," Samantha separates herself from this contested feminist issue. While she "didn't believe in having servants, she didn't mind dating a man who did." Samantha's abject stance on servitude reveals the unresolved nature of this debate within feminist discussions. Her statement reveals that she sees the value in having servants, but might experience discomfort in hiring servants for herself. The man she's dating, Harvey Turkell, is described as a real estate giant "who made a killing in the market, turning Chelsea sweatshops into condos for the upwardly trendy." Sum, his Asian domestic servant, initially appears servile and polite. However, when Harvey is absent, we learn that Sum is quite the opposite. Sum's English speaking improves and she is no longer restricted in her physical range of motion. Sum has been putting on an act to appear complicit with her boss's classist and racist views of her.

Samantha is abused by Sum and is falsely accused of assaulting her. While Sum cries to "Mr. Harvey," Samantha rolls her eyes in disgust. She did not have to feel discomfort about hiring servants after all. Sum was actually exploiting Mr. Harvey, not the other way around. Writing class exploitation in a way that subverts and distorts the reality of domestic servitude upholds the ruling class's ability to condone the power they maintain over manipulative domestics. . . .

Class exploitation is not only addressed among poor, immigrant women. *Sex and the City* also explores how class differences affect relationships between men and women.

When Miranda the lawyer dates Steve the bartender, class issues emerge sooner than later. At first, it seems that the issue is purely financial. How can a couple resolve the nontraditional dynamic of a woman earning and possessing more money than a man? "She was so crazy about him that she let him take her out to dinner, but only to the places he could afford." Miranda cares deeply for Steve and will not allow him to spend beyond his means. Later described as "yuppie guilt," Miranda seems to harbor conflicting feelings about her class privilege. She wants Steve to spend and live within his means, but she also wants the same for herself. She fails to acknowledge and accept what she truly feels, and this results in guilt. Miranda and her friends discuss the impact of class differences on her relationship:

MIRANDA: None of this matters to me. I just don't want it to matter to him. It's like when single men have a lot of money, it works to their advantage. But when a single woman has money, it's like a problem you have to deal with. It's ridiculous. I wanna enjoy my success, not apologize for it.

SAMANTHA: Bravo, honey, bravo.

CHARLOTTE: But you're talking about more than a difference in income. You're talking about a difference in background and education. This guy is working class.

MIRANDA: Working class?

CARRIE: It's the millennium sweetie. We don't say things like "working class."

CHARLOTTE: But you're trying to pretend that we live in a classless society, and we don't. ("Old Dogs, New Dicks," episode 21)

In this instance, Charlotte is the voice of reason. Miranda wants to liberate herself from the discomfort she believes is the result of money mattering to Steve. In this instance, the problem has been misdiagnosed. Miranda has judged Steve negatively regarding his working-class background. In one scene, Miranda visits Steve's apartment for the first time. She is visibly constricted in her reaction. She scans the apartment and sees his corduroy suit. What remains unsaid is that Miranda thinks his apartment is trash and his suit is inappropriate for her upcoming company dinner party. Miranda believes that as is, Steve is inadequate, prompting her to take him shopping for a new suit.

Miranda offers to pay, but Steve will not let her. While up to this point, Miranda would not allow Steve to spend beyond his means, she has now decided that she "does not want to apologize for her success." Steve's credit card is denied, and he becomes upset with Miranda's attempts to pay for the suit. However, he does take the suit home by charging some on credit, writing a check, and paying in cash. Steve would later return the suit, explaining that he could not afford it. He ends the relationship with Miranda on the evening they were to attend the party. Miranda believes that Steve has ended the relationship because she's [being] punished for her success. When Miranda decided that Steve was not good enough and needed to be changed, it marked the beginning of the end.

Sex and the City tells us that Miranda and Steve's class difference is the cause of their breakup. This is not true. Their relationship is going well until Miranda fails to admit to herself that she is ashamed of Steve and exploits his economic status in order to avoid feeling guilty and to further uphold her social position. Miranda and Steve eventually will resume their friendship, have a child together, and get married. This, of course, occurs after Steve gains class mobility by becoming a business owner.

Sex and the City, with its mass-based popularity and appeal, projects powerful images to audiences. When we fail to critically read and reread media presentations, we run the risk of internalizing and reproducing our own oppression.

WORKS CITED

"Boy, Girl, Boy, Girl." Episode 34. Writ. Jenny Bicks. Dir. Pam Thomas. *Sex and the City*. Home Box Office. 25 June 2000.

"The Caste System." Episode 22. Writ. Darren Star. Dir. Allison Anders. *Sex and the City*. Home Box Office. 8 Aug. 1999.

"Cock-a-Doodle-Doo." Episode 48. Writ. Michael Patrick King. Dir. Allen Coulter. *Sex and the City*. Home Box Office. 15 Oct. 2000.

"Evolution." Episode 23. Writ. Cindy Chupak. Dir. Pam Thomas. *Sex and the City*. Home Box Office. 15 Aug. 1999.

Labi, Nadya. "Girl Power." *Time*. (1998).

Lorde, Audre. (2000). "Age, Race, Class, and Sex: Women Redefining Women." *Feminist Theory: A Reader*. Ed. Wendy Kolmar and Frances Bartkowski. Mountain View: Mayfield, 2003. 288–93.

Maddox, Garry. "Is *Sex and the City* Gay?" *Sydney Morning Herald Online*. 9 Feb. 2004. 5 July 2006, www.smh.com.au/articles/2004/02/09/1076175068807.html?from=storyrhs.

Morgenstern, Joy. "Myths of Bisexuality." *Off Our Backs*. May–June (2004): 46–48.

"No Ifs, Ands, or Butts." Episode 35. Writ. Michael Patrick King. Dir. Nacole Holofcenter. *Sex and the City*. Home Box Office. 9 July 2000.

"Old Dogs. New Dicks." Episode 21. Writ. Jenny Bicks. Dir. Alan Taylor. *Sex and the City*. Home Box Office. 1 Aug. 1999.

"What Goes Around Comes Around." Episode 47. Writ. Darren Starr. Dir. Allen Coulter. *Sex and the City*. Home Box Office. 8 Oct. 2000.

Yudkin, Marcia. "Transsexualism and Women: A Critical Perspective." *Feminist Studies* 4.3 (1978): 97–106.

45

Racist Stereotyping in the English Language

ROBERT B. MOORE

LANGUAGE AND CULTURE

An integral part of any culture is its language. Language not only develops in conjunction with a society's historical, economic and political evolution; it also reflects that society's attitudes and thinking. Language not only *expresses* ideas and concepts but actually *shapes* thought.[1] If one accepts that our dominant white culture is racist, then one would expect our language—an indispensable transmitter of culture—to be racist as well. Whites, as the dominant group, are not subjected to the same abusive characterization by our language that people of color receive. Aspects of racism in the English language that will be discussed in this essay include terminology, symbolism, politics, ethnocentrism, and context.

SOURCE: From Paula S. Rothenberg, ed., *Racism and Sexism: An Integrated Study* (New York: St. Martin's Press, 1988), pp. 269–79. Reprinted by permission of the Council on Interracial Books for Children c/o Lawrence Jordan Literary Agency, 345 West 121st Street, NY, NY 10027.

Before beginning our analysis of racism in language we would like to quote part of a TV film review which shows the connection between language and culture.[2]

> Depending on one's culture, one interacts with time in a very distinct fashion. One example which gives some cross-cultural insights into the concept of time is language. In Spanish, a watch is said to "walk." In English, the watch "runs." In German, the watch "functions." And in French, the watch "marches." In the Indian culture of the Southwest, people do not refer to time in this way. The value of the watch is displaced with the value of "what time it's getting to be." Viewing these five cultural perspectives of time, one can see some definite emphasis and values that each culture places on time. For example, a cultural perspective may provide a clue to why the negative stereotype of the slow and lazy Mexican who lives in the "Land of Mañana" exists in the Anglo value system, where time "flies," the watch "runs" and "time is money."

A SHORT PLAY ON "BLACK" AND
"WHITE" WORDS

Some may blackly (angrily) accuse me of trying to blacken (defame) the English language, to give it a black eye (a mark of shame) by writing such black words (hostile). They may denigrate (to cast aspersions; to darken) me by accusing me of being blackhearted (malevolent), of having a black outlook (pessimistic, dismal) on life, of being a blackguard (scoundrel)—which would certainly be a black mark (detrimental fact) against me. Some may black-brow (scowl at) me and hope that a black cat crosses in front of me because of this black deed. I may become a black sheep (one who causes shame or embarrassment because of deviation from the accepted standards), who will be blackballed (ostracized) by being placed on a blacklist (list of undesirables) in an attempt to blackmail (to force or coerce into a particular action) me to retract my words. But attempts to blackjack (to compel by threat) me will have a Chinaman's chance of success, for I am not a yellow-bellied Indian-giver of words, who will whitewash (cover up or gloss over vices or crimes) a black lie (harmful, inexcusable). I challenge the purity and innocence (white) of the English language. I don't see things in black and white (entirely bad or entirely good) terms, for I am a white man (marked by upright firmness) if there ever was one. However, it would be a black day when I would not "call a spade a spade," even though some will suggest a white man calling the English language racist is like the pot calling the kettle black. While many may be niggardly (grudging, scanty) in their support, others will be honest and decent—and to them I say, that's very white of you (honest, decent).

The preceding is of course a white lie (not intended to cause harm), meant only to illustrate some examples of racist terminology in the English language.

OBVIOUS BIGOTRY

Perhaps the most obvious aspect of racism in language would be terms like "nigger," "spook," "chink," "spic," etc. While these may be facing increasing social disdain, they certainly are not dead. Large numbers of white Americans continue to utilize these terms. "Chink," "gook," and "slant-eyes" were in common usage among U.S. troops in Vietnam. An NBC nightly news broadcast, in February 1972, reported that the basketball team in Pekin, Illinois, was called the "Pekin Chinks" and noted that even though this had been protested by Chinese Americans, the term continued to be used because it was easy, and meant no harm. Spiro Agnew's widely reported "fat Jap" remark and the "little Jap" comment of lawyer John Wilson, during the Watergate hearings, are surface indicators of a deep-rooted Archie Bunkerism.

Many white people continue to refer to Black people as "colored," as for instance in a July 30, 1975, *Boston Globe* article on a racist attack by whites on a group of Black people using a public beach in Boston. One white person was quoted as follows:

> We've always welcomed good colored people in South Boston but we will not tolerate radical blacks or Communists. . . . Good colored people are welcome in South Boston, black militants are not.

Many white people may still be unaware of the disdain many African Americans have for the term "colored," but it often appears that whether used intentionally or unintentionally, "colored" people are "good" and "know their place," while "Black" people are perceived as "uppity" and "threatening" to many whites. Similarly, the term "boy" to refer to African American men is now acknowledged to be a demeaning term, though still in common use. Other terms such as "the pot calling the kettle black" and "calling a spade a spade" have negative racial connotations but are still frequently used, as for example when President Ford was quoted in February 1976 saying that even though Daniel Moynihan had left the U.N., the U.S. would continue "calling a spade a spade."

COLOR SYMBOLISM

The symbolism of white as positive and black as negative is pervasive in our culture, with the black/white words used in the beginning of this essay only one of many aspects. "Good guys" wear white hats and ride white horses, "bad guys" wear black hats and ride black horses. Angels are white, and devils are black. The definition of *black* includes "without any moral light or goodness, evil, wicked, indicating disgrace, sinful," while that of *white* includes "morally pure, spotless, innocent, free from evil intent."

A children's TV cartoon program, *Captain Scarlet*, is about an organization called Spectrum, whose purpose is to save the world from an evil extraterrestrial force called the Mysterons. Everyone in Spectrum has a color name—Captain Scarlet, Captain Blue, etc. The one Spectrum agent who has been mysteriously taken over by the Mysterons and works to advance their evil aims is Captain Black. The person who heads Spectrum, the good organization out to defend the world, is Colonel White.

Three of the dictionary definitions of white are "fairness of complexion, purity, innocence." These definitions affect the standards of beauty in our culture, in which whiteness represents the norm. "Blondes have more fun" and "Wouldn't you really rather be a blonde" are sexist in their attitudes toward women generally, but are racist white standards when applied to third world women. A 1971 *Mademoiselle* advertisement pictured a curly-headed, ivory-skinned woman over the caption, "When you go blonde go all the way," and asked: "Isn't this how, in the back of your mind, you always wanted to look? All wide-eyed and silky blonde down to there, and innocent?" Whatever the advertising people meant by this particular woman's innocence, one must remember that "innocent" is one of the definitions of the word white. This standard of beauty when preached to all women is racist. The statement "Isn't this how, in the back of your mind, you always wanted to look?" either ignores third world women or assumes they long to be white.

Time magazine in its coverage of the Wimbledon tennis competition between the black Australian Evonne Goolagong and the white American Chris Evert described Ms. Goolagong as "the dusky daughter of an Australian sheepshearer," while Ms. Evert was "a fair young girl from the middle-class groves of Florida." *Dusky* is a synonym of "black" and is defined as "having dark skin; of a dark color; gloomy; dark; swarthy." Its antonyms are "fair" and "blonde." *Fair* is defined in part as "free from blemish, imperfection, or anything that impairs the appearance, quality, or character: pleasing in appearance, attractive; clean; pretty; comely." By defining Evonne Goolagong as "dusky," *Time* technically defined her as the opposite of "pleasing in appearance; attractive; clean; pretty; comely."

The studies of Kenneth B. Clark, Mary Ellen Goodman, Judith Porter and others indicate that this persuasive "rightness of whiteness" in U.S. culture affects children before the age of four, providing white youngsters with a false sense of superiority and encouraging self-hatred among third world youngsters.

ETHNOCENTRISM OR FROM A
WHITE PERSPECTIVE

Some words and phrases that are commonly used represent particular perspectives and frames of reference, and these often distort the understanding of the reader or listener. David R. Burgest[3] has written about the effect of using the terms "slave" or "master." He argues that the psychological impact of the statement referring to "the master raped his slave" is different from the impact of the same statement substituting the words: "the white captor raped an African woman held in captivity."

> Implicit in the English usage of the "master-slave" concept is ownership of the "slave" by the "master," therefore, the "master" is merely abusing his property (slave). In reality, the captives (slave[s]) were African individuals with human worth, right and dignity and the term "slave" denounces that human quality thereby making the mass rape of African women by white captors more acceptable in the minds of people and setting a mental frame of reference for legitimizing the atrocities perpetuated against African people.

The term slave connotes a less than human quality and turns the captive person into a thing. For example, two McGraw-Hill Far Eastern Publishers textbooks (1970) stated, "At first it was the slaves who worked the cane and they got only food for it. Now men work cane and get money." Next time you write about slavery or read about it, try transposing all "slaves" into "African people held in captivity," "Black people forced to work for no pay" or "African people stolen from their families and societies." While it is more cumbersome, such phrasing conveys a different meaning.

PASSIVE TENSE

Another means by which language shapes our perspective has been noted by Thomas Greenfield,[4] who writes that the achievements of Black people—and Black people themselves—have been hidden in

> the linguistic ghetto of the passive voice, the subordinate clause, and the "understood" subject. The seemingly innocuous distinction (between active/passive voice) holds enormous implications for writers and speakers. When it is effectively applied, the rhetorical impact of the passive voice—the art of making the creator or instigator of action totally disappear from a reader's perception—can be devastating.

For instance, some history texts will discuss how European immigrants came to the United States seeking a better life and expanded opportunities, but will note that "slaves *were brought* to America." Not only does this omit the destruction of African societies and families, but it ignores the role of northern merchants and southern slaveholders in the profitable trade in human beings. Other books will state that "the continental railroad *was built*," conveniently omitting information about the Chinese laborers who built much of it or the oppression they suffered.

Another example. While touring Monticello, Greenfield noted that the tour guide

made all the black people of Monticello disappear through her use of the passive voice. While speaking of the architectural achievements of Jefferson in the active voice, she unfailingly shifted to passive when speaking of the work performed by Negro slaves and skilled servants.

Noting a type of door that after 166 years continued to operate without need for repair, Greenfield remarks that the design aspect of the door was much simpler than the actual skill and work involved in building and installing it. Yet his guide stated: "Mr. Jefferson designed these doors . . ." while "the doors were installed in 1809." The workers who installed those doors were African people whom Jefferson held in bondage. The guide's use of the passive tense enabled her to dismiss the reality of Jefferson's slaveholding. It also meant that she did not have to make any mention of the skills of those people held in bondage.

POLITICS AND TERMINOLOGY

"Culturally deprived," "economically disadvantaged" and "underdeveloped" are other terms which mislead and distort our awareness of reality. The application of the term "culturally deprived" to third world children in this society reflects a value judgment. It assumes that the dominant whites are cultured and all others without culture. In fact, third world children generally are bicultural, and many are bilingual, having grown up in their own culture as well as absorbing the dominant culture. In many ways, they are equipped with skills and experiences which white youth have been deprived of, since most white youth develop in a monocultural, monolingual environment. Burgest[5] suggests that the term "culturally deprived" be replaced by "culturally dispossessed," and that the term "economically disadvantaged" be replaced by "economically exploited." Both these terms present a perspective and implication that provide an entirely different frame of reference as to the reality of the third world experience in U.S. society.

Similarly, many nations of the third world are described as "underdeveloped." These less wealthy nations are generally those that suffered under colonialism and neo-colonialism. The "developed" nations are those that exploited their resources and wealth. Therefore, rather than referring to these countries as "underdeveloped," a more appropriate and meaningful designation might be "over exploited." Again, transpose this term next time you read about "underdeveloped nations" and note the different meaning that results.

Terms such as "culturally deprived," "economically disadvantaged" and "underdeveloped" place the responsibility for their own conditions on those being so described. This is known as "Blaming the Victim."[6] It places responsibility for poverty on the victims of poverty. It removes the blame from those in power who benefit from, and continue to permit, poverty.

Still another example involves the use of "non-white," "minority" or "third world." While people of color are a minority in the U.S., they are part of the vast majority of the world's population, in which white people are a distinct minority. Thus, by utilizing the term minority to describe people of color in the U.S., we can lose sight of the global majority/minority reality—a fact of some importance in the increasing and interconnected struggles of people of color inside and outside the U.S.

To describe people of color as "non-white" is to use whiteness as the standard and norm against which to measure all others. Use of the term "third world" to describe all people of color overcomes the inherent bias of "minority" and "non-white." Moreover, it connects the struggles of third world people in the U.S. with the freedom struggles around the globe.

The term third world gained increasing usage after the 1955 Bandung Conference of "non-aligned" nations, which represented a third force outside of the two world superpowers. The "first world" represents the United States, Western Europe and their sphere of influence. The "second world" represents the Soviet Union and its sphere. The "third world" represents, for the most part, nations that were, or are, controlled by the "first world" or West. For the most part, these are nations of Africa, Asia and Latin America.

"LOADED" WORDS AND NATIVE AMERICANS

Many words lead to a demeaning characterization of groups of people. For instance, Columbus, it is said, "discovered" America. The word *discover* is defined as "to gain sight or knowledge of something previously unseen or unknown; to discover may be to find some existent thing that was previously unknown." Thus, a continent inhabited by millions of human beings cannot be "discovered." For history books to continue this usage represents a Eurocentric (white European) perspective on world history and ignores the existence of, and the perspective of, Native Americans. "Discovery," as used in the Euro-American context, implies the right to take what one finds, ignoring the rights of those who already inhabit or own the "discovered" thing.

Eurocentrism is also apparent in the usage of "victory" and "massacre" to describe the battles between Native Americans and whites. *Victory* is defined in the dictionary as "a success or triumph over an enemy in battle or war; the decisive defeat of an opponent." *Conquest* denotes the "taking over of control by the victor, and the obedience of the conquered." *Massacre* is defined as "the unnecessary, indiscriminate killing of a number of human beings, as in barbarous warfare or persecution, or for revenge or plunder." *Defend* is described as "to ward off attack from; guard against assault or injury; to strive to keep safe by resisting attack."

Eurocentrism turns these definitions around to serve the purpose of distorting history and justifying Euro-American conquest of the Native American homelands. Euro-Americans are not described in history books as invading Native American lands, but rather as defending *their* homes against "Indian" attacks. Since European communities were constantly encroaching on land already occupied, then a more honest interpretation would state that it was the Native Americans who were "warding off," "guarding" and "defending" their homelands.

Native American victories are invariably defined as "massacres," while the indiscriminate killing, extermination and plunder of Native American nations by Euro-Americans is defined as "victory." Distortion of history by the choice of "loaded" words used to describe historical events is a common racist practice. Rather than portraying Native Americans as human beings in highly defined and complex societies, cultures and civilizations, history books use such adjectives as "savages," "beasts," "primitive," and "backward." Native people are referred to as "squaw," "brave," or "papoose" instead of "woman," "man," or "baby."

Another term that has questionable connotations is *tribe*. The *Oxford English Dictionary* defines this noun as "a race of people; now applied especially to a primary aggregate of people in a primitive or barbarous condition, under a headman or chief." Morton Fried,[7] discussing "The Myth of Tribe," states that the word "did not become a general term of reference to American Indian society until the nineteenth century. Previously, the words commonly used for Indian populations were 'nation' and 'people.'" Since "tribe" has assumed a connotation of primitiveness or backwardness, it is suggested that the use of "nation" or "people" replace the term whenever possible in referring to Native American peoples.

The term *tribe* invokes even more negative implications when used in reference to American peoples. As Evelyn Jones Rich[8] has noted, the term is "almost always used to refer to third world people and it implies a stage of development which is, in short, a put-down."

"LOADED" WORDS AND AFRICANS

Conflicts among diverse peoples within African nations are often referred to as "tribal warfare," while conflicts among the diverse peoples within European countries are never described in such terms. If the rivalries between the Ibo and the Hausa and Yoruba in Nigeria are described as "tribal," why not the rivalries between Serbs and Slavs in Yugoslavia, or Scots and English in Great Britain, Protestants and Catholics in Ireland, or the Basques and the Southern Spaniards in Spain? Conflicts among African peoples in a particular nation have religious, cultural, economic and/or political roots. If we can analyze the roots of conflicts among European peoples in terms other than "tribal warfare," certainly we can do the same with African peoples, including correct reference to the ethnic groups or nations involved. For example, the terms "Kaffirs," "Hottentot" or "Bushmen" are names imposed by white Europeans. The correct names are always those by which a people refer to themselves. (In these instances Xhosa, Khoi-Khoin and San are correct.[9])

The generalized application of "tribal" in reference to Africans—as well as the failure to acknowledge the religious, cultural and social diversity of African peoples—is a decidedly racist dynamic. It is part of the process whereby Euro-Americans justify, or avoid confronting, their oppression of third world peoples. Africa has been particularly insulted by this dynamic, as witness the pervasive "darkest Africa" image. This image, widespread in Western culture, evokes an Africa covered by jungles and inhabited by "uncivilized," "cannibalistic," "pagan," "savage" peoples. This "darkest Africa" image avoids the geographical reality. Less than 20 percent of the African continent is wooded savanna, for example. The image also ignores the history of African cultures and civilizations. Ample evidence suggests this distortion of reality was developed as a convenient rationale for the European and American slave trade. The Western powers, rather than exploiting, were civilizing and Christianizing "uncivilized" and "pagan savages" (so the rationalization went). This dynamic also served to justify Western colonialism. From Tarzan movies to racist children's books like *Doctor Dolittle* and *Charlie and the Chocolate Factory*, the image of "savage" Africa and the myth of "the white man's burden" has been perpetuated in Western culture.

A 1972 *Time* magazine editorial lamenting the demise of *Life* magazine, stated that the "lavishness" of *Life*'s enterprises included "organizing safaris into darkest Africa." The same year, the *New York Times'* C. L. Sulzberger wrote that Africa has "a history as dark as the skins of many of its people." Terms such as "darkest Africa," "primitive," "tribe" ("tribal") or "jungle," in reference to Africa, perpetuate myths and are especially inexcusable in such large circulation publications.

Ethnocentrism is similarly reflected in the term "pagan" to describe traditional religions. A February 1973 *Time* magazine article on Uganda stated, "Moslems account for only 500,000 of Uganda's 10 million people. Of the remainder, 5,000,000 are Christians and the rest pagan." *Pagan* is defined as "Heathen, a follower of a polytheistic religion; one that has little or no religion and that is marked by a frank delight in and uninhibited seeking after sensual pleasures and material goods." *Heathen* is defined as "Unenlightened; an unconverted member of a people or nation that does not acknowledge the God of the Bible. A person whose culture or enlightenment is of an inferior grade, especially an irreligious person." Now, the people of Uganda, like almost all Africans, have serious religious beliefs and practices. As used by Westerners, "pagan" connotes something wild, primitive and inferior—another term to watch out for.

The variety of traditional structures that African people live in are their "houses," not "huts." A *hut* is "an often small and temporary dwelling of simple construction." And to describe Africans as "natives" (noun) is derogatory terminology—as in, "the natives are restless." The dictionary definition of *native* includes: "one of a people inhabiting a territorial area at the time of its discovery or becoming familiar to a foreigner; one belonging to a people having a less complex civilization." Therefore, use of "native," like use of "pagan" often implies a value judgment of white superiority.

QUALIFYING ADJECTIVES

Words that would normally have positive connotations can have entirely different meanings when used in a racial context. For example, C. L. Sulzberger, the columnist of the *New York Times*, wrote in January 1975, about conversations he had with two people in Namibia. One was the white South African administrator of the country and the other a member of SWAPO, the Namibian liberation movement. The first is described as "Dirk Mudge, who as senior elected member of the administration is a kind of acting Prime Minister. . . ." But the second person is introduced as "Daniel Tijongarero, an intelligent Herero tribesman who is a member of SWAPO. . . ." What need was there for Sulzberger to state that Daniel Tijongarero is "intelligent"? Why not also state that Dirk Mudge was "intelligent"—or do we assume he wasn't?

A similar example from a 1968 *New York Times* article reporting on an address by Lyndon Johnson stated, "The President spoke to the well-dressed Negro officials and their wives." In what similar circumstances can one imagine a reporter finding it necessary to note that an audience of white government officials was "well-dressed"?

Still another word often used in a racist context is "qualified." In the 1960s white Americans often questioned whether Black people were "qualified" to hold public office, a question that was never raised (until too late) about white officials like Wallace, Maddox, Nixon, Agnew, Mitchell, et al. The question of qualifications has been raised even more frequently in recent years as white people question whether Black people are "qualified" to be hired for positions in industry and educational institutions. "We're looking for a qualified Black" has been heard again and again as institutions are confronted with affirmative action goals. Why stipulate that Blacks must be "qualified," when for others it is taken for granted that applicants must be "qualified"?

SPEAKING ENGLISH

Finally, the depiction in movies and children's books of third world people speaking English is often itself racist. Children's books about Puerto Ricans or Chicanos often connect poverty with a failure to speak English or to speak it well, thus blaming the victim and ignoring the racism which affects third world people regardless of their proficiency in English. Asian characters speak a stilted English ("Honorable so and so" or "Confucius say") or have a speech impediment ("roots or ruck," "very solly," "flied lice"). Native American characters speak another variation of stilted English ("Boy not hide. Indian take boy."), repeat certain Hollywood-Indian phrases ("Heap big" and "Many moons") or simply grunt out "Ugh" or "How." The repeated use of these language characterizations functions to make third world people seem less intelligent and less capable than the English-speaking white characters.

WRAP-UP

A *Saturday Review* editorial[10] on "The Environment of Language" stated that language

> . . . has as much to do with the philosophical and political condition-
> ing of a society as geography or climate. . . . People in Western cul-
> tures do not realize the extent to which their racial attitudes have been
> conditioned since early childhood by the power of words to ennoble
> or condemn, augment or detract, glorify or demean. Negative lan-
> guage infects the subconscious of most Western people from the time
> they first learn to speak. Prejudice is not merely imparted or superim-
> posed. It is metabolized in the bloodstream of society. What is needed
> is not so much a change in language as an awareness of the power
> of words to condition attitudes. If we can at least recognize the
> underpinnings of prejudice, we may be in a position to deal with the
> effects.

To recognize the racism in language is an important first step. Consciousness of the influence of language on our perceptions can help to negate much of that influence. But it is not enough to simply become aware of the effects of racism in conditioning attitudes. While we may not be able to change the language, we can definitely change our usage of the language. We can avoid using words that degrade people. We can make a conscious effort to use terminology that reflects a progressive perspective, as opposed to a distort-ing perspective. It is important for educators to provide students with opportu-nities to explore racism in language and to increase their awareness of it, as well as learning terminology that is positive and does not perpetuate negative human values.

NOTES

1.   Simon Podair, "How Bigotry Builds Through Language," *Negro Digest*, March 1967.

2.   Jose Armas, "Antonia and the Mayor: A Cultural Review of the Film," *The Journal of Ethnic Studies*, Fall 1975.

3.   David R. Burgest, "The Racist Use of the English Language," *Black Scholar*, Sept. 1973.

4.   Thomas Greenfield, "Race and Passive Voice at Monticello," *Crisis*, April 1975.

5.   David R. Burgest, "Racism in Everyday Speech and Social Work Jargon," *Social Work*, July 1973.

6.   William Ryan, *Blaming the Victim*, Pantheon Books, 1971.

7.   Morton Fried, "The Myth of Tribe," *National History*, April 1975.

8.   Evelyn Jones Rich, "Mind Your Language," *Africa Report*, Sept./Oct. 1974.

9.   Steve Wolf, "Catalogers in Revolt Against LC's Racist, Sexist Headings," *Bulletin of Interracial Books for Children*, Vol. 6, Nos. 3&4, 1975.

10. "The Environment of Language," *Saturday Review*, April 8, 1967.

REFERENCES

Also see:

Roger Bastide, "Color, Racism and Christianity," *Daedalus*, Spring 1967.

Kenneth J. Gergen, "The Significance of Skin Color in Human Relations," *Daedalus*, Spring 1967.

Lloyd Yabura, "Towards a Language of Humanism," *Rhythm*, Summer 1971.

UNESCO, "Recommendations Concerning Terminology in Education on Race Questions," June 1968.

46

Crimes Against Humanity

WARD CHURCHILL

If nifty little "pep" gestures like the "Indian Chant" and the "Tomahawk Chop" are just good clean fun, then let's spread the fun around, shall we?

During the past couple of seasons, there has been an increasing wave of controversy regarding the names of professional sports teams like the Atlanta "Braves," Cleveland "Indians," Washington "Redskins," and Kansas City "Chiefs." The issue extends to the names of college teams like Florida State University "Seminoles," University of Illinois "Fighting Illini," and so on, right on down to high school outfits like the Lamar (Colorado) "Savages." Also involved have been team adoption of "mascots," replete with feathers, buckskins, beads, spears, and "warpaint" (some fans have opted to adorn themselves in the same fashion), and nifty little "pep" gestures like the "Indian Chant" and "Tomahawk Chop."

A substantial number of American Indians have protested that use of native names, images, and symbols as sports team mascots and the like is, by definition, a virulently racist practice. Given the historical relationship between Indians and non-Indians during what has been called the "Conquest of America," American Indian Movement leader (and American Indian Anti-Defamation Council founder) Russell Means has compared the practice to contemporary Germans naming their soccer teams the "Jews," "Hebrews," and "Yids," while adorning

SOURCE: From Ward Churchill, "Crimes Against Humanity," *Z Magazine* 6 (March 1993): 43–47. Reprinted by permission of the author.

their uniforms with grotesque caricatures of Jewish faces taken from the Nazis' anti-Semitic propaganda of the 1930s. Numerous demonstrations have occurred in conjunction with games—most notably during the November 15, 1992 match-up between the Chiefs and Redskins in Kansas City—by angry Indians and their supporters.

In response, a number of players—especially African Americans and other minority athletes—have been trotted out by professional team owners like Ted Turner, as well as university and public school officials, to announce that they mean not to insult but to honor native people. They have been joined by the television networks and most major newspapers, all of which have editorialized that Indian discomfort with the situation is "no big deal," insisting that the whole thing is just "good, clean fun." The country needs more such fun, they've argued, and "a few disgruntled Native Americans" have no right to undermine the nation's enjoyment of its leisure time by complaining. This is especially the case, some have argued, "in hard times like these." It has even been contended that Indian outrage at being systematically degraded—rather than the degradation itself—creates "a serious barrier to the sort of intergroup communication so necessary in a multicultural society such as ours."

Okay, let's communicate. We are frankly dubious that those advancing such positions really believe their own rhetoric, but, just for the sake of argument, let's accept the premise that they are sincere. If what they say is true, then isn't it time we spread such "inoffensiveness" and "good cheer" around among *all* groups so that *everybody* can participate *equally* in fostering the round of national laughs they call for? Sure it is—the country can't have too much fun or "intergroup involvement"—so the more, the merrier. Simple consistency demands that anyone who thinks the Tomahawk Chop is a swell pastime must be just as hearty in their endorsement of the following ideas—by the logic used to defend the defamation of American Indians—should help us all really start yukking it up.

First, as a counterpart to the Redskins, we need an NFL team called "Niggers" to honor Afro-Americans. Half-time festivities for fans might include a simulated stewing of the opposing coach in a large pot while players and cheerleaders dance around it, garbed in leopard skins and wearing fake bones in their noses. This concept obviously goes along with the kind of gaiety attending the Chop, but also with the actions of the Kansas City Chiefs, whose team members—prominently including black team members—lately appeared on a poster looking "fierce" and "savage" by way of wearing Indian regalia. Just a bit of harmless "morale boosting," says the Chiefs' front office. You bet.

So that the newly formed Niggers sports club won't end up too out of sync while expressing the "spirit" and "identity" of Afro-Americans in the above fashion, a baseball franchise—let's call this one the "Sambos"—should be formed. How about a basketball team called the "Spearchuckers"? A hockey team called the "Jungle Bunnies"? Maybe the "essence" of these teams could be depicted by images of tiny black faces adorned with huge pairs of lips. The players could appear on TV every week or so gnawing on chicken legs and spitting watermelon seeds at one another. Catchy, eh? Well, there's "nothing to be upset about," according to those who love wearing "war bonnets" to the Super

Bowl or having "Chief Illiniwik" dance around the sports arenas of Urbana, Illinois.

And why stop there? There are plenty of other groups to include. Hispanics? They can be "represented" by the Galveston "Greasers" and San Diego "Spics," at least until the Wisconsin "Wetbacks" and Baltimore "Beaners" get off the ground. Asian Americans? How about the "Slopes," "Dinks," "Gooks," and "Zipperheads?" Owners of the latter teams might get their logo ideas from editorial page cartoons printed in the nation's newspapers during World War II: slant-eyes, buck teeth, big glasses, but nothing racially insulting or derogatory, according to the editors and artists involved at the time. Indeed, this Second World War–vintage stuff can be seen as just another barrel of laughs, at least by what current editors say are their "local standards" concerning American Indians.

Let's see. Who's been left out? Teams like the Kansas City "Kikes," Hanover "Honkies," San Leandro "Shylocks," Daytona "Dagos," and Pittsburgh "Polacks" will fill a certain social void among white folk. Have a religious belief? Let's all go for the gusto and gear up the Milwaukee "Mackerel Snappers" and Hollywood "Holy Rollers." The Fighting Irish of Notre Dame can be rechristened the "Drunken Irish" or "Papist Pigs." Issues of gender and sexual preferences can be addressed through creation of teams like the St. Louis "Sluts," Boston "Bimbos," Detroit "Dykes," and the Fresno "Fags." How about the Gainesville "Gimps" and Richmond "Retards," so the physically and mentally impaired won't be excluded from our fun and games?

Now, don't go getting "overly sensitive" out there. None of this is demeaning or insulting, at least not when it's being done to Indians. Just ask the folks who are doing it, or their apologists like Andy Rooney in the national media. They'll tell you in fact they *have* been telling you—that there's been no harm done, regardless of what their victims think, feel, or say. The situation is exactly the same as when those with precisely the same mentality used to insist that Step 'n' Fetchit was okay, or Rochester on the *Jack Benny Show*, or Amos and Andy, Charlie Chan, the Frito Bandito, or any of the other cutesy symbols making up the lexicon of American racism. Have we communicated yet?

Let's get just a little bit real here. The notion of "fun" embodied in rituals like the Tomahawk Chop must be understood for what it is. There's not a single non-Indian example used above which can be considered socially acceptable in even the most marginal sense. The reasons are obvious enough. So why is it different where American Indians are concerned? One can only conclude that, in contrast to the other groups at issue, Indians are (falsely) perceived as being too few, and therefore too weak, to defend themselves effectively against racist and otherwise offensive behavior.

Fortunately, there are some glimmers of hope. A few teams and their fans have gotten the message and have responded appropriately. Stanford University, which opted to drop the name "Indians" from Stanford, has experienced no resulting drop-off in attendance. Meanwhile, the local newspaper in Portland, Oregon, recently decided its long-standing editorial policy prohibiting use of racial epithets should include derogatory team names. The Redskins, for instance, are now referred to as "the Washington team," and will continue to be described in this way until the franchise adopts an inoffensive moniker (newspaper sales in Portland have suffered no decline as a result).

Such examples are to be applauded and encouraged. They stand as figurative beacons in the night, proving beyond all doubt that it is quite possible to indulge in the pleasure of athletics without accepting blatant racism into the bargain.

NUREMBERG PRECEDENTS

On October 16, 1946, a man named Julius Streicher mounted the steps of a gallows. Moments later he was dead, the sentence of an international tribunal composed of representatives of the United States, France, Great Britain, and the Soviet Union having been imposed. Streicher's body was then cremated, and—so horrendous were his crimes thought to have been—his ashes dumped into an unspecified German river so that "no one should ever know a particular place to go for reasons of mourning his memory."

Julius Streicher had been convicted at Nuremberg, Germany, of what were termed "Crimes Against Humanity." The lead prosecutor in his case—Justice Robert Jackson of the United States Supreme Court—had not argued that the defendant had killed anyone, nor that he had personally committed any especially violent act. Nor was it contended that Streicher had held any particularly important position in the German government during the period in which the so-called Third Reich had exterminated some 6,000,000 Jews, as well as several million Gypsies, Poles, Slavs, homosexuals, and other untermenschen (subhumans).

The sole offense for which the accused was ordered put to death was in having served as publisher/editor of a Bavarian tabloid entitled *Der Sturmer* during the early-to-mid 1930s, years before the Nazi genocide actually began. In this capacity, he had penned a long series of virulently anti-Semitic editorials and "news" stories, usually accompanied by cartoons and other images graphically depicting Jews in extraordinarily derogatory fashion. This, the prosecution asserted, had done much to "dehumanize" the targets of his distortion in the mind of the German public. In turn, such dehumanization had made it possible—or at least easier—for average Germans to later indulge in the outright liquidation of Jewish "vermin." The tribunal agreed, holding that Streicher was therefore complicit in genocide and deserving of death by hanging.

During his remarks to the Nuremberg tribunal, Justice Jackson observed that, in implementing its sentences, the participating powers were morally and legally binding themselves to adhere forever after to the same standards of conduct that were being applied to Streicher and the other Nazi leaders. In the alternative, he said, the victorious allies would have committed "pure murder" at Nuremberg—no different in substance from that carried out by those they presumed to judge—rather than establishing the benchmark for justice" which was intended.

Yet in the United States of Robert Jackson, the indigenous American Indian population had already been reduced, in a process which is ongoing to this day, from perhaps 12.5 million in the year 1500 to fewer than 250,000 by the beginning of the 20th century. This was accomplished, according to official sources, "largely through the cruelty of [Euro-American] settlers," and an informal but clear governmental policy which had made it an articulated goal to "exterminate these red vermin," or at least whole segments of them.

Bounties had been placed on the scalps of Indians—any Indians—in places as diverse as Georgia, Kentucky, Texas, the Dakotas, Oregon, and California, and had been maintained until resident Indian populations were decimated or disappeared altogether. Entire peoples such as the Cherokee had been reduced to half their size through a policy of forced removal from their homelands east of the Mississippi River to what were then considered less preferable areas in the West.

Others, such as the Navajo, suffered the same fate while under military guard for years on end. The United States Army had also perpetrated a long series of wholesale massacres of Indians at places like Horseshoe Bend, Bear River, Sand Creek, the Washita River, the Marias River, Camp Robinson, and Wounded Knee.

Through it all, hundreds of popular novels—each competing with the next to make Indians appear more grotesque, menacing, and inhuman—were sold in the tens of millions of copies in the U.S. Plainly, the Euro-American public was being conditioned to see Indians in such a way as to allow their eradication to continue. And continue it did until the Manifest Destiny of the U.S.—a direct precursor to what Hitler would subsequently call Le bens raum politik (the politics of living space)—was consummated.

By 1900, the national project of "clearing" Native Americans from their land and replacing them with "superior" Anglo-American settlers was complete: the indigenous population had been reduced by as much as 98 percent while approximately 97.5 percent of their original territory had "passed" to the invaders. The survivors had been concentrated, out of sight and mind of the public, on scattered "reservations," all of them under the self-assigned "plenary" (full) power of the federal government. There was, of course, no Nuremberg-style tribunal passing judgment on those who had fostered such circumstances in North America. No U.S. official or private citizen was ever imprisoned—never mind hanged—for implementing or propagandizing what had been done. Nor had the process of genocide afflicting Indians been completed. Instead, it merely changed form.

Between the 1880s and the 1980s, nearly half of all Native American children were coercively transferred from their own families, communities, and cultures to those of the conquering society. This was done through compulsory attendance at remote boarding schools, often hundreds of miles from their homes, where native children were kept for years on end while being systematically "deculturated" (indoctrinated to think and act in the manner of Euro-Americans rather than Indians). It was also accomplished through a pervasive foster home and adoption program—including "blind" adoptions, where children would be permanently denied information as to who they were/are and where they'd come from—placing native youths in non-Indian homes.

The express purpose of all this was to facilitate a U.S. governmental policy to bring about the "assimilation" (dissolution) of indigenous societies. In other words, Indian cultures as such were to be caused to disappear. Such policy objectives are directly contrary to the United Nations 1948 Convention on Punishment and Prevention of the Crime of Genocide, an element of international law arising from the Nuremberg proceedings. The forced "transfer of the children" of a targeted "racial, ethnical, or religious group" is explicitly prohibited as a genocidal activity under the Convention's second article.

Article II of the Genocide Convention also expressly prohibits involuntary sterilization as a means of "preventing births among" a targeted population. Yet, in 1975, it was conceded by the U.S. government that its Indian Health Service (IHS), then a subpart of the Bureau of Indian Affairs (BIA), was even then conducting a secret program of involuntary sterilization that had affected approximately 40 percent of all Indian women. The program was allegedly discontinued, and the IHS was transferred to the Public Health Service, but no one was punished. In 1990, it came out that the IHS was inoculating Inuit children in Alaska with Hepatitis-B vaccine. The vaccine had already been banned by the World Health Organization as having a demonstrated correlation with the HIV-Syndrome which is itself correlated to AIDS. As this is written, a "field test" of Hepatitis-A vaccine, also HIV-correlated, is being conducted on Indian reservations in the northern plains region.

The Genocide Convention makes it a "crime against humanity" to create conditions leading to the destruction of an identifiable human group, as such. Yet the BIA has utilized the government's plenary prerogatives to negotiate mineral leases "on behalf of" Indian peoples paying a fraction of standard royalty rates. The result has been "super profits" for a number of preferred U.S. corporations. Meanwhile, Indians, whose reservations ironically turned out to be in some of the most mineral-rich areas of North America, which makes us, the nominally wealthiest segment of the continent's population, live in dire poverty.

By the government's own data in the mid-1980s, Indians received the lowest annual and lifetime per capita incomes of any aggregate population group in the United States. Concomitantly, we suffer the highest rate of infant mortality, death by exposure and malnutrition, disease, and the like. Under such circumstances, alcoholism and other escapist forms of substance abuse are endemic in the Indian community, a situation which leads both to a general physical debilitation of the population and a catastrophic accident rate. Teen suicide among Indians is several times the national average.

The average life expectancy of a reservation-based Native American man is barely 45 years; women can expect to live less than three years longer.

Such itemizations could be continued at great length, including matters like the radioactive contamination of large portions of contemporary Indian Country, the forced relocation of traditional Navajos, and so on. But the point should be made: genocide, as defined in international law, is a continuing fact of day-to-day life (and death) for North America's native peoples. Yet there has been—and is—only the barest flicker of public concern about, or even consciousness of, this reality. Absent any serious expression of public outrage, no one is punished and the process continues.

A salient reason for public acquiescence before the ongoing holocaust in Native North America has been a continuation of the popular legacy, often through more effective media. Since 1925, Hollywood has released more than 2,000 films, many of them rerun frequently on television, portraying Indians as strange, perverted, ridiculous, and often dangerous things of the past. Moreover, we are habitually presented to mass audiences one-dimensional, devoid of recognizable human motivations and emotions; Indians thus serve as props, little more. We have thus been thoroughly and systematically dehumanized.

Nor is this the extent of it. Everywhere, we are used as logos, as mascots, as jokes: "Big-Chief" writing tablets, "Red Man" chewing tobacco, "Winnebago" campers, "Navajo" and "Cherokee" and "Pontiac" and "Cadillac" pickups and automobiles. There are the Cleveland "Indians," the Kansas City "Chiefs," the Atlanta "Braves" and the Washington "Redskins" professional sports teams—not to mention those in thousands of colleges, high schools, and elementary schools across the country—each with their own degrading caricatures and parodies of Indians and/or things Indian. Pop fiction continues in the same vein, including an unending stream of New Age manuals purporting to expose the inner works of indigenous spirituality in everything from pseudo-philosophical to do-it-yourself styles. Blond yuppies from Beverly Hills amble about the country claiming to be reincarnated 17th century Cheyenne Ushamans ready to perform previously secret ceremonies.

In effect, a concerted, sustained, and in some ways accelerating effort has gone into making Indians unreal. It is thus of obvious importance that the American public begin to think about the implications of such things the next time they witness a gaggle of face-painted and war-bonneted buffoons doing the "Tomahawk Chop" at a baseball or football game. It is necessary that they think about the implications of the grade-school teacher adorning their child in turkey feathers to commemorate Thanksgiving. Think about the significance of John Wayne or Charlton Heston killing a dozen "savages" with a single bullet the next time a western comes on TV. Think about why Land-o-Lakes finds it appropriate to market its butter with the stereotyped image of an "Indian princess" on the wrapper. Think about what it means when non-Indian academics profess—as they often do—to "know more about Indians than Indians do themselves." Think about the significance of charlatans like Carlos Castaneda and Jamake Highwater and Mary Summer Rain and Lynn Andrews churning out "Indian" bestsellers, one after the other, while Indians typically can't get into print.

Think about the real situation of American Indians. Think about Julius Streicher. Remember Justice Jackson's admonition. Understand that the treatment of Indians in American popular culture is not "cute" or "amusing" or just "good, clean fun."

Know that it causes real pain and real suffering to real people. Know that it threatens our very survival. And know that this is just as much a crime against humanity as anything the Nazis ever did. It is likely that the indigenous people of the United States will never demand that those guilty of such criminal activity be punished for their deeds. But the least we have the right to expect—indeed, to demand—is that such practices finally be brought to a halt.

47

Media Magic
Making Class Invisible

GREGORY MANTSIOS

Of the various social and cultural forces in our society, the mass media is arguably the most influential in molding public consciousness. Americans spend an average twenty-eight hours per week watching television. They also spend an undetermined number of hours reading periodicals, listening to the radio, and going to the movies. Unlike other cultural and socializing institutions, ownership and control of the mass media is highly concentrated. Twenty-three corporations own more than one-half of all the daily newspapers, magazines, movie studios, and radio and television outlets in the United States.[1] The number of media companies is shrinking and their control of the industry is expanding. And a relatively small number of media outlets is producing and packaging the majority of news and entertainment programs. For the most part, our media is national in nature and single-minded (profit-oriented) in purpose. This media plays a key role in defining our cultural tastes, helping us locate ourselves in history, establishing our national identity, and ascertaining the range of national and social possibilities. In this essay, we will examine the way the mass media shapes how people think about each other and about the nature of our society.

The United States is the most highly stratified society in the industrialized world. Class distinctions operate in virtually every aspect of our lives, determining the nature of our work, the quality of our schooling, and the health and safety of our loved ones. Yet remarkably, we, as a nation, retain illusions about living in an egalitarian society. We maintain these illusions, in large part, because the media

SOURCE: From Paula Rothenberg, ed., *Race, Class, and Gender in the United States: An Integrated Study*, 4th ed. (New York: St. Martin's Press, 1998). Reprinted with permission of the author.

hides gross inequities from public view. In those instances when inequities are revealed, we are provided with messages that obscure the nature of class realities and blame the victims of class-dominated society for their own plight. Let's briefly examine what the news media, in particular, tells us about class.

ABOUT THE POOR

The news media provides meager coverage of poor people and poverty. The coverage it does provide is often distorted and misleading.

The Poor Do Not Exist

For the most part, the news media ignores the poor. Unnoticed are forty million poor people in the nation—a number that equals the entire population of Maine, Vermont, New Hampshire, Connecticut, Rhode Island, New Jersey, and New York combined. Perhaps even more alarming is that the rate of poverty is increasing twice as fast as the population growth in the United States. Ordinarily, even a calamity of much smaller proportion (e.g., flooding in the Midwest) would garner a great deal of coverage and hype from a media usually eager to declare a crisis, yet less than one in five hundred articles in the *New York Times* and one in one thousand articles listed in the *Readers Guide to Periodic Literature* are on poverty. With remarkably little attention to them, the poor and their problems are hidden from most Americans.

When the media does turn its attention to the poor, it offers a series of contradictory messages and portrayals.

The Poor Are Faceless

Each year the Census Bureau releases a new report on poverty in our society and its results are duly reported in the media. At best, however, this coverage emphasizes annual fluctuations (showing how the numbers differ from previous years) and ongoing debates over the validity of the numbers (some argue the number should be lower, most that the number should be higher). Coverage like this desensitizes us to the poor by reducing poverty to a number. It ignores the human tragedy of poverty—the suffering, indignities, and misery endured by millions of children and adults. Instead, the poor become statistics rather than people.

The Poor Are Undeserving

When the media does put a face on the poor, it is not likely to be a pretty one. The media will provide us with sensational stories about welfare cheats, drug addicts, and greedy panhandlers (almost always urban and Black). Compare these images and the emotions evoked by them with the media's treatment of middle-class (usually white) "tax evaders," celebrities who have a "chemical dependency," or wealthy businesspeople who use unscrupulous means to "make a profit." While the behavior of the more affluent offenders is considered an "impropriety" and a deviation from the norm, the behavior of the poor is considered repugnant, indicative of the poor in general, and worthy of our indignation and resentment.

The Poor Are an Eyesore

When the media does cover the poor, they are often presented through the eyes of the middle class. For example, sometimes the media includes a story about community resistance to a homeless shelter or storekeeper annoyance with panhandlers. Rather than focusing on the plight of the poor, these stories are about middle-class opposition to the poor. Such stories tell us that the poor are an inconvenience and an irritation.

The Poor Have Only Themselves to Blame

In another example of media coverage, we are told that the poor live in a personal and cultural cycle of poverty that hopelessly imprisons them. They routinely center on the Black urban population and focus on perceived personality or cultural traits that doom the poor. While the women in these stories typically exhibit an "attitude" that leads to trouble or a promiscuity that leads to single motherhood, the men possess a need for immediate gratification that leads to drug abuse or an unquenchable greed that leads to the pursuit of fast money. The images that are seared into our mind are sexist, racist, and classist. Census figures reveal that most of the poor are white not Black or Hispanic, that they live in rural or suburban areas not urban centers, and hold jobs at least part of the year.[2] Yet, in a fashion that is often framed in an understanding and sympathetic tone, we are told that the poor have inflicted poverty on themselves.

The Poor Are Down on Their Luck

During the Christmas season, the news media sometimes provides us with accounts of poor individuals or families (usually white) who are down on their luck. These stories are often linked to stories about soup kitchens or other charitable activities and sometimes call for charitable contributions. These "Yule time" stories are as much about the affluent as they are about the poor: they tell us that the affluent in our society are a kind, understanding, giving people— which we are not.* The series of unfortunate circumstances that have led to

---

*American households with incomes of less than $10,000 give an average of 5.5 percent of their earning to charity or to a religious organization, while those making more than $100,000 a year give only 2.9 percent. After changes in the 1986 tax code reduced the benefits of charitable giving, taxpayers earning $500,000 or more slashed their average donation by nearly one-third. Furthermore, many of these acts of benevolence do not help the needy. Rather than provide funding to social service agencies that aid the poor, the voluntary contributions of the wealthy go to places and institutions that entertain, inspire, cure, or educate wealthy Americans—art museums, opera houses, theaters, orchestras, ballet companies, private hospitals, and elite universities. (Robert Reich, "Secession of the Successful," *New York Times Magazine*, February 17, 1991, p. 43.)

impoverishment are presumed to be a temporary condition that will improve with time and a change in luck.

Despite appearances, the messages provided by the media are not entirely disparate. With each variation, the media informs us what poverty is not (i.e., systemic and indicative of American society) by informing us what it is. The media tells us that poverty is either an aberration of the American way of life (it doesn't exist, it's just another number, it's unfortunate but temporary) or an end product of the poor themselves (they are a nuisance, do not deserve better, and have brought their predicament upon themselves).

By suggesting that the poor have brought poverty upon themselves, the media is engaging in what William Ryan has called "blaming the victim."[3] The media identifies in what ways the poor are different as a consequence of deprivation, then defines those differences as the cause of poverty itself. Whether blatantly hostile or cloaked in sympathy, the message is that there is something fundamentally wrong with the victims—their hormones, psychological make up, family environment, community, race, or some combination of these—that accounts for their plight and their failure to lift themselves out of poverty.

But poverty in the United States is systemic. It is a direct result of economic and political policies that deprive people of jobs, adequate wages, or legitimate support. It is neither natural nor inevitable: there is enough wealth in our nation to eliminate poverty if we chose to redistribute existing wealth or income. The plight of the poor is reason enough to make the elimination of poverty the nation's first priority. But poverty also impacts dramatically on the non-poor. It has a dampening effect on wages in general (by maintaining a reserve army of unemployed and underemployed anxious for any job at any wage) and breeds crime and violence (by maintaining conditions that invite private gain by illegal means and rebellion-like behavior, not entirely unlike the urban riots of the 1960s). Given the extent of poverty in the nation and the impact it has on us all, the media must spin considerable magic to keep the poor and the issue of poverty and its root causes out of the public consciousness.

ABOUT EVERYONE ELSE

Both the broadcast and the print news media strive to develop a strong sense of "we-ness" in their audience. They seek to speak to and for an audience that is both affluent and like-minded. The media's solidarity with affluence, that is, with the middle and upper class, varies little from one medium to another. Benjamin DeMott points out, for example, that the *New York Times* understands affluence to be intelligence, taste, public spirit, responsibility, and a readiness to rule and "conceives itself as spokesperson for a readership awash in these qualities."[4] Of course, the flip side to creating a sense of "we," or "us," is establishing a perception of the "other." The other relates back to the faceless, amoral, undeserving, and inferior "underclass." Thus, the world according to the news media is divided between the "underclass" and everyone else. Again the messages are often contradictory.

The Wealthy Are Us

Much of the information provided to us by the news media focuses attention on the concerns of a very wealthy and privileged class of people. Although the concerns of a small fraction of the populace, they are presented as though they were the concerns of everyone. For example, while relatively few people actually own stock, the news media devotes an inordinate amount of broadcast time and print space to business news and stock market quotations. Not only do business reports cater to a particular narrow clientele, so do the fashion pages (with $2,000 dresses), wedding announcements, and the obituaries. Even weather and sports news often have a class bias. An all news radio station in New York City, for example, provides regular national ski reports. International news, trade agreements, and domestic policies issues are also reported in terms of their impact on business climate and the business community. Besides being of practical value to the wealthy, such coverage has considerable ideological value. Its message: the concerns of the wealthy are the concerns of us all.

The Wealthy (as a Class) Do Not Exist

While preoccupied with the concerns of the wealthy, the media fails to notice the way in which the rich as a class of people create and shape domestic and foreign policy. Presented as an aggregate of individuals, the wealthy appear without special interests, interconnections, or unity in purpose. Out of public view are the class interests of the wealthy, the interlocking business links, the concerted actions to preserve their class privileges and business interests (by running for public office, supporting political candidates, lobbying, etc.). Corporate lobbying is ignored, taken for granted, or assumed to be in the public interest. (Compare this with the media's portrayal of the "strong arm of labor" in attempting to defeat trade legislation that is harmful to the interests of working people.) It is estimated that two-thirds of the U.S. Senate is composed of millionaires.[5] Having such a preponderance of millionaires in the Senate, however, is perceived to be neither unusual nor antidemocratic; these millionaire senators are assumed to be serving "our" collective interests in governing.

The Wealthy Are Fascinating and Benevolent

The broadcast and print media regularly provide hype for individuals who have achieved "super" success. These stories are usually about celebrities and superstars from the sports and entertainment world. Society pages and gossip columns serve to keep the social elite informed of each others' doings, allow the rest of us to gawk at their excesses, and help to keep the American dream alive. The print media is also fond of feature stories on corporate empire builders. These stories provide an occasional "insider's" view of the private and corporate life of industrialists by suggesting a rags to riches account of corporate success. These stories tell us that corporate success is a series of smart moves, shrewd acquisitions, timely mergers, and well thought out executive suite shuffles. By painting the upper class in a positive light, innocent of any wrongdoing (labor leaders and union organizations usually get the opposite treatment), the media assures us that wealth and power are benevolent. One person's capital accumulation is presumed to be good for all. The elite, then, are portrayed as investment wizards, people of special talent and skill, who even their victims (workers and consumers) can admire.

The Wealthy Include a Few Bad Apples

On rare occasions, the media will mock selected individuals for their personality flaws. Real estate investor Donald Trump and New York Yankees owner George Steinbrenner, for example, are admonished by the media for deliberately seeking publicity (a very un-upper class thing to do); hotel owner Leona Helmsley was caricatured for her personal cruelties; and junk bond broker Michael Milkin was condemned because he had the audacity to rob the rich. Michael Parenti points out that by treating business wrongdoings as isolated deviations from the socially beneficial system of "responsible capitalism," the media overlooks the features of the system that produce such abuses and the regularity with which they occur. Rather than portraying them as predictable and frequent outcomes of corporate power and the business system, the media treats abuses as if they were isolated and atypical. Presented as an occasional aberration, these incidents serve not to challenge, but to legitimate, the system.[6]

The Middle Class Is Us

By ignoring the poor and blurring the lines between the working people and the upper class, the news media creates a universal middle class. From this perspective, the size of one's income becomes largely irrelevant: what matters is that most of "us" share an intellectual and moral superiority over the disadvantaged. As *Time* magazine once concluded, "Middle America is a state of mind."[7] "We are all middle class," we are told, "and we all share the same concerns": job security, inflation, tax burdens, world peace, the cost of food and housing, health care, clean air and water, and the safety of our streets. While the concerns of the wealthy are quite distinct from those of the middle class (e.g., the wealthy worry about investments, not jobs), the media convinces us that "we [the affluent] are all in this together."

The Middle Class Is a Victim

For the media, "we" the affluent not only stand apart from the "other"—the poor, the working class, the minorities, and their problems—"we" are also victimized by the poor (who drive up the costs of maintaining the welfare roles), minorities (who commit crimes against us), and by workers (who are greedy and drive companies out and prices up). Ignored are the subsidies to the rich, the crimes of corporate America, and the policies that wreak havoc on the economic well-being of middle America. Media magic convinces us to fear, more than anything else, being victimized by those less affluent than ourselves.

The Middle Class Is Not a Working Class

The news media clearly distinguishes the middle class (employees) from the working class (i.e., blue collar workers) who are portrayed, at best, as irrelevant, outmoded, and a dying breed. Furthermore, the media will tell us that the hardships faced by blue collar workers are inevitable (due to progress), a result of bad luck (chance circumstances in a particular industry), or a product of their own doing (they priced themselves out of a job). Given the media's presentation of reality, it is hard to believe that manual, supervised, unskilled, and semiskilled workers actually represent more than 50 percent of the adult working popula-tion.[8] The working class, instead, is relegated by the media to "the other."

In short, the news media either lionizes the wealthy or treats their interests and those of the middle class as one in the same. But the upper class and the middle class do not share the same interests or worries. Members of the upper class worry about stock dividends (not employment), they profit from inflation and global militarism, their children attend exclusive private schools, they eat and live in a royal fashion, they call on (or are called upon by) personal physicians, they have few consumer problems, they can escape whenever they want from environmental pollution, and they live on streets and travel to other areas under the protection of private police forces.[*][9]

The wealthy are not only a class with distinct life-styles and interests, they are a ruling class. They receive a disproportionate share of the country's yearly income, own a disproportionate amount of the country's wealth, and contribute a disproportionate number of their members to governmental bodies and decision-making groups—all traits that William Domhoff, in his classic work *Who Rules America*, defined as characteristic of a governing class.[10]

This governing class maintains and manages our political and economic structures in such a way that these structures continue to yield an amazing pro-portion of our wealth to a minuscule upper class. While the media is not above referring to ruling classes in other countries (we hear, for example, references to Japan's ruling elite),[11] its treatment of the news proceeds as though there were no such ruling class in the United States.

Furthermore, the news media inverts reality so that those who are working class and middle class learn to fear, resent, and blame those below, rather than those above them in the class structure. We learn to resent welfare, which accounts for only two cents out of every dollar in the federal budget (approximately $10 billion) and provides financial relief for the needy,[**] but learn little about the $11 billion the federal government spends on individuals with incomes in excess of

---

[*]The number of private security guards in the United States now exceeds the number of public police officers. (Robert Reich, "Secession of the Successful," *New York Times Magazine*, February 17, 1991, p. 42.)

[**]A total of $20 billion is spent on welfare when you include all state funding. But the average state funding also comes to only two cents per state dollar.

$100,000 (not needy),[12] or the $17 billion in farm subsidies, or the $214 billion (twenty times the cost of welfare) in interest payments to financial institutions.

Middle-class whites learn to fear African Americans and Latinos, but most violent crime occurs within poor and minority communities and is neither inter-racial[†] nor interclass. As horrid as such crime is, it should not mask the destruction and violence perpetrated by corporate America. In spite of the fact that 14,000 innocent people are killed on the job each year, 100,000 die prematurely, 400,000 become seriously ill, and 6 million are injured from work-related accidents and diseases, most Americans fear government regulation more than they do unsafe working conditions.

Through the media, middle-class—and even working-class—Americans learn to blame blue collar workers and their unions for declining purchasing power and economic security. But while workers who managed to keep their jobs and their unions struggled to keep up with inflation, the top 1 percent of American families saw their average incomes soar 80 percent in the last decade.[13] Much of the wealth at the top was accumulated as stockholders and corporate executives moved their companies abroad to employ cheaper labor (56 cents per hour in El Salvador) and avoid paying taxes in the United States. Corporate America is a world made up of ruthless bosses, massive layoffs, favoritism and nepotism, health and safety violations, pension plan losses, union busting, tax evasions, unfair competition, and price gouging, as well as fast buck deals, financial speculation, and corporate wheeling and dealing that serve the interests of the corporate elite, but are generally wasteful and destructive to workers and the economy in general.

It is no wonder Americans cannot think straight about class. The mass media is neither objective, balanced, independent, nor neutral. Those who own and direct the mass media are themselves part of the upper class, and neither they nor the ruling class in general have to conspire to manipulate public opinion. Their interest is in preserving the status quo, and their view of society as fair and equitable comes naturally to them. But their ideology dominates our society and justifies what is in reality a perverse social order—one that perpetuates unprecedented elite privilege and power on the one hand and widespread deprivation on the other. A mass media that did not have its own class interests in preserving the status quo would acknowledge that inordinate wealth and power undermines democracy and that a "free market" economy can ravage a people and their communities.

† In 92 percent of the murders nationwide the assailant and the victim are of the same race (46 percent are white/white, 46 percent are black/black), 5.6 percent are black on white, and 2.4 percent are white on black. (FBI and Bureau of Justice Statistics, 1985–1986, quoted in Raymond S. Franklin, *Shadows of Race and Class*, University of Minnesota Press, Minneapolis, 1991, p. 108.)

NOTES

1. Martin Lee and Norman Solomon, *Unreliable Sources,* Lyle Stuart (New York, 1990), p. 71. See also Ben Bagdikian, *The Media Monopoly,* Beacon Press (Boston, 1990).

2. Department of Commerce, Bureau of the Census, "Poverty in the United States: 92," *Current Population Reports, Consumer Income,* Series *P60–185,* pp. xi, xv, 1.

3. William Ryan, *Blaming the Victim,* Vintage (New York, 1971).

4. Benjamin Demott, *The Imperial Middle,* William Morrow (New York, 1990), p. 123.

5. Fred Barnes, "The Zillionaires Club," *The New Republic,* January 29, 1990, p. 24.

6. Michael Parenti, *Inventing Reality,* St. Martin's Press (New York, 1986), p. 109.

7. *Time,* January 5, 1979, p. 10.

8. Vincent Navarro, "The Middle Class—A Useful Myth," *The Nation,* March 23, 1992, p. 1.

9. Charles Anderson, *The Political Economy of Social Class,* Prentice Hall (Englewood Cliffs, N.J., 1974), p. 137.

10. William Domhoff, *Who Rules America,* Prentice Hall (Englewood Cliffs, N.J., 1967), p. 5.

11. Lee and Solomon, *Unreliable Sources,* p. 179.

12. *Newsweek,* August 10, 1992, p. 57.

13. *Business Week,* June 8, 1992, p. 86.

48

Who(Se) Am I?

The Identity and Image of Women in Hip-Hop

IMANI PERRY

. . . BLACK WOMEN'S BODIES IN HIP-HOP VIDEOS

In the last few years of the 20th century, the visual image of black women in hip-hop rapidly deteriorated into one of widespread sexual objectification and degradation. For years before, hip-hop had been accused of misogyny—critics

SOURCE: Dines, Gail, and Jean M. Humez, eds. 2003. *Gender, Race, Class, in Media: A Text-Reader,* 2nd ed. Copyright © 2003 by Sage Publications. Reprinted by permission.

often citing the references to women as bitches and "hoes." But it is also true that hip-hop was often scapegoated, being no more misogynistic than American popular culture in general, although perhaps peppered with less polite language. But in the late years of the 20th century, hip-hop took a particularly pernicious turn, which is not only full of sexist assertions but threatens patriarchal impact.

It seemed to happen suddenly. Every time you turned on BET or MTV there was a disturbing music video. Black men rapped surrounded by dozens of black and Latina women dressed in swimsuits, or scantily clad in some other fashion. Video after video was the same, each one more objectifying than the next. Some were in strip clubs, some at the pool, beach, hotel rooms, but the recurrent theme was dozens of half-naked women.

This was a complex kind of sexist message as well. Its attack on black female identity was multifaceted. First, and most obviously, the women are commodified. They appear in the videos quite explicitly as property, not unlike the luxury cars, Rolex watches, and platinum and diamond medallions that were also featured. The male stars of the videos do not get these legions of women because of charisma or sexual prowess. Rather, they are able to buy them because they are wealthy. The message is not, "I am a Don Juan," but instead, "I am rich and these are my spoils." Not only are the women commodified, but so is sex as a whole.

Moreover, the women are often presented as vacuous, doing nothing but swaying around seductively. Their eyes are averted from the camera, thereby allowing the viewer to have a voyeuristic relationship to them. Or they look at the camera, eyes fixed in seductive invitation, mouth slightly open. Extremely rare are any signs of thought, humor, irony, intelligence, anger, or any other emotion.

Even the manner in which the women dance is a signal of cultural destruction. Black American dance is "discursive" (in that sexuality is usually combined with humor and the body is used to converse with other moving bodies). The women who appear in these videos are usually dancing in a two-dimensional fashion, a derivative but unintellectual version of black dance, more reminiscent of symbols of pornographic male sexual fantasy than of the ritual, conversational, and sexual traditions of black dance. Despite all the gyrations of the video models, their uninterested wet-lipped languor stands in sharp contrast to (for example) the highly sexualized "boodie dancing" of the Deep South (which features polyrhythmic rear end movement, innuendo, and sexual bravado).

This use of black women in the music videos of male hip-hop artists often makes very clear reference to the culture of strip clubs and pornography. Women dance around poles; porn actresses and exotic dancers are often featured in the videos and they bring the movement-based symbols of their trades with them. The introduction of porn symbols into music videos is consistent with a larger movement that began in the late 1990s, in which pornographic imagery, discourses, and themes began to enter American popular culture. Powerful examples may be found in the *Howard Stern Show*, E! Entertainment television, and daytime talk shows. Stars of pornographic films attain mainstream celebrity, exotic dancers are routine talk show guests, and the public face of lesbianism becomes not a matter of the sexual preference of women, but the sexual consumption and fantasy life of men. The videos are an appropriate companion piece to this wider trend. Although the music videos are male centered in that they assume a heterosexual male viewer who will appreciate the images of sexually available young women, it is clear that young women watch them as well. The messages such videos send to young women are instructions on how to be sexy and how to look in order to capture the attention of men with wealth and charisma. Magazines geared toward young women have given such instructions on how women should participate in their own objectification for decades. However, never before has a genre completely centralized black women in this process.

The beauty ideal for black women presented in these videos is as impossible to achieve as the waif-thin models in *Vogue* magazine are for white women. There is a preference for lighter-complexioned women of color, with long and straight or loosely curled hair. Hair that hangs slick against the head when wet as the model emerges out of a swimming pool (a common video image) is at a premium too. Neither natural tightly curled hair nor most coarse relaxed hair becomes slick, shining, and smooth when wet. It is a beauty ideal that contrasts sharply to the real hair of most black women. When brown-skinned or dark-skinned women appear in the videos, they always have hair that falls well below shoulder length, despite the fact that the average length of black women's natural hair in the United States today is 4 to 6 inches, according to renowned black hairstylist John Atchison.

The types of bodies that the camera shots linger on are specific. The videos have assimilated the African American ideal of a large rotund behind, but the video ideal also features a very small waist, large breasts, and slim shapely legs and arms. Often while the camera features the faces of lighter-complexioned women it will linger on the behinds of darker women, implying the same thing as the early 1990s refrain from Sir Mix a Lot's "Baby Got Back" that lauded the face of a woman from Los Angeles and the behind of a woman from Oakland. That is, the ideal is a high-status face combined with a highly sexualized body (which is often coded as the body of a poor or working-class woman).[1] Color is aligned with class and women are "created" (i.e., through weaves, pale makeup, and camera filters) and valued by how many fantasy elements have been pieced together in their bodies.

THE IMPACT OF THE IMAGE

Although one might argue that the celebration of the rotund behind signals an appreciation of black women's bodies, the image taken as a whole indicates how difficult a beauty ideal this is to attain for anyone. A small percentage of women, even black women, have such "Jessica Rabbit" (the voluptuous cartoon character from the 1990s film *Who Framed Roger Rabbit?*) proportions. As journalist

Tomika Anderson wrote for *Essence* magazine, "In movies, rap songs and on television, we're told that the attractive, desirable and sexy ladies are the ones with 'junk in their trunks.' And even though this might seem ridiculous, some of us actually listen to (and care about) these obviously misogynistic subliminal messages—just as we are affected by racialized issues like hair texture and skin tone.[2]

Americans have reacted with surprise and abundant social scientific data that show that black girls are the social group who score highest on self-esteem assessments and tend to have much better body images than white girls. Although these differences in esteem and body image are to a large extent attributable to cultural differences with black girls having been socialized to see beauty in strong personality characteristics and grooming rather than in particular body types, I believe the media play a role as well. White girls are inundated with images of beauty that are impossible for most to attain: sheets of blond air, very thin bodies, large breasts, no cellulite, small but round features, high cheekbones. Over the years, black women have been relatively absent from public images of beauty, an exclusion that may have saved black girls from aspiring to impossible ideals. But with the recent explosion of objectified and highly idealized images of black women in music videos, it is quite possible that the body images and even self-esteem of black girls will begin to drop, particularly as they move into adolescence and their bodies come under scrutiny. Many of the music videos feature neighborhood scenes, which include children. In them, little black girls are beautiful. They laugh, smile, play Double Dutch, and more. They are full of personality, and they are a cultural celebration. Their hair is plaited, twisted or curled, and adorned with colorful ribbons that match their outfits in characteristic black girl grooming style. And yet the adult women are generally two dimensional and robbed of personality. Is this what puberty is supposed to hold for these girls?

A FEMINIST RESPONSE?

In such troubling moments, we should all look for a gender critical voice, in the world, in ourselves. Where do we find a response to this phenomenon that will compellingly argue against such characterizations of black women, a hip-hop feminism? There has been a feminist presence in hip-hop since the 1980s. From Salt n Pepa to Queen Latifah to MC Lyte and others, there is a feminist legacy in hip-hop, and hip-hop feminism continues to exist despite the widespread objectification of black female bodies. We can find numerous examples of feminist and antisexist songs in hip-hop and hip-hop soul. Mary J. Blige, Lauryn Hill, Destiny's Child, Missy Elliot, Erykah Badu, and others all have their individual manners of representing black female identity and self-definition.

Missy transgresses gender categories with her man-tailored suits and her frequent presence as narrator of the action in the music videos of male hip-hop artists, an extremely rare location for a woman. Missy is a large woman who presents a glamorous and stylish image but never is presented in an objectifying manner. She uses bizarreness to entice rather than being a sexpot (appearing in one video in an outfit that resembled a silver balloon before a fun-house mirror). Although Missy Elliot may not be distinctive for brilliant rhyming, she has a noteworthy acumen for making hit songs as a producer and rapper, and she consistently maintains her personal dignity.

Alicia Keys, one of the crop of singer songwriters who fit into the hip-hop nation, also presents an image that contrasts sharply with the video models. The classically trained pianist, who has claimed Biggie Smalls and Jay Z among her music influences, appeared in her first music video for the song "Fallin" in a manner that was stylish and sexy but decidedly not self-exploiting. Her hair in cornrows, wearing a leather jacket and fedora, she sings with visible bluesy emotion. She describes repeatedly falling in love with a man who is not good for her. In the music video, Keys travels by bus to visit the man in prison. This element is an important signifier of hip-hop sensibilities, as it is the one art form that consistently engages with the crisis of black imprisonment and considers imprisoned people as part of its community. As she rides in the bus, she gazes at women prisoners working in a field outside the window. . . . The women on the bus riding to visit men in prison mirror the women outside of the bus, who are prison laborers. This visual duality is a commentary on the problem of black female imprisonment, a problem that is often overlooked in discussions about the rise of American imprisonment and black imprisonment in particular. It makes reference both to the fact that many black women are the mates of men who are imprisoned and to the reality that many black women wind up in prison because of being unwittingly or naively involved with men who participate in illegal activities. These social ills are poignantly alluded to in the video by a close-up of a stone-faced woman in prison clothing with a single tear rolling down her cheek.

Another critical example of a black feminist space in the hip-hop world is found in singer songwriter India.Arie. A young brown-skinned and dreadlocked woman, she burst upon the music scene with her song and companion music video "Video," which is a critique of the image of women in videos. In the refrain, Arie tells listeners that she's not the type of woman who appears in music videos, that her body type is not that of a supermodel but nevertheless she loves herself without hesitation.

Similar lyrics assert that value is found in intelligence and integrity rather than expensive clothes, liquor, and firearms. The video celebrates Arie, who smiles and dances and pokes fun at the process of selecting girls for music videos. She rides her bicycle into the sunshine with her guitar strapped across her shoulder. Arie refuses to condemn artists who present a sexy image but has stated that she will not wear a skirt above calf length on stage and that she will do nothing that will embarrass her family. Musically, although her sound is folksy soul, she does understand her work as being related to hip-hop. "I'm trying to blend acoustic and hip-hop elements," India explains. "I used the most acoustic-sounding drum samples, to have something loud enough to compete with other records, but to keep the realistic, softer feel.[3]

More than the compositional elements, Arie understands her work as inflected with hip-hop sensibilities. She says:

> I don't define hip-hop the way a record company would. The thread that runs through both my music and hip-hop is that it's a very precise expression of my way of life. It's like blues; it's [a] very real and honest output of emotion into a song. Because of that legacy, my generation now has an opportunity to candidly state our opinions. That's what my album is about. I just wanna be me.[4]

Arie's definition of hip-hop as honest self-expression is true to the ideology that was at the heart of hip-hop at its beginnings and that continues to be a concept professed to by multitudes of hip-hop artists. However, that element of hip-hop is in tension with the process of celebrity creation. The "honest" words in hip-hop exist in a swamp of image making. It is not enough to examine the clear and simple feminist presences in hip-hop; we must consider the murkier ones as well. When it comes to feminist messages, often the words and language of a hip-hop song may have feminist content but the visual image may be implicated in the subjugation of black women. Unlike the individualistic and expressive visuals we have of Arie, Keys, or Elliot, other artists are often marketed in a manner that is quite similar to the way in which objectified video models are presented.

TENSIONS BETWEEN TEXTS

Women hip-hop artists who are self-consciously "sexy" in their appearance, style, and words have a much more difficult road in carving out a feminist space in hip-hop than performers such as Elliot, Keys, and Arie. This is because the language of sexiness is also the language of sexism in American popular culture in general, and in hip-hop videos in particular. . . . [W]hen the women who articulate subjectivity are increasingly presented in visual media as objects rather than subjects, as they are now, then their statement to the world is ambiguous at best, and at worst the feminist message of their word is undermined.

The space a musical artist occupies in popular culture is multitextual. Lyrics, interviews, music, and videos together create a collage, often finely planned, out of which we are supposed to form impressions. But the texts may be in conflict with one another. Lil Kim, the much discussed, critiqued, and condemned nasty-talking bad girl of hip-hop, is a master of shock appeal. Her outfits often expose her breasts, her nipples covered by sequined pasties that are color coordinated with the rest of her attire. Despite Kim's visual and lyrical vulgarity, many of her critics admit to finding her endearing. She is known by her interviewers to be sweet-natured and generous. But Lil Kim is a contradiction because although she interviews as vulnerable and sweet, she raps with a hardness adored by her fans. She has an impressive aggressive sexual presence, and she has often articulated through words a sexual subjectivity along with an in-your-face camera presence. However, as Kim has developed as an entertainer it is clear that her image is complicit in the oppressive language of American cinematography with regards to women's sexuality. She has adopted a "Pamela Anderson in brown skin" aesthetic, calling on pornographic tropes, but losing the subversiveness that was sometimes apparent in her early career. . . .

It is a delicate balance, but it is important to distinguish between sexual explicitness and internalized sexism. Although many who have debated the image of female sexuality have put "explicit" and "self-objectifying" on one side, and "respectable" and "covered-up" on the other, that is a flawed means of categorization. The nature of sexual explicitness is important to consider, and will be increasingly important as more nuanced images will present themselves. There is a creative possibility for explicitness to be liberatory because it may expand the confines of what women are allowed to say and do. We just need to refer to the history of blues music, which is full of raunchy, irreverent, and transgressive women artists, for examples. However, the overwhelming prevalence of the Madonna/whore dichotomy in American culture means that any woman who uses explicit language or images in her creative expression is in danger of being symbolically cast into the role of whore regardless of what liberatory intentions she may have, particularly if she doesn't have complete control over her image.

Let us turn to other examples to further explore the tensions between text and visual image in women's hip-hop. . . . Eve is one of the strongest feminist voices in hip-hop. She rhymes against domestic violence and for women's self-definition and self-reliance. She encourages women to hold men in their lives accountable for behavior that is disrespectful or less than loving. Yet the politics of Eve's image are conflicted. She has appeared in music videos for songs on which she has collaborated with male hip-hop artists. Those videos are filled with the stock legions of objectified video models. Eve is dressed provocatively and therefore validates the idea of attractiveness exemplified by the models. But she is distinguished from these women because she is the star. She is dignified and expressive while they are not. Her distinction from the other women supports their objectification. She is the exception that makes the rule, and it is her exceptionalism that allows her to have a voice. . . . In fact, a number of women hip-hop artists, who claim to be the only woman in their crews, to be the one who can hang with the fellas, are making arguments through their exceptionalization that justify the subjugation of other women, even the majority of women.

Moreover, both Eve and Lil Kim often speak of the sexual power they have as being derived from their physical attractiveness to men. It is therefore a power granted by male desire, rather than a statement of the power of female sexual desire. Although neither artist has completely abandoned the language of empowering female subjectivity in her music, any emphasis on power granted through being attractive in conventional ways in this media language limits the feminist potential of their music. In one of the songs in which Eve most explicitly expresses desire, "Gotta Man," it is a desire for a man that is rooted in his ability to be dominant. She describes him as "the only thug in the hood who is wild enough to tame me"[5] and therefore she is "The Shrew," willingly stripped of her defiant power by a sexual union. Instead of using her aggressive tongue to challenge prevailing sexist sexual paradigms, she affirms them by saying that she simply needs a man who is stronger than most, stronger than she is, to bring everything back to normal.

The tensions present in hip-hop through the interplay of the visual and the linguistic, and the intertextuality of each medium, are various. Even Lauryn Hill, often seen as the redeemer of hip-hop due to her dignified, intellectually challenging, and spiritual lyricism, has a complicated image. As a member of the Fugees, she was often dressed casually, in baggy yet interesting clothes, thoroughly rooted in hip-hop style. It seems to be no accident that Lauryn Hill became a celebrity, gracing the covers of *British GQ, Harper's Bazaar,* and numerous other magazines, only when her sartorial presentation changed. Her skirts got shorter and tighter, her cleavage more pronounced, and her dreadlocks longer. When she began to sport an alternative style that nevertheless had mainstream acceptability, she was courted by high-end designers such as Armani. As Lauryn's image became more easily absorbable into the language of American beauty culture, her celebrity grew. She even appeared on the cover of *Sophisticates Black Hair Styles and Care Guide,* a black beauty magazine in which natural hair is at best relegated to a couple of small pictures of women with curly afros or afro weaves, while the vast majority of photos are of women with long straight weaves and relaxers. She was certainly one of the few *Sophisticates* cover models ever to have natural hair and the only with locks. (Interestingly, the silhouette of the locks was molded into the shape of shoulder-length relaxed hair.) . . . [I]t is important to note that Lauryn became widely attractive when her silhouette, thin body and big hair, matched that of mainstream beauty. So even as Lauryn has been treated as the symbol of black women's dignity and intelligence in hip-hop (and rightfully so given her brilliant lyricism), she too was pulled into the sexist world of image making. Although she has made some public appearances since cutting off her long hair, getting rid of the make-up, and returning to baggy clothes, about her has noticeably dropped. . . .

I used the examples of Lil Kim, Eve, [and] Lauryn Hill, all very distinct artists, to draw attention to the kinds of tensions that might exist between a feminist content in hip-hop lyrics and the visual image of that artist. To further illustrate this point, let us now turn to a comparison that offers a dramatic example of the relationship between visual images and the message of musical texts.

COMPARATIVE READINGS OF THE
CREOLE PROSTITUTE

In 2001, a remake of the 1975 LaBelle classic "Lady Marmalade" hit the airwaves. Twenty-six years after it was first recorded, it once again became a hit. The 2001 version was performed by a quartet of successful young female artists, pop sensation Christina Aguilera, R&B singers Pink and Mya, and rapper Lil Kim. Recorded for the soundtrack of the movie *Moulin Rouge*, a postmodern rendering of the famous Parisian cabaret circa 1899, the song served as a fantastic commercial for the film. And with all those popular songstresses, it was a surefire moneymaker. The cultural impact of the most recent version of "Lady Marmalade," however, was quite distinct from that of the original.

The original version of the song was sung by a trio of young black women who had recently shed their super sweet name "The Bluebells," a fourth member, their bouffant hairdos, and their chiffon gowns for a more radical image as LaBelle. Patti LaBelle sang the lead on the song penned by Kenny Nolan and produced by Allen Toussaint. She told a fable about a Creole prostitute in New Orleans, Lady Marmalade. Through the rhythm of her voice, Patti was able to transmit Lady Marmalade's strut and attitude. Marmalade turned her conservative john's world upside down, and thereby robbed her exploiter of some of his power. The song, with the racy lyrics Voulez vous couchez avec moi, ce soir?" was provocative and yet melancholy. And despite the fact that LaBelle's members purportedly didn't know the meaning of the French lyrics when they recorded the song, the song had a feminist sensibility about it. This was due to Patti's vocal interpretation and the visual presentation of all of LaBelle. They were telling a story of the past in which a woman found a little subversive power, but the storytellers themselves were contemporary women, futuristic even. LaBelle were rock glam stars and they stood outside of standard paradigms of female sexuality and objectification. They were somehow women's movement women, black power women, and transgressive women at once.

Patti brought the listener of the song to a corner in Storyville, the historic sex district in New Orleans. She told the story of a sister there, and made that listener feel the energy and melancholy of the Creole prostitute, sympathize with her, recognize her power. Yet Patti escaped being cast into Marmalade's position herself. In 2001, the singers of the song "Lady Marmalade" did not tell Marmalade's legendary story—rather, they became her. That process of embodying the Creole prostitute occurred largely through the visual representation of the song in the music video, which received a huge amount of airplay on MTV and BET, and one live performance.

In the video, the four women are attired in vintage style elaborate lingerie and dance about in rooms that look like the images of bordello boudoirs we have seen in film before. This embodiment had no subversive elements but instead was a glamorization of a turn-of-the-century image of prostitution. Moreover, the diversity and hybridity of the artists are exploited for the sake of the sexual fantasy. There is Christina, blond and blue eyed, Latina and Irish, who sings with the trills and moans of black gospel tradition. There is Mya, whose café au lait skin and long curling hair remind us of the song's description of Lady Marmalade's appearance. There is Pink, a white woman with the whiskey voice of a black blues singer, and Lil Kim, the brown-skinned rapper with the blond wig and blue contact lenses. Their hybridity is used for no more interesting purpose than the reification of the well-worn and generations-old image of the whore at the racial crossroads, the lascivious and tragic mulatto who is defined by her sexuality. Lil Kim is the one of the group who ultimately reveals that these women are embodiments of Lady Marmalade. She raps the story in the first person rather than the third. She tells listeners that she and her "sisters" are about the business of using their sexualities in exchange for material goods from men.

The racial and gender politics of the video are supportive of historic racist imagery of women of color. *Moulin Rouge* is filled with white stars, yet "Lady Marmalade," the song that introduces the movie, is racialized as black, being a hip-hop and R&B creation. The song advertises the film but its blackness is not central to the film's imagination. While the singers reflect the status of the movie's star as a courtesan, dancer, and singer, the politics of race and sex automatically locate them in a lower status position than that of the star of the film. The degradation of black female space simultaneously with the use of black discourses to define mainstream sexuality is nothing new in American culture. The film is not alone in playing that game. But that this version hearkens so closely to the fate of the Creole woman in relationship to white women in Louisiana history, socially defined as prostitute rather than lady, through everything from anti-miscegenation laws and quadroon balls to cultures of concubinage, is particularly troubling. It claims to be, and symbolically is, 1899 all over again.

Ironically, the earlier version of the song was produced by a man, and the later version was produced by a black woman, hip-hop phenom Missy Elliot, discussed earlier. Despite her own feminist presentation, she participates in this subjugating image, in a manner analogous to the way in which Eve participates in the subjugation of other women in music videos in which she appears. Elliot's image in the video as narrator/madam in the video is free from objectification, but supports the objectification of her fellow artists. Moreover, Elliot's own arguably feminist presence is trumped by the subjugated presence of the others.

THE COLONIZER AND COLONIZED

Novelist and cultural critic Toni Cade Bambara had great insight into the race and gender politics of American media. She reminded us in her essay "Language and the Writer" that

> the creative imagination has been colonized. The global screen has been colonized. And the audience—readers and viewers—is in bondage to an industry. It has the money, the will, the muscle, and the propaganda machine oiled up to keep us all locked up in a delusional system—as to even what America is.[6]

Musical artists are cultural actors, but those backed by record labels are hardly independent actors. In music videos and photo layouts they exist within what Bambara has described as colonized space, particularly around race and gender. In a context in which a short tight dress, and a camera rolling up the body, lingering on behinds and breasts, has particular very charged meanings with regard to gender and personal value, we must ask, How powerful are words that intend to contradict such objectification? How subversive are revolutionary words in a colonized visual world full of traditional gender messages? . . .

Often, language, even aggressive liberatory language, becomes nearly powerless in the face of the powerful discourse of the visual within the texts of music videos.

So then we ask, How should we read these artists who have feminist voices and sexist images? If they are linguistic proponents of women's power, subjectivity, and black feminism, why do they participate in creating such conflicting visual textual representations? First, it is important to acknowledge that in a society with such strong hegemonies of race, class, gender, and sexuality, virtually all of us, regardless of how committed we are to social justice and critical thinking, are conflicted beings. We want to be considered attractive even though we understand how attractiveness is racialized, gendered, and classed in our society, and how the designation often affirms structures of power and domination. Separating out healthy desires to be deemed attractive from those desires for attractiveness that are complicit in our oppression is challenging. Similarly, we want to be successful, but success is often tied to race, class, gender, and body politics that implicitly affirm the oppression of others. We support the status quo in order to succeed within it, despite our better judgment.

These tensions exist within the artist as much as within the average citizen, and we should therefore be cautious in our judgment of the artists. However, even if we were insensitive to these internal conflicts as they might exist in famous hip-hop artists, the artists still should not be considered solely responsible for the tensions between their words and image. The reality is that the "realness" in popular hip-hop and R&B stars is as much an illusion as it is real. Their public images are constructed by teams more often than by themselves. The conflicted images we see from some "feminist talking, sexpot walking" hip-hop artists may be as much a sign of a conflict between their agendas and those of the record companies, stylists, video directors, and so forth, as a sign of internal conflict. Each artist is a corporate creation—pun intended.

PROPERTY AND SUBVERSIVE POTENTIAL

As a college student, I met the black woman filmmaker Julie Dash. I had recently seen her short film *Illusions* and her landmark feature *Daughters of the Dust*. Excited by her work and thrilled to meet her, I gushed about how I wanted to "do what she did." She warned me that if I wrote a screenplay I'd better direct the movie myself if I wanted the substance to be intact once it was completed. In giving me this warning she was testifying as to the degrees of ownership of art, the interaction between words and image, and the importance of black female self-articulation in a colonized media.

Despite the powerful hand of corporate interests in hip-hop, it is a music that has sustained a revolutionary current with respect to consumer culture, albeit one that is increasingly fragile. This revolutionary current exists in the underground communities of unsigned artists (rappers, or MCs, as well as poets or "spoken word" artists) who push forward creative development without corporate involvement. They are cultural workers and artists in the organic sense, and proprietors of their own images. Analogous to independent filmmakers, local underground artists are a good source when we seek feminist and other politically progressive messages in hip-hop. However, most of the contemporary hip-hop audience has little access to underground artists. It is now overwhelmingly, albeit not exclusively, a recorded art form. Therefore, as cultural readers we should consider what the scope of power is for artists who are signed to record labels.

One clear location of authorial power exists in their ownership of their copyrighted lyrics. The owned lyrics are an asserted property right that competes with the concept that the artist herself is a "property." For women artists who have written and copyrighted their own lyrics, the lyrics might be one of their only areas of control, and it may be where we can best find their intended political messages. Looking to the distinction between the copyrighted material owned by artists and the music videos owned by the record companies, we have some indication of what particular political tensions face a given artist.

CONCLUSION: POSSIBILITIES FOR DISSENT

We know that the politics of the artist are often neutralized by the image made by the record company. Although a famous person has a legally cognizable property interest in his or her public image as a whole, when she consents to making a music video, she grants the record company the use of her image. She allows for the creation of a product that features her as a product and that in turn encourages the sale of her words and music. Perhaps there is a clue in that web to how artists might regain subversive power through language. If their words were not simply liberatory and progressive but also critically engaged, mocked, or challenged the very images that made the artist into a celebrity, the words of the artist might not be dwarfed by the image. Instead they might latch onto the image, shift its meaning, and bring it closer to being owned by the artist. Imagine an artist looking lustily into the camera while critiquing the gaze she is giving you, or discussing the sexism implicit in the sexy dress she is wearing. Although this strategy might simply give rise to further conflicted images, there is the possibility that it would force the listener to critically read the image. Certainly, in earlier periods of hip-hop groups such as a Tribe Called Quest and De La Soul often embedded in their music strong critiques of the music industry to which they understood themselves to be "enslaved" as commodities. Such critiques played a role in their success at being popular, political, and authentic groups, and they provide a useful model for a feminist voice in hip-hop.

There are surely a number of other strategies as well that might be employed by women hip-hop artists who seek innovative modes of feminist articulation and self-definition in an arena dominated by corporate interests and sexism. Hip-hop is an art form that has consistently been engaged in innovation, improvisation, and reinterpretation. It is therefore neither unreasonable nor naive to anticipate a new generation of feminist voices in hip-hop that will respond in increasingly sophisticated and complex ways to a sexist and racist society.

NOTES

1. There are many hip-hop lyrics that identify the voluptuous body with women who live in housing projects or from the 'hood. As well, the assumption of lighter-complexioned black women being of higher socioeconomic status or greater sexual desirability is a longstanding aspect of black American culture. Although this cultural phenomenon was challenged in the late civil rights era, it flourishes in the images that appear in many television shows, movies, and books and in the tendency of black male movie stars, musicians, and athletes to choose very light complexioned spouses if they marry black women.

2. Tomika Anderson, "Nothing Butt the Truth" *Essence,* November 2001, 116.

3. India. Arie interview, www.mtv.com.

4. India. Arie interview, www.mtv.com.

5. Eve, "Gotta Man," Ruffryders Interscope Records (2000).

6. Toni Cade Bambara. "Language and the Writer," *Deep Sightings and Rescue Missions: Fiction, Essays and Conversations*, ed. by Toni Morrison (PantheonNew York, 1996), 140.

49

Can Education Eliminate Race, Class, and Gender Inequality?

ROSLYN ARLIN MICKELSON AND STEPHEN SAMUEL SMITH

INTRODUCTION

Parents, politicians, and educational policy makers share the belief that a "good education" is *the* meal ticket. It will unlock the door to economic opportunity and thus enable disadvantaged groups or individuals to improve their lot dramatically.[1] This belief is one of the assumptions that has long been part of the American Dream. According to the putative dominant ideology, the United States is basically a meritocracy in which hard work and individual effort are rewarded, especially in financial terms.[2] Related to this central belief are a series of culturally enshrined misconceptions about poverty and wealth. The central one is that poverty and wealth are the result of individual inadequacies or strengths rather than the results of the distributive mechanisms of the capitalist economy. A second misconception is the belief that everyone is the master of her or his own fate. The dominant ideology assumes that American society is open and competitive, a place where an individual's status depends on talent and motivation, not inherited position, connections, or privileges linked to ascriptive characteristics like gender or race. To compete fairly, everyone must have access to education free of the fetters of family background, gender, and race. Since the middle of this century, the reform policies of the federal government have been designed, at least officially, to enhance individuals' opportunities to acquire education. The question we will explore in this essay is whether expanding educational opportunity is enough to reduce the inequalities of race, social class, and gender which continue to characterize U.S. society.

We begin by discussing some of the major educational policies and programs of the past forty-five years that sought to reduce social inequality through expanding equality of educational opportunity. This discussion highlights the success and failures of programs such as school desegregation, compensatory education, Title IX, and job training. We then focus on the barriers these programs face in actually reducing social inequality. Our point is that inequality is so deeply rooted in the structure and operation of the U.S. political economy, that, at best, educational reforms can play only a limited role in ameliorating such inequality. In fact, there is considerable evidence that indicates that, for

SOURCE: Printed by permission of the authors

poor and many minority children, education helps legitimate, if not actually reproduce, significant aspects of social inequality in their lives. Finally, we speculate about education's potential role in individual and social transformation.

First, it is necessary to distinguish among equality, equality of opportunity, and equality of educational opportunity. The term *equality* has been the subject of extensive scholarly and political debate, much of which is beyond the scope of this essay. Most Americans reject equality of life conditions as a goal, because it would require a fundamental transformation of our basic economic and political institutions, a scenario most are unwilling to accept. As Ralph Waldo Emerson put it, "The genius of our country has worked out our true policy—opportunity."

The distinction between equality of opportunity and equality of outcome is important. Through this country's history, equality has most typically been understood in the former way. Rather than a call for the equal distribution of money, property, or many other social goods, the concern over equality has been with equal opportunity in pursuit of these goods. In the words of Jennifer Hochschild, "So long as we live in a democratic capitalist society—that is, so long as we maintain the formal promise of political and social equality while encouraging the practice of economic inequality—we need the idea of equal opportunity to bridge that otherwise unacceptable contradiction."[3] To use a current metaphor: If life is a game, the playing field must be level; if life is a race, the starting line must be in the same place for everyone. For the playing field to be level, many believe education is crucial because it gives individuals the wherewithal to compete in the allegedly meritocratic system. In America, then, *equality* is really understood to mean *equality of opportunity*, which itself hinges on *equality of educational opportunity*.

THE SPOTTY RECORD OF FEDERAL EDUCATIONAL REFORMS

In the past [fifty] years, a series of educational reforms initiated at the national level has been introduced into local school systems. All of the reforms aimed to move education closer to the ideal of equality of educational opportunity. Here we discuss several of these reforms, and how the concept of equality of educational opportunity has evolved. Given the importance of race and racism in U.S. history, many of the federal education policies during this period attempted to redress the most egregious forms of inequality based on race.

School Desegregation

Although American society has long claimed to be based on equality of opportunity, the history of race relations suggests the opposite. Perhaps the most influential early discussion of this disparity was Gunnar Myrdal's *An American Dilemma*, published in 1944. The book vividly exposed the contradictions between the ethos of freedom, justice, equality of opportunity and the actual experiences of African Americans in the United States.[4]

The links among desegregation, expanded educational opportunity, and the larger issue of equality of opportunity are very clear from the history of the desegregation movement. This movement, whose first phase culminated in the 1954 *Brown* decision outlawing *de jure* segregation in school, was the first orchestrated attempt in U.S. history to directly address inequality of educational opportunity. The NAACP strategically chose school segregation to be the camel's nose under the tent of the Jim Crow (segregated) society. That one of the nation's foremost civil rights organizations saw the attack on segregated schools as the opening salvo in the battle against society-wide inequality is a powerful example of the American belief that education has a pivotal role in promoting equality of opportunity.

Has desegregation succeeded? This is really three questions: First, to what extent are the nation's schools desegregated? Second, have desegregation efforts enhanced students' academic outcomes? Third, what are the long-term outcomes of desegregated educational experiences?

Since 1954, progress toward the desegregation of the nation's public schools has been uneven and limited. Blacks experienced little progress in desegregation until the mid-1960s when, in response to the civil rights movement, a series of federal laws, executive actions, and judicial decisions resulted in significant gains, especially in the South. Progress continued until 1988, when the effects of a series of federal court decisions and various local and national political developments precipitated marked trends toward the resegregation of Black students. Nationally, in 1994–1995, 33 percent of Black students attended majority White schools compared with the approximately 37 percent who attended majority White schools for much of the 1980s.[5]

Historically, Latinos were relatively less segregated than African Americans. However, from the mid-1960s to the mid-1990s there was a steady increase in the percentage of the Latino students who attended segregated schools. As a result, education for Latinos is now more segregated than it is for Blacks.

Given the long history of legalized segregation in the south, it is ironic that the South's school systems are now generally the country's most *de*segregated, while those in the northeast are the most intensely segregated. However, even desegregated schools are often resegregated at the classroom level by tracking or ability grouping. There is a strong relationship between race and social class, and racial isolation is often an outgrowth of residential segregation and socioeconomic background.

Has desegregation helped to equalize educational outcomes? A better question might be which desegregation programs under what circumstances accomplish which goals? Evidence from recent desegregation research suggests that, overall, children benefit academically and socially from well-run programs. Black students enjoy modest academic gains, while the academic achievement of White children is not hurt, and in some cases is helped, by desegregation. In school systems which have undergone desegregation efforts, the racial gap in educational outcomes has generally been reduced, but not eliminated.

More important than short-term academic gains are the long-term consequence of desegregation for Black students. Compared to those who attended
racially isolated schools, Black adults who experienced desegregated education
as children are more likely to attend multiracial colleges and graduate from
them, work in higher-status jobs, live in integrated neighborhoods, assess their
abilities more realistically when choosing an occupation, and to report assess
their abilities more realistically when choosing an occupation, and to report
interracial friendships.[6]

Despite these modest, but positive, outcomes, in the last decade of the twentieth century, most American children attend schools segregated by race, ethnicity, and social class. Consequently, [fifty] years of official federal interventions
aimed at achieving equality of educational opportunity through school desegregation have only made small steps toward achieving that goal; children from different race and class backgrounds continue to receive segregated and, in many
respects, unequal educations. . . .

Title IX

Title IX of the 1972 Higher Education Act is the primary federal law prohibiting
sex discrimination in education. It states, "No person in the United States shall,
on the basis of sex, be excluded from participation in, be denied benefit of, or be
subjected to discrimination under any program or activity receiving Federal
financial assistance." Until Title IX's passage, gender inequality in educational
opportunity received minimal legislative attention. The act mandates gender
equality of treatment in admission, courses, financial aid, counseling services,
employment, and athletics.

The effect of Title IX upon college athletics has been especially controversial. While women constitute 53 percent of undergraduates, they are only
37 percent of college athletes. This is undoubtedly due to the complex interaction
between institutional practices and gender-role socialization over the life course.
Certainly, the fact that the vast majority of colleges spend much more money on
recruiting and scholarships for male athletes contributes to the disparities.

In spring 1997, the U.S. Supreme Court refused to review a lower court's
ruling in *Brown v. Cohen* that states in essence that Title IX requires universities
to provide equal athletic opportunities for male and female students regardless of
cost. Courts have generally upheld the following three-pronged test for compliance: (1) the percentage of athletes who are females should reflect the percentage
of students who are female; (2) there must be a continuous record of expanding
athletic opportunities for females athletes; and (3) schools must accommodate the
athletic interests and abilities of female students. As of the ruling, very few universities were in compliance with the law.

Gender discrimination exists in other areas of education where it takes a variety of forms. For example, in K–12 education official curricular materials frequently feature a preponderance of male characters. Male and female characters
typically exhibit traditional gender roles. Vocational education at the high school

and college level remains gender-segregated to some degree. School administrators at all levels are overwhelmingly male although most teachers in elementary and secondary schools are female. In higher education, the situation is more complex. Faculty women in academia are found disproportionately in the lower ranks, are less likely to be promoted, and continue to earn less than their male colleagues.

Like the laws and policies aimed at eliminating race differences in school processes and outcomes, those designed to eliminate gender differences in educational opportunities have, at best, only narrowed them. Access to educational opportunity in the United States remains unequal for people of different gender, race, ethnic, and socioeconomic backgrounds.

EQUALITY OF EDUCATIONAL OPPORTUNITY
AND EQUALITY OF INCOME

Despite the failures of . . . many programs to eliminate the inequality of educational opportunity over the past 45 years, there is one indicator of substantial progress: measured in median years, the gap in educational attainment between Blacks and Whites, and between males and females, has all but disappeared. In 1997, the median educational attainment of most groups was slightly more than twelve years. In the 1940s, by contrast, White males and females had a median educational attainment of just under nine years, African American males about five years, and African American women about six.

However, the main goal of educational reform is not merely to give all groups the opportunity to receive the same quality and quantity of education. According to the dominant ideology, the ultimate goal of these reforms is to provide equal educational opportunity in order to facilitate equal access to jobs, housing, and various other aspects of the American dream. It thus becomes crucial to examine whether the virtual elimination of the gap in educational attainment has been accompanied by a comparable decrease in other measures of inequality.

Of the various ways inequality can be measured, income is one of the most useful. Much of a person's social standing and access to the good things in life depends on his or her income.[7] Unfortunately, the dramatic progress in narrowing the gap in educational attainment has not been matched by a comparable narrowing of the gap in income inequality. Median individual earnings by race and gender indicate that White men still earn significantly more than any other group. Black men trail White men, and all women earn significantly less than all men. Even when occupation, experience, and level of education are controlled, women earn less than men, and Black men earn less than White men. It is only Black and White women with comparable educational credentials in similar jobs who earn about the same.

The discrepancy between the near elimination of the gap in median educational attainment and the ongoing gaps in median income is further evidence that addressing the inequality of educational opportunity is woefully in sufficient for addressing broader sources of inequality throughout society.

This discrepancy can be explained by the nature of the U.S. political economy. The main cause of income inequality is the structure and operation of U.S. capitalism, a set of institutions which scarcely have been affected by the educational reforms discussed earlier. Greater equality of educational opportunity has not led to a corresponding decrease in income inequality because educational reforms do not create more good-paying jobs, affect gender-segregated and racially segmented occupational structures, or limit the mobility of capital either between regions of the country or between the United States and other countries. For example, no matter how good an education White working-class or minority youth may receive, it does nothing to alter the fact that thousands of relatively good paying manufacturing jobs have left northern inner cities for northern suburbs, the sunbelt, or foreign countries.

Many argue that numerous service jobs remain or that new manufacturing positions have been created in the wake of this capital flight. But these pay less than the departed manufacturing jobs, are often part-time or temporary, and frequently do not provide benefits. Even middle-class youth are beginning to fear the nature of the jobs which await them once they complete their formal education. Without changes in the structure and operation of the capitalist economy, educational reforms alone cannot markedly improve the social and economic position of disadvantaged groups. This is the primary reason that educational reforms do little to affect the gross social inequalities that inspired them in the first place.

BEYOND ATTAINMENT: THE PERSISTENCE OF EDUCATIONAL INEQUALITY

Educational reforms have not led to greater overall equality for several additional reasons. While race and gender gaps in educational attainment have narrowed considerably, educational achievement remains highly differentiated by social class, gender, and race. Many aspects of school processes and curricular content are deeply connected to race, social class, and gender inequality. But gross measures of educational outputs, such as median years of schooling completed, mask these indicators of inequality.

Not all educational experiences are alike. Four years of public high school in Beverly Hills are quite different from four years in an inner-city school. Family background, race, and gender have a great deal to do with whether a person goes to college and which institution of higher education she or he attends. The more privileged the background, the more likely a person is to attend an elite private university.

For example, according to Jacobs, women trail men slightly in representation in high status institutions of higher education because women are less likely to attend engineering programs and are more likely to be part-time students (who are themselves more likely to attend lower status institutions such as community colleges). Gender segregation in fields of study remains marked, with women less likely than men to study in scientific and mathematical fields. Furthermore, there is substantial race and ethnic segregation between institutions of higher education. Asian-Americans and Latinos are more segregated from Whites than are African-Americans. Whites and Asian-Americans are more likely to attend higher status universities than are African-Americans and Latinos.[8]

These patterns of race and gender segregation in higher education have direct implications for gender and race gaps in occupational and income attainment. Math and science degree recipients are more likely to obtain more lucrative jobs. A degree from a state college is not as competitive as one from an elite private university. Part of the advantage of attending more prestigious schools comes from the social networks to which a person has access and can join.

Another example of persistent inequality of educational opportunity is credential inflation. Even though women, minorities, and members of the working class now obtain higher levels of education than they did before, members of more privileged social groups gain even higher levels of education. At the same time, the educational requirements for the best jobs (those with the highest salaries, benefits, agreeable working conditions, autonomy, responsibility) are growing. Those with the most education from the best schools tend to be the top candidates for the best jobs. Because people from more privileged backgrounds are almost always in a better position to gain these desirable educational credentials, members of the working class, women, and minorities are still at a competitive disadvantage. Due to the dynamics of credential inflation, educational requirements previously necessary for the better jobs and now within the reach of many dispossessed groups are inadequate and insufficient in today's labor market. The credential inflation process keeps the already privileged one step (educational credential) ahead of the rest of the job seekers.[9]

One additional aspect of the persistent inequalities in educational opportunities concerns what sociologists of education call the hidden curriculum. This concept refers to two separate but related processes. The first is that the content and process of education differ for children according to their race, gender, and class. The second is that these differences reflect and thus help reproduce the inequalities based on race, gender, and class that characterize U.S. society as a whole.

One aspect of the hidden curriculum is the formal curriculum's ideological content. Anyon's work on U.S. history texts demonstrates that children from more privileged backgrounds are more likely to be exposed to rich, sophisticated, and complex materials than are their working-class counterparts.[10] Another aspect of the hidden curriculum concerns the social organization of the school and the classroom. Some hidden curriculum theorists suggest that tracking, ability grouping, and conventional teacher-centered classroom interactions contribute to the reproduction of the social relations of production at the workplace. Lower-track classrooms are disproportionately filled with working-class and minority students. Students in lower tracks are more likely than those in higher tracks to be assigned repetitive exercises with low levels of cognitive challenge. Lower-track students are likely to work individually and to lack classroom experience with problem solving or other independent, creative activities. Such activities are more conducive to preparing students for working-class jobs than for professional and managerial positions. Correspondence principle theorists argue that educational experiences from preschool to high school are designed to differentially prepare students for their ultimate positions in the work force, and that a student's placement in various school programs is strongly related to her or his race and class origin. Critics charge that the correspondence principle has been applied in too deterministic and mechanical a fashion. Evidence abounds of student resistance to class, gender, and race differentiated education.[11] This is undoubtedly why so many students drop out or graduate from high school with minimal levels of literacy and formal skills. Nonetheless, hidden curriculum theory offers a compelling contribution to explanations of how and why school processes and outcomes are so markedly different according to the race, gender, and social class of students.

CONCLUSION

In this [essay] we have argued that educational reforms alone cannot reduce inequality. Nevertheless, education remains important to any struggle to reduce inequality. Moreover, education is more than a meal ticket; it is intrinsically worthwhile and crucially important for the survival of democratic society. Many of the programs discussed in this essay contribute to the enhancement of individuals' cognitive growth and thus promote important nonsexist, nonracist attitudes and practices. Many of these programs also make schools somewhat more humane places for adults and children. Furthermore, education, even reformist liberal education, contains the seeds of individual and social transformation. Those of us committed to the struggle against inequality cannot be paralyzed by the structural barriers that make it impossible for education to eliminate inequality. We must look upon the schools as arenas of struggle against race, gender, and social class inequality.

NOTES

1. This essay draws on an article by Roslyn Arlin Mickelson that appeared as "Education and the Struggle Against Race, Class and Gender Inequality," *Humanity and Society 11*(4) (1987): 440–64.

2. Ascertaining whether a set of beliefs constitutes the dominant ideology in a particular society involves a host of difficult theoretical and empirical questions. For this reason we use the term *putative dominant ideology*. For discussion of these questions, see Nicholas Abercrombie et al., *The Dominant Ideology Thesis* (London: George Allen & Unwin, 1980); James C. Scott, *Weapons of the Weak* (New Haven: Yale University Press, 1985); Stephen Samuel Smith, "Political Acquiescence and Beliefs About State Coercion" (unpublished Ph.D. dissertation, Stanford University, 1990).

3. Jennifer Hochschild, "The Double-Edged Sword of Equal Educational Opportunity." Paper presented at the meeting of the American Education Research Association, Washington, D.C., April 22, 1987.

4. Gunnar Myrdal, *An American Dilemma: The Negro Problem and Modern Democracy* (New York: Harper & Row, 1944).

5. Gary Orfield, Mark D. Bachmeier, David R. James, and Tamela Eitle, "Deepening Segregation in American Public Schools" (Cambridge, MA: Harvard Project on School Desegregation, 1997).

6. Amy Stuart Wells and Robert L. Crain, "Perpetuation Theory and the Long-Term Effects of School Desegregation," *Review of Educational Research* 64(4) (1994): 531–55.

7. To be sure, income does not measure class-based inequality, but there is a positive correlation between income and class. Income has the additional advantage of being easily quantifiable. Were we to use another measure of inequality, e.g., wealth, the disjuncture between it and increases in educational attainment would be even larger. Although the distribution of wealth in U.S. society has remained fairly stable since the Depression, the gap between rich and poor increased in the 1980s and 1990s. Although accurate data are difficult to obtain, a 1992 study by the Federal Reserve found that in 1989 the top one-half of one percent of households held 29 percent of the wealth held by all households.

8. Jerry A. Jacobs, "Gender and Race Segregation Between and Within Colleges" (paper presented at the Eastern Sociological Society, Boston, MA, April 1996).

9. Randall Collins, *The Credential Society* (New York: Academic Press, 1979).

10. Jean Anyon, "Social Class and the Hidden Curriculum of Work," *Journal of Education* 162(1) (1980): 67–92; Jean Anyon, "Social Class and School Knowledge," *Curriculum Inquiry 10* (1981): 3–42.

11. Samuel Bowles and Herbert Gintis, *Schooling in Capitalist America: Educational Reform and the Contradictions of Economic Life* (New York: Basic Books, 1976); Roslyn Arlin Mickelson, "The Case of the Missing Brackets: Teachers and Social Reproduction," *Journal of Education 169*(2) (1987): 78–88.

50

Why Segregation Matters

Poverty and Educational Inequality

GARY ORFIELD AND CHUNGMEI LEE

INTRODUCTION

Much of the discussion about school reform in the U.S. in the past two decades has been about racial inequality. Former President Bush promised that the No Child Left Behind Act and the expansion of high stakes testing to high schools can end the "soft racism of low expectations." Yet a disproportionate number of the schools being officially labeled as persistent failures and facing sanctions under this program are segregated minority schools. Large city school systems are engaged in massive efforts to break large segregated high poverty high schools into small schools, hoping that it will create a setting better able to reduce inequality, while others claim that market forces operating through charter schools and private schools could end racial inequalities even though both of these are even more segregated than public schools and there is no convincing evidence for either of these claims. More and more of the still standing court orders and plans for desegregated schools are being terminated or challenged in court, and the leaders of the small number of high achieving segregated schools in each big city or state are celebrated. The existence of these schools is being used to claim that we can have general educational success within the existing context of deepening segregation.[1] Clearly the basic assumption is that separate schools can be made equal and that we need not worry about the abandonment of the movement for integration whose history was celebrated so extensively last year on the 50th anniversary of the *Brown* decision even as the schools continued to resegregate. There has been a continuous pattern of deepening segregation for black and Latino students now since the 1980s.

What if this basic assumption is wrong? What if the Supreme Court was correct a half century ago in its conclusion that segregated schools were "inherently unequal"? What if Martin Luther King's many statements about how segregation harms both the segregator and the segregated, drastically limits opportunity, and does not provide the basis for building a successful interracial society are correct? What if the Supreme Court's sweeping conclusion in the 2003 University of Michigan case that there is compelling evidence that diversity

SOURCE: From *Why Segregation Matters: Poverty and Educational Inequality* by Gary Orfield and Chungmei Lee, January 2005, The Civil Rights Project at University of California, Los Angeles. Reprinted by permission.

improves the education of all students is true and applies with even greater force to public schools?

If, however, it is wrong to assume that segregation is irrelevant and policies that ignore that fact simply punish the victims of segregation because they fail to take into account many of the causes of the inequality, then current policy is being built on a foundation that cannot produce the desired results and may even compound the existing inequalities. We believe this to be true. Segregated schools are unequal and there is very little evidence of any success in creating "separate but equal" outcomes on a large scale.

One of the common misconceptions over the issue of resegregation of schools is that many people treat it as simply a change in the skin color of the students in a school. If skin color were not systematically linked to other forms of inequality, it would, of course, be of little significance for educational policy. Unfortunately that is not and never has been the nature of our society. Socioeconomic segregation is a stubborn, multidimensional and deeply important cause of educational inequality. U.S. schools are now 41 percent nonwhite and the great majority of the nonwhite students attend schools which now show substantial segregation. Levels of segregation for black and Latino students have been steadily increasing since the 1980s, as we have shown in a series of reports.[2] Achievement scores are strongly linked to school racial composition and so is the presence of highly qualified and experienced teachers.[3] The nation's shockingly high dropout problem is squarely concentrated in heavily minority high schools in big cities.[4] The high level of poverty among children, together with many housing policies and practices which exclude poor people from most communities, mean that students in inner city schools face isolation not only from the white community but also from middle class schools. Minority children are far more likely than whites to grow up in persistent poverty. Since few whites have direct experience with concentrated poverty schools, it is very important to examine research about its effects. . . .

Martin Luther King understood the nature of racial inequality and campaigned against segregation, discrimination and poverty. Dr. King died more than a third of a century ago and with his death the civil rights movement lost its central voice and focus and faced a strengthening movement toward preservation of the status quo. With the passage of time and changing political leadership we have seen sweeping policy reversals, rising segregation, especially in the South and West, and a loss of understanding of the reasons for Dr. King's crusades against racial separation. Certainly there was nothing about Dr. King that held that black institutions were bad—he was the proud pastor of an overwhelmingly black church of great influence and power and a proud graduate of the preeminent black college for men, Morehouse in Atlanta. Segregation was evil in his mind not because of skin color but because it almost always led to unequal opportunities, given the realities of American society, and because it produced both ignorance and damaging racial stereotypes in the minds of both the segregated and the segregators. Segregation was a basic structure that subordinated and limited opportunities for nonwhite children. Dr. King advocated not only plans that brought minority children into previously segregated white schools but much deeper transformations in which segregated schools became truly integrated with equal treatment and respect for all groups of students.

Segregation was never just a black-white problem, never just a Southern problem, and never just a racial problem, but in the initial struggle in the South of the mid-twentieth century that was clearly the focus. By the time Dr. King organized his last movement, the Poor People's Campaign, his approach was clearly multiracial, with a deepening emphasis on poverty as well as racial discrimination. Speaking ten days before he died, King spoke of his conviction that it was "absolutely necessary now to deal massively and militantly with the economic problem. . . . So the grave problem facing us is the problem of economic deprivation, with the syndrome of bad housing and poor education and improper health facilities all surrounding this basic problem."[5] Had he not been assassinated shortly before that movement came to Washington, perhaps the link between racial and economic isolation would be better understood as would the profound impact of double segregation (often triple segregation for immigrant children who are also isolated by language in their schools).

The civil rights movement was never about sitting next to whites, it was about equalizing opportunity. If high poverty schools are systematically unequal and segregated minority schools are almost always high poverty schools, it is much easier to understand both the consequences of segregation and the conditions that create the possibility of substantial gains in desegregated classes. At a time when the racial achievement gaps remain substantial and desegregation orders are being challenged, it is particularly important to understand the pattern that is developing and to think seriously about how to address it. . . .

POVERTY SEGREGATION AND RACIAL INEQUALITY

In the South of Dr. King's time, the world was largely black and white, apart from sections of Texas and Florida. The civil rights movement was largely understood at its peak as a movement to end discrimination against blacks. The black student enrollment of the country was many times larger than the Latino enrollment and the Asian enrollment was still almost insignificant.

Immigration has transformed American schools as the number of black students grew slowly, the number of Latinos and Asian students exploded, and white enrollment continuously declined as a proportion of the total. Enrollment statistics for the 2002–3 school year show the multiracial nature of our nation's public schools: Latinos are now the largest minority group at 18 percent, closely followed by black students at 17 percent. Together, these two groups are now more than a third of the total student population. In the West and South, the two most populous regions, with 54 percent of the nation's public school students, blacks and Latinos comprise at least 30 percent of the student population in most of the states. In many areas, Latinos and Asians are making their presence felt in previously biracial environments. Asians now outnumber black students in the Western region, stretching from the Rockies to the Pacific coast. In the Northeast, the share of Latino students rivals the share of black students. In these two regions, students have greater potential to attend multiracial schools than do their peers in the Midwest and Border states, where whites comprise 70 percent or more of the public school population.

White shares of enrollment have been shrinking and Latino shares rapidly climbing for a third of a century,[6] and both birth and immigration statistics strongly suggest that they will continue. Thus, the importance of understanding the conditions affecting nonwhite children becomes more important every year. These changes will also require us to think about race and racial/ethnic isolation in a much richer multiracial context.

The national enrollment statistics show the continuation of the historic concentration of Latinos in the West, where they make up 35 percent of all students, and blacks in the South, where they account for 27 percent of total enrollment (Table 1). In the West there are five times as many Latino as black students and slightly more Asians than blacks. In the other regions there are more blacks than Latinos but the numbers are changing. Almost one-fifth of the students in the South are now Latino as are one-eighth of the students in the Northeast. There is now clear evidence of large secondary Latino migrations into areas where the minority population has historically been overwhelmingly black, such as Georgia and North Carolina.

Due to the severe white residential isolation in outlying suburbs,[7] white students are the least likely group to attend truly multiracial schools (Table 2).

T A B L E 1   **Public School Enrollments by Race/Ethnicity and Region, 2002–03**

Region	Total Enrollment	%White	%Black	%Latino	%Asian	%Native American
West	11,086,700	48	7	35	8	2
Border	3,518,342	70	21	4	2	3
Midwest	9,850,818	75	15	7	3	1
South	13,880,097	51	27	19	2	0
Northeast	8,296,140	66	16	13	5	0
Alaska	134,364	59	5	4	6	26
Hawaii	183,829	20	2	5	72	1
Bureau of Indian Affairs	46,126	0	0	0	0	100
U.S. Total	46,996,416	59	17	18	4	1

TABLE 2   Racial Composition of Schools Attended by the Average Student of Each Race, 2002–03

| Percent Race in Each School | Racial Composition of School Attended by Average: | | | | |
	White Student	Black Student	Latino Student	Asian Student	Native American Student
%White	78	30	28	45	44
%Black	9	54	12	12	7
%Latino	8	13	54	20	11
%Asian	3	3	5	22	2
%Native American	1	1	1	1	36
Total	100	100	100	100	100

In contrast, although black and Latino students attend schools with a majority of students from their own groups, the average black and Latino students attend more diverse schools than do their white peers. The average black student attends a school where one-eighth of the students are Latino, and the average Latino student is in a school with a similar fraction of blacks. Black and Latino students attend schools, on average, that are 30 percent or less white. The contrast in terms of contact with their own groups are most extreme for whites and Asians. Whites are most isolated within their own racial group—attending schools where almost four-fifths of the students are white. In contrast, Asians are least isolated within their own racial group—with only about one-fifth Asian classmates. Asians attend the most diverse schools of all, with 45 percent white, 12 percent black, and 20 percent Latino students.

Enrollment in Predominantly Minority, Intensely Segregated Minority and Extremely Segregated Minority Schools[8]

Because of the severe isolation of students in their own racial groups, particularly of white students, black and Latino students attend predominantly minority schools in disproportionate numbers. Twice as many black and Latino students as white students attend predominantly minority (>50% minority) schools and three times as many attend intensely segregated schools (>90–100% minority). About 1.4 million black students and close to a million Latino students attend schools that are almost all minority (99–100% minority) compared to less than ten thousand white students. While it is true that, by definition, the majority of students in these intensely segregated and extremely segregated minority schools

T A B L E 3   Percent of Students from Each Racial Group in 50–100%, 90–100%, and 99–100% Minority Schools, 2002–03

Race	Type of School		
	50–100% Minority Schools	90–100% Minority Schools	99–100% Minority Schools
Black	73	38	18
Latino	77	38	11
White	12	1	0
Asian	56	15	1
Native American	52	27	16

(90% and 99% minority respectively) would be minority students, it is also clear that there are many more black and Latino students attending these schools than overall enrollment numbers should suggest. To get a better sense of how segregated the public schools are, we must examine the percentages of each group attending schools with different racial compositions.

More than three-quarters (77%) of Latino students attend majority minority schools, closely followed by black students at 73 percent (Table 3). Asian and Native American students attend these schools in substantial numbers. Despite the fact that these two groups *together* comprise just 5 percent of total public school enrollment, *more than half* of each group attend majority minority schools. In contrast, less than 12 percent of white students attend these majority minority schools, and less than 1 percent white students attend overwhelmingly minority schools (90–100% and 99–100% minority schools). More than a third of all black and Latino students attend 90–100 percent schools, closely followed by Native American students at 27 percent. More than 10 percent of each group attend 99–100 percent minority schools. Overall, black students experience even more segregation than do their Native American peers.

Enrollment in Predominantly, Intensely Segregated, and Extremely Segregated White Schools

Close to 90 percent of white students attend schools that are at least half white (Table 4). Given the racial composition of our nation's public schools—where close to 60 percent of the students are white—this is to be expected. However, 2 out of every 5 white students attend schools that are 90–100 percent white. This reflects substantial concentration of white students in certain areas, such as the suburbs of our nation. In contrast, only about 2 percent black and Latino students, 6 percent Asian, and 8 percent Native American students attend these overwhelmingly white schools.

T A B L E 4   **Percent of Students from Each Racial Group in 50–100%, 90–100%, and 99–100% White Schools, 2002–03**

Race	Type of School		
	50–100% Minority Schools	90–100% Minority Schools	99–100% Minority Schools
Black	28	2	0
Latino	23	2	0
White	88	41	5
Asian	45	6	0
Native American	49	8	0

THE POVERTY DIMENSION IN SEGREGATION

Many who belittle the desegregation movement tend to assume that integration-ists are too preoccupied by issues of race and that it is absurd to suppose that changing the color of a student's classmates would make any real difference. Desegregation, they claim, is not only virtually irrelevant to school reform but it is also insulting to suggest that there is something wrong about an all-black or all-Latino school. Many cite as examples minority schools that despite all odds were able to provide quality education to the students. The implication is that the civil rights leaders had an incorrect, simplistic and even racially paternal-istic theory that does not work and detracts attention from more important goals.

Segregation has never just been by race: segregation by race is systematically linked to other forms of segregation, including segregation by socioeconomic status, by residential location, and increasingly by language. Since the 1970s, there has been a gradual decline of white families in large metropolitan centers as they moved to suburbs or small cities, leaving a large concentration of black and Latino students in central cities.[9] . . .

Their communities usually reflect conditions of distress—housing inade-quacy and decay, weak and failing infrastructure, and critical lack of mentors and shortage of jobs—all of which adversely affect inner city children's educa-tional success. Isolated in the inner cities, high poverty schools must also struggle with the challenges posed by enrolling a student body lacking health and proper nutrition, violence in the form of crime and gangs, and unstable home environ-ments. Furthermore, the stigma experienced by people living in these communi-ties often feeds back into a vicious cycle of stagnation and unequal opportunity. Middle class, and even many low income whites, can expect their children to attend low poverty schools. In contrast, even middle income minority families often end up in neighborhoods and schools with high poverty concentrations because of housing discrimination and other forces that perpetuate and exacer-bate segregated residential patterns.[10]

The simplification of segregation into purely a racial issue ignores the fact that schools tend to reflect and intensify the racial stratification in society. Desegregation efforts aim at breaking the pernicious link between the two by taking a black and Latino student from a high poverty school to a middle class school that often has better resources, more qualified teachers, tougher academic competition and access to more developed social networks. . . .

NATIONAL TRENDS

The racial differences in exposure to poverty are striking (Table 5). The average white and Asian student attends schools with the lowest shares of poor students. The average black and Latino student attends schools in which close to half the students are poor, more than twice the exposure of whites to poor students. The average Native American student experienced the biggest increase in exposure to poor students, from 31 percent to 38 percent in 2002.

. . . Black and Latino students are more than three times as likely whites to be in high poverty schools and 12 times as likely to be in schools where almost everyone is poor. These are major consequences of residential and educational segregation.

Dropouts, "Dropout Factories"—Relationship with Poverty and Race Segregation

Several studies have shown the link between segregation by poverty, race, and academic performance.[11] . . . Nationally, Asians have the highest graduation rate at 77 percent, followed by 75 percent of white students. In contrast, a little more than half of all black, Latino, and Native American students graduated on time in 2001. . . .

T A B L E 5   **Percent Poor* in Schools Attended by the Average Student, by Race and Year**

Percent Poor	White Student	Black Student	Latino Student	Asian Student	Native American Student
1996–97	19	43	46	29	31
1998–99	20	39	44	26	35
2000–01	19	45	44	26	31
2002–03	23	49	48	27	39

SOURCE: 1996–97; 1998–99; 2000–01; 2002–03 NCES Common Core of Data Public School Universe

*These numbers include both students eligible for and receiving free and reduced lunch. Unlike the Census Bureau's poverty measure, it does not include low income students not receiving subsidized lunches; thus it is likely to underestimate the level of exposure to poor students.

Across the nation, the huge problem of minority high school dropouts is concentrated in a few hundred high schools where a huge proportion of the students never finish, called "dropout factories" by Johns Hopkins researcher Robert Balfanz.[12] These high schools are overwhelmingly poor and nonwhite and, apart from the South, they are very largely urban. Though much more attention has been devoted in recent years to test scores, dropping out is, of course, the ultimate failure for a student in the post-industrial economy—a failure that usually causes deep and irreversible life-long damage to a student and his future family.

The 24 largest central cities together enroll more than 4.5 million of the public school population. These districts are so heavily minority that except for one district, more than 70 percent of the black and Latino students in these districts attend predominantly majority minority schools (50–100% minority schools), and in 20 districts, more than 90 percent of black students attend these schools. In Dallas, El Paso, and Santa Ana, 100 percent of Latino students attend schools that are predominantly minority, and in 15 districts, more than 90 percent of Latino students attend schools where more than half of their peers are minority students. [These districts], for the most part, are high poverty districts. Among these large districts, all of those with the lowest high school completion rates are central city systems with very high levels of segregation. The cities with the lowest completion rates are among the country's largest: New York City, Los Angeles, and Chicago. More than three-quarters of the students in these schools are minority. Sadly included in this group is the first city that experienced full urban desegregation outside the South, Denver, which is also the city in which the right of Latino students to desegregated education was established in a decision on the city school by the U.S. Supreme Court.

The Case for Desegregation

There is clear evidence that experience with diversity produces both short- and long-term advantages in terms of intellectual and social development. These findings strongly suggest that exposure to more desegregated settings can break the tendency for racial segregation to become self-perpetuating for all students later in life.[13] Furthermore, students of all races who are exposed to integrated educational settings feel much more comfortable about their ability to live and work among people of diverse racial and ethnic backgrounds.[14]

NOTES

1. Thernstrom, A. and Thernstrom, S. (2003). *No excuses: Closing the racial gap in learning*. New York: Simon & Schuster.

2. The most recent is Orfield, G. and Lee, C. (2004). *Brown at 50: King's dream or Plessy's nightmare?* Cambridge, MA: The Civil Rights Project at Harvard University.

3. Natriello, G., McDill, E. L. & Pallas, A. M. (1990). *Schooling disadvantaged children: Racing against catastrophe*. New York, NY: Teachers College Press; Schellenberg, S.

(1999). Concentration of poverty and the ongoing need for Title I. In G. Orfield and E. DeBray. (Eds.), *Hard Work for Good Schools (pp. 130–146)*: Cambridge, MA: The Civil Rights Project at Harvard University; Lee, C. (2004). *Racial segregation and educational outcomes in metropolitan Boston*. Cambridge: The Civil Rights Project at Harvard University.

4.   Balfanz, R. and Legter, N. (January, 2001). *How many central city high schools have a severe dropout problem, where are they located, and who attends them?* Paper presented at Dropouts in America Conference, Cambridge, MA.

5.   Washington, J. ed. (1986). *Testament of hope: The essential writings and speeches of Martin Luther King, Jr.*, New York: Harper Collins Publishers, p. 672.

6.   See Orfield, G. and Lee, C. (2004). *Brown at 50: King's dream or Plessy's nightmare?* Cambridge, MA: The Civil Rights Project at Harvard University.

7.   Multiracial schools are schools where there are at least three groups with 10 percent or more representation in the student population.

8.   Predominantly minority schools are schools that are over 50% minority, intensely segregated minority schools are more than 90% minority schools, and extremely segregated minority schools enroll over 99% minority students. These terms are also used for segregated white schools: predominantly white (>50% white), intensely segregated white (>90% white), and extremely segregated white (>99%white) schools.

9.   Hauser, R., Simmons, S. and Pager, D. (2004). High school dropout, race/ethnicity, and social background from the 1970s to the 1990s. In G. Orfield, (Ed.). *Dropouts in America: Confronting the graduation rate crisis*. Cambridge: Harvard Education Press.

10.   Harris, D., and McArdle, N. (2004). *More than money: The spatial mismatch between where minorities can afford to live and where they actually reside*. Cambridge, MA: The Civil Rights Project at Harvard University; Bradford, C. (2002). *Risk or race? Racial disparities and the subprime refinance market*. Washington, DC: Center for Community Change; Yinger, J. (1995). *Closed doors, opportunities lost: The continuing costs of housing discrimination*. New York: Russell Sage Foundation.

11.   Orfield, G., Losen, D., Wald, J., and Swanson, C. (2004). *Losing our future: How minority youth are being left behind by the graduation rate crisis*. Cambridge, MA: The Civil Rights Project at Harvard University. Contributors: Urban Institute, Advocates for Children of New York, and The Civil Society Institute; Balfanz, R. and Legters, N. (2003). Weak promoting power, minority concentration, and high schools with high dropout rates in urban America: A multiple cohort analysis of the 1990s using the Common Core of Data." Prepared for Making Dropouts Visible conference at Teachers College, Columbia University.

12.   Balfanz, R. and Legters, N. (2004). Locating the dropout crisis: Which high schools produce the nation's dropouts? In Gary Orfield, ed. *Dropouts in America: Confronting the Graduation Rate Crisis*. Cambridge, MA: Harvard Education Press.

13.   See Wells, A. S., and Crain, R. L. (1994). Perpetuation theory and the long–term effects of school desegregation. *Review of Educational Research, 64*, 531–555; Braddock, J. H. and McPartland, J. (1989). Social-psychological processes that perpetuate racial segregation: The relationship between school and employment segregation." *Journal of Black Studies.* 19(3): 267–289.

14.   *The Impact of Racial and Ethnic Diversity on Educational Outcomes: Cambridge, MA School District*. The Civil Rights Project at Harvard University, January 2002.

51

How a Scholarship Girl
Becomes a Soldier

The Militarization of Latina/o Youth in
Chicago Public Schools

GINA M. PÉREZ

L ike many working-class communities, Chicago Puerto Ricans have had a
complicated relationship with the United States military, an institution that
represents, in a particularly visible way, Puerto Rico's unresolved political status,
one that is contested by community activists. Recent broad-based political mobilization in both Chicago and New York City demanding the end of naval operations on the island of Vieques, as well as a smaller, less visible presence of island
residents supporting the navy's role on the island, is just one example of the complex responses Puerto Ricans have to the military. The military also is an institution that is understood to be (and is often successfully used as) an important avenue
of social mobility for families and households with limited resources and opportunities available to them, particularly for young women who often are sources of
productive and reproductive labor critical to a household's survival. . . .

In what follows, I draw on ethnographic and interview data to discuss how
Puerto Rican and Latina/o youth in Chicago are implicated in an increasingly
militarized world. Military programs like the Junior Reserve Officer Training
Program (JROTC) are not new (it was founded in 1916) and have long targeted
Puerto Rican youth. In recent years, however, these programs have operated
with new intensity, appearing more frequently in urban schools with an explicit
goal of targeting populations deemed "at risk" in "less affluent large urban
schools." It is not surprising that this "at risk" group includes young men who
are allegedly in danger of falling into gangs or drug use. What is alarming, however, is the extent to which these programs are now targeting young women
who are deemed "at risk" because of the possibility that they might become
unwed teenage mothers. The gendered dynamics through which militarization
is occurring on the homefront are in need of critical attention if we are to
understand how poor and working-class youth and their families struggle and
strategize to make ends meet. . . .

SOURCE: Identities: *Global Studies in Culture and Power*, 13:53–72. Copyright © 2006.
Reprinted by permission of the Taylor & Francis Group.

CHICAGO: A GLOBAL LATINO CITY

With more than three-quarters of a million Latina/o residents, Chicago is home to the third largest and one of the most diverse Latina/o populations in the country. It also is the only place where large numbers of Mexicans and Puerto Ricans of several generations live, marry, and work side by side. . . .

Chicago Latinas/os, however, are also some of the city's most impoverished residents, with nearly twenty-four percent of Latinas/os living in poverty. Chicago Puerto Ricans are the poorest of all Chicago Latinos. With limited employment opportunities, primarily in the low-wage service sector, Puerto Ricans in Chicago index the highest poverty rates in the city at 33.8 percent, compared to thirty-three percent for the city's African-American families.[1] In contrast with Cubans and South Americans, Puerto Ricans and Mexicans have lower educational and average income levels and are concentrated in the low-wage service sector and as operatives.[2] Moreover, studies by the Latino Institute, The Chicago Urban League, and Northern Illinois University demonstrate that none of the jobs requiring only a high school diploma—and even many of those demanding some post-secondary education—pay a living wage for a family with dependent children.[3] According to the Latino Institute (1994), more than seventy-five percent of Puerto Ricans are employed in those sectors of the economy. In short, while transnational investment and the loss of manufacturing jobs in the Chicago area have rendered Latinos in general much poorer than a decade earlier, Puerto Ricans specifically remain the most economically disadvantaged group in all of Chicago. It is no wonder that many Latina/o youth consider military service as one of the few opportunities that will guarantee employment, provide them with marketable skills, and serve as a vehicle for achieving economic stability.

JROTC, MILITARY SERVICE, AND
MAKING ENDS MEET

Beginning at a very early age, military service becomes one of the most appealing and seemingly sure options for many poor and working-class Latina/o families. Bombarded by television advertisements, visits by recruiters in junior high and high schools, and the prospect of receiving JROTC programs' financial incentives (both immediate and promises of money for the future), poor children and particularly children of color quickly come to consider the military—rather than college—as an employment venue after high school. Money, promises of free education, training, discipline, honor, and respect seduce young men and women into the military ranks.

Chicago schools provide particularly fertile ground for cultivating military careers after high school. Nearly eighty-five percent of Chicago public school students come from low-income families and are either African American or Latina/o.[4] They also are increasingly enmeshed in what Pauline Lipman has identified as "stratified academic programs," whereby African-American and Latina/o high school students attend schools with "limited offerings of advanced courses and new vocational academies, basic skills transitional high schools" or public military academies rather than the "[n]ew academically selective magnet schools and programs, mainly located in largely white upper-income and/or gentrifying neighborhoods" (Lipman 2003: 81). The Chicago public schools, for example, lead the nation with more than 10,000 students participating in a wide range of expanding public school military programs. Chicago is home to two school-wide military academies—Chicago Military Academy and Carver Military Academy—both affiliated with the United States Army and located on Chicago's South Side; eight military academies within regular high schools (four affiliated with the Army, two with the Marines, and one each with the Navy and the Air Force); a middle school military academy in the predominately Mexican neighborhood of Little Village; and part-time JROTC in forty-three Chicago high schools and after-school programs at seventeen middle schools.[5] . . .

In Chicago, JROTC is part of the Education to Careers (ETC) program, whose mission is to "equip students to successfully transition into postsecondary education, advanced training, and the workplace." . . . High school students may participate in JROTC to satisfy the Chicago public high school career education requirement or, in some schools, to satisfy the physical education requirement for graduation. JROTC also offers many extracurricular activities, including honor guard, competitions, service, field trips, and opportunities to visit military installations. Many students explain that it is precisely these curricular and extra-curricular offerings that make JROTC appealing. Almost all of the young Latina/o and African-American JROTC participants I interviewed in one Chicago high school also cited "being treated with respect" (especially while wearing the cadet uniform) both as a reason for joining, as well as one of the greatest advantages of participating in, JROTC.

Students' concern with respect is no small matter, since most reside in poor and working-class (and, often, slowly gentrifying) neighborhoods regarded in local media as dangerous and are enmeshed in racialized policing practices aimed at containing suspect youth.[6] Latina/o and African-American youth are painfully aware of how their bodies are read. Thus, wearing a military uniform is, perhaps, one way of negotiating the racialized systems of surveillance that not only operate within their neighborhoods, but also within their own schools. . . .

Even when young people are not in formal high schools, the military continues to shape many of their educational and employment opportunities in other ways. For example, when nineteen-year-old Marvin Polanco[7] entered a G.E.D. class I taught at a Puerto Rican cultural center in Chicago, he regaled me with stories about his experience in Lincoln's Challenge, a military boarding school affiliated with the Illinois National Guard.[8] Shortly after dropping out of Clemente High School, Marvin's mother signed him up for the five-month program, which promised to instill discipline, prepare him for the G.E.D., and pay him $2,200 after successfully graduating from the school. "But they tricked my mom," Marvin told me solemnly one day after class. He never passed the G.E.D., a failure he attributes to the fact that they required him to exercise, run, and engage in other physical training immediately before the exam, leaving him exhausted and unable to finish the test. Marvin received less than $500 when he finished the program. "They try to brainwash you there," he explained another day, referring to the deeply militarized atmosphere and training each cadet receives while in residence. Between the physical discipline imposed by the officers and the danger of getting jumped by gang members also participating in the program, Marvin said he often was afraid he wouldn't get out alive. After a few months in the program, he ran away from the school and returned home, begging his mother not to send him back to the school and promising to get his G.E.D. somewhere in Chicago.

Many of my male G.E.D. students wanted to go into the military once they completed their G.E.D. When they would approach me about this, I encouraged them to consider enrolling in one of Chicago's city colleges and eventually transferring to either University of Illinois–Chicago (UIC) or Northeastern University, which both had programs specifically targeting and supporting incoming students with their G.E.D. With the help of the Centro's counselors and staff, we put many students in touch with representatives from these universities, invited them on a tour of one of the city colleges, and helped organize a campus visit to a small liberal arts college to whet their educational appetites. Michael, one of the Centro's counselors, was extremely invested in pushing our students to consider college. As a G.E.D. graduate himself and current student at Northeastern, he knew firsthand the strong push toward military service and vocational training, and he worked tirelessly with me to provide the students with alternatives.

Considering the military as an option, however, infected even the brightest of students. Nineteen-year-old Eddie Vélez, for example, a smart student who delighted in being a good student and a math wiz after years of being told he was not, arrived late to G.E.D. class one day, announcing that he met with an army recruiter who explained to him how the G.I. Bill would pay for his college education. Eddie was seriously considering this option and his enthusiasm spread to Frankie, a sixteen-year-old student who was also interested in joining the military. I was furious and gathered the class together and asked them to think about why the military aggressively recruited them and other poor people of color and not the mostly white affluent Northwestern University students I also taught on the city's northern suburbs. Initially stunned, my students responded passionately about the "honor of serving your country," telling me about their fathers, grandfathers, uncles, and brothers who had all served time in the armed forces. They were angry with me for being so upset and critical of what for many of them had been small steps toward social and economic mobility. "My father and *grandfather* were in the army," Eddie insisted. "It's a good job. And that's what I need. I need to get out of my house and *away* from the drugs and get into a good place. I can't concentrate when I'm at home, you know. They're smokin', drinkin', and I can't concentrate. Even if I go to college, I still have to go *home*. Why do you think I work so many hours and *like* to come to school? I need to get out of that environment." The class was silent, and a few students nodded their heads.

The idea of using the military as a haven or as a way to avoid troubled households and communities is not lost on those who run JROTC, as well as some of the programs' staunchest advocates. . . . The idea of the benefits JROTC programs provide for "inner-city" and "at-risk" youth is so entrenched that the program's Pentagon funding [was] expected to increase more than fifty percent of its current budget of $215 million in 2001 to $326 million by 2004.[9] Much of this money goes to local school districts eager to implement JROTC programs, since doing so guarantees the infusion of federal funds that many proponents argue can be a strategy for broadening school curriculum and securing more money for already financially strapped schools. . . .

GENDER, SEXUALITY, AND THE QUEST
FOR AUTONOMY

Young men, of course, are not alone in considering the military as a viable educational and employment option. Increasingly, young girls swell the JROTC ranks, a phenomenon that allegedly demonstrates the military's egalitarian gender roles. . . . In Chicago public schools, young women outnumber young men in Army JROTC programs by a slim margin; they also comprise nearly half of all JROTC program participants, with Latinas and African-American women constituting the overwhelming majority.

Perhaps the appeal of more egalitarian gender relations partially explains young women's interest in JROTC. Ethnographic evidence, however, suggests that young girls' decisions to enter JROTC programs are slightly more complicated. Latina adolescents, for example, emphasize how household economic imperatives, as well as parental pressure and encouragement from friends and kin, influence their ideas about JRTOC and military service. Participating in JROTC's extracurricular programs may also offer freedom and autonomy otherwise unavailable to some young girls who are expected to abide by culturally prescribed norms of behavior requiring them to be in the home. Some parents support their daughter's participation in JROTC because of the program's promise to cultivate in its participants the values of structure, discipline, and honor. As parents grow increasingly concerned with their adolescent daughter's sexuality, JROTC participation can be regarded as another way of preserving their daughter's chastity and honor, while simultaneously reinforcing their critical role in their household's economic survival.

Thus, while military involvement allegedly keeps young men out of gangs, for young women it allegedly functions as a way to mitigate unwed teen pregnancy, providing necessary discipline for "unruly," "unpredictable" women's sexuality. Chicana and Puerto Rican scholars have demonstrated how culturally prescribed notions of female chastity and virginity among Latinas frequently result in family policing of young unmarried women's bodies by observing them, being vigilant of their movements and activities, and designating the household as the appropriate place for young women (Souza 2002; Trujillo

1991; Zavella 1997). These concerns with female chastity and purity also characterize poor and working-class *puertorriqueñas* whose families (usually mothers) not only monitor young women's sexual behavior, but who also depend on them for the economic survival of their households.

As Caridad Souza (2002) has shown in her research among working-class Puerto Rican women in Queens, young women's reproductive work is critical to households whose survival depends on family and community solidarity through the exchange of goods, services, and the pooling of resources across multiple kin and non-kin households. In this context, young women are expected to be *en la casa* (inside the home or family) for important material and cultural reasons: Without their reproductive work, adult women—mothers, grandmothers, and aunts—are easily overwhelmed by (and perhaps unable to meet) the reproductive demands within their own households and kin networks. Being *una muchacha de la casa* also means abiding by the cultural norms of respectability, chastity, and family honor valued by the community. To be *de la calle* (outside the home or literally "on the streets") is to be transgressive, sexually promiscuous, and dangerous (Souza 2002: 35).

For young Latinas, therefore, participating in JROTC programs offers both a way for them to contribute to their household economies as well as the possibility of creating a space for them to exercise some autonomy, giving them "legitimate" and "respectable" reasons to be outside of the home in ways that conform to cultural expectations of young women. In this way, young Puerto Rican women creatively construct ways of fulfilling their culturally prescribed gendered responsibilities, although they do so by participating in a hyper-masculinized social context that regards female sexuality as a problem in need of discipline and control (Enloe 2000). Seventeen-year-old Jasmín Rodríguez's story provides a glimpse into these competing demands and dreams that shape young women's thinking of JROTC and their sense of obligation to their families.

Throughout middle and high school, Jasmín has been a model student, consistently making honor roll, being ranked at the top of her high school class, and actively participating in school-sponsored community projects, writing contests, and youth-leadership programs. Jasmín has also been the recipient of summer scholarship programs targeting young women of color to attend summer-long programs at prestigious universities like Princeton and Cornell. She is an avid reader with a keen sense of social justice and inequality, characteristics that led her to announce at the age of ten that she was going to go to Harvard Law School one day so she could help make the world a better place. As a junior at one of Chicago's public high schools, she scored well on her PSATs and began receiving letters from four-year colleges and universities, whetting her desire to go away to college, to be the first woman in her family to do so. Certainly, how she is going to pay for college weighs heavily on her mind, but with the assurance of her high school counselors and mentors, Jasmín was confident that she would be able to get enough money from scholarships and financial aid to pay for college without having to rely on her mother for help.

Jasmín's enthusiasm for college has spread to her younger sister, Myrna, who also talks excitedly about college. Together with their college degrees, they talk confidently that they will soon be financially secure enough to help support their mother, Aida Rodríguez, and the rest of their family [who] barely survive month to month on the pooled resources of Aida's salary as a high school clerk and their older sister's and brother's part-time jobs as a cashier and a security guard at O'Hare Airport. Both Jasmín and Myrna believe college is a sure way to economic security and to helping their mother achieve her dream of finally owning a modest home for their family one day. Jasmín's determination to go to college has been such a defining feature of her personality that I was amazed when, in the Spring 2002, she told me she had signed up for the military. When she made the announcement at her nephew's first birthday party, she glowed with excitement and even seemed to take pleasure in the shocked response her decision provoked. When I asked her why she had changed her mind and why she was now considering the military, she repeated several times, "I just had to do it. It was the right decision. I'm really going to do this *and* go to college. And JROTC will help pay." When I assured her that based on her grades and family's financial need she would be able to pay for college without relying on money from military service, she replied, "Yeah, I know, but this is just another way to be sure I can pay for college. And they also give you money now. And we really need it."

Aida was equally confused by her daughter's decision. Aida takes enormous pride in Jasmín's accomplishments, and the walls and shelves in their small second story apartment, covered with mostly Jasmín's diplomas, awards, and trophies recognizing her academic achievements, are a clear sign of Aida's pride in and admiration for her daughter. Jasmín's decision, therefore, was tremendously distressing to Aida, who could not explain why Jasmín wanted to join JROTC and serve in the military after graduation. Aida had encouraged her oldest daughter, Milly, to participate in JROTC in high school and Aida even hoped Milly would go into the military after graduation, assuring her that doing so would benefit her financially and educationally. Aida feared that Milly wouldn't have the same kinds of options Jasmín had and military service, she assured me on many occasions, was also a way to keep Milly "out of trouble" and to prevent her from making the same mistakes Aida made as a young girl, namely becoming a teenaged single mother. Like many Puerto Rican mothers, Aida worried that Milly and her other daughters would *meterse las patas* (to mess up, and in this case, to get pregnant) as teenagers and she continuously strategized ways of protecting her daughters, including contemplating sending Milly to Puerto Rico when she was fourteen years old. Jasmín, however, was different and carried the hope and responsibility of being the first woman in her family to go away and graduate from college.

Jasmín understood her mother's hopes for her. But she was also painfully aware of her family's precarious economic situation that became even more tenuous with the birth of Milly's son, who brought an incredible amount of joy to the family. When I interviewed Jasmín about her decision later that summer, she provided a narrative filled with concerns with money, the need for security, and a desire to protect and serve both her nation and her family. She explained how one morning she was approached by the JROTC sergeant at her high school to take the ASVAB. His praise of her high scores, his relationship with her older sister, and his promise of $2,000 if she participated in a summer boot camp that would prepare her for her service in the army reserves all highlight a critical feature of successful JROTC programs: Their strength derives from a trusting and respectful relationship between the school commanding officer and the cadets. In June 2004 I observed, for example, the ease with which young men and women cadets interacted with their commanding officer, and his sensitive understanding of each cadet's struggles and hopes for the future. Many of the cadets I interviewed had siblings or cousins who also participated in JROTC, and one young man, Johnny, explained how he convinced his younger sister to join. When I asked him why, Johnny jokingly offered that since he is older and has "rank" in his unit, his younger sister would have to "show me respect" and defer to him in ways that she refused to do when at home. While Johnny admitted this was only one reason for persuading his sister to participate, his comments speak to the role of kinship and social networks in expanding the program's ranks. Over time, JROTC commanders meet these family members and friends and are able to draw on those networks not only to support his cadets, but also to approach young people who otherwise may not consider JROTC or military service after high school. It was precisely these networks that facilitated Jasmín's conversations with one of the JROTC officers in the school. She explains, "He was pretty convincing. He said he knew my family since he recruited my sister . . . and I saw it as a precaution just in case I didn't get any scholarship or financial help for college. I could do a lot with this money. My family has never been rich, and that money seems like so much. . . . I had my senior fees coming up and I can't expect my mother to pay for it. My senior trip, my prom. And it feels really good to have the money."

When I asked her if there were other reasons for her decision, she admitted that the events of 9/11 were also important.

> I have family in New York . . . and I sure didn't feel protected and I felt like [by joining the military] I would be protecting those I love. And even those I don't know, I felt like I could be doing something to help. . . . The women in my family have had to be protected by bad men [all their lives] and I don't want to have to do that. I could protect myself and not have to depend on others, but [my friends and family] can depend on me. Protecting people and standing up for what you believe in and being a good example of what the U.S. can be. I think our armed forces embody that.

The desire to feel protected and to equip oneself with important skills and financial resources to ensure independence fills Jasmín's narrative of education, her future, and her family indebtedness. They are also themes emerging from many Chicago Puerto Rican women's life histories and explain, in part, Aida's ambivalent embrace of military service for her daughters. For both Aida and Jasmín, the military's promise of money is clearly a means to ultimately achieve a better education and financial security. . . .

Jasmín's decision, however, is also informed by her quest for autonomy and power that will allow the women in her family to depend on her rather than have to be "protected by bad men." Military service provides some possibilities not only for her, but specifically the women in her family whose options are often circumscribed because of limited education and low-wage employment. Young women's participation in JROTC, therefore, is informed by social and economic need, although public officials are often quick to seize upon the deeply gendered concern with sexuality to promote the benefits of military service for young women. Public officials, for example, deploy the language of security and protection to advance the idea that military service protects "at risk" youth from danger and instills critical values allegedly absent in poor families, namely discipline, honor, and the value of hard work. Some JROTC officers and elected officials share parental concern with young women's sexuality and suggest that JROTC might be a new way to battle teen pregnancy. In Jackson, Mississippi, for example, state senator Robert Johnson recently applauded the increasing number of women entering into JROTC programs saying, "Jackson has more unwed mothers than just about any city of its size in the nation. We're talking about second- and third-generation single parents. The people criticizing JROTC are not the people living in these communities, because if they were, they would know that the people making the biggest difference, doing the most grass-roots work, are people with military backgrounds.

Regardless of whether the senator's statement is true, what is striking about his analysis is the explicit way in which he and others connect women's military service with sexual discipline, and how poor and working families also see these benefits of military service as well as the possibility it provides for sustaining precarious household economies and leading the way out of poverty. There are clearly a number of success stories of how this has happened, and these narratives are part of a larger American dream ideology that continues to animate impoverished Latina/o families.

CONCLUSION

With high poverty rates, low levels of educational attainment, and high drop-out rates among Latinas/os, military service becomes an appealing option for many Latinas/os whose life chances and economic options are increasingly circumscribed. These realities are not lost on military recruiters and the Department of Defense and the Pentagon, who have increased spending for JROTC programs in urban schools and target Latinos for military enlistment with the goal of boosting "the Latino numbers in the military." . . . Like their male counterparts, young Latinas are particularly vulnerable to military appeals to their sense of family obligation as well as notions of pride, discipline, loyalty, honor, and citizenship as they consider military service as one of many economic strategies to make ends meet. This discussion, therefore, is an attempt to analyze Latinas/os' choice to participate in military programs within a broader cultural and political-economic context. What is needed, however, is more historically informed ethnographic research that engages with these complex issues of military service, poverty, race, gender, and notions of citizenship and patriotism in a way that honors the real lived experiences and struggles of our informants, but also challenges conventional notions of community based on militarized (and masculinist) notions of home and nation, to build, instead, communities of justice, equal opportunity, and solidarity.

NOTES

1. Latino Institute (1995).

2. John Betancur et al. (1993) point out that despite Cubans' and South Americans' economic success, their wages according to the 1980 census still approximated those of Puerto Ricans and Mexicans rather than that of whites. The 1990 census, however, paints a very different picture, emphasizing the growing gap between Puerto Ricans and other Latino groups in terms of average incomes, employment rates, and poverty levels.

3. See Chicago Urban League et al. (1994, 1995a, 1995b).

4. Chicago Public Schools, CPS at a Glance, February 2005. Available at www.cps.k.12.il.us/AtAGlance.html. The three full-time military academies, Chicago Military Academy, Carver Area High School, and Austin Community Academy, all opened since 1999 (www.cps.k12.il.us/AtAGlance.html).

5. Ana Beatriz Cholo, Military Marches into Middle Schools, *Chicago Tribune*, 26 July 2002; Chicago Public Schools, JROTC Program Book, nd; Education to Careers (ETC) FY 2003–2004, available at www.cps.k12.il.us/AboutCPS/Financial_Information/FY2004_Final/CPS_Unit/Education/Education_to_Careers.pdf.

6. Elsewhere (Pérez 2002), I have documented how Latina/o youth (especially those in rapidly gentrifying neighborhoods) are implicated in the policing of urban space aimed at curbing, for example, gang activity. Although Chicago's anti-loitering ordinance was declared unconstitutional in 1999, some scholars and activists have highlighted the "ongoing attempts to legalize harassment and street sweeps of youth," particularly youth of color who are regarded as dangerous and who allegedly "need to be locked up or removed from public space" (Lipman 2003: 95).

7. All names are pseudonyms.

8. Lincoln's Challenge is a federally funded youth program for at-risk youth between the ages of sixteen and eighteen whose quasi-military training in "discipline, esprit-de-corps, leadership, and teamwork is producing rapid and effective change in individual behaviors and attitude." Lincoln's Challenge is the largest National Guard Youth Challenge program in the nation (www.lincolnschallengeacademy.org/challenge/challenge.htm).

9. Racism and Conscription in the JROTC, *Peace Review*, September 2002; Class Warfare, *Time*, 4 March 2002, p. 50.

REFERENCES

American Friends Service Committee 1999. *Trading Books for Soldiers: The True Cost of JROTC*. Philadelphia, PA: AFSC.

Betancur, John J., Teresa Cordova, and Maria de los Angeles Torres 1993. Economic Restructuring and the Process of Incorporation of Latinos into the Chicago Economy. In *Latinos in a Changing U.S. Economy*. Rebecca Morales and Frank Bonilla, eds. New York: Sage. Pp. 109–132.

Chicago Urban League, Latino Institute, and Northern Illinois University 1994. The Changing Economic Standing of Minorities and Women in the Chicago Metropolitan Area, 1970–1990. Chicago, IL: Chicago Urban League, Final Report.

Chicago Urban League, Latino Institute, and Northern Illinois University 1995a. When the Job Doesn't Pay: Contingent Workers in the Chicago Metropolitan Area. In *The Working Poor Project*. Chicago, IL: Chicago Urban League.

Chicago Urban League, Latino Institute, and Northern Illinois University 1995b. Jobs That Pay: Are Enough Jobs Available in Metropolitan Chicago? In *The Working Poor Project*. Chicago, IL: Chicago, Urban League.

Enloe, Cynthia 2000. *Maneuvers: The International Politics of Militarizing Women's Lives*. Berkeley: University of California Press.

Latino Institute 1994. *A Profile of Nine Latino Groups in Chicago*. Chicago IL: Latino Institute.

Latino Institute 1995. *Facts on Chicago's Puerto Rican Population*. Chicago IL: Latino Institute.

Lipman, Pauline 2003. Cracking Down: Chicago School Policy and the Regulation of Black and Latino Youth. In *Education as Enforcement: The Militarization and Corporatization of Schools*. Kenneth J. Saltman and David A. Gabbard, eds. New York and London: Routledge Falmer. Pp. 81–101.

Pérez, Gina 2002. The other "Real World": Gentrification and the social construction of place in Chicago. *Urban Anthropology 3* (1): 37–67.

Souza, Caridad 2002. Sexual identities of young Puerto Rican mothers. *Dialogo 6* Winter/Spring: 33–39.

Trujillo, Carla 1991. Chicana Lesbians: Fear and Loathing in the Chicano Community. In *Chicana Critical Issues*. Norma Alarcon, Rafaela Castro, Emma Perez, Beatriz Pesquera, Adaljiza Sosa Riddel, and Patricia Zavella, eds. Berkeley: Third World Woman Press. Pp. 117–126.

Zavella, Patricia 1997. "Playing with Fire": The Gendered Constructions of Chicana/Mexicana Sexuality. In *The Gender/Sexuality Reader: Culture, History, Political Economy*. Micaela di Leonardo and Roger Lancaster, eds. New York: Routledge. Pp. 392–408.

52

Life at the Top in America Isn't Just Better, It's Longer

JANNY SCOTT

Jean G. Miele's heart attack happened on a sidewalk in Midtown Manhattan in May 2004. He was walking back to work along Third Avenue with two colleagues after a several-hundred-dollar sushi lunch. There was the distant rumble of heartburn, the ominous tingle of perspiration. Then Miele, an architect, collapsed onto a concrete planter in a cold sweat.

Will L. Wilson's heart attack came four days earlier in the bedroom of his brownstone in Bedford-Stuyvesant in Brooklyn. He had been regaling his fiancée with the details of an all-you-can-eat dinner he was beginning to regret. Wilson, a Consolidated Edison office worker, was feeling a little bloated. He flopped onto the bed. Then came a searing sensation, like a hot iron deep inside his chest.

Ewa Rynczak Gora's first signs of trouble came in her rented room in the noisy shadow of the Brooklyn-Queens Expressway. It was the Fourth of July. Gora, a Polish-born housekeeper, was playing bridge. Suddenly she was sweating, stifling an urge to vomit. She told her husband not to call an ambulance; it would cost too much. Instead, she tried a home remedy: salt water, a double dose of hypertension pills, and a glass of vodka.

Architect, utility worker, maid: heart attack is the great leveler, and in those first fearful moments, three New Yorkers with little in common faced a single common threat. But in the months that followed, their experiences diverged. Social class—that elusive combination of income, education, occupation, and wealth—played a powerful role in Miele's, Wilson's, and Gora's struggles to recover.

Class informed everything from the circumstances of their heart attacks to the emergency care each received, the households they returned to, and the jobs they hoped to resume. It shaped their understanding of their illness, the support they got from their families, their relationships with their doctors. It helped define their ability to change their lives and shaped their odds of getting better.

Class is a potent force in health and longevity in the United States. The more education and income people have, the less likely they are to have and die of heart disease, strokes, diabetes, and many types of cancer. Upper-middle-class Americans live longer and in better health than middle-class Americans,

SOURCE: The *New York Times*, May 16, 2005. Copyright © 2005 by the New York Times. All rights reserved. Reprinted by permission.

who live longer and better than those at the bottom. And the gaps are widening, say people who have researched social factors in health.

As advances in medicine and disease prevention have increased life expectancy in the United States, the benefits have disproportionately gone to people with education, money, good jobs, and connections. They are almost invariably in the best position to learn new information early, modify their behavior, take advantage of the latest treatments, and have the cost covered by insurance.

Many risk factors for chronic diseases are now more common among the less educated than the better educated. Smoking has dropped sharply among the better educated, but not among the less. Physical inactivity is more than twice as common among high school dropouts as among college graduates. Lower-income women are more likely than other women to be overweight, though the pattern among men may be the opposite.

There may also be subtler differences. Some researchers now believe that the stress involved in so-called high-demand, low-control jobs further down the occupational scale is more harmful than the stress of professional jobs that come with greater autonomy and control. Others are studying the health impact of job insecurity, lack of support on the job, and employment that makes it difficult to balance work and family obligations.

Then there is the issue of social networks and support, the differences in the knowledge, time, and attention that a person's family and friends are in a position to offer. What is the effect of social isolation? Neighborhood differences have also been studied: How stressful is a neighborhood? Are there safe places to exercise? What are the health effects of discrimination?

Heart attack is a window on the effects of class on health. The risk factors—smoking, poor diet, inactivity, obesity, hypertension, high cholesterol, and stress—are all more common among the less educated and less affluent, the same group that research has shown is less likely to receive cardiopulmonary resuscitation, to get emergency room care, or to adhere to lifestyle changes after heart attacks.

"In the last twenty years, there have been enormous advances in rescuing patients with heart attack and in knowledge about how to prevent heart attack," said Ichiro Kawachi, a professor of social epidemiology at the Harvard School of Public Health. "It's like diffusion of innovation: whenever innovation comes along, the well-to-do are much quicker at adopting it. On the lower end, various disadvantages have piled onto the poor. Diet has gotten worse. There's a lot more work stress. People have less time, if they're poor, to devote to health maintenance behaviors when they are juggling two jobs. Mortality rates even among the poor are coming down, but the rate is not anywhere near as fast as for the well-to-do. So the gap has increased."

Bruce G. Link, a professor of epidemiology and socio-medical sciences at Columbia University, said of the double-edged consequences of progress: "We're creating disparities. It's almost as if it's transforming health, which used to be like fate, into a commodity. Like the distribution of BMWs or goat cheese."

THE BEST OF CARE

Jean Miele's advantage began with the people he was with on May 6, when the lining of his right coronary artery ruptured, cutting off the flow of blood to his sixty-six-year-old heart. His two colleagues were knowledgeable enough to dismiss his request for a taxi and call an ambulance instead.

And because he was in Midtown Manhattan, there were major medical centers nearby, all licensed to do the latest in emergency cardiac care. The emergency medical technician in the ambulance offered Miele a choice. He picked Tisch Hospital, part of New York University Medical Center, an academic center with relatively affluent patients, and passed up Bellevue, a city-run hospital with one of the busiest emergency rooms in New York.

Within minutes, Miele was on a table in the cardiac catheterization laboratory, awaiting angioplasty to unclog his artery—a procedure that many cardiologists say has become the gold standard in heart attack treatment. When he developed ventricular fibrillation, a heart rhythm abnormality that can be fatal within minutes, the problem was quickly fixed.

Then Dr. James N. Slater, a fifty-four-year-old cardiologist with some twenty-five thousand cardiac catheterizations under his belt, threaded a catheter through a small incision in the top of Miele's right thigh and steered it toward his heart. Miele lay on the table, thinking about dying. By 3:52 p.m., less than two hours after Miele's first symptoms, his artery was reopened and Slater implanted a stent to keep it that way.

Time is muscle, as cardiologists say. The damage to Miele's heart was minimal.

Miele spent just two days in the hospital. His brother-in-law, a surgeon, suggested a few specialists. His brother, Joel, chairman of the board of another hospital, asked his hospital's president to call New York University. "Professional courtesy," Joel Miele explained later. "The bottom line is that someone from management would have called patient care and said, 'Look, would you make sure everything's okay?' "

Things went less flawlessly for Will Wilson, a fifty-three-year-old transportation coordinator for Con Ed. He imagined fleetingly that he was having a bad case of indigestion, though he had had a heart attack before. His fiancée insisted on calling an ambulance. Again, the emergency medical technician offered a choice of two nearby hospitals—neither of which had state permission to do angioplasty, the procedure Jean Miele received.

Wilson chose the Brooklyn Hospital Center over Woodhull Medical and Mental Health Center, the city-run hospital that serves three of Brooklyn's poorest neighborhoods. At Brooklyn Hospital, he was given a drug to break up the clot blocking an artery to his heart. It worked at first, said Narinder P. Bhalla, the hospital's chief of cardiology, but the clot re-formed.

So Bhalla had Wilson taken to the Weill Cornell Center of New York–Presbyterian Hospital in Manhattan the next morning. There, Bhalla performed angioplasty and implanted a stent. Asked later whether Wilson would have been better off if he had had his heart attack elsewhere, Bhalla said the most important issue in heart attack treatment was getting the patient to a hospital quickly.

But he added, "In his case, yes, he would have been better off had he been to a hospital that was doing angioplasty."

Wilson spent five days in the hospital before heading home on many of the same high-priced drugs that Miele would be taking and under similar instructions to change his diet and exercise regularly. After his first heart attack, in 2000, he quit smoking; but once he was feeling better, he stopped taking several medications, drifted back to red meat and fried foods, and let his exercise program slip.

This time would be different, he vowed: "I don't think I'll survive another one."

Ewa Gora's experience was the rockiest. First, she hesitated before allowing her husband to call an ambulance; she hoped her symptoms would go away. He finally insisted; but when the ambulance arrived, she resisted leaving. The emergency medical technician had to talk her into going. She was given no choice of hospitals; she was simply taken to Woodhull, the city hospital Will Wilson had rejected.

Woodhull was busy when Gora arrived around 10:30 p.m. A triage nurse found her condition stable and classified her as "high priority." Two hours later, a physician assistant and an attending doctor examined her again and found her complaining of chest pain, shortness of breath, and heart palpitations. Over the next few hours, tests confirmed she was having a heart attack.

She was given drugs to stop her blood from clotting and to control her blood pressure, treatment that Woodhull officials say is standard for the type of heart attack she was having. The heart attack passed. The next day, Gora was transferred to Bellevue, the hospital Jean Miele had turned down, for an angiogram to assess her risk of a second heart attack.

But Gora, who was fifty-nine at the time, came down with a fever at Bellevue, so the angiogram had to be canceled. She remained at Bellevue for two weeks, being treated for an infection. Finally, she was sent home. No angiogram was ever done.

COMFORTS AND RISKS

Jean Miele is a member of New York City's upper middle class. The son of an architect and an artist, he worked his way through college, driving an ice-cream truck and upholstering theater seats. He spent two years in the military and then joined his father's firm, where he built a practice as not only an architect but also an arbitrator and an expert witness, developing real estate on the side.

Miele is the kind of person who makes things happen. He bought a $21,000 house in the Park Slope section of Brooklyn, sold it about fifteen years later for $285,000, and used the money to build his current house next door, worth over

$2 million. In Brookhaven, on Long Island, he took a derelict house on a single acre, annexed several adjoining lots, and created what is now a four-acre, three-house compound with an undulating lawn and a fifteen-thousand-square-foot greenhouse he uses as a workshop for his collection of vintage Jaguars.

Miele's architecture partners occasionally joked that he was not in the business for the money, which to some extent was true. He had figured out how to live like a millionaire, he liked to say, even before he became one. He had worked four-day weeks for twenty years, spending long weekends with his family, sailing or iceboating on Bellport Bay and rebuilding cars.

Miele had never thought of himself as a candidate for a heart attack—even though both his parents had died of heart disease; even though his brother had had arteries unclogged; and even though he himself was on hypertension medication, his cholesterol levels bordered on high, and his doctor had been suggesting he lose weight.

He was a passionate chef who put great store in the healthfulness of fresh ingredients from the Mieles' vegetable garden or the greengrocers in Park Slope. His breakfasts may have been a cardiologist's nightmare—eggs, sausage, bacon, pastina with a poached egg—but he considered his marinara sauce to be healthy perfection: just garlic, oil, tomatoes, salt, and pepper.

He figured he had something else working in his favor: he was happy. He adored his second wife, Lori, twenty-three years younger, and their six-year-old daughter, Emma. He lived within blocks of his two sisters and two of his three grown children from his first marriage. The house regularly overflowed with guests, including Miele's former wife and her husband. He seemed to know half the people of Park Slope.

"I walk down the street and I feel good about it every day," Miele, a gregarious figure with twinkling blue eyes and a taste for worn T-shirts and jeans, said of his neighborhood. "And yes, that gives me a feeling of well-being."

His approach to his health was utilitarian. When body parts broke, he got them fixed so he could keep doing what he liked to do. So he had had disk surgery, rotator cuff surgery, surgery for a carpal tunnel problem. But he was also not above an occasional bit of neglect. In March 2004, his doctor suggested a stress test after Miele complained of shortness of breath. On May 6, the prescription was still hanging on the kitchen cabinet door.

An important link in the safety net that caught Miele was his wife, a former executive at a sweater manufacturing company who had stopped work to raise Emma but managed the Mieles' real estate as well. While Miele was still in the hospital, she was on the Internet, Googling stents.

She scheduled his medical appointments. She got his prescriptions filled. Leaving him at home one afternoon, she taped his cardiologist's business card to the couch where he was sitting. "Call Dr. Hayes and let him know you're coughing," she said, her fingertips on his shoulder. Thirty minutes later, she called home to check.

She prodded Miele, gently, to cut his weekly egg consumption to two, from seven. She found fresh whole wheat pasta and cooked it with turkey sausage and broccoli rabe. She knew her way around nutrition labels.

Lori Miele took on the burden of dealing with the hospital and insurance companies. She accompanied her husband to his doctor's appointments and retained pharmaceutical dosages in her head.

"I can just leave and she can give you all the answers to all the questions," Miele said to his cardiologist, Dr. Richard M. Hayes, one day.

"Okay, why don't you just leave?" Hayes said back. "Can she also examine you?"

With his wife's support, Miele set out to lose thirty pounds. His pasta consumption plunged to a plate a week from two a day. It was not hard to eat healthfully from the Mieles' kitchens. Even the "junk drawer" in Park Slope was stocked with things like banana chips and sugared almonds. Lunches in Brookhaven went straight from garden to table: tomatoes with basil, eggplant, corn, zucchini flower tempura.

At his doctor's suggestion, Miele enrolled in a three-month monitored exercise program for heart disease patients, called cardiac rehab, which has been shown to reduce the mortality rate among heart patients by 20 percent. Miele's insurance covered the cost. He even managed to minimize the inconvenience, finding a class ten minutes from his country house.

He had the luxury of not having to rush back to work. By early June, he had decided he would take the summer off, and maybe cut back his workweek when he returned to the firm.

"You know, the more I think about it, the less I like the idea of going back to work," he said. "I don't see any real advantage. I mean, there's money. But you've got to take the money out of the equation."

So he put a new top on his 1964 Corvair. He played host to a large family reunion, replaced the heat exchanger in his boat, and transformed the ramshackle greenhouse into an elaborate workshop. His weight dropped to 189 pounds, from 211. He had doubled the intensity of his workouts. His blood pressure was lower than ever.

Miele saw Hayes only twice in six months, for routine follow-ups. He had been known to walk out of doctors' offices if he was not seen within twenty minutes, but Hayes did not keep him waiting. The Mieles were swept into the examining room at the appointed hour. Buoyed by the evidence of Miele's recovery, they would head out to lunch in downtown Manhattan. Those afternoons had the feel of impromptu dates.

"My wife tells me that I'm doing fourteen-hour days," Miele mused one afternoon, slicing cold chicken and piling it with fresh tomatoes on toast. "She said, You're doing better now than you did ten years ago.' And I said, I haven't had sex in a week.' And she said, Well?'"

Just one unpleasant thing happened. Miele's partners informed him in late July that they wanted him to retire. It caught him off guard, and it hurt. He countered by taking the position that he was officially disabled and therefore entitled to be paid for a full year after he began his medical leave. "I mean, the guy has a heart attack," he said later. "So you get him while he's down?"

LUKEWARM EFFORTS TO REFORM

Will Wilson fits squarely in the city's middle class. His parents had been share-croppers who moved north and became a machinist and a nurse. He grew up in Bedford-Stuyvesant and had spent thirty-four years at Con Ed. He had an income of $73,000, five weeks' vacation, health benefits, a house worth $450,000, and plans to retire to North Carolina when he is fifty-five.

Wilson, too, had imagined becoming an architect. But there had been no money for college, so he found a job as a utility worker. By age twenty-two, he had two children. He considered going back to school, with the company's support, to study engineering. But doing shift work, and with small children, he never found the time.

For years he was a high-voltage cable splicer, a job he loved because it meant working outdoors with plenty of freedom and overtime pay. But on a snowy night in the early 1980s, a car skidded into a stanchion, which hit him in the back. A doctor suggested that Wilson learn to live with the pain instead of having disk surgery, as Jean Miele had done.

So Wilson became a laboratory technician, then a transportation coordinator, working in a cubicle in a low-slung building in Astoria, Queens, overseeing fuel deliveries for the company's fleet. Some people might think of the work as tedious, Wilson said, "but it keeps you busy."

"Sometimes you look back over your past life experiences and you realize that if you would have done something different, you would have been some-place else," he said. "I don't dwell on it too much because I'm not in a negative position. But you do say, Well, dag, man, I should have done this or that.' "

Wilson's health was not bad, but far from perfect. He had quit drinking and smoking, but had high cholesterol, hypertension, and diabetes. He was slim, five foot nine, and just under 170 pounds. He traced his first heart attack to his smoking, his diet, and the stress from a grueling divorce.

His earlier efforts to reform his eating habits were half-hearted. Once he felt better, he stopped taking his cholesterol and hypertension drugs. When his cardiologist moved and referred Wilson to another doctor, he was annoyed by what he considered the rudeness of the office staff. Instead of demanding courtesy or finding another specialist, Wilson stopped going.

By the time Dr. Bhalla encountered Wilson at Brooklyn Hospital, there was damage to all three main areas of his heart. Bhalla prescribed a half-dozen drugs to lower Wilson's cholesterol, prevent clotting, and control his blood pressure.

"He has to behave himself," Bhalla said. "He needs to be more compliant with his medications. He has to really go on a diet, which is grains, no red meat, no fat. No fat at all."

Wilson had grown up eating his mother's fried chicken, pork chops, and macaroni and cheese. He confronted those same foods at holiday parties and big events. There were doughnut shops and fried chicken places in his neighborhood; but Wilson's fiancée, Melvina Murrell Green, found it hard to find fresh produce and good fish.

"People in my circle, they don't look at food as, you know, too much fat in it," Wilson said. "I don't think it's going to change. It's custom."

At Red Lobster after his second heart attack, Green would order chicken and Wilson would have salmon—plus a side order of fried shrimp. "He's still having a problem with the fried seafood," Green reported sympathetically.

Whole grains remained mysterious. "That we've got to work on," she said. "Well, we recently bought a bag of grain something. I'm not used to that. We try to put it on the cereal. It's okay."

In August 2004, Green's blood pressure shot up. The culprit turned out to be a turkey chili recipe that she and Wilson had discovered: every ingredient except the turkey came from a can. She was shocked when her doctor pointed out the salt content. The Con Ed cafeteria, too, was problematic. So Wilson began driving to the Best Yet Market in Astoria at lunch to troll the salad bar.

Dr. Bhalla had suggested that Wilson walk for exercise. There was little open space in the neighborhood, so Wilson and Green often drove just to go for a stroll. In the fall of 2004 he entered a cardiac rehab program like Miele's, only less convenient. He would drive into Manhattan after work, during the afternoon rush, three days a week. He would hunt for on-street parking or pay too much for a space in a lot. Then a stranger threatened to damage Wilson's car in a confrontation over a free spot, so Wilson switched to the subway.

For a time, he considered applying for permanent disability. But Con Ed allowed him to return to work "on restrictions," so he decided to go back, with plans to retire in a year and a half. The week before he went back, he and Green took a seven-day cruise to Nassau. It was a revelation.

"Sort of like helped me to see there's a lot more things to do in life," he said. "I think a lot of people deny themselves certain things in life, in terms of putting things off, I'll do it later.' Later may never come."

IGNORING THE RISKS

Ewa Gora is a member of the working class. A bus driver's daughter, she arrived in New York City from Krakow in the early 1990s, leaving behind a grown son. She worked as a housekeeper in a residence for the elderly in Manhattan, making beds and cleaning toilets. She said her income was $21,000 to $23,000 a year, with health insurance through her union.

For $365 a month, she rented a room in a friend's Brooklyn apartment on a street lined with aluminum-sided row houses and American flags. She used the friend's bathroom and kitchen. She was in her seventh year on a waiting list for a subsidized one-bedroom apartment in the adjacent Williamsburg neighborhood. In the meantime, she had acquired a roommate: Edward Gora, an asbestos-removal worker newly arrived from Poland and ten years her junior, whom she met and married in 2003.

Like Jean Miele, Ewa Gora had never imagined she was at risk of a heart attack, though she was overweight, hypertensive, and a thirty-year smoker, and heart attacks had killed her father and sister. She had numerous health problems, which she addressed selectively, getting treated for back pain, ulcers, and so on, until the treatment became too expensive or inconvenient, or her insurance declined to pay.

"My doctor said, Ewa, be careful with cholesterol,' " recalled Gora, whose vestigial Old World sense of propriety had her dressed in heels and makeup for every visit to Bellevue. "When she said that, I think nothing; I don't care. Because I don't believe this touch me. Or I think she have to say like that because she doctor. Like cigarettes: she doctor, she always told me to stop. And when I got out of the office, lights up."

Gora had a weakness for the peak of the food pyramid. She grew up on her mother's fried pork chops, spare ribs, and meatballs—all cooked with lard—and had become a pizza, hamburger, and french fry enthusiast in the United States. Fast food was not only tasty but also affordable. "I eat terrible," she reported cheerily from her bed at Bellevue. "I like grease food and fast food. And cigarettes."

She loved the feeling of a cigarette between her fingers, the rhythmic rise and fall of it to her lips. Using her home computer, she had figured out how to buy Marlboros online for just $2.49 a pack. Her husband smoked, her friends all smoked. Everyone she knew seemed to love tobacco and steak.

Her life was physically demanding. She would rise at 6:00 a.m. to catch a bus to the subway, change trains three times, and arrive at work by 8:00 a.m. She would make twenty-five to thirty beds, vacuum, cart out trash. Yet she says she loved her life. "I think America is El Dorado," she said. "Because in Poland now is terrible; very little bit money. Here, I don't have a lot of, but I live normal. I have enough, not for rich life but for normal life."

The precise nature of Gora's illness was far from clear to her even after two weeks in Bellevue. In her first weeks home, she remained unconvinced that she had had a heart attack. She arrived at the Bellevue cardiology clinic for her first follow-up appointment imagining that whatever procedure had earlier been cancelled would then be done, that it would unblock whatever was blocked, and that she would be allowed to return to work.

Jad Swingle, a doctor completing his specialty training in cardiology, led Gora through the crowded waiting room and into an examining room. She clutched a slip of paper with words she had translated from Polish using her pocket dictionary:

"dizzy," "groin," "perspiration." Swingle asked her questions, speaking slowly. Do you ever get chest discomfort? Do you get short of breath when you walk?

She finally interrupted: "Doctor, I don't know what I have, why I was in hospital. What is this heart attack? I don't know why I have this. What I have to do to not repeat this?"

No one had explained these things, Gora believed. Or, she wondered, had she not understood? She perched on the examining table, ankles crossed, reduced by the setting to an oversize, obedient child. Swingle examined her, then said he would answer her questions "in a way you'll understand." He set about explaining heart attacks: the narrowed artery, the blockage, the partial muscle death.

Gora looked startled.
"My muscle is dead?" she asked.
Swingle nodded.
What about the procedure that was never done?
"I'm not sure an angiogram would help you," he said.
She needed to stop smoking, take her medications, walk for exercise, come back in a month.
"My muscle is still dead?" she asked again, incredulous.
"Once it's dead, it's dead," Swingle said. "There's no bringing it back to life."

Outside, Gora tottered toward the subway, fourteen blocks away, on pink high-heeled sandals in 89-degree heat. "My thinking is black," she said, uncharacteristically glum. "Now I worry. You know, you have hand? Now I have no finger."

If Jean Miele's encounters with the health care profession in the first months after his heart attack were occasional and efficient, Ewa Gora's were the opposite. Whereas he saw his cardiologist just twice, Gora, burdened by complications, saw hers a half-dozen times. Meanwhile, her heart attack seemed to have shaken loose a host of other problems.

A growth on her adrenal gland had turned up on a Bellevue CAT scan, prompting a visit to an endocrinologist. An old knee problem flared up; an orthopedist recommended surgery. An alarming purple rash on her leg led to a trip to a dermatologist. Because of the heart attack, she had been taken off hormone replacement therapy and was constantly sweating. She tore open a toe stepping into a pothole and needed stitches.

Without money or connections, moderate tasks consumed entire days. One cardiology appointment coincided with a downpour that paralyzed the city. Gora was supposed to be at the hospital laboratory at 8:00 a.m. to have blood drawn and back at the clinic at 1:00 p.m. In between, she wanted to meet with her boss about her disability payments. She had a 4:00 p.m. appointment in Brooklyn for her knee.

So at 7:00 a.m., she hobbled through the rain to the bus to the subway to another bus to Bellevue. She was waiting outside the laboratory when it opened.

Then she took a bus uptown in jammed traffic, changed buses, descended into the subway at Grand Central Terminal, rode to Times Square, found service suspended because of flooding, climbed the stairs to Forty-second Street, manuevered through angry crowds hunting for buses, and found another subway line.

She reached her workplace an hour and a half after leaving Bellevue; if she had had the money she could have made the trip in twenty minutes by cab. Her boss was not there. So she returned to Bellevue and waited until 2:35 p.m. for her one o'clock appointment. As always, she asked Dr. Swingle to let her return to work. When he insisted she have a stress test first, a receptionist gave her the first available appointment—seven weeks away.

Meanwhile, Gora was trying to stop smoking. She had quit in the hospital, then returned home to a husband and a neighbor who both smoked. To be helpful, her husband smoked in the shared kitchen next door. He was gone most of the day, working double shifts. Alone and bored, Gora started smoking again, then called Bellevue's free smoking cessation program and enrolled.

For the next few months, she trekked regularly to "the smoking department" at Bellevue. A counselor supplied her with nicotine patches and advice, not always easy for her to follow: stay out of the house; stay busy; avoid stress; satisfy oral cravings with, say, candy. The counselor suggested a support group, but Gora was too ashamed of her English to join. Even so, over time her tobacco craving waned.

There was just one hitch: Gora was gaining weight. To avoid smoking, she was eating. Her work had been her exercise and now she could not work. Dr. Swingle suggested cardiac rehab, leaving it up to Gora to find a program and arrange it. Gora let it slide. As for her diet, she had vowed to stick to chicken, turkey, lettuce, tomatoes, and low-fat cottage cheese. But she got tired of that. She began sneaking cookies when no one was looking—and no one was.

She cooked separate meals for her husband, who was not inclined to change his eating habits. She made him meatballs with sauce, liver, soup from spare ribs. Then one day she helped herself to one of his fried pork chops, and was soon eating the same meals he was. As an alternative to eating cake while watching television, she turned to pistachios, and then ate a pound in a single sitting.

Cruising the 99 Cent Wonder store in Williamsburg, where the freezers were filled with products like Budget Gourmet Rigatoni with Cream Sauce, she pulled down a small package of pistachios: two and a half servings, thirteen grams of fat per serving. "I can eat five of these," she confessed, ignoring the nutrition label. Not servings. Bags.

Heading home after a trying afternoon in the office of the apartment complex in Williamsburg, where the long-awaited apartment seemed perpetually just out of reach, Gora slipped into a bakery and emerged with a doughnut, her first since her heart attack. She found a park bench where she had once been accustomed to reading and smoking. Working her way through the doughnut, confectioners' sugar snowing onto her chest, she said ruefully, "I miss my cigarette."

She wanted to return to work. She felt uncomfortable depending on her husband for money. She worried that she was becoming indolent and losing her English. Her disability payments, for which she needed a doctor's letter every month, came to just half of her $331 weekly salary. Once, she spent hours searching for the right person at Bellevue to give her a letter, only to be told to come back in two days.

The copayments on her prescriptions came to about eighty dollars each month. Unnerving computer printouts from the pharmacist began arriving: "Maximum benefit reached." She switched to her husband's health insurance plan. Twice, Bellevue sent bills for impossibly large amounts of money for services her insurance was supposed to cover. Both times she spent hours traveling into Manhattan to the hospital's business office to ask why she had been billed. Both times a clerk listened, made a phone call, said the bill was a mistake, and told her to ignore it.

When the stress test was finally done, Dr. Swingle said the results showed she was not well enough to return to full-time work. He gave her permission for part-time work, but her boss said it was out of the question. By November, four months after her heart attack, her weight had climbed to 197 pounds from 185 in July. Her cholesterol levels were stubbornly high and her blood pressure was up, despite drugs for both.

In desperation, Gora embarked upon a curious, heart-unhealthy diet clipped from a Polish-language newspaper. Day 1: two hardboiled eggs, one steak, one tomato, spinach, lettuce with lemon and olive oil. Another day: coffee, grated carrots, cottage cheese, and three containers of yogurt. Yet another: just steak. She decided not to tell her doctor. "I worry if he don't let me, I not lose the weight," she said.

UNEVEN RECOVERIES

Nearly a year after his heart attack, Jean Miele was, remarkably, better off. He had lost thirty-four pounds and was exercising five times a week and taking subway stairs two at a time. He had retired from his firm on the terms he wanted. He was working from home, billing $225 an hour. More money in less time, he said. His blood pressure and cholesterol were low. "You're doing great," Dr. Hayes had said. "You're doing better than ninety-nine percent of my patients."

Will Wilson's heart attack had been a setback. His heart function remained impaired, though improved somewhat. At one checkup in the spring of 2005, his blood pressure and his weight had been a little high. He still enjoyed fried shrimp on occasion, but he took his medications diligently. He graduated from cardiac rehab with plans to join a health club with a pool. And he was looking forward to retirement.

Ewa Gora's life and health were increasingly complex. With Dr. Swingle's reluctant approval, she returned to work in November 2004. She had moved into the subsidized apartment in Williamsburg, which gave her her own kitchen and bathroom for the first time in seven years. But she began receiving menacing phone calls from a collection agency about an old bill her health insurance had not covered. Her husband, with double pneumonia, was out of work for weeks.

She had her long-awaited knee surgery in January 2005. But it left her temporarily unable to walk. Her weight hit two hundred pounds. When the diet failed, she considered another consisting largely of fruit and vegetables sprinkled with an herbal powder. Her blood pressure and cholesterol remained ominously high. She had been warned that she was now a borderline diabetic.

"You're becoming a full-time patient, aren't you?" Swingle remarked.

53

Intersections of Race, Class, and Gender in Public Health Interventions

AMY J. SCHULZ, NICHOLAS FREUDENBERG, AND JESSIE DANIELS

In recent years, scholars, practitioners, and activists have produced voluminous new evidence documenting the contributions of racial, socioeconomic, and gender inequalities to the disparities in health observed in the United States and elsewhere. While recording these differences is an important first step, progress toward the international goal of "health for all" requires moving beyond documentation to action. If we are to achieve greater social and health equity, we must better understand the social, cultural, and political processes that produce disparities in health. Only then will we be able to act effectively to reverse or mitigate those processes.

. . . We present two case studies in which we examine gender, race, and class in the social production of unequal health. We use these case studies to consider the potential for scholars, activists, community members, and practitioners to intervene to promote greater equity in health. The first case study examines the social context within which young men of color in New York

SOURCE: Amy J. Schulz and Leith Mullergs, eds. 2006. *Gender, Race, Class & Health.* p. 371–393. San Francisco, CA: Jossey-Bass Inc. Reprinted by permission of John Wiley & Sons, Inc.

City experience heightened risk of HIV, substance abuse, and incarceration. It describes the development of a targeted intervention (REAL MEN) designed to mitigate those processes. The second case describes the Mothers of East Los Angeles, a group of women and families who for the past twenty years have actively mobilized to resist a variety of environmental threats to health in their neighborhoods. These case studies highlight the dynamic interplay of gender, race, and class within the two local contexts—New York City and East Los Angeles—as they shape the conditions of people's lives and contribute to the stark inequalities in health that now characterize the United States.

INTERSECTIONAL ANALYSIS AND HEALTH

Our analysis emerges from several distinct bodies of literature. First, public health researchers have documented persistent relationships between social inequalities and health: people who have fewer economic resources, and those who command less social or political power, have poorer health. These relationships hold across multiple health outcomes and persist over time, even as the major causes of death have undergone a transition from infectious to chronic diseases. . . .

Second, beginning in the 1980s and inspired by the social movements of the 1960s and 1970s, African American and Latina feminist activists and scholars developed a critique of social analyses that focused on any single system of oppression—that is, race, or class, or gender. They argued that such an approach limits our understanding of the ways that these categories intersect to create and maintain inequities. . . . Within health, this approach posits that one cannot understand or reduce racial or gender disparities in health without examining the ways that institutionalized racism intersects with gender and class to shape educational and economic opportunities, risk of imprisonment and associated health risks, and opportunities to promote health and well-being. . . .

The third source for our analysis here is the literature on the evaluation of public health interventions. Health practitioners and officials judge their efforts by their success in reducing illness and death. Most interventions address a single health condition, work on one level of social organization, and focus on short-term improvements. Another strand of intervention, however, links health and social justice and depends on the force of social movements and political mobilizations to make broader changes in living conditions (Brown and others, 2004; Johnston, Larana, and Gusfield, 1994). In the United States in the past five decades, the labor, civil rights, women's, environmental, and gay and lesbian movements have all contributed to improved health and reductions in health disparities based on class, gender, or race.

In the two case studies, we use insights from these three bodies of literature as a lens to examine the potential of intersectional analyses to inform action to promote health and reduce disparities. Although both cases unfold in large urban areas —New York City and East Los Angeles—they differ on multiple other dimensions, including the gender and ethnicity of participants, the health issues of concern, and the roles of residents, professionals, and related social movements. . . .

RETURNING EDUCATED AFRICAN AMERICAN, LATINO, AND LOW-INCOME MEN TO ENRICH NEIGHBORHOODS

Health professionals in consultation with community members developed Returning Educated African American, Latino, and Low-Income Men to Enrich Neighborhoods (REAL MEN) as a short-term health intervention for young men leaving jail and returning to their communities. . . .

REAL MEN seeks to increase young men's chances of economic and social stability, and thus better health, by linking them to employment and educational opportunities on their release from jail. The program also seeks to engage participants in a critical examination of how dominant social constructions of masculinity and race influence not only the contexts that they encounter but also their own actions and health risks. In providing opportunities for young men to analyze these social processes, REAL MEN hopes to help them identify within their life circumstances and communities tools and opportunities that may enhance their chances of staying out of jail and protect their own health and the well-being of people they care about.

Structuring Pathways to Health and Disease in New York City

In the past three decades, criminal justice, poverty, and educational policies have dramatically diminished the life opportunities available to low-income urban young people of color. Disinvestment in public education, large class sizes, and poorly trained teachers contribute to high dropout rates. . . . Over the past decade, New York State has transferred almost $1 billion from public higher education to building, staffing, and filling prisons. . . . As a result of the loss of manufacturing jobs, poor educational preparation, and other factors, half a million men of African descent in New York City lack jobs (Levitan, 2003), and these limited employment opportunities lead many young men into the illegal economy, selling drugs or engaging in other criminal activity.

Easy availability of drugs and alcohol and limited availability of effective prevention and treatment programs in low-income communities contribute to high rates of substance abuse. . . .

These life circumstances influence the health of young men of color both directly and indirectly. Young men who have been in jail or prison have higher rates of HIV and other sexually transmitted infections (STIs) than comparable nonjailed populations (Malow and others, 1997), primarily due to unprotected heterosexual contact before and after incarceration. HIV infection and drug use increase significantly between adolescence and early adulthood, among both incarcerated and general populations (Staton and others, 1999; Stiffman, Dore, Cunningham, and Earls, 1995). Most young men leave jail, prison, or juvenile detention without exposure to effective HIV prevention programs or linkages to community services (Freudenberg, 2001).[1] A majority of those leaving jail are rearrested within a year of release, creating a revolving door that serves to increase HIV infection and drug use by further disrupting the lives of these young men, their partners, and their communities (LeBlanc, 2003). . . .

Gender, Race, and Class and the Foundations of REAL MEN

REAL MEN is grounded in an analysis of how social constructions of masculinity, race, and class create unequal life opportunities, provide frameworks young men use to negotiate within the context of those inequalities, and combine to create unequal opportunities for health. It recognizes that young men often leave jail with a commitment to changing their lives but that those intentions are often overwhelmed by the social, economic, and political contexts they encounter after release. Macrolevel factors (such as systematic racism, blocked educational opportunities, and media representations of young men of color as violent or misogynist), community-level factors (such as social support networks and educational opportunities), and microlevel factors (such as family disruptions and interactions with peer networks) shape young men's lives following release from jail and their chances for financial and social stability, substance use, HIV and other sexually transmitted infections, and for being victims or perpetrators of violence.

In contrast to the majority of public health programs that have attempted to take on issues of HIV infection, substance abuse, mental health, or violence among incarcerated men, REAL MEN was developed to address selected aspects of broader life circumstances and dynamics of race, class, and gender that can perpetuate or protect against health risks. Specifically, working in conjunction with community organizations, young men participating in the program are linked to high school, general equivalency diploma, literacy, and job readiness programs in their communities. As appropriate, they are also connected to local providers of substance abuse as well as physical and mental health services in an effort to support their transition back to the community. In addition to helping young men plan for release and make connections to community resources, REAL MEN offers them an opportunity to engage in a critical examination of masculinity and racial/ethnic identities and to explore how these social concepts influence health by, for example, influencing both social context and actions taken within those contexts.

The curriculum for this latter component of REAL MEN was developed by program staff in consultation with young men in the community.[2] It builds on a body of research that suggests that gendered conceptions of risk, trust, and power profoundly affect health and health behavior (see, for example, Aronson, Whitehead, and Baber, 2003; DiClemente and others, 2004). Furthermore, the intervention uses growing evidence that some components of racial identity, such as ethnic pride, may help reduce the degree to which psychosocial assaults of racism are injurious to adolescents' mental health, self-esteem, and academic achievement . . . and may also reduce drug use or sexual risk behavior (Belgrave, 2002; Brook and others, 1998; Harvey and Hill, 2004; Johnson, 2002).

Based on these sources, REAL MEN planners hypothesized a number of pathways through which conceptions of masculinity may shape decisions made by young men as they are leaving jail. For example, the belief that manhood requires multiple sexual partners or having sex when high may increase the vulnerability of young men and their partners to HIV, unwanted pregnancies, and other adverse outcomes. Young men who believe that they will be perceived by their peers as "less manly" for taking action to protect their health and the health of their partners may be less likely to do so. The use of violence or coercion in relationships may also be linked to conceptions of masculinity that emphasize male power. Taking risks, and exhibiting an absence of fear in the face of risks, is promoted through both mainstream and alternative culture as a desirable characteristic of men . . . and may influence young men's willingness to engage in risky behaviors.

In a series of workshops offered in jail and after release, young men are encouraged to consider their options related to drugs, relationships, sex, criminal activity, and jail. In role plays, discussions of popular films and hip-hop lyrics, and group exercises, participants examine their own and media notions of manhood. In all sessions, participants develop and consider alternative responses to situations they are likely to encounter with friends, sexual partners, employers, and family. REAL MEN also offers activities designed to reinforce racial pride and solidarity in order to bolster young men's protection against the injurious effects of racism, foster attitudes and behaviors that promote community wellbeing, and motivate young men to plan a better future. Following release from jail, participants spend two full days in more group sessions, sign up for educational and employment services at Friends of Island Academy, the program's community sponsor, and attend sessions with their parents or partners. . . .

Interventions such as REAL MEN illustrate both the potentials and limits of targeted efforts grounded in an intersectional analysis to promote health and health equity. Such efforts can be an important point of intervention to reduce or alleviate the harmful effects of structural inequalities on the health of young men as they leave the prison system. While by themselves they do not address the underlying processes that create health inequalities in the first place, in conjunction with broader social movements for change, they can contribute to social justice and health equity.

THE MOTHERS OF EAST LOS ANGELES

The Madres del Este de Los Angeles (MELA) was founded in 1984 by a group of Mexican American women and families living on Los Angeles's east side.[3] MELA's mission statement suggests a group identity forged at the intersection of ethnicity, class, and gender: "Not economically rich—but culturally wealthy;

not politically powerful—but socially conscious; not mainstream educated—but armed with the knowledge, commitment and determination that only a mother can possess."

With roots in Saul Alinsky's Industrial Areas Foundation, the Chicano movement of the 1960s (Hernandez, 1980; Marin, 1991), and women's volunteer organizing efforts with local Catholic churches (Pardo, 1998), MELA grew into a network that at its peak included more than four hundred families and mobilized over thirty-five hundred people. Over a twenty-year period, MELA successfully organized to address a series of controversial issues related to the environment and health in East Los Angeles neighborhoods. They mobilized successful efforts to block construction of a state prison in their community (1986), halt an oil pipeline that would have passed under a junior high school (1987), prevent the construction of a toxic waste incinerator in their neighborhood (1987), and stop a chemical company from building a plant to treat cyanide and other hazardous wastes across the street from a local high school (1989) (Pardo, 1998; Freudenberg, 2004). Establishing ties to a variety of environmental groups nationally, MELA was active in local, state, and national environmental coalitions; collaborated with universities to study environmental causes of children's health problems; and worked on water conservation, lead poisoning awareness, and graffiti abatement programs. . . .

Structuring Pathways to Health and Disease in East Los Angeles

Longstanding political disenfranchisement and racist residential and employment policies have placed communities of color in jobs and residential neighborhoods in which they experience greater exposure to environmental toxins (for example, garment sweatshops, deteriorated housing in older urban areas) and limited access to the political and economic resources necessary to influence zoning or enforce environmental regulations. . . .

East Los Angeles is no exception, and class and race oppression created the particular health and social problems that MELA organized to address over two decades: prisons, toxic waste incinerators, and oil pipelines under schools. . . .

The vulnerability of East LA residents to unwanted local land uses was evident following World War II when five freeways were constructed through East LA, uprooting thousands of residents, contributing noise and pollution, and shaping the political consciousness of community members (Escobedo, 1979). In her ethnography of MELA, Pardo notes "all but one of the core activists [of MELA] shared the experience of losing a home and having to relocate to make way for a freeway" (1998, p. 72). This shared economic and political vulnerability of the community contributed to a political consciousness oriented toward social justice. Pardo quotes MELA activist Juana Gutierez as saying: "One of the things that really upsets me is injustice, and we have seen a lot of that in our community. Especially before, because I believe our people used to be less aware, we didn't assert ourselves as much. In the 1950s they put up the freeways, and just like that they gave us notice that we had to move" (1998, p. 73).

Living conditions within the community and prevailing stereotypes about its predominantly Mexican American population shaped outsiders' perceptions of the neighborhood. These perceptions made the neighborhood a target for undesirable projects and discrimination and limited residents' political influence. These same conditions contributed to the emergence of a sense of shared racial and class identity, political consciousness, and collective mobilization that led eventually to the development of MELA.

Gender, Race, Class, and the Foundations of Community Mobilization

The history of community activism within East Los Angeles predates MELA by several decades. In the late 1940s, community residents worked with a field organizer for the Industrial Areas Foundation (IAF) to create the Community Service Organization (CSO). In East LA, the CSO organized "house meetings" that engaged both men and women in the development of mobilization strategies for more than a decade. By the 1950s, residents of East LA were mobilizing around poor city services, voter registration, police brutality, and inadequate infrastructure and housing discrimination (Rose, 1994; Pardo, 1998).

In the 1960s, the Chicano movement, which was active in East Los Angeles, emphasized development of Mexican American leadership (Acuña, 1984; Muñoz, 1989; Taylor, 2002). Movement goals included fostering a positive Chicano identity, fighting against racism, and promoting equal rights (Acuña, 1984; García, 1986; Romo, 1983). Hernandez (1980) notes that while both women and men were active in the movement, men held the majority of the leadership positions. As with many other social movements of the time, there is some evidence that gender frameworks were actively used to discourage women from leadership positions. For example, women activists who challenged men's leadership roles were subject to labelling as "women's libbers and lesbians"— explicitly gendered labelling intended to be pejorative, thus trivializing women and the issues they raised (Marin, 1991; Pardo, 1998).

In the 1970s, the United Neighborhood Organization (UNO) emerged in East Los Angeles, cosponsored by the Catholic church and with organizers trained by Saul Alinsky's IAF (Ortiz, 1984). UNO was grounded in theological perspectives that incorporated both personal faith and the transformation of society toward the goal of social justice (Ortiz, 1984). The first president of UNO, Gloria Chavez, began her community activities through the PTA and her children's school activities, eventually translating those skills into organizing for social justice (Pardo, 1998).

Although each of these social justice movements differed somewhat in its emphasis (on class, Chicano identity, liberation theology), each offered opportunities for participants to develop organizing skills and a shared political consciousness that valued participation toward the goal of social justice or equality. Many of the women who became active in MELA had histories of working with one or more of these organizations and developed leadership skills as well as networks with other community activists through those experiences (Pardo, 1998).

Intersections of gender, race, and class shaped educational and employment opportunities as well as community activism. Opportunities for employment for Mexican American men in East Los Angeles were primarily in physically demanding positions (Pardo, 1998). Many women did not work outside the home, due to a combination of the poor quality of jobs available to women with their educational backgrounds, their desire to devote time to their families, and their partners' ability to earn sufficient income to support the households. While men often returned home following long and physically exhausting work-days, women's activities as church volunteers, as well as many of their histories in CSO, the Chicano movement, and UNO, provided organizing and fundraising skills that transferred readily to organizing around other community issues. Furthermore, at-home mothers were often the only community members available to attend daytime public hearings regarding land use and environmental issues (Rysavy, 1998).

In addition to the opportunities to participate in community organizing afforded by their gendered family, work, and community locations, MELA activists appealed directly to women as mothers. For example, Elsa Lopez, MELA program director, described recruiting women, saying, "We ask them, 'Are you ready to defend and protect your family?' " (Rysavy, 1998). Women also used maternal frames to influence policymakers, claiming the moral force that can be marshaled through the symbolism of motherhood and the protection of children and simultaneously defusing efforts to discredit them with stigmatizing portrayals of women activists. Even as they used their identities as mothers to promote their organizing efforts, women were acutely aware that their activism sometimes tested more traditional gender frameworks (Pardo, 1998). Thus, although they drew on frameworks of motherhood in their mobilizing efforts, women nonetheless engaged in critiques that challenged stereotypical gendered categories.

MELA activists also recognized that intersecting images of race, gender, and class contributed to the environmental health threats that they experienced. Film and media representations of East LA as a dangerous and deteriorated community with gangs of young Mexican American men fed negative and stereotyped images. Such representations rendered invisible the long-term residents who were working to stabilize the community and raise their children in it, and fostered outsiders' perceptions that there was little in the community worth protecting from, say, a prison or an incinerator (Pardo, 1998). The women of MELA recognized that these symbolic representations played an important role in shaping decisions that threatened the health of their communities, and worked skillfully to challenge portrayals of their community as graffiti-filled and gang-ridden. By posing counterimages of stable working-class families committed to education and community and by working with local policymakers to curtail film crews' continued production of negative imagery (Acuña, 1984; Pardo, 1998), MELA resisted the perpetuation of social inequalities that compromised the health of East LA residents. . . .

PROMOTING EQUITY IN HEALTH: POTENTIALS AND LIMITATIONS OF INTERSECTIONAL ANALYSES

How can the intersectional analyses illustrated in these case histories inform efforts to improve health and reduce health inequities? We identify several themes that illustrate both potentials and limitations of intersectional analyses in efforts to understand the underpinnings of health inequalities and develop effective strategies to promote greater equality.

Race, Class, and Gender Shape Material Conditions and Symbolic Representations

By looking at the combined impact of class, race, and gender on health, scholars and activists can develop a more comprehensive understanding of the complex social processes that influence well-being. The case histories show how the social categories of class, race, and gender affect health through two linked pathways: the material conditions that structure opportunities for health and disease and the symbolic representations that influence consciousness, political mobilization, and policy. The cases also demonstrate the potential for intervention through these two pathways.

In the New York case, material conditions limited job and educational opportunities, pushed young men into the illegal economy, and put them at risk of substance abuse, HIV infection, violence, and incarceration. In East Los Angeles, these material conditions exposed families to toxic substances, reduced opportunities for income, and segregated Mexican Americans into resource-poor neighborhoods. In each community, through processes that unfolded locally, these living conditions worsened the health of community residents. Any investigation of these communities that used only a single lens of class, race, or gender would miss the powerful synergy of the three in creating the inequities in health that characterize America today.

The case histories also show how symbolic representations of class, race, and gender have real consequences for health. In New York City, media portrayals of young men of color contributed to policy decisions that redirected funds from schools to jails. Public perceptions of young men of color limited their access to social support from caring adults, increased their isolation, and sometimes made young men themselves feel they had to live up to these negative images. Together, these class, race, and gender perceptions reinforced and amplified the adverse health consequences of deteriorating living conditions.

In East Los Angeles, symbolic representations of the community reflected racialized, classed, and gendered images, such as gangs of dangerous young Mexican American men. These stereotypical images, promoted by the film industry, influenced outsiders' perceptions of the neighborhood and contributed to the political and economic disenfranchisement of the community. These in turn played a role in the loss of retail and other services within East Los Angeles and numerous efforts to locate noxious facilities in these neighborhoods.

The Transformative Power of Intersectional Analyses

The case studies also show how expanding the scope of analysis from class, race, or gender alone, and from properties of individuals to concepts that illuminate social relationships, can create new opportunities for intervention. REAL MEN helps participants consider multiple identities—as partners, friends, fathers, family providers, and people responsible for their community and ethnic group. By exploring the contradictions within these identities and the discrepancy between young men's aspirations and their life circumstances, REAL MEN seeks to mobilize young men for personal and community transformation. Selecting release from jail as the point of intervention, REAL MEN moved beyond many more narrowly focused HIV/AIDS prevention efforts to help young men expand their limited educational and employment opportunities and cope more effectively with a hostile policy and economic environment that put them in the path of HIV infection and substance abuse. Postulating that prevalent conceptions of masculinity could amplify the risk in this environment, they sought to connect young men to organizations and individuals that could help them overcome barriers to school and work while engaging them in a critical structural examination of race and gender. By intervening at a critical point in the lives of young men whose health is actively threatened by the social context in which they exist, REAL MEN may help them better protect their own health and the health of their partners and communities.

The second case study also illustrates how a blended class, race, and gender analysis can reveal opportunities for intervention. Throughout its twenty-year history, MELA effectively used community mobilization, social protest, and conflict strategies to contest decisions about the siting of a variety of environmental health hazards in their communities. These decisions reflected broader economic and political inequalities, and by focusing on social justice for women, low-income communities, and Mexican Americans, as well as on health, MELA was able to mobilize wider constituencies than may have been possible with an appeal to any single category. As in REAL MEN, the personal experiences of oppression, dislocation, and injustice sparked resistance and proactive engagement in the community. In MELA, these shared experiences contributed to a sense of collective identity, critical analysis of socially structured inequalities, and the development of effective community mobilization skills. Women's location within the community as housewives, activists, mothers, and church volunteers provided an additional pillar for sustained local mobilization.

Working with existing political structures, as well as through informal networks, MELA resisted and sought to transform images of their communities as part of a larger struggle to challenge the perpetuation of inequalities. At a more personal level, they also effectively used the legitimacy attached to social roles as mothers concerned with family and community to sustain their efforts for change, while at the same time critiquing and expanding those frameworks through their activism.

New Actors and New Settings for Action

Another benefit of examining the multiple pathways that contribute to ill health and disparities in health is that such analyses can bring new actors into change efforts and identify additional settings for intervention activities. Among the actors involved in both case studies are individuals and their families, health professionals, community organizations, social movement members, and public officials. Settings for action include institutions like jails, schools, and service providers; mass and community media; public spaces such as streets; and government events like public hearings or court cases.

In both REAL MEN and MELA, the intervention instigators framed the problem in such a way that actors could enter the activities through multiple routes: as individuals concerned about their own or their families' health, as activists wanting to fight injustice or improve their neighborhood, or as service providers wanting to do their jobs better, for example. By defining the initial problem as one that affected individuals and families in their class, gender, and racial categories and by addressing both health and social justice, the planners invited people to decide for themselves the mix of motivations that would move them to action. This approach contrasts sharply with many traditional public health programs that use health professionals' definition of a specific disease or risk behavior as the starting point for intervention.

In East Los Angeles, social movements were a dominant presence in MELA's activities. The Chicano, community justice, environmental, women's, and prisoners' rights movements all contributed to specific actions and provided opportunities for integration with campaigns on other community issues as well as integration with regional and national efforts. In REAL MEN, the civil rights, educational equity, and prison reform movements were part of the context, raising public awareness of the social justice issues and serving as visible role models for young men in jail. Despite these differences, social movements played a key role in both cases. They focused attention on the underlying class, race, and gender inequalities; linked limited local efforts to broader regional and national mobilizations; inspired hope; and forced public officials to consider the consequences of inaction. In MELA, the social movements provided concrete support for many activities, while in REAL MEN, they provided a more symbolic representation of an alternative to a future of oppression.

From Analysis to Action: The Limits of Local Interventions

While the case studies illustrate the potential for intersectional analyses to inform local intervention efforts, they also illuminate the limits of these efforts. As we have noted, inequalities in health reflect underlying social structures that have long sorted people by class, race, and gender and exposed them to unequal living conditions and social environments that amplify disease for those lacking in political power. Obviously, no single health intervention or political action can reverse generations or a lifetime of exposure to health-damaging forces. In different ways, both MELA and REAL MEN are relatively modest interventions and can expect only modest impact. . . .

By comparing the outcomes of interventions like REAL MEN and MELA to those of other more limited public health programs and to our larger vision of health and social justice for all, we can set limited, achievable goals that also embody optimism about the future. Finally, by linking health and social justice and by tackling the full range of oppression and potential for transformation embodied in class, race, and gender, activists and scholars can help to create new opportunities for improving well-being and reducing inequalities in health.

NOTES

1. Jails incarcerate those detained but not convicted, those convicted and sentenced to less than a year, and some parole violators. Prisons house people convicted and sentenced to more than a year. Juvenile detention facilities lock up those defined as juveniles, often for offenses that would not be considered criminal if they were of the age of majority. In New York State, those under age sixteen are classified as juveniles. REAL MEN serves only adolescents incarcerated in jails, p. 376.

2. The titles of the group sessions are: "Staying Free, Staying Healthy," "Being a Man in Today's World," "Sex in the Risk Zone," "My People, My Pride, *Mi Gente, Mi Orgullo*," "Drugs in Your Life," and "Getting the Information You Need to Stay Free and Healthy."

3. We thank Mary Pardo for her permission to draw on material from Pardo (1998) for the case study of Mothers of East Los Angeles.

REFERENCES

Acuña, R. F. *Community Under Siege: A Chronicle of Chicanos East of the Los Angeles River.* Los Angeles: Chicano Studies Research Center, University of California, 1984, *1945–1975.*

Aronson, R. E., Whitehead, T. L., and Baber, W. L. "Challenges to Masculine Transformation Among Urban Low-Income African American Males." *American Journal of Public Health,* 2003, *93,* 732–741.

Belgrave, F. Z. "Relational Theory and Cultural Enhancement Interventions for African American Adolescent Girls." *Public Health Reports,* 2002, *117*(suppl. 1), S76–S81.

Brook, J. S., and others. "Drug Use Among African Americans: Ethnic Identity as a Protective Factor." *Psychology Reports,* 1998, *83*(3 pt. 2), 1427–1446.

Brown, P., and others. "Embodied Health Movements: New Approaches to Social Movements in Health." *Sociology of Health and Illness,* 2004, *26*(1), 50–80.

DiClemente, R. J., and others. "Efficacy of an HIV Prevention Intervention for African American Adolescent Girls: A Randomized Controlled Trial." *Journal of the American Medical Association,* 2004, *292,* 171–179.

Escobedo, R. *Boyle Heights Community Plan.* Los Angeles: Department of City Planning, 1979.

Freudenberg, N. "Jails, Prisons and the Health of Urban Populations: Review of the Impact of the Correctional System on Community Health." *Journal of Urban Health,* 2001, *78,* 214–240.

Freudenberg, N. "Community Capacity for Environmental Health Promotion: Determinants and Implications for Practice." *Health Education and Behavior,* 2004, *31* (4), 472–490.

García, J., "Forjando Ciudad: The Development of a Chicano Political Community in East Los Angeles." Unpublished doctoral dissertation, University of California, Riverside, 1986.

Harvey, A. R., and Hill, R. B. "Africentric Youth and Family Rites of Passage Program: Promoting Resilience Among At-Risk African American Youth." *Social Work,* 2004, *49*(1), 65–74.

Hernandez, P. "Lives of Chicana Activists: The Chicano Student Movement in the United States." In M. Mora and A. R. Del Castillo (eds.), *Mexican Women in the United States.* Los Angeles: Chicano Studies Research Center, University of California, 1980.

Johnson, R. L. "The Relationships Among Racial Identity, Self Esteem, Sociodemographics, and Health Promoting Lifestyles." *Research and Theory for Nursing Practice,* 2002, *16*(3), 193–207.

Johnston, H., Larana, E., and Gusfield, J. R., Jr. (eds.) *New Social Movements: From Ideology to Identity.* Philadelphia: Temple University Press, 1994.

LeBlanc, A. *Random Families: Love Drugs, Trouble and Coming of Age in the Bronx.* New York: Scribner, 2003.

Levitan, M. *A Crisis of Black Male Employment, Unemployment and Joblessness in New York City.* New York: Community Service Society, 2003.

Malow, R. M. and others. "Psychosocial Predictors of HIV Risk Among Adolescent Offenders Who Abuse Drugs." *Psychiatric Services,* 1997, *48,* 185–187.

Marin, M. V. *Social Protest in an Urban Barrio: A Study of the Chicano Movement, 1966–1974.* New York: University Press of America, 1991.

Muñoz, C. *Youth, Identity, Power: The Chicano Movement.* London: Verso Press, 1989.

Ortiz, I. D. "Chicano Urban Politics and the Politics of Reform in the Seventies." *Western Political Quarterly,* 1984, *37*(4), 565–577.

Pardo, M. S. *Mexican American Women as Activists: Identity and Resistance in Two Los Angeles Communities.* Philadelphia: Temple University Press, 1998.

Romo, R. *East Los Angeles: History of a Barrio.* Austin: University of Texas Press, 1983.

Rose, M. "Gender and Civic Activism in Mexican American Barrios in California: The Community Service Organization 1947–1962." In J. Meyerowitz (ed.), *Not June Cleaver: Women and Gender in Postwar America.* Philadelphia: Temple University Press, 1994.

Rysavy, T. "Mothers for Eco-Justice." 1998. www.yesmagazine.com/6RxforEarth/rysavy.htm.

Staton, M., and others. "Risky Sex Behavior and Substance Use Among Young Adults" *Health and Social Work,* 1999, *24,* 147–154.

Stiffman, A. R., Dore, P., Cunningham, R. M., and Earls, F. "Person and Environment in HIV Risk Behavior Change Between Adolescence and Early Adulthood." *Health Education and Behavior,* 1995, *22,* 211–226.

Taylor, D. "The Influence of Race, Class and Gender on Activism and the Development of Environmental Discourses." Seattle, Wash.: U.S. Forest Service, 2002, www.fs.fed.us/pnw/pubs/gtr534.pdf.

54

The First Americans

American Indians

C. MATTHEW SNIPP

By the end of the nineteenth century, many observers predicted that American Indians were destined for extinction. Within a few generations, disease, warfare, famine, and outright genocide had reduced their numbers from millions to less than 250,000 in 1890. Once a self-governing, self-sufficient people, American Indians were forced to give up their homes and their land, and to subordinate themselves to an alien culture. The forced resettlement to reservation lands or the Indian Territory (now Oklahoma) frequently meant a life of destitution, hunger, and complete dependency on the federal government for material needs.

Today, American Indians are more numerous than they have been for several centuries. While still one of the most destitute groups in American society, tribes have more autonomy and are now more self-sufficient than at any time since the last century. In cities, modern pan-Indian organizations have been successful in making the presence of American Indians known to the larger community, and have mobilized to meet the needs of their people (Cornell 1988; Nagel 1986; Weibel-Orlando 1991). In many rural areas, American Indians and especially tribal governments have become increasingly more important and increasingly more visible by virtue of their growing political and economic power. The balance of this [reading] is devoted to explaining their unique place in American society.

THE INCORPORATION OF AMERICAN INDIANS

The current political and economic status of American Indians is the result of the process by which they were incorporated into Euro-American society (Hall 1989). This amounts to a long history of efforts aimed at subordinating an otherwise self-governing and self-sufficient people that eventually culminated in widespread economic dependency. The role of the U.S. government in this process can be seen in the five major historical periods of federal Indian relations:

SOURCE: From Silvia Pedraza and Rubén G. Rumbaut, eds., *Origins and Destinies: Immigration, Race, and Ethnicity in America.* (Belmont, CA: Wadsworth, 1996), pp. 390–403. Reprinted by permission.

removal, assimilation, the Indian New Deal, termination and relocation, and self-determination.

Removal

In the early nineteenth century, the population of the United States expanded rapidly at the same time that the federal government increased its political and military capabilities. The character of Indian-American relations changed after the War of 1812. The federal government increasingly pressured tribes settled east of the Appalachian Mountains to move west to the territory acquired in the Louisiana Purchase. Numerous treaties were negotiated by which the tribes relinquished most of their land and eventually were forced to move west.

Initially the federal government used bargaining and negotiation to accomplish removal, but many tribes resisted (Prucha 1984). However, the election of Andrew Jackson by a frontier constituency signaled the beginning of more forceful measures to accomplish removal. In 1830 Congress passed the Indian Removal Act, which mandated the eventual removal of the eastern tribes to points west of the Mississippi River, in an area which was to become the Indian Territory and is now the state of Oklahoma. Dozens of tribes were forcibly removed from the eastern half of the United States to the Indian Territory and newly created reservations in the west, a long process ridden with conflict and bloodshed.

As the nation expanded beyond the Mississippi River, tribes of the plains, southwest, and west coast were forcibly settled and quarantined on isolated reservations. This was accompanied by the so-called Indian Wars—a bloody chapter in the history of Indian-White relations (Prucha 1984; Utley 1963). This period in American history is especially remarkable because the U.S. government was responsible for what is unquestionably one of the largest forced migrations in history.

The actual process of removal spanned more than a half-century and affected nearly every tribe east of the Mississippi River. Removal often meant extreme hardships for American Indians, and in some cases this hardship reached legendary proportions. For example, the Cherokee removal has become known as the "Trail of Tears." In 1838, nearly 17,000 Cherokees were ordered to leave their homes and assemble in military stockades (Thornton 1987, p. 117). The march to the Indian Territory began in October and continued through the winter months. As many as 8,000 Cherokees died from cold weather and diseases such as influenza (Thornton 1987, p. 118).

According to William Hagan (1979), removal also caused the Creeks to suffer dearly as their society underwent a profound disintegration. The contractors who forcibly removed them from their homes refused to do anything for "the large number who had nothing but a cotton garment to protect them from the sleet storms and no shoes between them and the frozen ground of the last stages of their hegira. About half of the Creek nation did not survive the migration and the difficult early years in the West" (Hagan 1979, pp. 77–81). In the West, a band of Nez Perce men, women, and children, under the leadership of Chief

Joseph, resisted resettlement in 1877. Heavily outnumbered, they were pursued by cavalry troops from the Wallowa valley in eastern Oregon and finally captured in Montana near the Canadian border. Although the Nez Perce were eventually captured and moved to the Indian Territory, and later to Idaho, their resistance to resettlement has been described by one historian as "one of the great military movements in history" (Prucha 1984, p. 541).

Assimilation

Near the end of the nineteenth century, the goal of isolating American Indians on reservations and the Indian Territory was finally achieved. The Indian population also was near extinction. Their numbers had declined steadily throughout the nineteenth century, leading most observers to predict their disappearance (Hoxie 1984). Reformers urged the federal government to adopt measures that would humanely ease American Indians into extinction. The federal government responded by creating boarding schools and the allotment acts—both were intended to "civilize" and assimilate American Indians into American society by Christianizing them, educating them, introducing them to private property, and making them into farmers. American Indian boarding schools sought to accomplish this task by indoctrinating Indian children with the belief that tribal culture was an inferior relic of the past and that Euro-American culture was vastly superior and preferable. Indian children were forbidden to wear their native attire, to eat their native foods, to speak their native language, or to practice their traditional religion. Instead, they were issued Euro-American clothes, and expected to speak English and become Christians. Indian children who did not relinquish their culture were punished by school authorities. The curriculum of these schools taught vocational arts along with "civilization" courses.

The impact of allotment policies is still evident today. The 1887 General Allotment Act (the Dawes Severalty Act) and subsequent legislation mandated that tribal lands were to be allotted to individual American Indians, . . . and the surplus lands left over from allotment were to be sold on the open market. Indians who received allotted tribal lands also received citizenship, farm implements, and encouragement from Indian agents to adopt farming as a livelihood (Hoxie 1984, Prucha 1984).

For a variety of reasons, Indian lands were not completely liquidated by allotment, many Indians did not receive allotments, and relatively few changed their lifestyles to become farmers. Nonetheless, the allotment era was a disaster because a significant number of allottees eventually lost their land. Through tax foreclosures, real estate fraud, and their own need for cash, many American Indians lost what for most of them was their last remaining asset (Hoxie 1984).

Allotment took a heavy toll on Indian lands. It caused about 90 million acres of Indian land to be lost, approximately two-thirds of the land that had belonged to tribes in 1887 (O'Brien 1990). This created another problem that continues to vex many reservations: "checkerboarding." Reservations that were subjected to allotment are typically a crazy quilt composed of tribal lands, privately owned "fee" land, and trust land belonging to individual Indian families.

Checkerboarding presents reservation officials with enormous administrative problems when trying to develop land use management plans, zoning ordinances, or economic development projects that require the construction of physical infrastructure such as roads or bridges.

The Indian New Deal

The Indian New Deal was short-lived but profoundly important. Implemented in the early 1930s along with the other New Deal programs of the Roosevelt administration, the Indian New Deal was important for at least three reasons. First, signaling the end of the disastrous allotment era as well as a new respect for American Indian tribal culture, the Indian New Deal repudiated allotment as a policy. Instead of continuing its futile efforts to detribalize American Indians, the federal government acknowledged that tribal culture was worthy of respect. Much of this change was due to John Collier, a long-time Indian rights advocate appointed by Franklin Roosevelt to serve as Commissioner of Indian Affairs (Prucha 1984).

Like other New Deal policies, the Indian New Deal also offered some relief from the Great Depression and brought essential infrastructure development to many reservations, such as projects to control soil erosion and to build hydroelectric dams, roads, and other public facilities. These projects created jobs in New Deal programs such as the Civilian Conservation Corps and the Works Progress Administration.

An especially important and enduring legacy of the Indian New Deal was the passage of the Indian Reorganization Act (IRA) of 1934. Until then, Indian self-government had been forbidden by law. This act allowed tribal governments, for the first time in decades, to reconstitute themselves for the purpose of overseeing their own affairs on the reservation. Critics charge that this law imposed an alien form of government, representative democracy, on traditional tribal authority. On some reservations, this has been an on-going source of conflict (O'Brien 1990). Some reservations rejected the IRA for this reason, but now have tribal governments authorized under different legislation.

Termination and Relocation

After World War II, the federal government moved to terminate its long-standing relationship with Indian tribes by settling the tribes' outstanding legal claims, by terminating the special status of reservations, and by helping reservation Indians relocate to urban areas (Fixico 1986). The Indian Claims Commission was a special tribunal created in 1946 to hasten the settlement of legal claims that tribes had brought against the federal government. In fact, the Indian Claims Commission became bogged down with prolonged cases, and in 1978 the commission was dissolved by Congress. At that time, there were 133 claims still unresolved out of an original 617 that were first heard by the commission three decades earlier (Fixico 1986, p. 186). The unresolved claims that were still pending were transferred to the Federal Court of Claims.

Congress also moved to terminate the federal government's relationship with Indian tribes. House Concurrent Resolution (HCR) 108, passed in 1953, called for steps that eventually would abolish all reservations and abolish all special programs serving American Indians. It also established a priority list of reservations slated for immediate termination. However, this bill and subsequent attempts to abolish reservations were vigorously opposed by Indian advocacy groups such as the National Congress of American Indians. Only two reservations were actually terminated, the Klamath in Oregon and the Menominee in Wisconsin. The Menominee reservation regained its trust status in 1975 and the Klamath reservation was restored in 1986.

The Bureau of Indian Affairs (BIA) also encouraged reservation Indians to relocate and seek work in urban job markets. This was prompted partly by the desperate economic prospects on most reservations, and partly because of the federal government's desire to "get out of the Indian business." The BIA's relocation programs aided reservation Indians in moving to designated cities, such as Los Angeles and Chicago, where they also assisted them in finding housing and employment. Between 1952 and 1972, the BIA relocated more than 100,000 American Indians (Sorkin 1978). However, many Indians returned to their reservations (Fixico 1986). For some American Indians, the return to the reservation was only temporary; for example, during periods when seasonal employment such as construction work was hard to find.

Self-Determination

Many of the policies enacted during the termination and relocation era were steadfastly opposed by American Indian leaders and their supporters. As these programs became stalled, critics attacked them for being harmful, ineffective, or both. By the mid-1960s, these policies had very little serious support. Perhaps inspired by the gains of the Civil Rights movement, American Indian leaders and their supporters made "self-determination" the first priority on their political agendas. For these activists, self-determination meant that Indian people would have the autonomy to control their own affairs, free from the paternalism of the federal government.

The idea of self-determination was well received by members of Congress sympathetic to American Indians. It also was consistent with the "New Federalism" of the Nixon administration. Thus, the policies of termination and relocation were repudiated in a process that culminated in 1975 with the passage of the American Indian Self-Determination and Education Assistance Act, a profound shift in federal Indian policy. For the first time since this nation's founding, American Indians were authorized to oversee the affairs of their own communities, free of federal intervention. In practice, the Self-Determination Act established measures that would allow tribal governments to assume a larger role in reservation administration of programs for welfare assistance, housing, job training, education, natural resource conservation, and the maintenance of reservation roads and bridges (Snipp and Summers 1991). Some reservations also have their own police forces and game wardens, and can issue licenses and levy taxes. The

Onondaga tribe in upstate New York have taken their sovereignty one step further by issuing passports that are internationally recognized. Yet there is a great deal of variability in terms of how much autonomy tribes have over reservation affairs. Some tribes, especially those on large and well-organized reservations have nearly complete control over their reservations, while smaller reservations with limited resources often depend heavily on BIA services. . . .

CONCLUSION

Though small in number, American Indians have an enduring place in American society. Growing numbers of American Indians occupy reservation and other trust lands, and equally important has been the revitalization of tribal governments. Tribal governments now have a larger role in reservation affairs than ever in the past. Another significant development has been the urbanization of American Indians. Since 1950, the proportion of American Indians in cities has grown rapidly. These American Indians have in common with reservation Indians many of the same problems and disadvantages, but they also face other challenges unique to city life.

The challenges facing tribal governments are daunting. American Indians are among the poorest groups in the nation. Reservation Indians have substantial needs for improved housing, adequate health care, educational opportunities, and employment, as well as developing and maintaining reservation infrastructure. In the face of declining federal assistance, tribal governments are assuming an ever-larger burden. On a handful of reservations, tribal governments have assumed completely the tasks once performed by the BIA.

As tribes have taken greater responsibility for their communities, they also have struggled with the problems of raising revenues and providing economic opportunities for their people. Reservation land bases provide many reservations with resources for development. However, these resources are not always abundant, much less unlimited, and they have not always been well managed. It will be yet another challenge for tribes to explore ways of efficiently managing their existing resources. Legal challenges also face tribes seeking to exploit unconventional resources such as gambling revenues. Their success depends on many complicated legal and political contingencies.

Urban American Indians have few of the resources found on reservations, and they face other difficult problems. Preserving their culture and identity is an especially pressing concern. However, urban Indians have successfully adapted to city environments in ways that preserve valued customs and activities—pow-wows, for example, are an important event in all cities where there is a large Indian community. In addition, pan-Indianism has helped urban Indians set aside tribal differences and forge alliances for the betterment of urban Indian communities.

These alliances are essential, because unlike reservation Indians, urban American Indians do not have their own form of self-government. Tribal governments do not have jurisdiction over urban Indians. For this reason, urban Indians must depend on other strategies for ensuring that the needs of their community are met, especially for those new to city life. Coping with the transition to urban life poses a multitude of difficult challenges for many American Indians. Some succumb to these problems, especially the hardships of unemployment, economic deprivation, and related maladies such as substance abuse, crime, and violence. But most successfully overcome these difficulties, often with help from other members of the urban Indian community.

Perhaps the greatest strength of American Indians has been their ability to find creative ways for dealing with adversity, whether in cities or on reservations. In the past, this quality enabled them to survive centuries of oppression and persecution. Today this is reflected in the practice of cultural traditions that Indian people are proud to embrace. The resilience of American Indians is an abiding quality that will no doubt ensure that they will remain part of the ethnic mosaic of American society throughout the twenty-first century and beyond.

REFERENCES

Cornell, Stephen. 1988. *The Return of the Native: American Indian Political Resurgence.* New York: Oxford University Press.

Fixico, Donald L. 1986. *Termination and Relocation: Federal Indian Policy, 1945–1960.* Albuquerque, NM: University of New Mexico Press.

Hagan, William T. 1979. *American Indians.* Chicago, IL: University of Chicago Press.

Hall, Thomas D. 1989. *Social Change in the Southwest, 1350–1880.* Lawrence, KS: University Press of Kansas.

Hoxie, Frederick E. 1984. *A Final Promise: The Campaign to Assimilate the Indians, 1880–1920.* Lincoln, NE: University of Nebraska Press.

Nagel, Joanne. 1986. "American Indian Repertoires of Contention." Paper presented at the annual meeting of the American Sociological Association, San Francisco, CA.

O'Brien, Sharon. 1990. *American Indian Tribal Governments.* Norman, OK: University of Oklahoma Press.

Prucha, Francis Paul. 1984. *The Great Father.* Lincoln, NE: University of Nebraska Press.

Snipp, C. Matthew and Gene F. Summers. 1991. "American Indian Development Policies," pp. 166–180 in *Rural Policies for the 1990s,* edited by Cornelia Flora and James A. Christenson. Boulder, CO: Westview Press.

Sorkin, Alan L. 1978. *The Urban American Indian.* Lexington, MA: Lexington Books.

Thornton, Russell. 1987. *American Indian Holocaust and Survival: A Population History since 1942.* Norman, OK: University of Oklahoma Press.

Utley, Robert M. 1963. *The Last Days of the Sioux Nation.* New Haven: Yale University Press.

Weibel-Orlando, Joan. 1991. *Indian Country, L.A.* Urbana, IL: University of Illinois Press.

55

Policing the National Body

Sex, Race, and Criminalization

JAEL SILLIMAN

American politicians, eager to garner support from the large middle-class, promote "family values," endlessly debate issues of abortion, and outdo each other as champions of "working families" (read "middle-class" families). . . .

Essentially, what we have is an America deeply divided across class and race lines. This makes it possible for mainstream America—its politicians and media—to ignore or rarely address issues of poverty, criminalization, and race that are pressing for communities of color. Incarceration rates for people of color are disproportionately high and assaults and searches by police, the Immigration and Naturalization Service (INS), and border patrol forces are daily occurrences in communities of color. This aggressive law enforcement regime is increasingly accepted by the mainstream as the price to be paid for law and order. A lead article in the February 2001 edition of the *New York Times Magazine* marks this decisive shift in public attitude in New York City from a more libertarian, turbulent, and nonconformist city towards a greater acceptance of aggressive law enforcement.

. . . The new wisdom in New York City—one of the bastions of liberalism in the country—is "We no longer believe that to solve crime we have to deal with the root causes of poverty and racism; we now believe that we can reduce crime through good policing."[1]

Aggressive law enforcement policies and actions are devastating women of color and their communities. Though there is a strong and growing law enforcement accountability movement,[2] the women's movement in general has not seen state violence as a critical concern.[3] The mainstream reproductive rights movement, consumed with protecting the right to abortion, has failed to respond adequately to the policing, criminalization, and incarceration of large numbers of poor people and people of color. It has not sufficiently addressed cuts in welfare and immigrant services that have made one of the most fundamental reproductive rights—the right to have a child and to rear a family—most tenuous for a large number of people.

SOURCE: From Jael Silliman and Anannya Bhattacharjee, eds., *Policing the National Body: Race, Gender, and Criminalization* (Boston: South End Press, 2002), pp. x–xxvi. Reprinted by permission of the South End Press.

The mainstream reproductive rights movement, largely dominated by white women, is framed around choice: the choice to determine whether or not to have children, the choice to terminate a pregnancy, and the ability to make informed choices about contraceptive and reproductive technologies. This conception of choice is rooted in the neoliberal tradition that locates individual rights at its core, and treats the individual's control over her body as central to liberty and freedom. This emphasis on individual choice, however, obscures the social context in which individuals make choices, and discounts the ways in which the state regulates populations, disciplines individual bodies, and exercises control over sexuality, gender, and reproduction.[4]

The state regulates and criminalizes reproduction for many poor women through mandatory or discriminatory promotion of long-acting contraceptives and sterilization, and by charging pregnant women on drugs with negligence or child abuse. An examination of the body politics, the state's power of "regulation, surveillance and control of bodies (individual and collective)," elucidates the scope and venues through which the state regulates its populations "in reproduction and sexuality, in work and in leisure, in sickness and other forms of deviance and human difference."[5] . . .

Women of color have independently articulated a broad reproductive rights agenda embedded in issues of equality and social justice, while keenly tuned to the state's role in the reproduction and regulation of women's bodies. In an effort to protect their reproductive rights, they have challenged coercive population policies, demanded access to safe and accessible birth control, and asserted their right to economic and political resources to maintain healthy children. Through these demands, they move from an emphasis on individual rights to rights that are at once politicized and collectivized. African-American women leaders articulate the barriers to exercising their reproductive rights:

> Hunger and homelessness. Inadequate housing and income to provide
> for themselves and their children. Family instability. Rape. Incest.
> Abuse. Too young, too old, too sick, too tired. Emotional, physical,
> mental, economic, social—the reasons for not carrying a pregnancy to
> term are endless and varied.[6]

These leaders remind us that a range of individual and social concerns must be engaged to realize reproductive rights for all women. . . . A key concern among women of color and poor communities today is: the difficulty of maintaining families and sustaining community in the face of increasing surveillance and criminalization. . . . Particular communities and women within them are conceived and reproduced as threats to the national body, imagined as white and middle-class. . . .

Poor women and women of color are criminalized. Bhattacharjee elaborates upon the prison system, police, INS, and border patrol forces to illustrate how they routinely undermine and endanger women's caretaking, caregiving, and reproductive functions. She shows how systemic and frequent abuses of reproductive rights and threats to bodily integrity are often overlooked by narrow

definitions of reproductive rights and single-issue movements. . . . Greater coali-
tion-building efforts between the women's movement, the reproductive rights
movement, the immigrant rights movement, the violence against women movement, and the enforcement accountability movement are needed to break down
barriers and to ensure the safety and self-determination of women of color. . . .

CRIME AND PUNISHMENT IN THE UNITED STATES

The criminal justice system has become a massive machine for arrest, detention,
and incarceration. The events of September 11, 2001 have intensified this trend.
Citing "national security" and the "war against terrorism," President Bush has
furthered the power of the criminal justice system to arrest noncitizens and to
circumvent the court system. Immigrants and communities of color will bear
the brunt of the intensified assault on civil liberties. In 1998, on any given day,
there were approximately six million people under some form of correctional
authority. The number of people in American prisons is expected to surpass
two million by late 2001. Federal Judge U. W. Clemon, after a visit to the
Morgan County Jail in Alabama, wrote in a blistering ruling, "To say the
Morgan County Jail is overcrowded is an understatement. The sardine-can
appearance of its cell units more nearly resemble the holding units of slave ships
during the Middle Passage of the eighteenth century than anything in the
twenty-first century."[7] Imprisonment is the solution currently proffered for
drug offenses in minority communities and for the other social problems
spawned by poverty. As welfare and service programs are gutted, the only social
service available to many of America's poor is jail![8]

Despite this surge in incarceration rates, it is widely accepted that prisons
encourage recidivism, transform the occasional offender into a habitual delin-
quent,[9] fail to eliminate crime, and ignore the social problems that drive individuals to engage in illegal actions. French historian and social critic Michel
Foucault explains the political rationale behind what the terms the "production
of delinquency," and its usefulness for those in power:

> [T]he existence of a legal prohibition creates around it a field of illegal
> practices, which one manages to supervise, while extracting from it an
> illicit profit through elements, themselves illegal, but rendered manipu-
> lable by their organization in delinquency. . . . Delinquency represents a
> diversion of illegality for the illicit circuits of profit and power of the
> dominant class.[10]

The building and maintenance of policing and prison systems is politically
expedient and highly profitable. Prisons boost local economies. Fremont
County, Colorado, home to thirteen prisons, promotes itself as the Corrections
Capital of the World. In *Going Up the River,* Joseph Hallinan explains how prisons have become public works projects that require a steady flow of inmates
to sustain them.[11] Corporations engage in bidding wars to run prisons, and the

federal government boasts about saving money by contracting out prison management to the private sector.

People of color are disproportionately represented in the prison industrial complex. Bureau of Justice statistics indicate that in 1999, 46 percent of prison inmates were Black and 18 percent were Hispanic. In "Killing the Black Community," Judith Scully argues that the war on drugs is used to justify and exercise control over the Black community. She contends that the United States government has historically maintained control over the Black community by selectively enforcing the law, arbitrarily defining criminal behavior and incarceration, and failing to punish white people engaged in lawless acts against the Black community, as exemplified in the infamous Rodney King and Amadou Diallo verdicts. The war on drugs demonstrates how government officials employ legal tools as well as racial rhetoric and criminal theory to criminalize and destroy Black communities. Scully explores how the creation of drug-related crimes such as the "crack baby" demonizes Black motherhood and undermines Black childbearing. Her essay exposes the institutional links between Blackness, suspicion, and criminality.

As strongly as class and race biases determine the criminal justice system, so does gender bias. Since 1980, the number of women in state and federal correctional facilities has tripled.[12] Amnesty International figures indicate that the majority of the over 140,000 women in the American penitentiary system are Black, Latina, and poor women, incarcerated largely for petty crimes.[13] For the same offense, Black and Latina women are respectively eight and four times more likely to be incarcerated than white women. The United Nations special report on violence against women in U.S. state and federal prisons noted the trend in prison management that emphasizes punishment rather than rehabilitation and a widespread reduction in welfare and support services within the criminal justice system. . . .

The criminal justice system works collaboratively with government, corporate, and professional institutions to perform and carry out disciplinary functions deemed necessary to uphold the system of injustice. The recent exposés on racial profiling, discriminatory sentencing, and the compliance of hospitals, medical professionals, and private citizens in administering drug tests or reporting substance abuse among pregnant women are a few examples of the ways in which a range of actors are drawn in (sometimes reluctantly) to the surveillance and disciplinary system. For example, feminist lawyer Lynn Paltrow, executive director of National Advocates for Pregnant Women, describes how the Medical University Hospital in Charleston instituted a policy of reporting and facilitating the arrest of pregnant patients, overwhelmingly African-Americans, who tested positive for cocaine. African-American women were dragged out of the hospital in chains and shackles where the medical staff worked in collaboration with the prosecutor and police to see if the threat of arrest would deter drug use among pregnant women.[14]

Such violations are not going unchallenged. In *Ferguson vs. City of Charleston* (March 2001), a lawsuit engineered by Lynn Paltrow, the US Supreme Court agreed that Americans have the right, when they seek medical help, to expect that their doctor will examine them to provide diagnosis and treatment, and not search them to facilitate their arrest.[15] . . .

It is essential that we do not separate the more blatant forms of policing—videocameras within prisons that track every move of a prisoner, racial profiling, drug tests disproportionately administered in poor communities, and the raids on illegal immigrants crossing over to the United States—from the disciplinary apparatus being deployed across society. The wide use of differential forms of control and discipline is apparent in the ways in which the public has acquiesced to the policing and surveillance increasingly employed in everyday lives.

As a society we are no longer outraged when we hear that public schools in poor neighborhoods are routinely policed and equipped with metal detectors, and that students are hauled away in handcuffs for petty misdemeanors. The bodies being patrolled, segregated by race, determine the form of discipline applied. Perhaps this differential treatment explains why, in response to the spate of shootings in public schools across the country, parents in affluent neighborhoods have rallied and called for greater policing to ensure their children's safety. Middle-class parents invite policing to "protect their children." This contrasts sharply with the policing imposed in schools in poor communities and communities of color that criminalizes rather than protects. Though surveillance and policing differ according to whether they are there to protect or to criminalize, both kinds of interventions further extend state control over individual and collective bodies. The overt and insidious intrusions consolidate power in state and corporate entities.

BIOLOGICAL CONTROL

Women of color have been the target of biological control ideologies since the founding of America. In her book *Killing the Black Body,* Dorothy Roberts traces the history of reproductive rights abuses perpetrated on the Black community from slavery to the present. Roberts shows how control over reproduction is systematically deployed as a form of racial oppression and argues that the denial of Black reproductive autonomy serves the interests of white supremacy.[16] Others have documented the history of sterilization abuse in the Latina and Native American communities, indicating that population control has a long history in the United States. A Native American reproductive activist reports that on the Pine Ridge reservation today, pregnant women with drinking problems are put in jail, as it is the only holding place for them.[17] . . .

The potential to extend biological control has expanded exponentially. Advances in data collection and storage through the computer, Internet, and genetic revolutions have made surveillance systems more efficient and invasive. The nineteenth-century "panoptical gaze" made it possible for a prisoner to be seen at all times, and through that process the prisoner internalized surveillance.[18] The new technologies radically expand the ability to collect, process, and encode large amounts of information on ever smaller surfaces. This makes it possible for the human body to be manipulated and controlled in radically new ways—from within the body itself. This intensification takes control from a mental to a physiological realm.

Like policing, corporate intrusion into the private sphere is increasingly naturalized. A great deal of data on individuals is bought, sold, and traded. In this instance, it is information on the rich and middle-class that is particularly coveted. The Internet revolution has made it possible for corporate and security interests to track every move made by an individual on the web to determine a person's consumer preferences, interests, and purchases, in addition to getting credit card information. Such intrusive data collecting techniques regarding an individual's tastes and preferences for corporate niche marketing is rarely framed as a surveillance problem. Critics have discussed it sometimes as a privacy concern, but by and large these incursions are accepted as a market-driven intrusion into our life.

Emerging reproductive technologies, such as cloning, have the potential to blur the distinctions between genetically distinct and genetically determined individuals. This raises ethical questions regarding who would count as fully human with the attendant civil rights and liberties. Valerie Hartouni points to how standards of humanity get "partialized" (making some less human than others) in this process.[19] She fears that, in the current social context, such technologies will be used to manage and contain diversity and the proliferation of difference.[20] Other critics are concerned with the eugenic possibilities of cloning and similar practices that work on humans from the inside out. The commodification of human life and the disruption that such technologies could have on kinship structures and human relations are a source of grave concern.

SURVEILLANCE AND NATIONAL (IN)SECURITY

Whereas the prison and policing systems are supposed to protect the nation from dangers within, the military, Border Patrols, and INS are ostensibly designed to protect the public from danger and threats that emanate from outside the national body. The discretionary authority and budgets of these institutions have expanded exponentially since September 11, 2001. . . . An influx of immigrants, the rhetoric of explosive population growth in the Third World, and angry young men inside and outside the United States are manufactured as a threat to national security.[21]

Immigrants in the United States are constructed as a source of danger. This threat has been used to justify the allocation of billions of dollars to the enforcement programs of the INS that patrol the nation. At present, the INS has more armed agents with arrest power than any other federal law enforcement agency.[22] Mandatory detention provisions have made immigrants the fastest growing incarcerated population in the United States. Stringent controls and border security forces are positioned at strategic places along the United States–Mexico border. Military-style tactics and equipment result in immigrants undertaking more dangerous, isolated routes to cross over where the risks of death, dehydration, and assault are exponentially higher.

Immigrant rights organizations and the press record detailed accounts of immigrants risking their lives to make their way through tortuous terrain to find work in the United States. The Mojave Desert in Southern California with its inhospitable terrain has become one such death trap for immigrants.[23] This dangerous crossing is a dramatic example of the extreme risks that immigrants take to escape INS agents and provide adequately for their families.

Vigilante groups and private citizens in Arizona have taken it upon themselves to "help" patrol ranchers "hunt" Mexican undocumented immigrants. Jose Palafox reports on the roundup of over 3,000 undocumented immigrants on Roger and Don Barnett's 22,000-acre property in Douglas, Arizona, near the United States–Mexico border. Ranchers circulated a leaflet asking for volunteers to help patrol their land while having "Fun in the Sun." Immigrants were considered "fair game."[24]

While border patrol forces and private citizens tighten their grip on illegal immigration and regulate the mobility of poor workers, social services are being slashed for legal immigrants. The latter are often portrayed as a drain on national resources and a direct threat to low-wage workers in the United States, and female immigrants of color are particularly targeted for their family settlement and community building roles. . . .

Even though many studies have shown that immigrant labor contributes to the economy, there is a widely held perception that they represent a financial drain. Despite the proverbial anti-immigrant sentiment, the United States economy depends on cheap immigrant labor. Very often undocumented workers perform jobs or are made to engage in activities usually considered too low-paying or too risky for citizens and residents. It has been estimated that 45,000 to 50,000 women and children are trafficked annually into and across the United States for the sex industry, sweatshops, domestic labor, and agricultural work. The INS compares the trafficking in women and children with the drug and weapons smuggling industry. . . .

BEYOND ANALYSIS: BUILDING A MOVEMENT

The . . . Committee on Women, Population, and the Environment (CWPE) [is] a multiracial alliance of feminist activists, health practitioners, and scholars committed to promoting the social and economic empowerment of women in a context of global peace and justice. We work toward eliminating poverty, inequality, racism, and environmental degradation. A crucial feature of our work . . . is our identification and cultivation of a political common ground, given our widely ranging ethnic and national identities. . . .

CWPE asserts that the oppression of women, not their reproductive capacities, needs to be eliminated and calls for drug treatment for women with substance abuse problems, decent jobs, educational opportunities, and mental health and child-care services. It is the lack of these services that deny human dignity and exacerbate conditions of poverty, social status, and gender discrimination. . .

This [essay] is directed at activists, students, policy-makers, and scholars who seek to understand the connections between the criminalization of people of color and the poor, and social and population control. We seek to [foster] a dialogue that builds collaborations between social movements to move beyond single-issue and identity-based politics towards an inclusive political agenda across progressive movements.

NOTES

1. Quoting Myron Magnet, editor of *City Journal*.
2. The law enforcement agencies referred to include local and state police agencies; prison systems at the local, state and federal levels; the United States Border Patrol and Interior Enforcement Operations of the Immigration and Naturalization Service (INS); and the rapidly expanding INS detention system.
3. This is not true in other parts of the world. For example, state violence has been a critical issue for the contemporary women's movement in India. Radha Kumar in *The History of Doing* writes: "The issue of rape has been one that most contemporary feminist movements internationally have focused on, firstly because sexual assault is one of the ugliest and most brutal expressions of masculine violence towards women, because rape and the historical discourse around it reveal a great deal about the social relations of reproduction, and thirdly because of what it shows about the way in which the woman's body is seen as representing the community. In India, it has been the latter reason which has been the most dominant in the taking up of campaigns against rape." (Delhi: Kali for Women, 1993), 128.
4. Foucault, in *The History of Sexuality,* refers to this form of state control as "biopower."
5. Nancy Scheper-Hughes and Margaret Lock, "The Mindful Body: A Prolegomenon to Future Work in Medical Anthropology," *Medical Anthropology Quarterly,* Vol. 1 (1987), 6–41.
6. "We Remember: African American Women are for Reproductive Freedom" (1998); statement signed and distributed by leaders in the African-American community including Byllye Avery, Reverend Willie Barro, Donna Brazil, Shirley Chisholm, and Dorothy Heights in support of keeping abortion safe and legal.
7. David Firestone, "Alabama's Packed Jails Draw Ire of Courts, Again," *New York Times,* May 1, 2001, A1.
8. Eve Goldberg and Linda Evans, "The Prison Industrial Complex and the Global Economy," *Political Environments* (Fall 1999/Winter 2000), 47.
9. Dr. James Gilligan, a Harvard psychotherapist who has worked on prisons and recidivism, argues that punishing violence with imprisonment does not stop violence because it continues to replicate the patriarchal code. See *Violence: Reflections on a National Epidemic* (New York: Vintage, 1997).
10. Michel Foucault, *Discipline and Punish* (New York: Vintage, 1995), 280.

11. Joseph Hallinan, *Going Up The River: Travels in a Prison Nation* (New York: Random House, 2001).

12. Jennifer Yanco, "Breaking the Silence: Women and the Criminal Justice System," *Political Environments,* No. 7 (Fall 1999/Winter 2000), 20.

13. This document, "The UN Special Report on Violence Against Women" by Special Rapporteur Radhika Coomaraswamy, is available from United Nations Publications at www.un.org.

14. Lynn Paltrow, "Pregnant Drug Users, Fetal Persons, and the Threat to *Roe v. Wade*," *Albany Law Review,* Vol. *62*, No. 3 (1999), 1024.

15. This decision affirms the Fourth Amendment to the US Constitution that protects every American, including those who are pregnant and those with substance abuse problems, from warrantless, unreasonable searches.

16. Dorothy Roberts, *Killing the Black Body* (New York: Vintage, 1998).

17. Native American Health Education Resource Center, SisterSong Native Women's Reproductive Health and Rights Roundtable Report (Lake Andes, SD: January 2001), 17.

18. Foucault, *Discipline and Punish,* 195–228.

19. For a rich discussion of this set of issues, see Valerie Hartouni, "Replicating the Singular Self," in *Cultural Conceptions: On Reproduction Technologies and the Remaking of Life* (Minneapolis, MN: University of Minnesota Press, 1997), 110–132.

20. Ibid., 119.

21. For more on the subject see Betsy Hartmann, "Population, Environment, and Security: A New Trinity" in Jael Silliman and Ynestra King, eds., *Dangerous Intersections* (Cambridge, MA: South End Press, 1999), 1–23.

22. Maria Jimenez, "Legalization Then and Now: An Eighty Year History," *Network News* (Oakland, CA: National Network for Immigrant and Refugee Rights, Summer 2000), 6.

23. Ginger Thompson, "The Desperate Risk of Death in a Desert," *New York Times,* October 31, 2000, A12.

24. Jose Palafox, "Welcome to America: Arizona Ranchers Hunt Mexicans," *Network News* (Summer 2000), 4.

56

Rape, Racism, and the Law

JENNIFER WRIGGINS

The history of rape in this country has focused on the rape of white women by Black men. From a feminist perspective, two of the most damaging consequences of this selective blindness are the denials that Black women are raped and that all women are subject to pervasive and harmful sexual coercion of all kinds. . . .

THE NARROW FOCUS ON BLACK OFFENDER/
WHITE VICTIM RAPE

There are many different kinds of rape. Its victims are of all races, and its perpetrators are of all races. Yet the kind of rape that has been treated most seriously throughout this nation's history has been the illegal forcible rape of a white woman by a Black man. The selective acknowledgement of Black accused/ white victim rape was especially pronounced during slavery and through the first half of the twentieth century. Today a powerful legacy remains that permeates thought about rape and race.

During the slavery period, statutes in many jurisdictions provided the death penalty or castration for rape when the convicted man was Black or mulatto and the victim white. These extremely harsh penalties were frequently imposed. In addition, mobs occasionally broke into jails and courtrooms and lynched slaves alleged to have raped white women, prefiguring Reconstruction mob behavior.

In contrast to the harsh penalties imposed on Black offenders, courts occasionally released a defendant accused of raping a white woman when the evidence was inconclusive as to whether he was Black or mulatto. The rape of Black women by white or Black men, on the other hand, was legal; indictments were sometimes dismissed for failing to allege that the victim was white. In those states where it was illegal for white men to rape white women, statutes provided less severe penalties for the convicted white rapist than for the convicted Black one. In addition, common-law rules both defined rape narrowly and made it a difficult crime to prove. . . .

SOURCE: From Jennifer Wriggins, "Rape, Racism and the Law" *Harvard Women's Law Journal* 6 (Spring 1983):103–141. Reprinted by permission of *Harvard Journal of Law and Gender*.

After the Civil War, state legislatures made their rape statutes race-neutral, but the legal system treated rape in much the same way as it had before the war. Black women raped by white or Black men had no hope of recourse through the legal system. White women raped by white men faced traditional common-law barriers that protected most rapists from prosecution.

Allegations of rape involving Black offenders and white victims were treated with heightened virulence. This was manifested in two ways. The first response was lynching, which peaked near the end of the nineteenth century. The second, from the early twentieth century on, was the use of the legal system as a functional equivalent of lynching, as illustrated by mob coercion of judicial proceedings, special doctrinal rules, the language of opinions, and the markedly disparate numbers of executions for rape between white and Black defendants.

Between 1882 and 1946 at least 4,715 persons were lynched, about three-quarters of whom were Black. Although lynching tapered off after the early 1950s, occasional lynch-like killings persist to this day. The influence of lynching extended far beyond the numbers of Black people murdered because accounts of massive white crowds torturing, burning alive, and dismembering their victims created a widespread sense of terror in the Black community.

The most common justification for lynching was the claim that a Black man had raped a white woman. The thought of this particular crime aroused in many white people an extremely high level of mania and panic. One white woman, the wife of an ex-Congressman, stated in 1898, "If it needs lynching to protect woman's dearest possession from human beasts, then I say lynch a thousand times a week if necessary." The quote resonates with common stereotypes that Black male sexuality is wanton and bestial, and that Black men are wild, criminal rapists of white women.

Many whites accepted lynching as an appropriate punishment for a Black man accused of raping a white woman. The following argument made to the jury by defense counsel in a 1907 Louisiana case illustrates this acceptance:

> Gentlemen of the jury, this man, a nigger, is charged with breaking into the house of a white man in the nighttime and assaulting his wife, with the intent to rape her. Now, don't you know that, if this nigger had committed such a crime, he never would have been brought here and tried; that he would have been lynched, and if I were there I would help pull on the rope.[1]

It is doubtful whether the legal system better protected the rights of a Black man accused of raping a white woman than did the mob. Contemporary legal literature used the term "legal lynching" to describe the legal system's treatment of Black men. Well past the first third of the twentieth century, courts were often coerced by violent mobs, which threatened to execute the defendant themselves unless the court convicted him. Such mobs often did lynch the defendant if the judicial proceedings were not acceptable to them. A contemporary authority on lynching commented in 1934 that "the local sentiment which would make a lynching possible would insure a conviction in the courts." Even if the mob was not overtly pressuring for execution, a Black defendant accused of raping a white woman faced a hostile, racist legal system. State court submission to mob pressure is well illustrated by the most famous series of cases about interracial rape, the Scottsboro cases of the 1930s. Eight young Black men were convicted of what the Alabama Supreme Court called "a most foul and revolting crime," which was the rape of "two defenseless white girls." The defendants were summarily sentenced to death based on minimal and dubious evidence, having been denied effective assistance of counsel. The Alabama Supreme Court upheld the convictions in opinions demonstrating relentless determination to hold the defendants guilty regardless of strong evidence that mob pressure had influenced the verdicts and the weak evidence presented against the defendants. In one decision, that court affirmed the trial court's denial of a change of venue on the grounds that the mobs' threats of harm were not imminent enough although the National Guard had been called out to protect the defendants from mob executions. The U.S. Supreme Court later recognized that the proceedings had in fact taken place in an atmosphere of "tense, hostile, and excited public sentiment." After a lengthy appellate process, including three favorable Supreme Court rulings, all of the Scottsboro defendants were released, having spent a total of 104 years in prison.

In addition, courts applied special doctrinal rules to Black defendants accused of the rape or attempted rape of white women. One such rule allowed juries to consider the race of the defendant and victim in drawing factual conclusions as to the defendant's intent in attempted rape cases. If the accused was Black and the victim white, the jury was entitled to draw the inference, based on race alone, that he intended to rape her. One court wrote, "In determining the question of intention, the jury may consider social conditions and customs founded upon racial differences, such as that the prosecutrix was a white woman and defendant was a Negro man."[2] The "social conditions and customs founded upon racial differences" which the jury was to consider included the assumption that Black men always and only want to rape white women, and that a white woman would never consent to sex with a Black man.

The Georgia Supreme Court of 1899 was even more explicit about the significance of race in the context of attempted rape, and particularly about the motivations of Black men. It held that race may properly be considered "to rebut any presumption that might otherwise arise in favor of the accused that his intention was to obtain the consent of the female, upon failure of which he would abandon his purpose to have sexual intercourse with her."[3] Such a rebuttal denied to Black defendants procedural protection that was accorded white defendants. . . .

The outcome of this disparate treatment of Black men by the legal system was often the same as lynching—death. Between 1930 and 1967, thirty-six percent of the Black men who were convicted of raping a white woman were executed. In stark contrast, only two percent of all defendants convicted of rape involving other racial combinations were executed. As a result of such disparate treatment, eighty-nine percent of the men executed for rape in this country were Black. While execution rates for all crimes were much higher for Black men than for white men, the differential was most dramatic when the crime was the rape of a white woman.

The patterns that began in slavery and continued long afterwards have left a powerful legacy that manifests itself today in several ways. Although the death penalty for rape has been declared unconstitutional, the severe statutory penalties for rape continue to be applied in a discriminatory manner. A recent study concluded that Black men convicted of raping white women receive more serious sanctions than all other sexual assault defendants. A recent attitudinal study found that white potential jurors treated Black and white defendants similarly when the victim was Black. However, Black defendants received more severe punishment than white defendants when the victim was white.

The rape of white women by Black men is also used to justify harsh rape penalties. One of the few law review articles written before 1970 that takes a firm position in favor of strong rape laws to secure convictions begins with a long quote from a newspaper article describing rapes by three Black men, who at 3 a.m. on Palm Sunday "broke into a West Philadelphia home occupied by an eighty-year-old widow, her forty-four-year-old daughter and fourteen-year-old granddaughter," brutally beat and raped the white women, and left the grandmother unconscious "lying in a pool of blood." This introduction presents rape as a crime committed by violent Black men against helpless white women. It is an image of a highly atypical rape—the defendants are Black and the victims white, the defendants and victims are strangers to each other, extreme violence is used, and it is a group rape. Contemporaneous statistical data on forcible rapes reported to the Philadelphia police department reveals that this rape case was virtually unique.[4] Use of this highly unrepresentative image of rape to justify strict rape laws is consistent with recent research showing that it is a prevalent, although false, belief about rape that the most common racial combination is Black offender and white victim.[5]

Charges of rapes committed by Black men against white women are still surrounded by sensationalism and public pressure for prosecution. Black men seem to face a special threat of being unjustly prosecuted or convicted. One example is Willie Sanders.[6] Sanders is a Black Boston man who was arrested and charged with the rapes of four young white women after a sensational media campaign and intense pressure on the police to apprehend the rapist. Although the rapes continued after Sanders was incarcerated, and the evidence against him was extremely weak, the state subjected him to a vigorous twenty-month prosecution. After a lengthy and expensive trial, and an active public defense, he was eventually acquitted. Although Sanders was clearly innocent, he could have been convicted; he and his family suffered incalculable damage despite his acquittal. . . .

From slavery to the present day, the legal system has consistently treated the rape of white women by Black men with more harshness than any other kind of rape. . . .

This selective focus is significant in several ways. First, since tolerance of coerced sex has been the rule rather than the exception, it is clear that the rape of white women by Black men has been treated seriously not because it is coerced sex and thus damaging to women, but because it is threatening to white men's power over both "their" women and Black men. Second, in treating Black offender/white victim illegal rape much more harshly than all coerced sex experienced by Black women and most coerced sex experienced by white women, the legal system has implicitly condoned the latter forms of rape. Third, this treatment has contributed to a paradigmatic but false concept of rape as being primarily a violent crime between strangers where the perpetrator is Black and the victim white. Finally, this pattern is perverse and discriminatory because rape is painful and degrading to both Black and white victims regardless of the attacker's race.

THE DENIAL OF THE RAPE OF BLACK WOMEN

The selective acknowledgement of the existence and seriousness of the rape of white women by Black men has been accompanied by a denial of the rape of Black women that began in slavery and continues today. Because of racism and sexism, very little has been written about this denial. Mainstream American history has ignored the role of Black people to a large extent; systematic research into Black history has been published only recently. The experiences of Black women have yet to be fully recognized in those histories, although this is beginning to change. Indeed, very little has been written about rape from the perspective of the victim, Black or white, until quite recently. Research about Black women rape victims encounters all these obstacles.

The rape of Black women by white men during slavery was commonplace and was used as a crucial weapon of white supremacy. White men had what one commentator called "institutionalized access" to Black women. The rape of Black women by white men cannot be attributed to unique Southern pathology, however, for numerous accounts exist of northern armies raping Black women while they were "liberating" the South.

The legal system rendered the rape of Black women by any man, white or Black, invisible. The rape of a Black woman was not a crime. In 1859 the Mississippi Supreme Court dismissed the indictment of a male slave for the rape of a female slave less than 10 years old, saying:

> [T]his indictment can not be sustained, either at common law or under our statutes. It charges no offense known to either system. [Slavery] was unknown to the common law . . . and hence its provisions are inapplicable. . . . There is no act (of our legislature on this subject) which embraces either the attempted or actual commission of a rape by a slave on a female slave. . . . Masters and slaves can not be governed by the same a slave on a female slave. . . . Masters and slaves can not be governed by the same system or laws; so different are their positions, rights and duties.[7]

This decision is illuminating in several respects. First, Black men are held to lesser standards of sexual restraint with Black women than are white men with white women. Second, white men are held to lesser standards of restraint with Black women than are Black men with white women. Neither white nor Black men were expected to show sexual restraint with Black women.

After the Civil War, the widespread rape of Black women by white men persisted. Black women were vulnerable to rape in several ways that white women were not. First, the rape of Black women was used as a weapon of group terror by white mobs and by the Ku Klux Klan during Reconstruction. Second, because Black women worked outside the home, they were exposed to employers' sexual aggression as white women who worked inside the home were not.

The legal system's denial that Black women experienced sexual abuse by both white and Black men also persisted, although statutes had been made race-neutral. Even if a Black victim's case went to trial—in itself highly unlikely—procedural barriers and prejudice against Black women protected any man accused of rape or attempted rape. The racist rule which facilitated prosecutions of Black offender/white victim attempted rapes by allowing the jury to consider the defendant's race as evidence of his intent, for instance, was not applied where both persons were "of color and there was no evidence of their social standing."[8] That is, the fact that a defendant was Black was considered relevant only to prove intent to rape a white woman; it was not relevant to prove intent to rape a Black woman. By using disparate procedures, the court implicitly makes two assertions. First, Black men do not want to rape Black women with the same intensity or regularity that Black men want to rape white women. Second, Black women do not experience coerced sex in the sense that white women experience it.

These attitudes reflect a set of myths about Black women's supposed promiscuity which were used to excuse white men's sexual abuse of Black women. An example of early twentieth century assumptions about Black women's purported promiscuity was provided by the Florida Supreme Court in 1918. In discussing whether the prior chastity of the victim in a statutory rape case should be presumed subject to defendant's rebuttal or should be an element of the crime which the state must prove, the court explained that:

> What has been said by some of our courts about an unchaste female
> being a comparatively rare exception is no doubt true where the pop-
> ulation is composed largely of the Caucasian race, but we would blind
> ourselves to actual conditions if we adopted this rule where another race
> that is largely immoral constitutes an appreciable part of the population.[9]

Cloaking itself in the mantle of legal reasoning, the court states that most young white women are virgins, that most young Black women are not, and that unchaste women are immoral. The traditional law of statutory rape at issue in the above-quoted case provides that women who are not "chaste" cannot be raped. Because of the way the legal system considered chastity, the association of Black women with unchastity meant not only that Black women could not be victims of statutory rape, but also that they would not be recognized as victims of forcible rape.

The criminal justice system continues to take the rape of Black women less seriously than the rape of white women. Studies show that judges generally impose harsher sentences for rape when the victim is white than when the victim is Black. The behavior of white jurors shows a similar bias. A recent study found that sample white jurors imposed significantly lighter sentences on defendants whose victims were Black than on defendants whose victims were white. Black jurors exhibited no such bias.

Evidence concerning police behavior also documents the fact that the claims of Black rape victims are taken less seriously than those of whites. A . . . study of Philadelphia police processing decisions concluded that the differential in police decisions to charge for rape "resulted primarily from a lack of confidence in the veracity of Black complainants and a belief in the myth of Black promiscuity."

The thorough denial of Black women's experiences of rape by the legal system is especially shocking in light of the fact that Black women are much more likely to be victims of rape than are white women.[10] Based on data from national surveys of rape victims, "the profile of the most frequent rape victim is a young woman, divorced or separated, Black and poverty stricken." . . .

CONCLUSION

The legal system's treatment of rape both has furthered racism and has denied the reality of women's sexual subordination. It has disproportionately targeted Black men for punishment and made Black women both particularly vulnerable and particularly without redress. It has denied the reality of women's sexual subordination by creating a social meaning of rape which implies that the only type of sexual abuse is illegal rape and the only form of illegal rape is Black offender/white victim. Because of the interconnectedness of rape and racism, successful work against rape and other sexual coercion must deal with racism. Struggles against rape must acknowledge the differences among women and the different ways that groups other than women are disempowered. In addition, work against rape must go beyond the focus on illegal rape to include all forms of coerced sex, in order to avoid the racist historical legacy surrounding rape and to combat effectively the subordination of women.

NOTES

1. *State v. Petit*, 119 La., 44 So. (1907).
2. *McQuirter v. State*, 36 Ala., 63 So. 2d (1953).
3. *Dorsey v. State*, 108 Ga., 34 S.E. (1899).
4. Out of 343 rapes reported to the Philadelphia police, 3.3% involved Black defendants accused of raping white women; 42% involved complaints of stranger rape; 20.5% involved brutal beatings; 43% involved group rapes.

5. In answer to the question, "Among which racial combination do most rapes occur?" 48% of respondents stated Black males and white females, 3% stated white males and Black females, 16% stated Black males and Black females, 33% stated white males and white females. Recent victim survey data contradict this prevalent belief; more than four-fifths of illegal rapes reported to researchers were between members of the same race, and white/Black rapes roughly equaled Black/white rapes.

6. Suffolk Superior Court indictment (1980).

7. *George v. State*, 37 Miss. (1859).

8. *Washington v. State*, 38 Ga., 75 S.E. (1912).

9. *Dallas v. State*, 76 Fla., 79 So. (1918).

10. Recent data from random citizen interviews suggest that Black women are much more likely to be victims of illegal rape than are white women.

57

Race, Poverty and Disability

Three Strikes and You're Out! Or Are You?

PAMELA BLOCK, FABRICIO E. BALCAZAR, AND
CHRISTOPHER B. KEYS

INTRODUCTION

Throughout the twentieth century, theories of biology and culture presented images of race, class, and disability in terms of deficiency and dependence. Biological models represented certain ethnic and racial groups as genetically inferior. Cultural models represented these groups as trapped in an inescapable cycle of poverty. Both models represented people of color with disabilities as social victims and/or socially threatening. Policies and practices based on these theories projected pejorative images on low-income ethnic and racial groups, and on people with disabilities. Individuals belonging to more than one of these categories—or all three—were especially vulnerable to social stigma.

SOURCE: From Pamela Block, Balcazar, Fabricio E., Keys, Christopher B. "Race, Poverty, and Disability: Three Strikes and You're Out! Or Are You?" (*Social Policy*, Fall 2002.) Reprinted by permission of the authors.

Such images of biological and cultural pathology have been rejected by many modern theorists who have adopted a minority group model for the analysis of groups disenfranchised on the basis of race, class, gender, sexual orientation, and/or disability. In these formulations, racism, classism, ableism and other forms of prejudice create barriers that result in social and economic marginalization for members of disenfranchised groups. African Americans and Latinos with disabilities face multiple challenges of discrimination and social barriers related to race and class, as well as to disability. However, insufficient attention has been given to the interaction of these variables.

RACE, POVERTY AND DISABILITY AS
BIOLOGICAL PATHOLOGY

From the time of the early eugenics family studies until eugenics' recent revival in academic and popular discussion, biological theories have always intersected theories of race, class, and disability. In the early 1900s, eugenics was the primary ideological framework in which policies and practices were developed to manage marginalized populations. Eugenics, the science of the genetic improvement of the human race, was used to establish race and class distinctions as "natural" and incontrovertible. The dominance of the upper class was mandated by their superior genetic heritage; the poor remained in poverty because of their degenerate genes.

Many proponents of eugenics believed in a hierarchy of "lower" and "higher" races. These theorists sought a state-controlled program to improve the quality of the human gene pool by optimizing the breeding of elite white Anglo-Saxon Protestants (positive eugenics), and controlling the reproduction of "degenerates," "morons," and those of "inferior" racial or ethnic groupings (negative eugenics). Special attention was given to those of "mixed" heritage and to foreigners from places other than Western Europe. "Mongrel" families (mixed Caucasian, African-American and Native American racial heritage) were considered unstable and degenerate. The genetic inferiority of the new wave of immigrants was widely discussed. Alarmed at the supposed decrease in the birth-rate of white Anglo-Saxon Protestants, eugenicists feared "race suicide." One psychologist advocated that "morons" be allowed to serve in the army in World War I, so that fewer "normal men" would be wasted.

Eugenicists believed that racial degeneration would change the national character, subsuming a primarily Nordic-European population under a "tide" of unwanted immigrants. Adherents of the biological-pathology model developed standardized tests that they believed proved the inferior intelligence of African Americans and immigrants. The biological model highlighted multiple stigma as justification for regressive social policy. Some case studies showed how racial character, poverty, and disability interacted with other factors such as illicit pregnancy, prostitution, gambling, alcoholism, drug addiction, tuberculosis, and syphilis. When several of these factors coexisted, it was considered an indication of degeneration due to pathological inheritance. Policymakers and high profile health professionals in the United States responded by establishing programs for the institutionalization and sterilization of supposedly pathological populations. Hundreds of thousands of individuals considered "feeble-minded" were institutionalized, sometimes for life, in facilities that still exist today.

The extreme example of both positive and negative eugenics was Nazi Germany. Those with Aryan characteristics were encouraged to have large families, while thousands of others were involuntarily sterilized and millions of others were annihilated under a policy of racial purification. Advances in the science of genetics and the example of Nazi Germany served to discredit the eugenics movement in the United States. However, eugenicists continued to take an active role in many professions, including population science and social service provision for marginalized groups (e.g., people on welfare, psychiatric patients, and people with cognitive disabilities).

The term "eugenics" disappeared from professional and popular discourse when it was discredited and thus no longer considered an effective means of program implementation. However, decades after eugenics disappeared from official public policy, eugenicists continued to influence the social service system by advocating institutionalization and sterilization of people with developmental disabilities and sterilization of women on welfare. The legacy of treatment models developed by eugenicists in the early 1900s (such as segregation from the opposite sex and sterilization) is still apparent in practices, such as genetic counseling, which encourage the abortion of fetuses with disabilities like Down's Syndrome. It is also visible in the physical and organizational structures of mental institutions and in the psychological scars of people who have been institutionalized for long periods. Additionally, eugenics ideologies persisted through the subtle influence of cultural beliefs concerning marginalized groups, evidenced by contradictory representations of people with disabilities in literature, film, and television as social victims or social threats. People with disabilities are often portrayed as unable to control their sexuality and capable of erupting into random and unpredictable acts of violence. Society and those with disabilities are only considered safe when such people are locked behind the walls of an institution or when deceased.

RACE, POVERTY AND DISABILITY AS CULTURAL PATHOLOGY

In the 1930s, biological theories began to be replaced by the theories of culture promulgated by sociologists and anthropologists. Various views of poverty were presented: as cultural disintegration and deprivation; as inspiring pathological forms of cultural structure, i.e., the "culture of poverty"; or as a result of the loss or lack of culture.

Although recognizably distinct from earlier biological theories, the new cultural theories included regressive images of people living in poverty that were disturbingly close to earlier models. Whether the explanation was biological or cultural, these differences (e.g. to be poor, a racial minority, or to have a disability) were presented as a form of self-defeating and self-perpetuating pathology.

Whether defined as a pathological lack of culture, or a culture that was pathological, the people living in poverty were considered deficient by leading theorists and in public policy. For example, the concept of "culture of poverty" impacted Department of Labor and War On Poverty thinking in the 1960s.

The prevalence of the deficiency approach has also characterized societal perceptions of people with disabilities. For example, the 1994 reauthorization of the Developmental Disabilities Assistance and Bill of Rights Act (PL 103-230), defined developmental disabilities as a series of "substantial functional limitations" in "self care," "self-direction," "capacity for independent living," and "economic self sufficiency." These "substantial functional limitations" were also exhibited by the "white trash" in an earlier period, with biology thought to be the root cause. Like pathological genes, the culture of poverty was reproduced from generation to generation, all but impossible to escape. Although by the 1950s and 1960s eugenics had been largely discredited, its ideas unintentionally provided a model of cultural pathology to supplant eugenics notions of biological pathology. Thus the social problems exhibited by minorities, such as teen pregnancy, alcohol and drug abuse, welfare culture, serial monogamy or multiple partners, and destabilized families, could be explained through cultural, rather than biological deficiencies. Whether in biological or cultural terms, the contours of these pathology arguments were quite comparable, focusing on the identification of individual deficits and blaming the victim for having them.

FROM MINORITY GROUP MODEL TO AN EMPOWERMENT FRAMEWORK OF ANALYSIS

Immediately after their appearance, scholars and activists began challenging the concepts of the culture of poverty and cultural deprivation, protesting that the supposed traits of people living in poverty were contemptuous distortions based on middle-class biases. Urban anthropologists wrote ethnographic accounts of African-American communities highlighting and valorizing cultural differences. Community, social, and developmental psychologists suggested that so-called "deficits" in language and learning actually reflected cultural differences, not cultural deficiency. Activists protested an approach that blamed victims for their problems and ignored the strengths of their cultures.

Like the eugenics family studies, the culture of poverty thesis functioned to justify blaming the poor for their poverty and blaming social problems on innate deficiencies. Culture became an alternative to biological explanations for the inferiority, dependency and marginality of certain social groups. Individuals from marginalized groups became categorized as mentally retarded, not because of their inherent "incompetence," but due to racial and class bias within school and social service systems.

The contrary view claimed that social problems should be addressed through the elimination of unequal power relations and re-distribution of wealth and income. This social perspective on disability is based on a minority group model that has also been adopted to explain systematic exclusion on the basis of sex, race, and sexual orientation. This model functions to liberate individuals with disabilities from long-held societal prejudices and mistaken assumptions.

Identity politics grew out of collective movements for equality on the basis of gender, race, and sexual orientation. People with disabilities are now applying this principle to their own situations. However, within the minority group model, a single issue (such as race, gender, or sexual orientation) is usually the focus for identity formation. Other issues are considered secondary or are not considered at all. Perhaps this is a response to earlier biological and cultural models that tended to use a multi-issue approach to "prove" the pathological nature (and culture) of marginalized groups. In the process of identity formation, attempts to stay as far away from the earlier pejorative images of race and poverty have resulted in the inadvertent exclusion of individuals facing the "triple jeopardy" of race, poverty, and disability. Perhaps to overcome the pressures of marginalization and to rally a critical mass of support, it is more effective, especially initially, to focus on one common characteristic. But inevitably to some degree this single-focus strategy excludes those facing triple jeopardy and their multiple concerns.

Disability rights activists reject earlier models of disability, adopting instead a model that emphasizes overcoming social barriers rather than focusing on individual pathology. In contrast to static biological and cultural models, the minority group model is a dynamic view of disability resulting from the interaction of the individual and the environment. Disability-rights activists argue that social and cultural values encouraging discrimination or the segregation of people with disabilities must be changed. In this model, the emphasis shifts from the individual to the society, and from victim-blaming to strategies for social change. Education, organizing, a sense of community, and pride in identity replace earlier experiences of isolation and shame. Individuals can band together with others who face similar challenges and learn how to challenge and navigate existing systems to their advantage. A good example is the national advocacy organization of people with physical disabilities known as ADAPT (www.adapt.org). This organization originally focused on efforts to make public transportation accessible in major cities in the country. Since the passage of the Americans with Disabilities Act, the group has focused more on efforts to promote state and national legislation regarding needs for personal care attendants. This group tends to approach social change one-issue-at-a-time.

Earlier biological and cultural models focused on individual pathology, thus obscuring the significant social barriers faced by people with the multiple stigmas of race, poverty, and disability. With multiple stigmas (such as gender, race, and disability), opportunities for employment become increasingly limited and people become trapped in poverty. Disability, poverty, and minority status are linked and intensify the already negative relationship between economic status and the existence of a disability. For example, the majority of Latinos with disabilities live below the poverty level, and African Americans with disabilities are particularly disadvantaged. Biological and cultural theories cannot sufficiently explain the discrepancies between the disabled and non-disabled, minority and non-minority populations. People living with multiple stigmas have limited opportunities and resources available to them and face societal barriers and oppression that result in poverty and exclusion.

An empowerment framework, one that refers to the increased degree of control people can have over relevant aspects of their environment, better accounts for systemic inequalities faced by minorities with disabilities and attempts to overcome these obstacles.

The minority group model of disability must encompass the complex and multilayered identities held by people with disabilities. For some individuals, disability is only one of many stigmas they face. A disability-rights model that does not recognize the multiple barriers resulting from the "triple jeopardy" of race, poverty, and disability may inadvertently exclude a large number of people who deserve representation, many of whom are most likely to be left behind. The one-issue-at-a-time approach is an effective but narrow strategy for social change. An empowerment framework of analysis that incorporates an understanding of multifaceted issues can serve as a catalyst for helping people to overcome multiple stigmas and support their efforts to seek positive social change.

This framework encourages people from multiple constituencies to develop a common agenda based on shared unmet needs. It also promotes self-reliance and an increased awareness and understanding of the social forces that maintain oppression and discrimination. Empowered individuals or groups are more likely to challenge the status quo and to pursue remedy to the social inequality that characterizes the existence of low-income people of color with disabilities.

CONCLUSION

Throughout the twentieth century, several models have evolved to address the problems of people from marginalized groups, such as those living in poverty or who have disabilities. In the early decades of the twentieth century, a model of biological pathology was used to explain persistent social problems, and pathological inheritance was believed to cause mental and physical deficiency. In the decades following World War II, biological explanations gave way to models of cultural pathology. Cultural deprivation was related to mental and physical deficiency, resulting in deficits in language and learning. Both biological and cultural pathology models have been criticized for over-looking the importance of social and political inequality. The minority group model was developed in the 80s and 90s to explain the persistence of the social barriers faced by people with disabilities and other marginalized groups. Currently, an empowerment framework reconceptualizes the problem and specifies action sequences by which individuals from marginalized groups with multiple stigmas may gain the social, political, and economic support needed to overcome barriers to their full participation in society.

Early perceptions of people with disabilities as victims and threats have been increasingly replaced by images of social actors empowered to and capable of overcoming functional limitations and social barriers. Attempts continue to be made to better understand the complex interaction of multiple variables, such as disability, gender, race, and economic status. Developing theories of social change support the promotion of empowerment. Minorities with disabilities living in poverty can have the strength, understanding, and motivation to learn about their own situation and act to change it. They can develop a critical consciousness about their oppressed status, and then plan and take action to address the impact of negative societal attitudes on them and others in similar circumstances.

The barriers faced by people of color with disabilities living in poverty should not be underestimated. Racism, ableism, and poverty severely limit opportunity. Established organizations are usually unprepared to serve minority populations with multiple needs. For example, there is a growing population of inner-city minorities who have become permanently disabled as a result of violence. Service providers have difficulty meeting their needs. In the past, such situations were addressed through single-issue movements, and service provision that divided and dealt with each issue in isolation.

What is needed now is greater implementation of empowerment approaches that address the nexus of race, poverty, and disability. With critical awareness, support and social change, individuals and groups facing multiple stigmata need not be relegated to the sidelines of society. Essential to this important mission are more indigenous leaders and grassroots groups who have experience working on these different levels to meet the needs, affirm the rights and enhance the choices of people of color with disabilities living in low-income neighborhoods. Three strikes need not mean you're out, but rather playing a game with a greater degree of difficulty.

58

Women's Human Rights

It's About Time!

RITA ARDITTI

The advancement of women and the achievement of equality between women and men are a matter of human rights and a condition for social justice and should not be seen in isolation as a women's issue. They are the only way to build a sustainable, just and developed society. Empowerment of women and equality between women and men are prerequisites for achieving political, social, economic, cultural and environmental security among all peoples.

—Platform for Action, Article 41,
Fourth World Conference on Women, Beijing, 1995

At the first meeting of a five-day seminar on Women and Human Rights that I was leading for doctoral students, one of the two men attending expressed deep skepticism about the need to discuss Human Rights through the lens of women's experiences. Didn't the Universal Declaration of Human Rights apply to everybody? Didn't Article 1 of the Declaration state that "All human beings are born free and equal in dignity and rights?" So, what was the fuss all about? Why was Women and Human Rights even a topic worthy of exploration? As far as he was concerned, we would be wasting our time.

When I recovered from my surprise at his attitude I offered a "yes, but . . ." answer that got me more or less intact through that first session. For the rest of the seminar, however, his questions stayed with me, forcing me to clarify, expand, examine, and connect in new ways each single topic discussed. By the end of the seminar, I was happy to hear that he had changed his mind. He found the seminar to be a mind-expanding experience and expressed his appreciation for the new insights he gathered. I believe that the shock I had because of his bluntness at the beginning of the seminar forced me to dig deeper and do a better job. For that, I thank him.

SOURCE: Rita Arditti "Women's Human Rights: It's about Time!" (2005). Reprinted with the permission of the author

I hope that this article will be useful to others who, like him, have questions about the relevance of the topic and to those who already believe that women's rights must be an integral part of the human rights paradigm and want to convince others to join them in this belief.

THE UNIVERSAL DECLARATION OF
HUMAN RIGHTS

The Universal Declaration of Human Rights, adopted in 1948 by the United Nations starts by asserting that freedom, justice, and peace in the world depend upon the recognition of human dignity and the rights of all members of the human family. It then goes on to spell out a variety of political, civic, social, cultural, and economic rights in its 30 articles. Political and civic rights have received most of the attention in the West, while economic, social, and cultural rights have been seen with suspicion, as if not fully qualifying as human rights. For people in the global South, however, this distinction does not make sense and it is because of their unrelenting insistence that now official UN circles have started more and more to pay attention to those rights.

According to the Declaration, all human beings have "inherent dignity," are "born free and equal," are "endowed with reason and conscience," and are "entitled to all the rights and freedoms set forth in this Declaration." Rights are not earned or conferred; they are part and parcel of the human condition. We have those rights by the very fact of our existence, simply because we are born, and the respect for those rights is considered essential to create a free, just, and peaceful world. Moreover, human rights are seen as indivisible and interdependent—they constitute a whole where each right is linked to the others and necessary for the full realization of the principles embodied in the Declaration.

Yes, my learner was right, the Declaration (and later UN documents since 1948) in its final written form, applies to all human beings. However, since its inception, the Declaration has struggled with the issue of women and its inclusion in the human rights framework. . . .

ARE HUMAN RIGHTS ONLY FOR MEN?

When feminists started to question the lack of attention to women's rights in the human rights paradigm and asserted that women's rights were not those of a "special" interest group but belonged at the center of the human rights conversation, they identified at least three themes that threw some light on this critical issue. One was the separation between public and private life [that] is considered "normal" in most societies; the second was the hidden issues associated with the power differential that permeate the relations between men and women in the great majority of cultures; and the third was the gap between the ideals of the UN documents and its practical implementation by a largely male establishment.

The separation of the public and private spheres is a given in patriarchal societies and ensures the control of women by men within different groups and at all levels. The male, as head of the household, rules home life and the individual *right to privacy,* an important human right, is often interpreted as if in the familial sphere there should *not be* governmental or community interference. The *right to privacy,* which includes the right to choose with whom one associates and deals with all reproductive decisions, becomes instead the right of men to control their "private" families. In other words, the human rights of women are not protected in the home with the result that the intimate relations between men and women follow a rigid hierarchical pattern and are left out of the human rights framework. As for the public sphere, civil and political rights, which are concerned with the right to life, seem primarily directed at the protection of men in public life. As such they do not address the many life-threatening situations that women encounter all over the world, like infanticide, malnutrition, reproductive health hazards, illegal abortion, less access to health care, trafficking, forced prostitution, and many other forms of violence.

The second theme, that of the power differential between men and women, manifests itself in the persistent discrimination against women in practically all spheres of life, in the violence aimed at maintaining women in a subordinate role, and in the dismissal of that pattern as a private or cultural matter. Charlotte Bunch, among others, saw violence against women as a political issue, which resulted from the "structural relationships of power, domination and privilege between women and men in society."[1] Bunch located women's bodies as the physical territory of this political struggle, as demonstrated by the resistance to allow control of women's bodies to women and the plethora of laws and regulations that ensure the physical subjugation of women to men.

And on the topic of implementation of women's human rights by the international law-making institutions that developed and support the traditional human rights framework, feminists pointed out that these institutions are very male-dominated "rendering suspect the claim of the objectivity and universality of international human rights law."[2] For instance, the International Court of Justice (also known as the World Court) in The Hague, the principal judicial organ of the UN, with 15 elected judges, has only one woman member as of 2004. And the International Law Commission, established by the UN General Assembly in 1947 to promote the progressive development of international law and its codification, with 34 members, elected its very first woman member only in 2001. As for the UN Secretariat, which carries on the day-to-day work of the organization, women hold 32.7 percent of the positions at the senior level. All in all, there is an over-representation of men, which extends also to the committees that monitor the implementation of human rights treaties. Implementation is an especially important issue in order to make declarations and treaties truly effective in the lives of women.

Recognizing that women's rights are human rights is an ongoing and arduous process and an agenda for struggle. New themes and areas of activism continue to arise as a result of this awareness and contribute to deepen and widen the human rights conversation in the near and far future. In the next section I address some of the activism that has both propelled and emerged from the women's human rights agenda.

WOMEN'S TRANSNATIONAL ORGANIZING

For the past 25 years women activists from all over the world, but particularly from the global South, have played an increasingly important role in the recognition of the rights of women as part of the human rights framework. The First United Nations Decade for Women, which took place in 1976–1985, marked an explicit commitment to women's issues, and in the midpoint of the decade, in 1979, the General Assembly adopted the Convention for the Elimination of All Forms of Discrimination Against Women (CEDAW). This Convention expressly addresses violations of the human rights of women and it represents a truly major step in the struggle for the recognition of women's rights as human rights. Its preamble and 30 articles are widely and rightly seen as an international bill of rights for women. Its first article sets the tone of the document by defining discrimination against women as

> any distinction, exclusion, or restriction made on the basis of sex, which has the effect or purpose of impairing or nullifying the recognition, enjoyment, or exercise by women, irrespective of their marital status, on a basis of equality of men and women, of human rights and fundamental freedom in the political, economic, social, cultural, civil, or any other field.

From this first article flows the rest of the document, specifying concrete steps to end discrimination against women and requiring action in all fields to advance women's human rights. Significantly, the Convention is the "only human rights treaty [that] affirms the reproductive rights of women and targets culture and tradition as influential forces shaping gender roles and family relations."[3] Just to give an idea of some of its salient points: CEDAW requires the end of traffic in women and the exploitation of prostitutes; mandates equal rights of women and men regarding their nationality and that of their children; demands equal access for women to family benefits, credit, bank loans, sports, and cultural life; focuses on the problems of rural women; guarantees equality before the law and equal access to administer property; requires steps to ensure equality in marriage and family relations, etc. CEDAW should be read in its entirety to fully appreciate its comprehensiveness. CEDAW has been ratified by 179 countries as of October 2004.

Since the 1975 UN World Conference on Women in Mexico City, marking International Women's Year, women activists from all over the planet have converged not only at the UN Women's conferences in Copenhagen (1980), Nairobi (1985), and Beijing (1995), but also at the UN World conferences on

Environment and Development (Rio de Janeiro, 1992), Human Rights (Vienna, 1993), and Population and Development (Cairo, 1994), pressing for the recognition of women's rights as human rights in all spheres of life. Many voices have convincingly argued that thanks to the persistent work of these activists, the UN global conferences have successfully incorporated gender analysis into areas previously considered "gender-neutral" and that this "gendering the agenda"[4] has greatly contributed to the recognition of women's rights. In other words, gender has now become part and parcel of the important global conversations in all spheres of life, such as war, peace, health, development, militarism, security, globalization, etc.

In 1995, twenty years after the first Women's conference, 3,000 women participated in the official UN Fourth World Conference on Women in Beijing, China, and over 30,000 attended the parallel NGO (non-governmental organizations) forum. Out of this conference emerged another significant document, the Platform for Action (PFA), which acknowledges the continual barriers to women's empowerment and calls on governments, NGOs and the private sector to take action. It highlights 12 critical areas of concern [that] are interrelated, interdependent, and considered "high priority." They are worth listing because they present a succinct summary of the main issues that impede the incorporation of women as full members of society all over the world and provide a map that helps inspire us and remind us of what still needs to be done. They are:

- The persistent and increasing burden of poverty on women
- Inequalities and inadequacies in and unequal access to education and training
- Inequalities and inadequacies in and unequal access to health care and related services
- Violence against women
- The effects of armed or other kinds of conflict on women, including those living under foreign occupation
- Inequality in economic structures and policies, in all forms of productive activities and in access to resources
- Inequality between men and women in the sharing of power and decision-making at all levels
- Insufficient mechanisms at all levels to promote the advancement of women
- Lack of respect for and inadequate promotion and protection of the human rights of women
- Stereotyping of women and inequality in women's access to and participation in all communication systems, especially in the media
- Gender inequalities in the management of natural resources and in the safeguarding of the environment
- Persistent discrimination against and violation of the rights of the girl child

The Beijing conference was also the scenario of the emerging backlash that the success of the transnational women's movement has generated. Well-organized conservative right-wing groups from various countries (including the United States) framed women's rights as threats to family, nation, and God. This countermovement objected to issues such as reproductive rights and LGBT rights, and engaged in a vigorous defense of traditional marriage and family arrangements. Women's rights were portrayed as an attempt to discount and take over national and religious values. In spite of this backlash, the conference managed to produce the PFA, which was signed by more than 180 governments although several countries expressed reservations about language that seemed to support abortion or alternative family structures.

Furthermore, in addition to this backlash, globalization and the huge influence of the United States in the international scene have allowed for direct impact on the lives of women all over the world, making more necessary than ever the transnational organizing for women's rights. Consider for instance, the Global Gag Rule. President Bush's first action when he took power in 2001 was to stop all United States financing to the United Nations Population Fund (UNPF) allegedly because the Fund "promotes" abortion. This was undoubtedly an attempt to impose conservative U.S. policies on the rest of the world and is in direct violation of women's reproductive and sexual health as explicitly addressed in the UN Conference on Population and Development in Cairo in 1994. The Global Gag Rule "denies health organizations in countries that receive U.S. family planning monies the right to use their own, non-U.S funds to perform abortions (even where legal), provide abortion counseling or referrals or advocate to change abortion laws."[5] This has had huge consequences for the health and lives of women in the poorest parts of the world and has prompted a solidarity campaign started by two women in the United States, Lois Abraham and Jane Roberts, to raise $34 million, one dollar at a time, to make up the funds that have been denied to the UNPF.

One encouraging example of women's transnational organizing has been the case of the International Criminal Court (ICC) based in The Hague. The Rome Statute, the creating document for the ICC, raised the standards of responding to crimes against women by recognizing rape, sexual slavery, enforced prostitution, forced pregnancy, enforced sterilization, and sexual violence as war crimes and crimes against humanity. Of the 18 judges elected last year, seven are women. This is an unprecedented proportion in international law circles and reflects the successful work of the Women's Caucus for Gender Justice whose mission is "strengthening advocacy in Women's Human Rights and international justice."[6] . . .

WOMEN'S RIGHTS AND CULTURAL RELATIVISM

. . . It seems though that in practice, over and over again the universality of human rights is particularly challenged when applied to women. The resistance to women's rights often takes the form of adopting a cultural relativism viewpoint and claiming that women's position in the private sphere leaves women outside the human rights framework. Specific abuses of women such as sexual slavery, genital mutilation, forced marriage, systematic rape, violence in the home, discrimination at many levels, etc., continue unabated worldwide.

Many factors come into play in the discussion of universality and relativism particularly because of the history of Western colonialism, national resistance movements, and the role of women in transmitting culture. Women are seen in most societies as the bearers and reproducers of culture and as such they carry a special weight in terms of maintaining tradition and group identity. Equating women with culture often manifests itself in traditional legal systems (sometimes referred as "customary laws") that discriminate against women in family life—i.e., divorce, inheritance, marriage, child custody, property ownership, etc. How to deal with the tension between women's rights as articulated in CEDAW and various cultural and religious systems can be a frustrating matter and as such it has been the focus of intense discussions. One thing that has emerged from these conversations is the need to avoid the "arrogant gaze" of the outsider (i.e., the West) and be extremely aware of the strategies used when trying to modify or eradicate cultural practices that harm women. . . .

Clearly, cultural beliefs, attitudes, and values have often been used to justify the oppression of women. Justifications such as "this is how we do things, this is part of our culture," negate the fact that cultures are dynamic entities, that what is acceptable today is different from what was acceptable 100 years ago (i.e., slavery, foot-binding, widow-burning, etc.). Resistance to cultural change is hard to overcome but we cannot give up and accept practices that deny women basic human rights. Furthermore, extremist positions often present themselves as the "true" bearers of culture in spite of the fact that there are different ideologies and deep contradictions within the cultures themselves. Supporting local activists to engage in "internal dialogue," since women do not speak with only one voice, followed by external dialogue with the transnational women's movement, seems a prudent strategy, though at times admittedly a slow and frustrating one, in furthering the recognition of women's rights.

. . . Like the quote at the beginning of this article from the Beijing Platform of Action states, the advancement of women and the full recognition of our humanity is the only way to build a sustainable, just, and developed society. So, let's get to work!

NOTES

1. Bunch, Charlotte. 1990. "Women's Rights as Human Rights: Toward a Re-Vision of Human Rights." *Human Rights Quarterly.* 12, No. 4: 486–498.

2. Charlesworth, Hilary. 2002. "The Hidden Gender of International Law." 16 *Temp. Int'l & Comp. L.J. 93.*

3. www.un.org/womenwatch/daw/cedaw/cedaw/htm.

4. Friedman, Elizabeth Jay. 2003. "Gendering the Agenda: The Impact of the Transnational Women's Rights Movement at the UN conferences of the 1990s." *Women's Studies International Forum 26*, No. 4: 313–331.

5. Petchesky, Rosalind. 2004, June. www.radiofeminista.net./junio04/notas/
    cairo+10-ing3.htm. Washington, Joi. "The U.S., Rowing Against the Tide."
    National Women's Health Network. *29,* Issue 3: 3.

6. The Women's Caucus for Gender Justice has changed its name to "Women's
    Initiatives for Gender Justice" and their web address is: www.iccwomen.org. For
    more info on the ICC go to: www.icc-cpi.int and www.iccnow.org.

PART IV

Pulling It All Together

MARGARET L. ANDERSEN AND
PATRICIA HILL COLLINS

U pon reaching the end of this book, students often want to know "What can I do?" As the editors of this volume who compile and revise it every couple of years, we know there is not a simple answer to this question. Once people know about the social injustices brought about by race, class, and gender, they may feel overwhelmed by the possibility of changing society. Some may still think that social inequities happen to other people, not to them, and they tune out. We understand that developing and then acting on an intersectional analysis of race, class, and gender is a lot to expect from our readers, most of whom are undergraduate students. For some, reading this book will be the first time they have even thought about such things. Others will have experienced some of what is written about here, but perhaps they will not before have thought beyond the particulars of their own lives. We know that developing an inclusive perspective and then deciding what to do about it is a complex process—one without simple solutions or ways of thinking. Thus, we have developed this last section of the book to examine what it means to pull together a new way of thinking—and acting—that recognizes and engages the complexity of race, class, gender, and the various other social factors that together make up this complex system we have called a *matrix of domination*.

In Part IV we ask "What would things look like were you to take an intersectional view of society?" This includes having the vision to think comprehensively and from multiple points of view as well as analyzing the various contexts where people negotiate their way through the structures of society. Thus, we open the last section with a classic visionary piece by feminist Audre Lorde ("Age, Race, Class, and Sex: Women Redefining Difference"). Lorde was an

inspirational thinker, one of the earliest to make us to think about the connections among race, class, gender, sexuality, and other forms of oppression. We
include her article because of the comprehensive vision for thinking inclusively
that she provides. Lorde shows how, in a society marked by differences, we learn
incomplete and incorrect views of others and ourselves that separate and divide
us. Relating across those differences is critical to human survival.

Vision is important to imagining social justice, and that vision is now being articulated in new ways by younger generations of feminists, represented here by Daisy
Hernandez and Pandora K. Leong ("Feminism's Future: Young Feminists of Color
Take the Mic"). Hernandez and Leong grew up, like many of our readers, in a historical period that encouraged young women to achieve and be powerful. As
Hernandez and Leong argue, though, the ideals of feminism cannot be achieved
without embracing a race and class analysis that reaches out to women of color and
working-class women. In the context of Lorde's feminist vision, their article reminds
us of the importance of feminism being for all women, not just the most successful.

Is there something in your life that you care about so much that it would spur
you to work for social justice? Is it your family? your children? your faith? your
neighborhood? a concern for a social issue? or an ethical framework that requires
not just talk but action? Most people think that people who work for social justice
must be somehow extraordinary like Martin Luther King Jr. or other heroic figures.
But most people who engage in social activism are ordinary, everyday people who
decide to take action about something that touches their lives.

For Eisa Nefertari Ulen ("Tapping Our Strength") religious identity as a U.S.
Muslim is the foundation for her urging us to build cross-cultural, multi-ethnic
connections among people. In a time period where simplistic and stereotypical
views of people defined as "other" pervade the dominant culture, Ulen expresses
the need to embrace the diverse experiences and perspectives of people, while
simultaneously using that diversity to build a more inclusive and just society.

The effort to generate a more just society develops in many social contexts—
schools, homes, communities, churches, and other locales. In "'Whosoever' Is
Welcome Here: An Interview with Reverend Edwin C. Sanders II," Gary David
Comstock describes how African American minister Sanders challenged the exclusionary practices of Southern African American churches by building a new church
that included everyone. In this organization a small group of people came together
in Nashville, Tennessee, in search of a community of worship that did not exclude
anyone and was welcoming of everyone. From its small beginnings, the
Metropolitan Interdenominational Church grew and is welcoming of people of
different races, classes, genders, sexualities, and religious traditions. Possessing a
vision of equality and an ethical framework that helps people see how inequalities

of race, class, gender, and sexuality dehumanize us all helps make a social institution more just. This article also reminds us that churches, even ostensibly conservative ones, can be important sites of change.

Kim Fellner ("The Starbucks Paradox") uses a familiar workplace to talk about how and where diversity can be achieved, but she also illustrates how this process can be filled with contradictions. Starbucks is part of the new global economy. On the one hand, Starbucks constitutes a large, for-profit corporation. Anti-globalization initiatives typically criticize companies like Wal-Mart, Nike, McDonald's, Starbucks, and similar icons of American culture as examples of the worst of globalization. Yet, as Fellner went around the country, she noticed that Starbucks employees and customers seemed far more diverse by race and class than participants in the American anti-globalization movement. You might think about how Starbucks compares in this regard to other places you might frequent—perhaps Abercrombie & Fitch, a company known for hiring only young, White workers—a practice for which they were sued and fined $40 million in 2005. These examples show you how observing organizations around you can reveal the differing race, class, and gender structures in very familiar organizations and institutions.

Whether social actors are located inside the institutions they wish to change or whether they stand outside its boundaries, the strategies they select reflect the opportunities and constraints of each specific site. Working from within organizations can mean trying to change the institutional policies and practices that either overtly discriminate based on race, class, and gender or that have unjust outcomes. Other people and groups work outside formal social institutions to effect social change. The familiar boycotts, picketing, public demonstrations, leafleting, and other direct-action strategies long associated with social movements of all types typically constitute actions taken outside an institution, and designed to pressure it to change. Although activities such as these can be trivialized in the media, it is important to remember that direct action from outsider locations represents one important way to work for social justice.

Celene Krauss's article, "Women of Color on the Front Line," describes the challenges of working in a grassroots movement dedicated to changing environmental policy. Krauss tells how the majority of participants in the environmental justice movement are mothers who became involved out of concern for their children's health. They were neither employees of bureaucracies nor government officials. Her article highlights how women often become politically involved—through concerns over family, children, and quality of life in their neighborhoods. Krauss suggests that the individual and collective actions of ordinary people who are not policymakers or diversity trainers form the bedrock for many social movements. She shows how applying a race, class, and gender framework to all efforts for social change is important for grassroots movements and organizations that are marginalized by dominant institutions.

When people are shut out of dominant social institutions, they sometimes form new social structures and organizations. Adia M. Harvey ("Becoming Entrepreneurs: Intersections of Race, Class, and Gender at the Black Beauty Salon") shows how African American women have created their own social spaces—spaces where they can realize their own entrepreneurial talents. Harvey's research on Black-woman-owned beauty salons shows that people negotiate their way through social structures and, in doing so, generate new ways of being and new institutional and organizational forms. Although Harvey's research is unique to Black women entrepreneurs, negotiating your way through structures that may oppose you is something that all people do every day within the particular structures they face. Thus, Harvey's article is not only a strong illustration of how you can analyze social life from an intersectional analysis of race, class, and gender, it is also a reminder of how people actively construct their lives even in the face of various and multiple forms of oppression.

Throughout this book we have examined how entrenched power relations of race, class, and gender affect us all. Dominant social institutions use ideologies to obscure the individual and collective political actions of everyday individuals who oppose social injustices. Latinos, African Americans, Native Americans, new immigrant groups, women, gays and lesbians, and members of the poor and working class have long challenged the institutional structures that constrain them. Thus, social institutions are both sites that reproduce social inequalities of race, class, gender, ethnicity, and sexuality and sites where people have challenged these same injustices. Workplaces, families, schools, the media, the military, and the criminal justice system—some of the social institutions we examined in Part III—are all potential sites of change.

Working for a more just society requires looking beyond what already exists. We must learn to see beyond what is in order to imagine what is possible. For example, what type of environmental policies would result if White working-class women and women of color were central in the environmental movement's decision-making processes? If all religious institutions welcomed all people, how might families and communities be changed? How might immigrant communities be changed if all service workers were paid an adequate wage? If working-class people's need for affordable, reliable transportation were central to public policy, would more money be devoted to mass transit and less to highway repair? How might health care be organized if Native American women's ideas were central to planning? If men and women truly learned to work together, how might economic security be better provided for all? Thinking inclusively about race, class, and gender stimulates this type of vision. All sites of change contain emancipatory possibilities, if only we can learn to imagine them.

59

Age, Race, Class, and Sex
Women Redefining Difference

AUDRE LORDE

Much of Western European history conditions us to see human differences in simplistic opposition to each other: dominant/subordinate, good/bad, up/down, superior/inferior. In a society where the good is defined in terms of profit rather than in terms of human need, there must always be some group of people who, through systematized oppression, can be made to feel surplus, to occupy the place of the dehumanized inferior. Within this society, that group is made up of Black and Third World people, working-class people, older people, and women.

As a forty-nine-year-old Black lesbian feminist socialist mother of two, including one boy, and a member of an interracial couple, I usually find myself a part of some group defined as other, deviant, inferior, or just plain wrong. Traditionally, in american society, it is the members of oppressed, objectified groups who are expected to stretch out and bridge the gap between the actualities of our lives and the consciousness of our oppressor. For in order to survive, those of us for whom oppression is as american as apple pie have always had to be watchers, to become familiar with the language and manners of the oppressor, even sometimes adopting them for some illusion of protection. Whenever the need for some pretense of communication arises, those who profit from our oppression call upon us to share our knowledge with them. In other words, it is the responsibility of the oppressed to teach the oppressors their mistakes. I am responsible for educating teachers who dismiss my children's culture in school. Black and Third World people are expected to educate white people as to our humanity. Women are expected to educate men. Lesbians and gay men are expected to educate the heterosexual world. The oppressors maintain their position and evade responsibility for their own actions. There is a constant drain of energy which might be better used in redefining ourselves and devising realistic scenarios for altering the present and constructing the future.

Institutionalized rejection of difference is an absolute necessity in a profit economy which needs outsiders as surplus people. As members of such an economy, we have *all* been programmed to respond to the human differences

SOURCE: From Audre Lorde, *Sister Outsider* (Freedom, CA: Crossing Press, 1984), pp. 114–23. Reprinted by permission.

between us with fear and loathing and to handle that difference in one of three ways: ignore it, and if that is not possible, copy it if we think it is dominant, or destroy it if we think it is subordinate. But we have no patterns for relating across our human differences as equals. As a result, those differences have been misnamed and misused in the service of separation and confusion.

Certainly there are very real differences between us of race, age, and sex. But it is not those differences between us that are separating us. It is rather our refusal to recognize those differences, and to examine the distortions which result from our misnaming them and their effects upon human behavior and expectation.

Racism, the belief in the inherent superiority of one race over all others and thereby the right to dominance. Sexism, the belief in the inherent superiority of one sex over the other and thereby the right to dominance. Ageism. Heterosexism. Elitism. Classism.

It is a lifetime pursuit for each one of us to extract these distortions from our living at the same time as we recognize, reclaim, and define those differences upon which they are imposed. For we have all been raised in a society where those distortions were endemic within our living. Too often, we pour the energy needed for recognizing and exploring difference into pretending those differences are insurmountable barriers, or that they do not exist at all. This results in a voluntary isolation, or false and treacherous connections. Either way, we do not develop tools for using human difference as a springboard for creative change within our lives. We speak not of human difference, but of human deviance.

Somewhere, on the edge of consciousness, there is what I call a *mythical norm,* which each one of us within our hearts knows "that is not me." In america, this norm is usually defined as white, thin, male, young, heterosexual, christian, and financially secure. It is with this mythical norm that the trappings of power reside within this society. Those of us who stand outside that power often identify one way in which we are different, and we assume that to be the primary cause of all oppression, forgetting other distortions around difference, some of which we ourselves may be practicing. By and large within the women's movement today, white women focus upon their oppression as women and ignore differences of race, sexual preference, class, and age. There is a pretense to a homogeneity of experience covered by the word *sisterhood* that does not in fact exist.

Unacknowledged class differences rob women of each others' energy and creative insight. Recently a women's magazine collective made the decision for one issue to print only prose, saying poetry was a less "rigorous" or "serious" art form. Yet even the form our creativity takes is often a class issue. Of all the art forms, poetry is the most economical. It is the one which is the most secret, which requires the least physical labor, the least material, and the one which can be done between shifts, in the hospital pantry, on the subway, and on scraps of surplus paper. Over the last few years, writing a novel on tight finances, I came to appreciate the enormous differences in the material demands between poetry and prose. As we reclaim our literature, poetry has been the major voice of poor, working class, and Colored women. A room of one's own may be a necessity for writing prose, but so are reams of paper, a typewriter, and plenty of time. The actual requirements to produce the visual arts also help determine, along class lines, whose art is whose. In this day of inflated prices for material, who are our sculptors, our painters, our photographers? When we speak of a broadly based women's culture, we need to be aware of the effect of class and economic differences on the supplies available for producing art.

As we move toward creating a society within which we can each flourish, ageism is another distortion of relationship which interferes without vision. By ignoring the past, we are encouraged to repeat its mistakes. The "generation gap" is an important social tool for any repressive society. If the younger members of a community view the older members as contemptible or suspect or excess, they will never be able to join hands and examine the living memories of the community, nor ask the all important question, "Why?" This gives rise to a historical amnesia that keeps us working to invent the wheel every time we have to go to the store for bread.

We find ourselves having to repeat and relearn the same old lessons over and over that our mothers did because we do not pass on what we have learned, or because we are unable to listen. For instance, how many times has this all been said before? For another, who would have believed that once again our daughters are allowing their bodies to be hampered and purgatoried by girdles and high heels and hobble skirts?

Ignoring the differences of race between women and the implications of those differences presents the most serious threat to the mobilization of women's joint power.

As white women ignore their built-in privilege of whiteness and define *woman* in terms of their own experience alone, then women of Color become "other," the outsider whose experience and tradition is too "alien" to comprehend. An example of this is the signal absence of the experience of women of Color as a resource for women's studies courses. The literature of women of Color is seldom included in women's literature courses and almost never in other literature courses, nor in women's studies as a whole. All too often, the excuse given is that the literatures of women of Color can only be taught by Colored women, or that they are too difficult to understand, or that classes cannot "get into" them because they come out of experiences that are "too different." I have heard this argument presented by white women of otherwise quite clear intelligence, women who seem to have no trouble at all teaching and reviewing work that comes out of the vastly different experiences of Shakespeare, Molière, Dostoyevsky, and Aristophanes. Surely there must be some other explanation.

This is a very complex question, but I believe one of the reasons white women have such difficulty reading Black women's work is because of their reluctance to see Black women as women and different from themselves. To examine Black women's literature effectively requires that we be seen as whole people in our actual complexities—as individuals, as women, as human—rather than as one of those problematic but familiar stereotypes provided in this society in place of genuine images of Black women. And I believe this holds true for the literatures of other women of Color who are not Black.

The literatures of all women of Color re-create the textures of our lives, and many white women are heavily invested in ignoring the real differences. For as long as any difference between us means one of us must be inferior, then the recognition of any difference must be fraught with guilt. To allow women of Color to step out of stereotypes is too guilt provoking, for it threatens the complacency of those women who view oppression only in terms of sex.

Refusing to recognize difference makes it impossible to see the different problems and pitfalls facing us as women.

Thus, in a patriarchal power system where whiteskin privilege is a major prop, the entrapments used to neutralize Black women and white women are not the same. For example, it is easy for Black women to be used by the power structure against Black men, not because they are men, but because they are Black. Therefore, for Black women, it is necessary at all times to separate the needs of the oppressor from our own legitimate conflicts within our communities. This same problem does not exist for white women. Black women and men have shared racist oppression and still share it, although in different ways. Out of that shared oppression we have developed joint defenses and joint vulnerabilities to each other that are not duplicated in the white community, with the exception of the relationship between Jewish women and Jewish men.

On the other hand, white women face the pitfall of being seduced into joining the oppressor under the pretense of sharing power. This possibility does not exist in the same way for women of Color. The tokenism that is sometimes extended to us is not an invitation to join power; our racial "otherness" is a visible reality that makes that quite clear. For white women there is a wider range of pretended choices and rewards for identifying with patriarchal power and its tools.

Today, with the defeat of ERA, the tightening economy, and increased conservatism, it is easier once again for white women to believe the dangerous fantasy that if you are good enough, pretty enough, sweet enough, quiet enough, teach the children to behave, hate the right people, and marry the right men, then you will be allowed to co-exist with patriarchy in relative peace, at least until a man needs your job or the neighborhood rapist happens along. And true, unless one lives and loves in the trenches it is difficult to remember that the war against dehumanization is ceaseless.

But Black women and our children know the fabric of our lives is stitched with violence and with hatred, that there is no rest. We do not deal with it only on the picket lines, or in dark midnight alleys, or in the places where we dare to verbalize our resistance. For us, increasingly, violence weaves through the daily tissues of our living—in the supermarket, in the classroom, in the elevator, in the clinic and the schoolyard, from the plumber, the baker, the saleswoman, the bus driver, the bank teller, the waitress who does not serve us.

Some problems we share as women, some we do not. You fear your children will grow up to join the patriarchy and testify against you, we fear our children will be dragged from a car and shot down in the street, and you will turn your backs upon the reasons they are dying.

The threat of difference has been no less blinding to people of Color. Those of us who are Black must see that the reality of our lives and our struggle does not make us immune to the errors of ignoring and misnaming difference. Within Black communities where racism is a living reality, differences among us often seem dangerous and suspect. The need for unity is often misnamed as a need for homogeneity, and a Black feminist vision mistaken for betrayal of our common interests as a people. Because of the continuous battle against racial erasure that Black women and Black men share, some Black women still refuse to recognize that we are also oppressed as women, and that sexual hostility against Black women is practiced not only by the white racist society, but implemented within our Black communities as well. It is a disease striking the heart of Black nationhood, and silence will not make it disappear. Exacerbated by racism and the pressures of powerlessness, violence against Black women and children often becomes a standard within our communities, one by which manliness can be measured. But these woman-hating acts are rarely discussed as crimes against Black women.

As a group, women of Color are the lowest paid wage earners in america. We are the primary targets of abortion and sterilization abuse, here and abroad. In certain parts of Africa, small girls are still being sewed shut between their legs to keep them docile and for men's pleasure. This is known as female circumcision, and it is not a cultural affair as the late Jomo Kenyatta insisted, it is a crime against Black women.

Black women's literature is full of the pain of frequent assault, not only by a racist patriarchy, but also by Black men. Yet the necessity for and history of shared battle have made us, Black women, particularly vulnerable to the false accusation that anti-sexist is anti-Black. Meanwhile, womanhating as a recourse of the powerless is sapping strength from Black communities, and our very lives. Rape is on the increase, reported and unreported, and rape is not aggressive sexuality, it is sexualized aggression. As Kalamu ya Salaam, a Black male writer points out, "As long as male domination exists, rape will exist. Only women revolting and men made conscious of their responsibility to fight sexism can collectively stop rape."[1]

Differences between ourselves as Black women are also being misnamed and used to separate us from one another. As a Black lesbian feminist comfortable with the many different ingredients of my identity, and a woman committed to racial and sexual freedom from oppression, I find I am constantly being encouraged to pluck out some one aspect of myself and present this as the meaningful whole, eclipsing or denying the other parts of self. But this is a destructive and fragmenting way to live. My fullest concentration of energy is available to me only when I integrate all the parts of who I am, openly, allowing power from particular sources of my living to flow back and forth freely through all my different selves, without the restrictions of externally imposed definition. Only then can I bring myself and my energies as a whole to the service of those struggles which I embrace as part of my living.

A fear of lesbians, or of being accused of being a lesbian, has led many Black women into testifying against themselves. It has led some of us into destructive alliances, and others into despair and isolation. In the white women's communities, heterosexism is sometimes a result of identifying with the white patriarchy, a rejection of that interdependence between women-identified women which allows the self to be, rather than to be used in the service of men. Sometimes it reflects a die-hard belief in the protective coloration of heterosexual relationships, sometimes a self-hate which all women have to fight against, taught us from birth.

Although elements of these attitudes exist for all women, there are particular resonances of heterosexism and homophobia among Black women. Despite the fact that woman-bonding has a long and honorable history in the African and African-American communities, and despite the knowledge and accomplishments of many strong and creative women-identified Black women in the political, social and cultural fields, heterosexual Black women often tend to ignore or discount the existence and work of Black lesbians. Part of this attitude has come from an understandable terror of Black male attack within the close confines of Black society, where the punishment for any female self-assertion is still to be accused of being a lesbian and therefore unworthy of the attention or support of the scarce Black male. But part of this need to misname and ignore Black lesbians comes from a very real fear that openly women-identified Black women who are no longer dependent upon men for their self-definition may well reorder our whole concept of social relationships.

Black women who once insisted that lesbianism was a white woman's problem now insist that Black lesbians are a threat to Black nationhood, are consorting with the enemy, are basically un-Black. These accusations, coming from the very women to whom we look for deep and real understanding, have served to keep many Black lesbians in hiding, caught between the racism of white women and the homophobia of their sisters. Often, their work has been ignored, trivialized, or misnamed, as with the work of Angelina Grimke, Alice Dunbar-Nelson, Lorraine Hansberry. Yet women-bonded women have always been some part of the power of Black communities, from our unmarried aunts to the amazons of Dahomey.

And it is certainly not Black lesbians who are assaulting women and raping children and grandmothers on the streets of our communities.

Across this country, as in Boston during the spring of 1979 following the unsolved murders of twelve Black women, Black lesbians are spearheading movements against violence against Black women.

What are the particular details within each of our lives that can be scrutinized and altered to help bring about change? How do we redefine difference for all women? It is not our differences which separate women, but our reluctance to recognize those differences and to deal effectively with the distortions which have resulted from the ignoring and misnaming of those differences.

As a tool of social control, women have been encouraged to recognize only one area of human difference as legitimate, those differences which exist between women and men. And we have learned to deal across those differences with the urgency of all oppressed subordinates. All of us have had to learn to live or work or coexist with men, from our fathers on. We have recognized and negotiated these differences, even when this recognition only continued the old dominant/subordinate mode of human relationship, where the oppressed must recognize the masters' difference in order to survive.

But our future survival is predicated upon our ability to relate within equality. As women, we must root out internalized patterns of oppression within ourselves if we are to move beyond the most superficial aspects of social change. Now we must recognize differences among women who are our equals, neither inferior nor superior, and devise ways to use each others' difference to enrich our visions and our joint struggles.

The future of our earth may depend upon the ability of all women to identify and develop new definitions of power and new patterns of relating across difference. The old definitions have not served us, nor the earth that supports us. The old patterns, no matter how cleverly rearranged to imitate progress, still condemn us to cosmetically altered repetitions of the same old exchanges, the same old guilt, hatred, recrimination, lamentation, and suspicion.

For we have, built into all of us, old blueprints of expectation and response, old structures of oppression, and these must be altered at the same time as we alter the living conditions which are a result of those structures. For the master's tools will never dismantle the master's house.

As Paulo Freire shows so well in *The Pedagogy of the Oppressed*,[2] the true focus of revolutionary change is never merely the oppressive situations which we seek to escape, but that piece of the oppressor which is planted deep within each of us, and which knows only the oppressors' tactics, the oppressors' relationships.

Change means growth, and growth can be painful. But we sharpen self-definition by exposing the self in work and struggle together with those whom we define as different from ourselves, although sharing the same goals. For Black and white, old and young, lesbian and heterosexual women alike, this can mean new paths to our survival.

> We have chosen each other
> and the edge of each others battles
> the war is the same
> if we lose
> someday women's blood will congeal
> upon a dead planet
> if we win
> there is no telling
> we seek beyond history
> for a new and more possible meeting.[3]

NOTES

1. From "Rape: A Radical Analysis, An African-American Perspective" by Kalamu ya Salaam in *Black Books Bulletin*, vol. 6, no. 4 (1980).

2. Seabury Press, New York, 1970.

3. From "Outlines," unpublished poem.

60

Feminism's Future
Young Feminists of Color Take the Mic

DAISY HERNÁNDEZ AND PANDORA L. LEONG

When San Jose State University senior Erika Jackson tried to recruit fellow women of color for a new feminist group on campus, the overwhelming reply was the sneer: "white women." Those words were code for another term: racist.

Many women of color, like their Anglo counterparts, eschew the term "feminism" while agreeing with its goals (the right to an abortion, equality in job hiring, girls' soccer teams). But women of color also dismiss the label because the feminist movement has largely focused on the concerns of middle-class white women. This has been a loss for people of color. Likewise, it's a loss for the movement if it expects to grow: the U.S. Census projects that the Latino and Asian-American population is expected to triple by 2050.

The "browning" of America has yet to serve as a wakeup call for feminist organizers. Attempts to address the racism of the feminist movement have largely been token efforts without lasting effects. Many young women of color still feel alienated from a mainstream feminism that doesn't explicitly address race. One woman of color, who wishes not to be identified and is working with the March for Women's Lives, put it this way: "We're more than your nannies and outreach workers. We're your future."

Progressive movements have a long history of internal debates, but for feminists of color the question of racism and feminism isn't about theories. It's about determining our place in the movement. As the daughters of both the civil rights and feminist movements, we were bred on girlpower, identity politics, and the emotional and often financial ties to our brothers, fathers, aunties, and moms back home, back South, back in Pakistan, Mexico or other homelands. We live at the intersections of identities, the places where social movements meet, and it's here that our feminism begins.

ORGANIZATIONS AS OBSTACLES

Feminism in the United States has stagnated in part because it has largely neglected a class and race analysis. Feminism can't survive by helping women

SOURCE: *In These Times* 28, no. 12 (April 21, 2004): 32–33. Reprinted by permission.

climb the corporate ladder while ignoring cuts to welfare. Family and medical leave only matter if we have jobs with benefits. Feminism has to recruit beyond college campuses.

"If the message doesn't get broader, [communities of color] aren't going to open their arms," says Sang Hee Won of Family Planning Advocates in Albany, New York. "These issues don't resonate with an immigrant woman on the streets of New York City. I'm first generation. When I think of my parents, they have so many other things to think about. People are struggling with daily lives and it's especially hard to connect [traditional feminist] issues with their situation."

The priorities of national feminist organizations often are secondary to our daily struggles. Reproductive freedom has to include access to affordable health-care and the economic opportunities to provide for the children some of us do want to have. Likewise, it's jarring to see the word "policing" on a feminist Web site and be directed to information on gender equity in police departments without mention of police brutality.

For feminists of color to identify with the mainstream movement, national organizations need to address race explicitly. Women of color always have participated but largely have remained ignored. Organizations purport to be aware but don't hire, promote, or recognize women of color as leaders. Affinity groups and special projects remain ghettoized add-ons.

"[Feminist organizations] try and are well-intentioned," says Lauren Martin, a New York activist. "They talk a lot but don't do a lot. Organizations I've worked with talk a lot about being anti-racist. [There would be] lots of trainings and in-services, but [racist] incidents that occurred would be brushed under the rug."

"Their attitude is, 'I'm going to empower you. I'm going to teach you,'" says Alma Avila-Pilchman, program manager at ACCESS, a reproductive rights organization in Oakland, California. "When the truth is we already have that power. We need to use it. We need to be listened to."

CHANGE IN LEADERSHIP

The young feminists of color we interviewed called for the inclusion of women of color and low-income women in national campaigns—when the agenda is being set.

"Forming a real coalition means starting from the very beginning rather than the 'add and stir' approach," Martin says. "The beginning is when issues are defined. It doesn't work to tack our perspective on at the end and call it 'outreach.'"

Khadine Bennett, a board member of Third Wave Foundation, which supports the activism of young women, says that feminist organizations need to share their power. "Sometimes your organization is not the best one to carry out the work," she says. "Part of the mandate from funders should be to work with people of color organizations."

[More than 30 years after the first charges of racism against the movement, these young feminists believe progressive women of color need to be the leaders of national feminist groups. That the executive directors of these organizations and senior staff still are overwhelmingly white testifies to the movement's division. The professionalization of the nonprofit world has deepened this divide by internalizing corporate expectations and marginalizing the involvement of women who can't afford to work for free. In pursuit of mainstream acceptance, organizations are losing touch with the grassroots that could revive feminism. There needs to be a commitment to leadership development among women of color and low-income women that includes mentoring and training.]

SEEKING COMMON CAUSE

[The movement also should consider models already practiced by younger activists who actively seek out coalitions.] "The people I know are working around anti-violence including sexuality and antiwar work and anti-globalization," says Mina Trudeau, of Hampshire College's Civil Liberties and Public Policy Program. "Our feminism is about social justice."

[Election years are good moments to broaden an organization's agenda.] Last fall, Erika Jackson's feminist campus group organized against Proposition 54, which would have eliminated racial classifications in California. They were the first student organization to tackle the issue, and they didn't debate whether it was a feminist issue.

"Like with public health, we talked about how it affects women of color," she says.[A lack of racial classifications would hide the higher rate of low birth weight babies born to women of color.] The measure was defeated in the November 2003 state election.

Leah Lakshmi Piepzna-Samarasinha, a Toronto-based spoken-word artist, [sees race as a central part of the work she did in counseling women who have suffered from sexual abuse and racism.] "You can't deal with the abuse and not the colonialism," she says of her work with Native American women. Healing, she adds, can often mean reconnecting to cultural pride.

Avila-Pilchman has talked to women of color "who've given up on working with white women." However, she doesn't fall into that category." I don't think that all white women don't want to work with us. I can't think that. But how is it going to happen? When?"

These are questions the mainstream feminist movement must answer, and some are hopeful.

Trudeau observes, "There is new visioning. Maybe this happens at all different points, but at this time, we're conscious of our history and of where we want to go. I think there's some back and forth, an internal dialogue that will hopefully take us to a better place."

61

Tapping Our Strength

EISA NEFERTARI ULEN

I walk with women draped in full-length fabric. We swirl through the delicate smell of incense and oil filling air around the mosque. Even as the mad Manhattan streets overflow with noise, we sisters rustle past the crisp ease of brothers in pressed cotton tunics and loose-fitting pants, past tables of over-garments and woven caps, Arabic books, and Islamic tapes. Our scarves flap and wave in bright color or sober earth tones above an Upper East Side sidewalk that is transformed, every Friday, into a bazaar—the sandy souk reborn on asphalt.

Worshippers walk through a stone gate, along a path, and into a room unadorned yet filled with spiritual energy. Women and men lean to remove their shoes where rows of slippers, sandals, heels, and sneakers line the entrance hall. White walls bounce light onto the high ceiling. Men sit in rows along the carpeted floors, shifting in silence as we file past them, up the stairs and to a loft where other women sit and wait. "As salaam u alaikum." "Wa alaikum as salaam." "Kaifa halak—how are you, sister?" "Al humdilillah—praise be to God—I am well."

Soon the imam's voice begins to resonate in the hushed rooms. Contemplative quiet focuses communal piety. The cleric speaks in Arabic, then in English, building ideas about a complete way of life. About an hour later, when he concludes, a chanting song, lyrical poetry, calls the Muslims to prayer. We women stand tall, shoulder to shoulder, forming ranks, facing Mecca, kneeling down and then forward in complete submission to Allah, our faces just tapping the prayer rug.

And our ranks are growing. Islam is the second most popular religion in the world, with over one billion Muslims forming a global Umma (community) that represents about 23 percent of the world's population. In virtually every country of Western Europe, Islam is now the second religion after Christianity. There are approximately six million Muslims in the U.S. today, and about 60 percent or so are immigrants. About 30 percent of the remaining American-Muslim population are African-American, with U.S.-born Latinos and Asian-Americans making up the 10 percent difference. There are now more Muslims in the U.S. than Jews, and the numbers of new shahadas and Muslim immigrants continue to

SOURCE: *Shattering the Stereotypes: Muslim Women Speak Out*, edited by Fawzia Afzal-Khan, p. 42–50. Copyright 2005 by Eisa Nefertari Ulen. Reprinted with the permission of Olive Branch Books, an imprint of Interling Publishing Group, Inc.

rise. Islam is even changing the way the United States sounds, as the azan converges with church bells, calling Muslims to prayer five times a day.

Words of worship are filling the air with Arabic all across America. This country will increasingly need to explore gender, generation, politics, and plurality from an Islamic perspective. The veiled lives of American-Muslim women, so often garbled into still passivity, pulsate with social ramifications.

So how will pluralistic America shift and groan under the weight of this new diversity? What happens when uber-girrrl in spiked heels and spiked hair turns the corner on her urban street and peers into the wide eyes of a woman whose face is covered with cloth? What happens when uber-girrrl's daughter brings home a friend who has two mommies—and one daddy? Under what terms do we launch that dialogue of encounter? Are women who insist on wearing hijab unselfconsciously oppressed, or—particularly in the land that gave birth to wet T-shirt contests—are they performing daily acts of resistance by covering their hair? In the West, where long blond tresses signify a certain power through sexuality and set the standard for beauty, are veiled women the most daring revolutionaries? In workplaces, where anything less than full assimilation is dismissed, are women who quietly refuse to uncover actually storming the gates for our own liberation? Is liberation possible within the veil? American feminists would do well to engage these and other questions, and then again to engage what may seem the easy answers.

I am peering outward, clustered with sisters in scarves. And I am also aligned with women sporting spikes. I am a Muslim woman. I am also a womanist, a feminist rooted in the traditions of Sistah Alice Walker. I run those mad Manhattan streets, contribute my own voice to the cacophony; I also sit in focused silence, shifting space to embrace the presence of my many-hued sisters. I celebrate the sanctity of varicolored flesh, of difference, that is celebrated in Islam. And I am also an ardent advocate of Black Empowerment, of Uplift, of Pride—I am a Race Woman.

When non-Muslims ask how a progressive womanist sistah like me could convert to Islam, I tell them Muslim women inherited property, participated in public life, divorced their husbands, worked and controlled the money they earned, even fought on the battlefield—1,400 years ago. When Muslims ask how a woman who submits to Allah like me could still be feminist, I tell them the same thing and add that modern realities too often fall short of the Islamic ideal. The Qur'an was revealed because Arabs were burying their newborn daughters in the sand, because Indians were burning their wives for dowry, because Europeans were keeping closet mistresses in economic servitude, because Africans were mutilating female genitals, and because the Chinese word for woman is slave. Obviously, these forms of violence against women continue, very often at the hands of Muslim women and men. The presence of patriarchal, sanctioned assaults against women and girls anywhere in the global Ummah horrifies me, particularly because I recognize these atrocities as anti-Islamic.

Contrary to popular opinion among non-Muslims, the Qur'an rejects the sexist propaganda that Eve is the first sinner who tempted Adam and led him to perdition. Both are held responsible for their exploits in the garden. Muslims believe Islam is the perpetuation—the refinement—of monotheistic religion and admonishes the persecution of people on the basis of gender, race, and class. Non-Muslims often confuse sexist individuals or groups with an entire religious system—and a cross-racial, multicultural swath of the world's population gets entangled in the inevitable stereotypes.

This political irresponsibility is dangerous especially because Islam is the sustenance more and more women feel they need. Indeed, while American women were publicly calling for more foreplay a few decades ago, Islam sanctioned equal pleasure in spouses' physical relationship—again, 1,400 years ago. Muslim men actually receive Allah's blessing when they bring their wives to orgasm.

Knowing many Muslim women are cutting their daughters, I celebrate Islam and teach to raise awareness about the culturally manifested, pre-Islamic practice of female genital mutilation, which is falsely considered an Islamic practice worldwide. I feel the same passionate need for widespread truth and empowerment when I read about women across this country knifing their breasts and hips and faces in the name of Western-inspired beautification. Fellow feminists too often allow difference to impede a coalition based on these virtual duplications, on this cross-cultural torture. My Muslim sisters too often think feminism is a secular evil that would destroy the very foundations of our faith. This is all a waste. While we allow difference to divide us, women everywhere are steadily slicing into their own flesh—and into the flesh of emergent women around them.

Any concerted efforts to link and liberate on the part of American feminists must proceed from factual knowledge of the veiled "other." Immigrant Muslim women must begin to align themselves with non-Muslim-American women, even as they maintain their *deen* (religion) and their home culture. Increasing conversion in this country demands U.S. citizens interested in women's freedom begin to understand why women like myself have chosen Islam.

I became Muslim because the Qur'an made sense, because my mind and spirit connected. Islam is a thinking chick's religion. Education is more than just a privilege in Islam; it is a demand. Qur'anic exhortations to reflect and understand highlight each Muslim's duty to increase in knowledge, a key component of this *deen,* this religion, where science especially supports a better understanding of spirit. Islam makes no distinction between women and men and access to knowledge, though some men would deny women's Islamic right to education, just as men in America have historically denied women the opportunity to learn.

The more I think about feminism and Islam, the more compatibility emerges from the dust of difference, and the more potential I see. I want to reconcile the great gulf that all my sisters—American non-Muslims who can't get past the *hijab* and American Muslims who can't get past secularism—see when they peer (usually past) "the other." I understand my Muslim sisters' trepidation, because first wave feminism's relationship with African-Americans lacked a cohesion born of acute commitment and fell victim to white supremacist techniques. Likewise, second wave feminism fell victim to the science of divide and conquer. Daring individuals pushed past division, though. They leaped over the great gulfs system controllers contrived to separate abolition and suffrage, Black Power and Women's Lib, and embraced a decidedly universal freedom. I am still slightly shocked when Muslims and non-Muslims claim I cannot be both Muslim and feminist. I am leaping the great abyss dividing submission (to Allah) and resistance (to patriarchy) in this increasingly complex place called America. We must build bridges.

We must build cross-cultural, multi-ethnic bridges. Now, especially now, I ask my African-American sistahs to remember our legacy of domestic terrorism, of white sheets streaking by on horseback, of strange fruit, of Black men burned alive, of four little girls. Now, more than ever, we must remember the centuries of domestic terrorism in this country, but we must also remember this: that countless white women were our sisters in fighting the horror and pain their fathers and brothers wrought. We must remember this, too: We must remember the white men appalled by the terrorism of white supremacists, the white men who battled their own souls. We must allow these memories to help form a link connecting non-Muslims with Muslims in the country today. I am asking women to remember today. I am asking you.

I am ready to do this important new work. Islam fuels my momentum. I empower myself when I wash and wrap for prayer. I transform out of a space belonging to big city chaos and into a space conjuring inner peace. I renew. With the ritual Salaat, I generate serenity. I can create and channel strong energy as I pray.

Although I only cover for prayer, I deeply admire women who choose to outwardly manifest their connection to the Divine within. I want more non-Muslims to understand veiled Muslim women and respect them for celebrating Islamic creed, for resisting overwhelming economic forces in this country, for not succumbing to the images captured in high fashion gloss. By living in constant alignment with faith, they challenge the misogynist systems that compel too many Western women and girls to binge, purge, and starve themselves. For these pious sisters, plain cloth is the most meaningful accessory they could ever wear. To me, American-Muslim women who choose to cover undeniably act out real life resistance to the hyper-sexualization of girls and women in the West. In the context of consumerist America, women who cover express power of intellect over silhouette, of mind over matters of the flesh.

Because I move in non-Muslim circles, I hear too many of my fellow feminists focus on *hijab,* urging complete unveiling as the key to unleashing an authentic liberation. For them, scarves strangle any movement toward Muslim women's emancipation. I ask them to just imagine 1,400 years—generations—of women moving without bustles, hoops, garters, bustiers, corsets, zippers, pantyhose, buckles, belts, pins, and supertight micro-minis. The way I see it, Western men wear comfortable shoes and slacks while women are pinned, underwired, heeled, and buttoned to psychological death. We American women still strangle ourselves every day we get up and get dressed for work.

Ah, you say—but even in their loose garb, Muslim women are still, so, so . . . passive. But Muslim women are not silent, not sitting still. We do not require American pity. We take the very best America has to offer. We are moving our bodies. As American women wow the world with unleashed athletic excellence in the WNBA, women's soccer, bobsledding, and pole vaulting, Muslim-American women are running and kicking along that mainstream—in full *hijab* or not. For Muslim immigrants who hail from nations that denied women access to physical movement, this country has freed them to pilot their own bodies. Many American Muslims are destroying the cultural forces that chained them while remaining true to the essence of Islam.

I remember hearing Sister Ama Shabazz, a bi-racial Muslim educator and lecturer (her mother is Japanese-American and her father is African-American) urging a large group of Muslim women to take swimming classes and learn CPR to satisfy the Shariah (Islamic law) not to defy it. (Anecdotes can be so helpful sometimes.) A friend of mine, whose mother immigrated to this country from Colombia, wears long loose clothing to the gym three days a week, then washes to cover and go home. An African-American girlfriend of mine roller-blades through her Bedford Stuyvescent neighborhood, full *hijab* blowing through her own body's wind.

These women fiercely assert Islam even though they have felt American hands tug at their clothing, especially since 9/11. They are obviously Muslim even though non-Muslims hurl offensive epithets or gestures at them. They do not cower. "It takes a warrior to be a Muslim woman," says another friend, a New Yorker born and bred Dominican. I agree with her.

Yet there is so much promise in the future: I know two Muslim high school students of mixed Iraqi/Indian heritage who play tennis in traditional whites and have earned black belts in karate. One even coaches the boys' basketball team. Interestingly, their immigrant mother could not wear blue jeans because her father forbade it. Certainly some men are still using women to assert a political agenda via Islam. I recoil when I see young Muslim girls in full *hijab* while their brothers skip beside them in shorts and t-shirts. I think about the women I know who cover themselves and their daughters for the wrong reason, and then I remember I know some women who wear push-up bras for the same wrong reason: to please men.

We must recognize that the similarities in our oppression as women far outweigh the differences in the ways that oppression manifests. And to do that we must fuse our stories. Like African-American women who have fought to wear locks and braids and naturals on the job—or have fought to use relaxers without the ultra-righteous disparaging them as loser sell-outs—Muslim women have had to fight to wear *hijab* here. For everything from job security to an American passport photo, Muslim women have been asked to uncover. At root, we are all denied our right to represent an authentic self by these predominant cultural and social forces.

The last place we women of all faiths need to suffer the indignity of judgment based solely on outward appearance is in the company of other women. Right now, half of American non-Muslim women encourage other women to be free by being naked, and the other half desperately tries to get women and girls to cover up. Meanwhile, the men simply get dressed in the morning. Likewise, Muslim women who wear *hijab* are automatically considered unhappy, while men who wear turbans and long loose clothing are just considered Muslim.

Non-Muslim women need to stop telling Muslim women their traditional Islamic garb symbolizes oppression. Muslim women need to open themselves to coalitions with women in mini-skirts. Only then will we work successfully toward a world where all women can truly wear what they feel. Ultimately, societies grant men much more freedom in clothing. Perhaps this is the point from which our discussion should launch.

We must begin to think more critically, and honestly, about media representations of all women. While Muslim immigrants need to reconsider the East's portrayal of American women as loose and wild, feminists need to check their sweeping generalizations about the seemingly inherent violence and suppression the media projects as Islam.

Since the 1970s America has slowly shifted evil empire status from the former Soviet Union to the site of underground power, where black, slick, liquid energy fuels America's Middle East policy. Americans have been taught to fear the Arab world so that America can easily justify killing Arabs. Images of Middle Eastern men in the state of *jihad* demonized the people of an entire region. But the only legitimate Islamic war is a war waged in self-defense—the other guy must be the clear aggressor. And the direct translation of *jihad* is struggle, while the primary focus of that struggle is within. We must remember this as we watch our evening news, as we watch bombs fall from U.S. planes. We must remember this as we vote.

American feminists should not join the Pentagon and media in denigrating veiled women and our faith as archaic, out-of-touch, regressive. This is part of the propaganda of fear America needs to perpetuate in order to maintain world dominance. This country takes the very universal problem of sexism and often presents it as if it were exclusively a Muslim issue, as when non-Muslims degrade Islam for allowing men to marry up to four wives, even as American men practice their own kind of polygamy—via mistresses, madams, and baby's-mammas. Who are we to judge? After all, while the United States has never had a woman president, Pakistan, Turkey, and Bangladesh—all Muslim countries—have had female heads of state.

Of course, as Jane Smith of the Hartford Seminary says, "I think you'd have to be blind not to see things going on in the Islamic world—and in the name of Islam—that are not Islamic." Muslim women would do well to remember that the Hadith (sayings of Prophet Muhammad) have been interpreted to give men powerful social advantages over women, and that there are men, and women complicit in their own oppression, who use Islam as justification for misogyny.

Certainly the 9/11 attacks were not Islamic. Islam means peace. We greet each other with peace. Islam is no more violent than Christianity. Yet there have been American Christian networks formed to throw bombs—at abortion clinics. When have you ever heard the term Christian terrorist? We Americans do not profile white men with crew cuts, but a white man in a crew cut bombed the federal office building in Oklahoma City. Certainly we should not denigrate Christianity—and Christians—because of the few who would use their faith as a justification for violence. Why has it been easy for white Americans to turn on their own darker brothers?

Maybe we simply need to understand each other. Certainly America needs to begin the work necessary to understand Islam. El Hajj Malik El Shabazz said in his letter from Mecca, "America needs to understand Islam, because it is the one religion that erases from its society the race problem." Back at the Islamic Cultural Center in Manhattan, when the congregational Jummah prayer concludes, chatter fills the once hushed room as women and men prepare to leave. They step off the carpet, slip on their shoes, and the women readjust their *hijabs.* Vari-colored Indian and Pakistani sisters toss beautifully brilliant cloth sari-style. Olive-skinned Arab sisters check the pin securing the cotton scarves underneath their chins. Deep brown African sisters toss oversized lightweight cloth in their handbags, revealing their artfully wrapped gelees. And some women—of all colors and nationalities—take their *hijab* off completely, now that prayer has ended. This dynamic diversity might just be what the next wave of American feminism needs.

Muslim women and men are active forces in many different struggles, just as Western feminists struggle against misogynist forces. As a Muslim brother of mine once reminded me, race is just a smokescreen. Gender is just a smokescreen. Religion is just a smokescreen. These are tools of the oppressor used to separate and slay as he takes.

We have so much work to do. I chose Islam for the wonder of the word, because I believe in the five pillars of the faith, because I love Allah and justice. I have been blessed to bear witness to women's realities in what people think of as two different worlds, and I have seen that those realities are essentially the same.

I bear witness to the woman beaten by her lover in the street outside my Brooklyn apartment—and to the woman tied to a Nigerian whipping post. I bear witness to the woman forced to strip to survive in Atlanta—and the woman forced to cover to survive in Afghanistan. I bear witness to the ever-increasing legions of women caught in this country's prison industrial complex, often because of their associations with husbands and male lovers—and I bear witness to the women struggling against inequity in interpretations of the Shariah in Islamic courtrooms. How do we measure a veiled woman's pain? Does it weigh more or less than the trauma in an American woman's eyes? Should we compare and contrast the horror, brutality, and hard smack against a woman's cheek?

I simply ask that we warrior women, Muslim and non-Muslim, stand shoulder to shoulder, forming ranks, bending forward to carry all our sisters, Muslim and non-Muslim, tapping our collective strength.

62

"Whosoever" Is Welcome Here

An Interview with Reverend Edwin C. Sanders II

GARY DAVID COMSTOCK

Reverend Edwin C. Sanders II is the founding pastor of Metropolitan Interdenominational Church, which is located in a working class neighborhood of small houses in Nashville, Tennessee. Reverend Sanders is African American and the congregation is predominantly African American. Lesbian/bisexual/gay/transgendered people are welcome and encouraged to participate in the life of the church. I attended a packed Sunday morning service in June 1998 and interviewed Rev. Sanders in the afternoon.

GARY COMSTOCK: How did the church get started?

REVEREND SANDERS: I had been the Dean of the Chapel at Fisk University. I left in 1980 in a moment of controversy. A new president had come, and we weren't able to mesh. I left and had no where to go. I didn't have a plan. And instantly there were folks who were part of the chapel experience there at Fisk who wanted to organize a new church. I felt no spiritual interest whatsoever in organizing a new church, but about seven months later I felt like I clearly heard the voice saying, "This is something to do." I'm glad it worked out that way, because I think if I had done it directly after Fisk it would have been born out of a reaction and we probably wouldn't have developed the kind of identity, sense of mission, and direction the way we did. There were twelve people who came together and said they wanted to do this.

One of my good friends, Bill Turner, is a sociologist, and we were at Fisk together. Bill has a theory he advanced that institutions—and he built his theory around black institutions—cannot break out of the mold from which they were born. There was something about the way an institution is framed in its beginning, and no matter what you do

SOURCE: From Eric Brandt, ed., *Dangerous Liaisons: Blacks, Gays, and the Struggle for Equality* (New York: The New Press, 1999), pp. 142–57. Reprinted by permission of the author.

you don't escape it. I thought that was absurd, but in time I came to think that he had something. The congregation was a mix of white and black men and women from all kinds of denominational backgrounds, and that mix turned out to be significant because from the beginning people identified us as being inclusive at least across racial terms and definitely inclusive and equal in gender terms. . . .

In that original group we had also a young man—one of my very dear friends—who was gay. Don was living a bisexual lifestyle at that point, but mainly to keep up appearances for professional purposes. Another one of my dearest friends went through a major mental breakdown at the time we were starting the church. I felt it very important not to abandon him and to include him. So we had a guy going through major psychological problems, somebody who was gay, and we had the racial mix.

We did not have the class mix at the beginning. Most of the folks were associated in some way with the academic community. Nothing like we have now. Today we have an unbelievable mix of people. I mean there are people who are doctors, lawyers, dentists and business people, and we also have a lot of folks who are right off the street, blue collar workers, in treatment for drugs and alcohol, going through a lot of transition. And that mix has evolved. Although we said in the beginning that's what we wanted to be, in actuality the current mix goes beyond that of the original twelve members.

The presence of Don, the one gay black male in the original congregation, had a lot to do with our current mix of people, and it had a lot to do with how we got so involved with HIV/AIDS ministry. The church began in '81 and he died in '84. It's almost hilarious when I think about it because I'm so involved in HIV/AIDS work now, but when he died I remember he was real sick, we didn't know what was going on, and they told us he died of toxicosis. I remember saying what in the world is that? I researched it and found out that it had something to do with cat and bird droppings, but Don didn't have cats. It was AIDS. You'd hear people talking about this strange disease because then it was 1984, but what that meant for us was that we got involved before it became a publicly recognized and discussed issue. Don's presence and death immediately pushed us in ministering to and being responsive to folks with AIDS and in dealing with the issue of homosexuality.

The other thing I was going to tell you which is kind of funny is the name. I will never forget when I told one of my friends we were going to name the church Metropolitan

Interdenominational. He said are you sure you want to do that. I said what do you mean, you know, I was pretty naive. He said all the gay churches across the country are called Metropolitan churches. I said that's where I feel the Lord leads me, I feel Metropolitan. In my mind what that meant was that Nashville happens to have been the first metropolitan government in the United States. It's the first place where the county and city combined, so everything in and around Nashville is referred to as metropolitan government. But my friend said to me that everybody is going to instantly say you're the gay church. I said so be it and we went on with it. Like I said, there was a hand bigger than mine at work. . . .

COMSTOCK: What was the turnaround point for the class mix? How and when did it happen?

REV. SANDERS: It was real clear. Early on what happened was we got involved in prison ministry—actually directed a ministry called the Southern Prison Ministry for a while. Going in and out of the prisons we started to develop relationships that translated to folks coming out of prison and getting involved in the church. After I had done the prison ministry for a while it became crystal clear to me that 80 percent of the people I was dealing with in prison were there for alcohol and drug related issues. I started reassessing this whole issue of drugs and alcohol and decided that's an area we had to begin to focus our ministries. So, I got involved, did the training, and became certified as a counselor. I would venture to say that 25 percent of the people in this church are folks that I first met in treatment. Thirty to 35 plus percent of this congregation are folks who are in treatment from alcohol and drug use. This is a place where a lot of folks know they can come. We've got all these names, you know, the drug church, the AIDS church, we get those labels. But it's all right because that's what we do. We hit those themes a lot.

I've learned some things over the years that I pretty much hold to and this is one of them. . . . We don't say we are a black church, we don't say that we are a gay church or a straight church, we don't say that we are anything other than a church that celebrates our oneness in Christ. I'm convinced that has turned out to be the real key to being able to hold this diverse group of folks together. I must admit I'm a person that has a negative thing about the word diversities. I don't use it much. We don't celebrate our differences, we celebrate our oneness. That ends up being an avenue for a lot of folks being attracted and feeling comfortable here. New folks say I got here and I just felt like no one was looking at me strange, no one was treating me different, I was just able to be here. . . .

COMSTOCK: I was impressed by the informality of the service today. People seem to relax and fit in. The choir is not so focused on performance that they're not interacting with the congregation. And your own manner is informal. It's a style that lets people in. You aren't just saying all people are welcomed here, you actually do something that let's them feel at home. I was also struck by the openness of the windows—the plants inside and the view into the park. . . .

REV. SANDERS: . . . We have the sense of informality. People are arriving from the time we welcome the guests at the beginning until just after the sermon when there's maximum presence in the audience. People come in slow, and we give folks an opportunity to leave early. We even say it in the bulletin, we just say if you need to leave just leave quietly. And we know folks do. There's a lot of folks who want to hear the choir or they want to hear the sermon, but they don't want all the rest of it. So we do it that way. It works out. One of the real hooks for just about everyone at Metropolitan is the fellowship circle at the end. We actually have a few Jews and a couple of Muslims who worship here regularly. The Muslim family does a very interesting thing which helped me to appreciate the significance of the fellowship circle at the end. We offer communion—the Eucharist, the Lord's supper—every Sunday, and they stay until we get to that part of the service. Then they go outside or into the vestibule, but they come back in. I remember when Omar first started attending, I said to him it's interesting to me that you don't leave at that point. He said, "No, no, no, I wouldn't miss the fellowship circle." And it made me realize that the communion of the fellowship circle is more important than the bread and the wine. Probably the real communion is what we do when we stand there at the end and sing "We've Come Too Far to Turn Back Now." We do that every week. It's our theme. That's a very significant moment. I've heard people tell me when they have to be late that they rush to get here just to catch the end of it. One woman said to me it was enough for her if she just got here for the fellowship circle.

A lot of the informality is very intentional. For instance, the only thing at Metropolitan that's elevated is the altar. Nothing else is above ground level. None of the seating is differentiated. In most churches the ministers have different, higher, bigger chairs than everyone else. We don't do that. We sit in the same seats the other folks do. My choir director is always telling me we'd get better sound if we could elevate the back rows. I say no, everybody's got to be on the same level and got to sit in the same seats. We do a lot of symbolic things like that.

COMSTOCK: Do other clergy give you much flack for working with needle exchange, welcoming gay folks, and working on issues that they may see as too progressive?

REV. SANDERS: They do. But let me tell you something. I have been able to have a level of involvement with ministers in this community that probably has brought credibility to what we do. I would like to think that we have maintained our sense of integrity especially as it relates to our consistency in ministry. Folks tend to respect that, so even when they disagree, they also look more seriously and harder at what it is they're questioning. . . .

COMSTOCK: "Whosoever" is from John 3:16–17: "For God so loved the world that he gave his only Son, that whosoever believes in him should not perish but have eternal life." Has it been an expressed theme from the beginning?

REV. SANDERS: It's been our theme for the last seven or eight years. . . . I think we're growing in this inclusivity all the time. The language issue was big for us. We try to use inclusive language. If somebody else comes into the church that is not into that, it sticks out like a sore thumb. Does our inclusivity also mean that there is a real tolerance for folks who are perhaps not where we think folks should be in terms of issues like inclusive language, issues that relate to sexuality, gay and lesbian issues? I think the answer to that question is yes, you have to make room for those folks too. That's a real struggle. Another one of our little clichés, and we don't have a lot of them, is we say we try to be inclusive of all and alienating to none. Not being alienating is a real trick. It's amazing how easy you can alienate folks without realizing it, in ways that you're just not aware of. . . .

COMSTOCK: What would you do if some new people had trouble with cross dressing, transgendered, lesbian or gay people? How do you get them to stay and deal with it rather than leave?

REV. SANDERS: We tracked that issue a couple of times. Let me tell you what's happened. Most folk are here for a while before some of it settles in, before they start to notice everything. It's amazing to me how people get caught up in Metropolitan. They'll join and get involved and then they'll go through membership class, that's usually when it starts to hit them, that they say to themselves, "Oh oh, I'm seeing some stuff that I'm not sure about here." But what we've discovered is that what seems to help us more than anything else is that folks end up remembering what their initial experience had been when they first came here. We even have one couple, this is my favorite story to tell, who has talked about coming here on one of those Sundays when we were hitting the theme of inclusive hard. And they left here and said what kind of church is that? But then a couple of weeks later they said let's go back over there again. They came back and then they didn't come again for about two or three months. They were visiting other churches. They said when they got down to thinking about all the churches, they had felt warmth and connection here. They said you know that church really did kind of work; and they ended up coming back. When folks are here there's a warmth they feel, there is a connection they feel, there's a comfort zone in which they'll eventually deal with the issues that might be their point of difference. We've seen folks move in their thinking and there's some folks that have not been able to do it. I've got one young man—I really do think he just loves being here—who says, "I just can't fathom this gay thing. Why do you insist on it." I said, "You know, you're the one who's lifting this up."

As I said, we try to make sure that when any issue is brought up, it's in the course of things. It's not like we stop and have gay liberation day, just like we don't celebrate Black History month. We don't focus on or celebrate these identities or difference, but yet if you're around here you can't help but pick up on the church's support for these issues and differences. Today, for instance, we sang "Lift Up Your Voice and Sing," which is known as the black national anthem. But we call it "Our Song of Liberation," and I've tried to help people to understand that. . . .

I have a lot of divinity students who serve as my pastoral assistants, and most of them are women. This place has become a real refuge for black women in the ministry. There are not a lot of clergy opportunities for them, so I've tried to figure out how to incorporate them into the life and ministry of the church. They actually run a lot of our ministries. And most of them are extremely well prepared academically. They're more prepared than I am to do the work and they do it well. When they started coming we suddenly became a magnet, a place where there was this rush of folks out of divinity school to come and be a part of ministry here. At one level I probably seem like I'm extremely freewheeling and loose, but I'm probably a lot more intentional about how things are going than folks realize, especially as they relate to the focus of ministry here at the church. One of my real concerns was not having the time to orient the young ministers, to bring them into a full awareness of what makes this place click. What I have discovered is that if I get the inclusive piece established in the beginning with them, I don't have to worry about the rest of it as much. If they buy into understanding that inclusivity is at the heart of this church, I don't have to worry about keeping my eye on them all the time. . . .

COMSTOCK: You know that the example here of accepting and welcoming lesbians and gay men is an anomaly in the Church. You provide inspiration and hope, an example of something positive that is actually happening. Most gay people feel alienated from their churches, but I've found that African-American more often than white gay people emphasize the importance of religion in their family, community, and history and say that most of the pain and sadness in their lives centers around the Church. They claim that being rejected by the black church is especially devastating because this institution has been, and continues to be, the only place where they can take real refuge from the racism they experience in the society. Clearly that's not happening here at Metropolitan.

REV. SANDERS: We realized when we started doing our HIV/AIDS ministry that there are organizations that were established by gay white men who had done a good job of developing services, but we kept seeing there was something that was not happening for gay and lesbian African Americans. And we realized it was community. No matter how much they tried to get that in other communities, there is the whole thing of cultural comfort. They were looking for the context where there were people who looked like them, who were extensions of their family supporting them. It became clear to me that what we needed to do was to establish a place where people could literally come and where people had a sense of community. Consequently, what we call the Wellness Center here is more than anything else just a place where you can come and feel at home. It has sofas and tables and chairs. Folks sit around. It's a place where there is a community that is an insulating, supporting place to come to. And although we provide some direct services for those folks, more than anything else I think the greatest service we provide is a safe place, a comfort zone, a community, a way to be connected to community. So I know real clearly what you're talking about, and I've heard it spoken too. It's one of the real issues for African Americans who are gay and lesbian. So our simple response has been to try to create a space where folks will have a certain level of comfort.

The problem I often run into, which speaks to what you were asking about, is that I think the greatest trouble we have sometimes is folks in the gay and lesbian community want to celebrate their sense of life and lifestyle more than Metropolitan lends itself to. In the same way, I have folks who don't understand why we don't do more things that are clearly more defined as being Afrocentric. There's a strong Afrocentric movement now within religious circles, and I end up dealing with them the same way I do with the folks who want to have ways in which they celebrate being gay and lesbian. My greatest struggle ends up being with folks who want to lift that up more, and I say the only thing we lift up is the basis of our oneness. I think the Church has been a place where folks have not been able to find community, when they have been rejected. The African American church is a pretty conservative entity, always has been. It's probably even better at holding up the conventions of American Christianity than other institutions. Consequently folks can draw some pretty hard lines, and that's what we've been dealing with at Metropolitan. But the other side of the Black church is that it is known as being an institution of compassion. So at the same time you have folks giving voice to some conservative ideas and practices, you can also appeal to a tradition and practice of compassion. That's why in our HIV/AIDS ministry we've been able to bring other churches into the loop. We're trying to get thirty churches to be involved in doing education and intervention on HIV/AIDS. We've been able to engage them on the level of compassion. Once we get them involved at that level then, we see folks open their eyes to other issues. African American churches have been very effective in compassion ministry for years around issues of sickness, death and dying. What we're trying to do is get folks to put as much emphasis on living and not just being there to minister to sickness and dying.

Another thing is the contradictions in the Church. The unspoken message that says it's all right for you to be here, just don't say anything, just play your little role. You can be in the choir, you can sit on the piano bench, but don't say you're gay. We had an experience here in Nashville which I'm sure could be evidenced in other places in the world. You know how a few years ago in the community of male figure skaters there was a series of deaths related to HIV/AIDS. The same thing happened a few years ago with musicians in the black churches. At one point here in Nashville there were six musicians who died of AIDS. In every instance it was treated with a hush. Nobody wanted to deal with the fact that all of these men were gay black men, and yet they'd been there leading the music for them. It's that contradiction where folks say yeah you're here but don't say anything about who you really are, don't be honest and open about yourself. I believe that the way in which you get the Church to respond is to continue to force the issue in terms of the teachings of Christ, to be forthright in seeing how the issue is understood in relationship to Jesus Christ. One thing I've learned about dealing with inner city African Americans is that you have to bring it home for them in a way that has some biblical basis. I'm always challenged by this, but I'm always challenging them to find a place where Jesus ever rejected anyone. I don't think anyone can find it. I don't think there's anybody that Jesus did not embrace.

63

The Starbucks Paradox

KIM FELLNER

It was November 1999 in Seattle and the U.S. global justice movement had taken to the streets. Suddenly, there was a crash not 20 feet from where I stood, and the Starbucks window collapsed in a hail of glass. "Zowie," I said to

SOURCE: Kim Fellner, "The Starbucks Paradox," by Kim Fellner. *Colorlines* (Spring 2004 v. 7, i1). Reprinted by permission of Colorlines Magazine, www.colorlines.com.

my husband Alec, "I sure hate the WTO and capitalism, but that's our coffee store."

I've been pondering the contradictions ever since, watching my progressive colleagues devour the latest episode of *Friends,* gleefully shop for real estate, and sip their lattes. Political ideology notwithstanding, we were avid participants in popular culture and petty capital. How could we so glibly demonize that which we so cheerfully consumed?

And as I went around the country, I couldn't help notice that both the employees and habituès of Starbucks seemed far more diverse by race and class than the American anti-globalization movement. I wanted to know, was big, by definition, bad? Was Starbucks' touted commitment to "values" just a cynical ploy to complement its branding and market share?

Few people argued that coffee was inherently evil like bombs or SUVs. Rather, Starbucks stood accused of: buying coffee at prices that couldn't sustain the farmers; purchasing from farms that degraded the environment; causing neighborhoods to gentrify and small cafes to wither; and representing the mega-branding that's killing small businesses and homogenizing the world.

I frankly like having Starbucks at the airport, and at strip malls in strange cities. I wouldn't mind independently-owned coffee shops instead, but Starbucks is usually what's there. Moreover, progressives have tended to romanticize small businesses; yet many sweatshops in this country have been small, family-owned enterprises, and that didn't benefit those who worked there. As a rule, racial minorities have fared better in larger institutions. Was Starbucks doing right by race? What about class and politics?

What, specifically, is wrong with this emblematic corporation? And is anything right? It seems to me that the movement is old enough to make some distinctions.

THE VIEW FROM HEADQUARTERS

The mermaid from the Starbucks logo peers coyly from the top of the refurbished Sears warehouse in Seattle that serves as company headquarters. Although one union official claims this is cultishness gone amok, I can't help seeing it as a humorous and engaging design feature.

The people I meet are also humorous and engaging. They include Paula Boggs, executive vice president, general counsel and secretary; Wanda Herndon, senior VP, Worldwide Public Affairs; David Pace, executive VP, Partner Resources; and Sandra Taylor, senior VP, Corporate Social Responsibility. While they are polished spokespeople for the Starbucks mission and policies, they don't really strike me as cult material. What does strike me is that three of the four are African American women. Women comprise barely 13 percent of Fortune 500 general counsel and women of color merely 1.6 percent of all corporate officer positions. At Starbucks, more than 34 percent of the top officers (vice president or higher) are women and/or people of color. And although Starbucks is cagey about its statistics, they're clearly far ahead of the miserable norm. "Certainly, at the top of this company, diversity in the form of race rings loud," says Boggs, the general counsel. "I would think, fairly, we have some work to do when it comes to middle management, but we're on the right path."

"You will see diversity by age, sex, race, sexual orientation, family status throughout the organization," David Pace asserts. "It holds an equal place in our guiding principles, just like respect and dignity, just like customer satisfaction."

Those principles are spelled out in the company's six-point mission statement, which is central to the way the company defines itself and acculturates its employees. Yet Pace is aware that, "Our guiding statement is probably not that different from people who are involved in scandals. The difference is the commitment to it."

MAKING GOOD ON EIGHTH STREET

Tawana Green, a lively 28-year-old African American, Washington, DC native with unruly dreads, believes in the Starbucks mission. She started with the company the same year as the Seattle protests and now, four years later, is the manager of the new Starbucks on SE Eighth Street, two blocks from my home on Capitol Hill.

It's the opportunity Green was looking for. "I've been on my own since I was 17," she relates, "and started work right out of high school. I did the office thing, but found myself sleeping at the computer. So I went into retail where I could work with people."

She was impressed by the full benefits for anyone working 20 hours or more, almost unheard of in the world of retail. In addition to a generous health, life, and disability insurance package, the company offers: a 401K with a match; Starbucks "bean stock," which reflects the annual success of the company and is allocated among employees based on hours worked; discounts on Starbucks public stock; and to top it off, a free pound of coffee every week.

She's hired a total of 15 workers, and all but one survived the first month. Visiting on different shifts, it's clear that she's assembled mostly people of color from the community; they start at $7.50 an hour plus tips, and most of them work enough hours to qualify for the health benefits. Although some Starbucks workers claim they have a difficult time getting those hours, the company says that more than two-thirds of all employees qualify.

As I listen to Green and take in the familiar décor, the comfortable chairs and small merchandise displays, I find myself thinking about what journalist/activist Naomi Klein describes as the branding phenomenon, where what's on sale is not just a product but an image and a lifestyle: not just coffee, but "The Starbucks Experience." Klein hates the clone-like nature of the stores and what she sees as the cynical manipulation of the consumer. And I honestly know what she means. Starbucks is putting in a bid to be a Pleasantville version of the village square.

Then I look at Green, her co-workers, and the customers, and it doesn't look like such a terrible thing. "My goal is to provide a great place to hang out and also give back to the community," Tawana Green tells me. "I'm not saying it's easy. But here I am, 28, with a high school education, and I'm a store manager with a piece of the company. Where else could I get this opportunity? It doesn't get much better than this!"

UNIFORM MISERY

However, for Sandra Evans, an employee at Cintas—the mega-company producing uniforms, laundry, and cleaning supply products with which Starbucks contracts—it doesn't get much worse. Evans, a 57-year-old African American, works 40 hours a week sorting shirts at the Cintas plant in Aspen, Pennsylvania. She also works 20 hours at the local K-Mart. "I've got two jobs, and I'm still struggling," she wryly notes. "This company is making millions, and people can't even pay their electric bills." She loves to walk, entertain, and cook, but "I'm just too busy working." And while she drinks a cup of coffee a day, she's never had a Starbucks, "although I hear they're excellent."

Driven by grievances of unpaid and forced overtime, discrimination and unsafe working conditions, she and her fellow workers are engaged in a campaign to join UNITE, the Union of Needletrades, Industrial and Textile Employees. Evans is also among a group of workers who recently filed an EEOC class action complaint against Cintas for gender and race discrimination. "When I got hired two years ago, they told me $8.50 was the starting rate. Then they hired an 18-year-old white girl, right out of high school, and started her at $9.50. Recently a Hispanic woman started, and they're paying her $8.25." Evans reports there's only one black supervisor at her plant; at another plant, they do not hire any blacks as drivers, even if they are well-qualified, and some plants are stratified by gender as well.

"It's a sad situation," Evans sighs. "I just hope they can be a fair company, hire people at the same rate, and a decent rate, and give people of color a chance to move up. It's not just about me, and it's not just about money. It's about finally treating people with respect."

VOICE VS. VOTE

"I've met Howard twice," Tawana Green tells me, easily referring to Starbucks board chair Howard Schultz. At a company leadership conference, "he told us, 'we take care of the partners, the partners take care of the customers, and the customers take care of the business.'"

Green avoids the word "employees" in favor of the company lexicon, "partners," and quickly corrects herself the one time the e-word slips in. However, a number of Starbucks workers in Canada and the U.S. have begged to differ on the partner/employee question.

Jef Keighley, a national representative for the Canadian Automobile Workers, is bitter when he describes his experiences organizing Starbucks in British Columbia, where the union is in a drawn-out contract negotiation covering 10 Vancouver stores. "We used to have 12 stores," he says, "but the company has had a hand in organizing decertifications at two of those stores, even selecting and paying for the lawyer. We've been at the Labor Board for a year and a half."

Keighley admires Starbucks' tactical and PR sophistication, but dismisses any sincerity around the mission statement as "an absolute crock." He's particularly exercised about the bean stock benefit. "Most would be better off with the money in their paychecks," he asserts. "Our committee knows they're only getting the sizzle while Starbucks keeps the steak."

In his own book, Howard Schultz—a Brooklyn boy made good—clearly expresses a belief that benevolent management should make unions superfluous. It's almost as though the desire to have outside representation is seen as a personal affront that hurts Schultz's feelings more than his profit margins.

UNITE says it has yet to hear from Starbucks about Cintas. But Starbucks General Counsel Paula Boggs insists that, "We take any allegation regarding our suppliers very seriously. We have instituted a supplier code of conduct, and we do look for partners that share our values. Is it perfect, always? No. But it's certainly worked in the main for us."

BRINGING IT ALL BACK HOME

So, what about the most common grievances against Starbucks?

A crucial concern on the global level has been the survival of coffee farmers and the environmental sustainability of production. Coffee is second only to petroleum on the world commodities market, and the price is at catastrophic lows of 40 to 50 cents per pound, threatening roughly 250 million small farmers. There seems little question that pressure from human rights and environmental groups in the mid-'90s spurred Starbucks' commitments in Third World coffee-growing countries. They instituted coffee-sourcing guidelines, and have negotiated long-term contracts and created direct relationships with suppliers to stabilize the income of the farmers. In 2002, they paid an average of $1.20 per pound up to a high of about $1.41. Steve Coats of the US/Labor and Education in the Americas Project (US/LEAP) suggests that, "They don't move any faster than you push them. They may purchase only 1 percent of the coffee, but they can set the standards and tone." However, both US/LEAP and Global Exchange acknowledge that Starbucks has made significant changes in its policies and practices.

There is also a pervasive belief that Starbucks is driving out the independents, but evidence is slim. As I recall, in the distant past of 15 years ago, most coffeehouses were near universities and in Italian neighborhoods. Cheap coffee was found at the local diner, and good coffee was an oxymoron. Are we really saying that there should be no coffee-houses in urban communities of color? Are we arguing against a decent latte at the airport? It may be an unnecessary luxury. But a sign of evil conspiracy? I'm not convinced.

Nor is Starbucks equivalent to Wal-Mart; it has not forced a downward spiral in either prices or wages, it is arguably the best employer in its sector, and it has not turned old main streets into ghost towns. However, Starbucks *has* joined McDonald's and Wal-Mart in hyper-aggressive expansion, especially abroad, thus setting itself up as a symbol of the Americanization of the world.

At the community level, though, it's often hard for people to understand why the store where they shop and work has been targeted. "I remember being in a demonstration where a McDonald's or Starbucks window, I forget which, was shattered," says Colin Rajah, a long-time global justice organizer, currently a program associate at the National Network of Immigrant and Refugee Rights. "We could see the staff, people of color, running toward the back of the store and cowering behind the counter, really scared. I was in a group of young global justice activists of color, and we immediately found ourselves identifying with the people inside the store. And we realized that tactic was a mistake.

"Sure we have to look at the larger picture and how the capitalist system works, but if these entities provide a service to the community, and are not abusing their workers, there's something to be said for that as well. What we should be looking for is the democratization of the process."

I couldn't really get an answer at Starbucks headquarters on how big was too big. "I guess too big is all relative," says Starbucks VP Wanda Herndon, "but big is not necessarily bad if you're doing good things." Mostly, she feels misunderstood by the protesters, who have continued to sporadically damage stores around the country. "The way I felt was that the people who were protesting didn't understand who Starbucks really is, the company, the culture, the people," she explains. "But because our siren makes good news, we became a symbol of all that went wrong with WTO.

"We've always tried to balance profitability with benevolence, and I think that's a concept that's hard for people to grasp. You want to be profitable, but you realize you need to give back if you want to be a part of the communities in which you do business."

As it turns out, Starbucks appears improved on trade, good on race, and a model on health care for part-timers. Amazingly, they have only recently acquired their first government relations staffer, and it remains to be seen whether they will parlay their position into fighting nationally for civil rights, the environment, or national health care.

"People who go into corporate management didn't sign up to be civil servants," notes global justice organizer Liz Butler. "But increasingly, the crucial decisions are being made in board rooms, and we need them to take on that role."

And, at Starbucks, there's some hope of getting through; after all, some of the top management were the anti-war activists of the '70s. Meanwhile, I'm certain that Tawana Green would care about the fate of Sandra Evans, though I wonder whether the company VP's will feel the same. I'm not holding my breath that they'll ever prefer union representation to benevolence. But I'd like to think that they will not be down with the race and gender discrimination of a Cintas.

As Liz Butler puts it, "McDonald's never says it's a good neighbor, but Starbucks does. It's our job to hold them to it." Perhaps then Starbucks won't be doing business with Sandra Evans' oppressor, and activists won't be throwing bricks through Tawana Green's window.

64

Women of Color on the Front Line

CELENE KRAUSS

Toxic waste disposal is a central focus of women's grass-roots environmental activism. Toxic waste facilities are predominantly sited in working-class and low-income communities and communities of color, reflecting the disproportionate burden placed on these communities by a political economy of growth that distributes the costs of economic growth unequally. Spurred by the threat that toxic wastes pose to family health and community survival, female grass-roots activists have assumed the leadership of community environmental struggles. As part of a larger movement for environmental justice, they constitute a diverse constituency, including working-class housewives and secretaries, rural African American farmers, urban residents, Mexican American farm workers, and Native Americans.

These activists attempt to differentiate themselves from what they see as the white, male, middle-class leadership of many national environmental organizations. Unlike the more abstract, issue-oriented focus of national groups, women's focus is on environmental issues that grow out of their concrete, immediate experiences. Female blue-collar activists often share a loosely defined ideology of environmental justice and a critique of dominant social institutions and mainstream environmental organizations, which they believe do not address the broader issues of inequality underlying environmental hazards. At the same time, these activists exhibit significant diversity in their conceptualization of toxic waste issues, reflecting different experiences of class, race, and ethnicity.

SOURCE: From Robert D. Bullard, ed., *Unequal Protection: Environmental Justice and Communities of Color* (San Francisco: Sierra Club Books, 1994), pp. 256–71. Reprinted by permission.

This [essay] looks at the ways in which different working-class women formulate ideologies of resistance around toxic waste issues and the process by
which they arrive at a concept of environmental justice. Through an analysis of
interviews, newsletters, and conference presentations, I show the voices of white,
African American, and Native American female activists and the resources that
inform and support their protests. What emerges is an environmental discourse
that is mediated by subjective experiences and interpretations and rooted in the
political truths women construct out of their identities as housewives, mothers,
and members of communities and racial and ethnic groups.

THE SUBJECTIVE DIMENSION OF
GRASS-ROOTS ACTIVISM

Grass-roots protest activities have often been trivialized, ignored, and viewed as
self-interested actions that are particularistic and parochial, failing to go beyond a
single-issue focus. This view of community grass-roots protests is held by most
policymakers as well as by many analysts of movements for progressive social
change.

In contrast, the voices of blue-collar women engaged in protests regarding
toxic waste issues tell us that single-issue protests are about more than the single
issue. They reveal a larger world of power and resistance, which in some measure
ends up challenging the social relations of power. This challenge becomes visible
when we shift the analysis of environmental activism to the experiences of
working-class women and the subjective meanings they create around toxic
waste issues.

In traditional sociological analysis, this subjective dimension of protest has
often been ignored or viewed as private and individualistic. Feminist theory,
however, helps us to see its importance. For feminists, the critical reflection on
the everyday world of experience is an important subjective dimension of social
change. Feminists show us that experience is not merely a personal, individualistic concept. It is social. People's experiences reflect where they fit in the social
hierarchy. Thus, blue-collar women of differing backgrounds interpret their
experiences of toxic waste problems within the context of their particular cultural histories, starting from different assumptions and arriving at concepts of
environmental justice that reflect broader experiences of class and race.

Feminist theorists also challenge a dominant ideology that separates the
"public" world of policy and power from the "private" and personal world of
everyday experience. By definition, this ideology relegates the lives and concerns
of women relating to home and family to the private, nonpolitical arena, leading
to invisibility of their grass-roots protests about issues such as toxic wastes. As
Ann Bookman has noted in her important study of working-class women's community struggles, women's political activism in general, and working-class political life at the community level in particular, remain "peripheral to the historical
record . . . where there is a tendency to privilege male political activity and labor

activism."[1] The women's movement took as its central task the reconceptualiza-tion of the political itself, critiquing this dominant ideology and constructing a new definition of the political located in the everyday world of ordinary women rather than in the world of public policy. Feminists provide a perspective for making visible the importance of particular, single-issue protests regarding toxic wastes by showing how ordinary women subjectively link the particulars of their private lives with a broader analysis of power in the public sphere.

Social historians such as George Rudé have pointed out that it is often diffi-cult to understand the experience and ideologies of resistance because ordinary working people appropriate and reshape traditional beliefs embedded within working-class culture, such as family and community. This point is also relevant for understanding the environmental protests of working-class women. Their protests are framed in terms of the traditions of motherhood and family; as a result, they often appear parochial or even conservative. As we shall see, how-ever, for working-class women, these traditions become the levers that set in motion a political process, shaping the language and oppositional meanings that emerge and providing resources for social change.

Shifting the analysis of toxic waste issues to the subjective experience of ordi-nary women makes visible a complex relationship between everyday life and the larger structures of public power. It reveals the potential for human agency that is hidden in a more traditional sociological approach and provides us with a means of seeing "the sources of power which subordinated groups have created."[2]

The analysis presented in this [essay] is based on the oral and written voices of women involved in toxic waste protests. Interviews were conducted at envi-ronmental conferences such as the First National People of Color Environmental Leadership Summit, Washington, D.C., 1991, and the World Women's Congress for a Healthy Planet, Miami, Florida, 1991, and by telephone. Additional sources include conference presentations, pamphlets, books, and other written materials that have emerged from this movement. This research is part of an ongoing comparative study that will examine the ways in which experiences of race, class, and ethnicity mediate women's environmental activ-ism. Future research includes an analysis of the environmental activism of Mexican American women in addition to that of the women discussed here.

TOXIC WASTE PROTESTS AND THE
RESOURCE OF MOTHERHOOD

Blue-collar women do not use the language of the bureaucrat to talk about envi-ronmental issues. They do not spout data or marshal statistics in support of their positions. In fact, interviews with these women rarely generate a lot of discussion about the environmental problem per se. But in telling their stories about their protest against a landfill or incinerator, they ultimately tell larger stories about their discovery or analysis of oppression. Theirs is a political, not a technical, analysis.

Working-class women of diverse racial and ethnic backgrounds identify the toxic waste movement as a women's movement, composed primarily of mothers. Says one woman who fought against an incinerator in Arizona and subsequently worked on other anti-incinerator campaigns throughout the state, "Women are the backbone of the grass-roots groups; they are the ones who stick with it, the ones who won't back off." By and large, it is women, in their traditional role as mothers, who make the link between toxic wastes and their children's ill health. They discover the hazards of toxic contamination: multiple miscarriages, birth defects, cancer deaths, and so on. This is not surprising, as the gender-based division of labor in a capitalist society gives working-class women the responsibility for the health of their children.

These women define their environmental protests as part of the work that mothers do. Cora Tucker, an African American activist who fought against uranium mining in Virginia and who now organizes nationally, says:

> It's not that I don't think that women are smarter, [she laughs] but I think that we are with the kids all day long. . . . If Johnny gets a cough and Mary gets a cough, we try to discover the problem.

Another activist from California sums up this view: "If we don't oppose an incinerator, we're not doing our work as mothers."

For these women, family serves as a spur to action, contradicting popular notions of family as conservative and parochial. Family has a very different meaning for these women than it does for the middle-class nuclear family. Theirs is a less privatized, extended family that is open, permeable, and attached to community. This more extended family creates the networks and resources that enable working-class communities to survive materially given few economic resources. The destruction of working-class neighborhoods by economic growth deprives blue-collar communities of the basic resources of survival; hence the resistance engendered by toxic waste issues. Working-class women's struggles over toxic waste issues are, at root, issues about survival. Ideologies of motherhood, traditionally relegated to the private sphere, become political resources that working-class women use to initiate and justify their resistance. In the process of protest, working-class women come to reject the dominant ideology, which separates the public and private arenas.

Working-class women's extended network of family and community serves as the vehicle for spreading information and concern about toxic waste issues. Extended networks of kinship and friendship become political resources of opposition. For example, in one community in Detroit, women discovered patterns of health problems while attending Tupperware parties. Frequently, a mother may read about a hazard in a newspaper, make a tentative connection between her own child's ill health and the pollutant, and start telephoning friends and family, developing an informal health survey. Such a discovery process is rooted in what Sara Ruddick has called the everyday practice of mothering.[3] Through their informal networks, they compare notes and experiences and develop an oppositional knowledge used to resist the dominant knowledge of experts and the decisions of government and corporate officials.

These women separate themselves from "mainstream" environmental organizations, which are seen as dominated by white, middle-class men and concerned with remote issues. Says one woman from Rahway, New Jersey: "The mainstream groups deal with safe issues. They want to stop incinerators to save the eagle, or they protect trees for the owl. But we say, what about the people?"

Another activist implicitly criticizes the mainstream environmental groups when she says of the grass-roots Citizens' Clearinghouse for Hazardous Wastes:

> Rather than oceans and lakes, they're concerned about kids dying.
> Once you've had someone in your family who has been attacked by the
> environment—I mean who has had cancer or some other disease—you
> get a keen sense of what's going on.

It is the traditional, "private" women's concerns about home, children, and family that provide the initial impetus for blue-collar women's involvement in issues of toxic waste. The political analyses they develop break down the public-private distinction of dominant ideology and frame a particular toxic waste issue within broader contexts of power relationships.

THE ROLE OF RACE, ETHNICITY, AND CLASS

Interviews with white, African American, and Native American women show that the starting places for and subsequent development of their analyses of toxic waste protests are mediated by issues of class, race, and ethnicity.

White working-class women come from a culture in which traditional women's roles center on the private arena of family. They often marry young; although they may work out of financial necessity, the primary roles from which they derive meaning and satisfaction are those of mothering and taking care of family. They are revered and supported for fulfilling the ideology of a patriarchal family. And these families often reflect a strong belief in the existing political system. The narratives of white working-class women involved in toxic waste issues are filled with the process by which they discover the injustice of their government, their own insecurity about entering the public sphere of politics, and the constraints of the patriarchal family, which, ironically prevent them from becoming fully active in the defense of their family, especially in their protest. Their narratives are marked by a strong initial faith in "their" government, as well as a remarkable transformation as they become disillusioned with the system. They discover "that they never knew what they were capable of doing in defense of their children."

For white working-class women, whose views on public issues are generally expressed only within family or among friends, entering a more public arena to confront toxic waste issues is often extremely stressful. "Even when I went to the PTA," says one activist, "I rarely spoke. I was so nervous." Says another: "My views have always been strong, but I expressed them only in the family. They were not for the public." A strong belief in the existing political system is characteristic of these women's initial response to toxic waste issues. Lois Gibbs, whose involvement in toxic waste issues started at Love Canal, tells us, "I believed if I had a problem I just had to go to the right person in government and he would take care of it."

Initially, white working-class women believe that all they have to do is give the government the facts and their problem will be taken care of. They become progressively disenchanted with what they view as the violation of their rights and the injustice of a system that allows their children and family to die. In the process, they develop a perspective of environmental justice rooted in issues of class, the attempt to make democracy real, and a critique of the corporate state. Says one activist who fought the siting of an incinerator in Sumter County, Alabama: "We need to stop letting economic development be the true God and religion of this country. We have to prevent big money from influencing our government."

A recurring theme in the narratives of these women is the transformation of their beliefs about government and power. Their politicization is rooted in the deep sense of violation, betrayal, and hurt they feel when they find that their government will not protect their families. Lois Gibbs sums up this feeling well:

> I grew up in a blue-collar community. We were very into democracy. There is something about discovering that democracy isn't democracy as we know it. When you lose faith in your government, it's like finding out your mother was fooling around on your father. I was very upset. It almost broke my heart because I really believed in the system. I still believe in the system, only now I believe that democracy is of the people and by the people, that people have to move it, it ain't gonna move by itself.

Echoes of this disillusionment are heard from white blue-collar women throughout the country. One activist relates:

> We decided to tell our elected officials about the problems of incineration because we didn't think they knew. Surely if they knew that there was a toxic waste dump in our county they would stop it. I was politically naive. I was really surprised because I live in an area that's like the Bible Belt of the South. Now I think the God of the United States is really economic development, and that has got to change.

Ultimately, these women become aware of the inequities of power as it is shaped by issues of class and gender. Highly traditional values of democracy and motherhood remain central to their lives. But in the process of politicization through their work on toxic waste issues, these values become transformed into resources of opposition that enable women to enter the public arena and challenge its legitimacy. They justify their resistance as a way to make democracy real and to protect their children.

White blue-collar women's stories are stories of transformations: transformations into more self-confident and assertive women; into political activists who challenge the existing system and feel powerful in that challenge; into wives and mothers who establish new relationships with their spouses (or get divorced) and new, empowering relationships with their children as they provide role models of women capable of fighting for what they believe in.

African American working-class women begin their involvement in toxic waste protests from a different place. They bring to their protests a political awareness that is grounded in race and that shares none of the white blue-collar women's initial trust in democratic institutions. These women view government with mistrust, having been victims of racist policies throughout their lives. Individual toxic waste issues are immediately framed within a broader political context and viewed as environmental racism. Says an African American activist from Rahway, New Jersey:

> When they sited the incinerator for Rahway, I wasn't surprised. All you have to do is look around my community to know that we are a dumping ground for all kinds of urban industrial projects that no one else wants. I knew this was about environmental racism the moment that they proposed the incinerator.

An African American woman who fought the siting of a landfill on the South Side of Chicago reiterates this view: "My community is an all-black community isolated from everyone. They don't care what happens to us." She describes her community as a "toxic doughnut":

> We have seven landfills. We have a sewer treatment plant. We have the Ford Motor Company. We have a paint factory. We have numerous chemical companies and steel mills. The river is just a few blocks away from us and is carrying water so highly contaminated that they say it would take seventy-five years or more before they can clean it up.

This activist sees her involvement in toxic waste issues as a challenge to traditional stereotypes of African American women. She says, "I'm here to tell the story that all people in the projects are not lazy and dumb!"

Some of these women share experiences of personal empowerment through their involvement in toxic waste issues. Says one African American activist:

> Twenty years ago I couldn't do this because I was so shy. . . . I had to really know you to talk with you. Now I talk. Sometimes I think I talk too much. I waited until my fifties to go to jail. But it was well worth it. I never went to no university or college, but I'm going in there and making speeches.

However, this is not a major theme in the narratives of female African American activists, as it is in those of white blue-collar women. African American women's private work as mothers has traditionally extended to a more public role in the local community as protectors of the race. As a decade of African American feminist history has shown, African American women have historically played a central role in community activism and in dealing with issues of race and economic injustice. They receive tremendous status and recognition from their community. Many women participating in toxic waste protests have come out of a history of civil rights activism, and their environmental protests, especially in the South, develop through community organizations born during the civil rights movement. And while the visible leaders are often male, the base of the organizing has been led by African American women, who, as Cheryl Townsend Gilkes has written, have often been called "race women," responsible for the "racial uplift" of their communities.[4]

African American women perceive that traditional environmental groups only peripherally relate to their concerns. As Cora Tucker relates:

> This white woman from an environmental group asked me to come down to save a park. She said that they had been trying to get black folks involved and that they won't come. I said, "Honey, it's not that they aren't concerned, but when their babies are dying in their arms they don't give a damn about a park." I said, "They want to save their babies. If you can help them save their babies, then in turn they can help you save your park." And she said, "But this is a real immediate problem." And I said, "Well, these people's kids dying is immediate."

Tucker says that white environmental groups often call her or the head of the NAACP at the last minute to participate in an environmental rally because they want to "include" African Americans. But they exclude African Americans from the process of defining the issues in the first place. What African American communities are doing is changing the agenda.

Because the concrete experience of African Americans' lives is the experience and analysis of racism, social issues are interpreted and struggled with within this context. Cora Tucker's story of attending a town board meeting shows that the issue she deals with is not merely the environment but also the disempowerment she experiences as an African American woman. At the meeting, white women were addressed as Mrs. So-and-So by the all-white, male board. When Mrs. Tucker stood up, however, she was addressed as "Cora":

> One morning I got up and I got pissed off and I said, "What did you call me?" He said, "Cora," and I said, "The name is Mrs. Tucker." And I had the floor until he said "Mrs. Tucker." He waited five minutes before he said "Mrs. Tucker." And I held the floor. I said, "I'm Mrs. Tucker," I said, "Mr. Chairman, I don't call you by your first name and I don't want you to call me by mine. My name is Mrs. Tucker. And when you want me, you call me Mrs. Tucker." It's not that—I mean it's not like you gotta call me Mrs. Tucker, but it was the respect.

In discussing this small act of resistance as an African American woman, Cora Tucker is showing how environmental issues may be about corporate and state power, but they are also about race. For female African American activists, environmental issues are seen as reflecting environmental racism and linked to other social justice issues, such as jobs, housing, and crime. They are viewed as part of a broader picture of social inequity based on race. Hence, the solution articulated in a vision of environmental justice is a civil rights vision—rooted in the everyday experience of racism. Environmental justice comes to mean the need to resolve the broad social inequities of race.

The narratives of Native American women are also filled with the theme of environmental racism. However, their analysis is laced with different images. It is a genocidal analysis rooted in the Native American cultural identification, the experience of colonialism, and the imminent endangerment of their culture. A Native American woman from North Dakota, who opposed a landfill, says:

> Ever since the white man came here, they keep pushing us back, taking our lands, pushing us onto reservations. We are down to 3 percent now, and I see this as just another way for them to take our lands, to completely annihilate our races. We see that as racism.

Like that of the African American women, these women's involvement in toxic waste protests is grounded from the start in race and shares none of the white blue-collar women's initial belief in the state. A Native American woman from southern California who opposed a landfill on the Rosebud Reservation in South Dakota tells us:

> Government did pretty much what we expected them to do. They supported the dump. People here fear the government. They control so many aspects of our life. When I became involved in opposing the garbage landfill, my people told me to be careful. They said they anni-hilate people like me.

Another woman involved in the protest in South Dakota describes a government official's derision of the tribe's resistance to the siting of a landfill:

> If we wanted to live the life of Mother Earth, we should get a tepee and live on the Great Plains and hunt buffalo.

Native American women come from a culture in which women have had more empowered and public roles than is the case in white working-class culture.

Within the Native American community, women are revered as nurturers. From childhood, boys and girls learn that men depend on women for their survival. Women also play a central role in the decision-making process within the tribe. Tribal council membership is often equally divided between men and women; many women are tribal leaders and medicine women. Native American religions embody a respect for women as well as an ecological ethic based on values such as reciprocity and sustainable development: Native Americans pray to Mother Earth, as opposed to the dominant culture's belief in a white, male, Anglicized representation of divinity.

In describing the ways in which their culture integrates notions of environmentalism and womanhood, one woman from New Mexico says:

> We deal with the whole of life and community; we're not separated, we're born into it—you are it. Our connection as women is to the Mother Earth, from the time of our consciousness. We're not environmentalists. We're born into the struggle of protecting and preserving our communities. We don't separate ourselves. Our lifeblood automatically makes us responsible; we are born with it. Our teaching comes from a spiritual base. This is foreign to our culture. There isn't even a word for dioxin in Navajo.

In recent years, Native American lands have become common sites for commercial garbage dumping. Garbage and waste companies have exploited the poverty and lack of jobs in Native American communities and the fact that Native American lands, as sovereign nation territories, are often exempt from local environmental regulations. In discussing their opposition to dumping, Native American women ground their narratives in values about land that are inherent in the Native American community. They see these projects as violating tribal sovereignty and the deep meaning of land, the last resource they have. The issue, says a Native American woman from California, is

> protection of the land for future generations, not really as a mother, but for the health of the people, for survival. Our tribe bases its sovereignty on our land base, and if we lose our land base, then we will be a lost people. We can't afford to take this trash and jeopardize our tribe.
>
> If you don't take care of the land, then the land isn't going to take care of you. Because everything we have around us involves Mother Earth. If we don't take care of the land, what's going to happen to us?

In the process of protest, these women tell us, they are forced to articulate more clearly their cultural values, which become resources of resistance in helping the tribe organize against a landfill. While many tribal members may not articulate an "environmental" critique, they well understand the meaning of land and their religion of Mother Earth, on which their society is built.

CONCLUSION

The narratives of white, African American, and Native American women involved in toxic waste protests reveal the ways in which their subjective, particular experiences lead them to analyses of toxic waste issues that extend beyond the particularistic issue to wider worlds of power. Traditional beliefs about home, family, and community provide the impetus for women's involvement in these issues and become a rich source of empowerment as women reshape traditional language and meanings into an ideology of resistance. These stories challenge traditional views of toxic waste protests as parochial, self-interested, and failing to go beyond a single-issue focus. They show that single-issue protests are ulti-mately about far more and reveal the experiences of daily life and resources that different groups use to resist. Through environmental protests, these women challenge, in some measure, the social relations of race, class, and gender.

These women's protests have different beginning places, and their analyses of environmental justice are mediated by issues of class and race. For white blue-collar women, the critique of the corporate state and the realization of a more genuine democracy are central to a vision of environmental justice. The defini-tion of environmental justice that they develop becomes rooted in the issue of class. For women of color, it is the link between race and environment, rather than between class and environment, that characterizes definitions of environ-mental justice. African American women's narratives strongly link environment justice to other social justice concerns, such as jobs, housing, and crime. Environmental justice comes to mean the need to resolve the broad social inequities of race. For Native American women, environmental justice is bound up with the sovereignty of the indigenous peoples.

In these women's stories, their responses to particular toxic waste issues are inextricably tied to the injustice they feel as mothers, as working-class women, as African Americans, and as Native Americans. They do not talk about their pro-tests in terms of single issues. Thus, their political activism has implications far beyond the visible, particularistic concern of a toxic waste dump site or the siting of a hazardous waste incinerator.

NOTES

1. Sandra Morgen, " 'It's the Whole Power of the City Against Us!': The Development of Political Consciousness in a Women's Health Care Coalition," in *Women and the Politics of Empowerment,* eds. Ann Bookman and Sandra Morgen (Philadelphia: Temple University Press, 1988), p. 97.

2. Sheila Rowbotham, *Women's Consciousness, Man's World* (New York: Penguin, 1973).

3. See Sara Ruddick, *Maternal Thinking: Towards a Politics of Peace* (New York: Ballantine, 1989).

4. Cheryl Townsend Gilkes, "Building in Many Places: Multiple Commitments and Ideologies in Black Women's Community Work," in *Women and the Politics of Empowerment,* op. cit.

65

Becoming Entrepreneurs

Intersections of Race, Class, and Gender at the Black Beauty Salon

A great deal of research has applied the concept of intersectionality to understand the ways race, class, and gender affect individuals. Feminist researchers
have been particularly effective in promoting scholarship that examines race,
gender, and class as interlocking categories of oppression, wherein the experiences of some minority groups (e.g., Black women) must be understood as a
consequence of racial, gendered, and sometimes class-based inequality (Chafetz
1997; Collins 1990; King 1988). For these women, understanding their experiences involves more than just knowing how race and gender lead to inequality
but mandates an understanding of how race is gendered and gender is racialized
(Browne and Misra 2003). It is this totality that researchers seek to explore.

Accepting the premise that the intersections of race, gender, and class affect
virtually all aspects of life, feminist researchers examine these intersections in
social arenas such as the workplace, community organizations, media, and others
(Collins 1990; Jewell 1993; Rollins 1985). Studies that address interacting
oppressions in the workplace often focus on the ways that race, gender, and
class shape minority women's access to certain jobs, compensation for work,
and the perception of these women's suitability for various occupations
(Browne 1999; Higginbotham 1997). These intersecting oppressions often relegate minority women into the bottom of the labor queue, where economic stability is precarious and they are over-represented in low-skill, low-status work
(Browne and Kennelly 1999; Reskin and Roos 1990).

This study builds on existing studies of intersectionality and work by exploring the ways the intersections of race, gender, and class shape working-class
Black women's experiences with entrepreneurship. While research has demonstrated that intersecting oppressions are likely to channel working-class Black
women into occupations like janitorial, food, or health service work, little is
known about how these interacting inequalities affect Black women's entrepreneurial activity. This study shifts the center to focus on a group whose

SOURCE: *Gender & Society*, Vol. 19 No. 6, December 2005: 789–808. © 2005
Sociologists for Women in Society. Reprinted by permission of Sage Publications.

entrepreneurial activity may differ structurally and experientially from the entrepreneurs who are normally the subject of this type of research. . . .

THEORETICAL FRAMEWORK

. . . The legal and social gains of the 1960s have had an enormous impact on Black women's occupational opportunities; nonetheless, structural, entrenched problems still remain for African American women in the workplace. Institutional and individual discrimination, the glass ceiling, and hostile workplace cultures are still realities that constrain occupational opportunities for many Black women. For working Black women, the relocation of jobs to suburbs, rise of the technology industry, and lack of educational opportunities also leave limited work options, and those that exist tend to be low paying with low prestige and little authority, job security, or benefits.

Faced with constrained opportunities in the labor market, business ownership can become an appealing pathway to economic support for some Black women (Smith 1992). Indeed, the increase in Black women's entrepreneurial activity speaks to its growing popularity as an alternative to the paid labor force. Between the years 1997 and 2004, the numbers of Black women entrepreneurs grew 33 percent, with recent estimates suggesting that there are now approximately 414,472 firms owned by African American women (Center for Women's Business Research 2004). This study focuses on an often understudied segment of business owners—Black working-class women—to explore how their entrepreneurial activity is influenced by race, gender, and class.

Black Women's Work in the Hair Industry

Structural disadvantages in the labor market have made entry into many fields virtually impossible for Black women, but some fields have been easier to access than others. For instance, Black women have a long history of work within the hair industry dating back to the early twentieth century. Black women in the late 1800s and early 1900s sold hair products in Black communities. Probably the most notable and successful of these early entrepreneurs was Madame C. J. Walker, the inventor of the straightening comb and first African American millionaire. Unlike other mostly white entrepreneurs at the time who promoted Black hair care products as ways to eliminate "kinky, snarly, ugly, curly hair," Madame Walker established her fortune by marketing her products as ones that would promote healthy hair growth and also employed other African American women as sales associates (Rooks 1996, 35).

Madame Walker's success in the hair industry reflects structural barriers that heightened the appeal of entrepreneurship in the hair industry for many African American women. With racial segregation a common practice, and women's occupational opportunities severely constrained, the intersection of race and gender meant that Black women were effectively excluded from most jobs. In addition, in keeping with the custom of racial segregation, whites in the beauty industry often refused to service Black customers. As such, the beauty industry was distinctive as a niche that Black women could enter relatively easily, pursue entrepreneurship, and maintain access to a relatively untapped consumer base. Their entrepreneurial work in this area reflects the assertion that minority and women's entrepreneurship often has elements of middleman minority characteristics. In this particular case, women in the beauty industry did not follow the typical middleman minority route of selling goods and services such as hair products to the masses. They did, however, function as middleman minorities in that these entrepreneurs provided a service to groups with whom elites were loath to interact.

Boyd (2000) argues that the combination of racial oppression and labor market disadvantage during the great depression in fact led Black women to pursue entrepreneurship. As the economy declined and employment became increasingly scarce, many Black women were dismissed from positions they had held and increasingly relied on domestic service for necessary wages. However, some women gravitated to the beauty industry to supplement their earnings because it was one of few available options for additional income and because a demand for these services came from other Black women, who perceived that maintaining a well-groomed appearance was a necessity for securing employment. Labor market disadvantage, economic depression, and racial discrimination made entrepreneurship in the hair industry highly appealing for African American women during this time.

Today, African American women have more occupational opportunities than ever before, and many are moving into the private sector as well as into economically rewarding fields of entrepreneurship that were unavailable in the past, such as entertainment and consulting. Nonetheless, Black women's contemporary entrepreneurship in the hair industry remains contextualized by pervasive structural issues that they face in the occupational sphere, including discrimination, wage inequality, glass ceilings, and occupational sex segregation.

However, these studies address only how intersectionality shapes Black women's entrepreneurship at structural levels. While it is important to address how institutionalized racism, sexism, and classism channel Black women into certain occupational niches, equally important are the ways these interlocking factors shape the more micro-level, interactional aspects of entrepreneurship. This study explores, at the level of social interaction, how intersectionality affects Black women's work practices, patterns of business ownership, and experiences as entrepreneurs.

DATA COLLECTION

To assess how race, gender, and class shaped the entrepreneurial activity of working-class Black women, I conducted interviews with 11 Black female hair salon owners in a major city in the mid-Atlantic United States. The women I interviewed owned their businesses for time periods ranging from 2 to 15 years. All interviews took place in the women's shops and generally lasted one hour to an hour and a half. The majority of interviews were tape-recorded, with the exception of three women who were uncomfortable being recorded. In these cases I took copious notes.

The women ranged in age from 25 to 50, and all had worked as stylists before eventually becoming salon owners. I contacted respondents through a snowball method. Four of the respondents had worked as stylists for other owners included in this study before eventually becoming owners themselves. Two women were sole operators, meaning they hired no employees. Of the other nine women, the number of employees ranged from 2 to 13 stylists. Nine owners had children; of these nine, their children ranged in age from toddlers to adulthood. The locations of the salons covered a wide range, including working-class neighborhoods, middle-class areas on the outskirts of the city, and urban, lower-class neighborhoods. Similarly, clienteles ranged in income but did not vary widely with regard to race and gender. Some owners had a handful of white or Asian customers, but at each shop, the majority of the clientele were Black women. The location of the shop tended to mirror the social class of the clientele, for example, shops in middle-class neighborhoods usually had mostly middle-class women as a customer base.

All of the owners interviewed here hailed from working-class backgrounds. . . . As members of the working class, these women had been previously employed as sales associates, fast food workers, or telemarketers, positions that generated annual incomes of about $25,000 or less annually. They did not own their homes, although six had owned cars, and none had amassed any other forms of wealth or educational attainment past high school (although all owners completed cosmetology school). Furthermore, none of these women were raised in families where the combined household incomes reached the middle-class or upper-middle-class levels Sernau describes of $40,000 to $75,000 per year. . . .

FINDINGS

Rather than describing the ways certain aspects of entrepreneurship were shaped by race or gender or class, I have chosen instead to pinpoint key aspects of the women's entrepreneurial activity to highlight how these various facets were shaped by some combination of these social constructs. In this way, a much clearer picture is visible of how one or more of these issues determine the contours of working-class Black women's entrepreneurship. The intersections of race, class, and gender (or some combination of these) are most clearly revealed in two areas of working-class Black women's entrepreneurship: the process of becoming entrepreneurs and relationships with stylists.

Becoming Entrepreneurs

In some ways, the women's decision to become entrepreneurs reflects the findings of existing studies of gender and entrepreneurial activity. Specifically, for many women in this study, entrepreneurship is a gendered choice in that it allows them to balance the demands of work and family. The desire to balance work and family is an important factor that often distinguishes women's motivations for entrepreneurial activity from men. In this regard, then, Black women's initial choice to become entrepreneurs is shaped by gendered factors that appeal to many other women regardless of race or class. Many women stated that as salon owners, it was much easier for them to meet the challenges of working full-time and raising children. They were able to set their own hours and could structure their time to devote adequate attention to children and work.

Chandra, for example, is 25 with a one-year-old daughter. Chandra's salon is located in what she acknowledges is a "bad location" because of the rampant drug trade and related criminal activity nearby. She has owned her salon for nearly two years and employs two other stylists, one of whom is her best friend, Sarah. In response to a question about why she decided to open her own salon, Chandra stated, "I can honestly tell you, the most rewarding part [of owning a salon] is knowing that [because] I do have a small baby, I can come and go as I please, and I don't have to answer to anybody." In contrast to Chandra, Danette, 35, owns a salon located on the outskirts of the city in a quiet, middle-class, tree-lined neighborhood. Like many of the women interviewed for this study, Danette's salon is spacious, with a bulletin board announcing upcoming community events and African American art lining the walls. Danette also indicates that part of the lure of entrepreneurship was the opportunity it offered her to balance work and family. She too spoke of bringing her son to the salon when he was a toddler and allowing him to entertain himself in the shop while she worked: "Everybody loved [my son], he was cute. . . . Everybody loved him so much that he used to be in the hair, have hair everywhere, and they would tell me, 'Oh, just leave him alone. . . . And then when he got older he would come in Saturday night and clean up for me, things like that." The decision to become a business owner because it offers the advantage of balancing work and family thus reflects one particular way in which entrepreneurial decisions are influenced by gender. However, aspects of intersectionality are present here too vis-à-vis the unspoken fact that as working-class women, professional child care may be too costly for them to consider it a financially viable option. . . .

The decision to become entrepreneurs in the beauty industry represents a choice that clearly reflects gendered norms, given that the hair salon capitalizes on and profits from societal messages about gender and appearance. Perhaps more important, the appreciation for "making women beautiful" was frequently cited as one of the motivations that led them to this line of work. A number of owners stated that they loved their work: they loved working with hair and enjoyed the ability to "make women look good." Maxine is 34 and owns a small, sparsely furnished salon on the ground level of a four-story walk-up building. Her shop is located in a middle-class neighborhood adjacent to a major university. Prior to opening her salon, she was a stylist at a shop located a few doors down from the one she now owns. She was compelled to enter the hair industry because "I've always loved working with hair, and I had been doing it since I was a teenager. The best part of this work is fixing people up, making them look nice; they smile, say thank you. It makes me feel good." The decision to become entrepreneurs in the hair industry itself indicates how gender shapes entrepreneurial activity. It is clear that these women's motivations for entrepreneurship in this field reflect the gendered notion that it is appropriate for women to devote attention, time, and money to meeting dominant standards of beauty.

However, Black women's entrepreneurial work is in some ways shaped by gendered factors that might be applicable to other women entrepreneurs; in other ways, their initial pathway to entrepreneurship is specifically delineated by the intersection of race and gender. Gender ideology demands that women should be attractive, but overlapping racial messages often insist that Black women do not meet beauty ideals. Interestingly, the social significance attached to women's hair exposes this contradiction. Hair has long been considered a symbol of feminine beauty and attractiveness, but long, straight, blonde hair has been epitomized as the female ideal. Instead, Black women's hair in its natural state is more likely to be short, tightly curled, and dark—the opposite of the ideal.

For Black women, the social significance of hair reflects a "gendered racism" that has very particular and specific implications. The racialized standard of beauty can produce a conundrum for African American women in a society where hair is unequivocally connected to beauty and adherence to beauty ideals is linked to "access to rewards such as employment, income, and self-esteem" (St. Jean and Feagin 1998, 74). While many women experience contradictions and tensions that stem from the symbolic significance of hair, race and gender intersect to shape these issues in a manner that reinforces mainstream culture's messages of African American women's inferiority (hooks 1992; Weitz 2001; Wilson and Russell 1996).

Gendered racism labels Black women's hair unattractive and unappealing; consequently, a large consumer base exists for Black women who work in the hair industry. As such, many of the women interviewed became salon owners with full confidence that there was a market for their services. Their awareness of this market and their ability to tap into it stemmed both from prior experience as stylists and from their first-hand familiarity with the ways in which gendered racism shapes Black women's feelings about their hair. Greta, a tall, confident woman in her thirties, runs a small but cozy salon in the same neighborhood as Maxine's. She worked alone for a year before hiring two stylists to join her in her shop. Greta states that she never worries about losing business because "[Black women] will go get our hair done. Even when I had to work alone, I had customers, because we will get our hair done. We'll take the rent money to go get our hair done! We don't believe in looking crazy." Greta's remarks indicate her familiarity with her customers' attitudes about hair and the importance, for some Black women, of making sure hair is professionally styled even if it comes at the expense of other necessities. Interestingly, she also likens not having professionally done hair to "looking crazy," further emphasizing the perception that Black women's hair needs professional styling to be considered appealing. . . .

Coming from working-class backgrounds also shapes these women's entrepreneurial activity in a way that differs from middle- and upper-class women's decisions to enter into self-employment. Budig (2001) finds that many upper-class women engage in entrepreneurship as a response to the glass ceiling they experience in the workplace. Entrepreneurship thus becomes a way for them to avoid or minimize the gender discrimination they face as paid workers. Similarly, Davies-Netzley (2000) cites this as one of the primary factors that leads white and Latina women into entrepreneurship. While these women are able to attain middle- or upper-class status (e.g., cars, homes, benefits), they are often motivated to become entrepreneurs as they recognize that institutionalized and individual workplace discrimination will eventually limit (or has already limited) their continued upward mobility.

The working-class Black women interviewed for this study, however, face different circumstances. Prior to becoming salon owners, these women worked primarily in the low-skill, low-wage service jobs in which Black women are disproportionately represented: fast food, factory work, telemarketing, filing clerks, and sales associates. Entrepreneurship, then, did not provide a way to continue upward mobility as much as it was a step toward upward mobility. Unlike middle- and upper-class women who became entrepreneurs to avoid a glass ceiling, the women interviewed here became entrepreneurs as a route to securing basic financial stability that middle- and upper-class women already enjoy. While class plays a role in both groups of women's entrepreneurial work, it shapes working-class women's advent into entrepreneurship differently than middle- and upper-class women's.

Furthermore, working-class salon owners' economic needs intersected with gender-based ones in that many sought economic stability to meet familial responsibilities. Chandra has owned her salon for nearly two years and explicitly stated that financial considerations were why she made the move from working as a stylist at a chain salon to opening her own establishment, but she concedes that familial concerns were important as well: "I have a baby, and at Supercuts, you have to split like 50 percent with them, and they take out for the shampoo, they take out for the conditioner, so I said, I'm not going to work for them, because I was barely making $500 every two weeks. Being here, I can make $500 in a day, or in two days, and it's mine." Chandra's statement underscores how becoming an entrepreneur was shaped by her class position as well as by racialized, gendered factors. As an African American woman, Chandra's hardships in the labor market likely reflect the complex interplay of gender and race. As a working-class mother, economic stability is absolutely essential, to the point where the deduction of shampoo and other supplies from her wages meant that she was not earning enough to support her family. Salon ownership, therefore, offers Chandra a way to sidestep some forms of racial and gendered discrimination and to achieve the economic stability necessary to provide for her family. . . .

Relationships with Stylists: The Helping Ideology

The relationship between Black women salon owners and the stylists they hire is the other key area of entrepreneurship that is shaped by the interplay of race, gender, and class. Salon owners interviewed for this study typically hired 1 to 3 stylists to work in their shops. Miranda and Lola, both in their mid-thirties, are the only two owners interviewed who work alone. In contrast, Lana employed 14 stylists, more than any other owner interviewed for this study. All the owners operated salons in which they charged stylists booth rent. Under this system, stylists pay a monthly rent to owners for the use of space at the salon. Stylists are then permitted to keep all money from business transactions but are not guaranteed any benefits such as medical insurance, dental insurance, retirement plans, or sick leave. Although stylists are technically independent contractors, they must follow the rules and guidelines established by the salon owner, and either party can terminate the working relationship.

The relationship between owners and stylists is unique in a number of respects. Possibly most striking is the ideology of help and support that shapes, to varying degrees, owners' relationships with stylists. Owners generally profess a willingness to assist stylists in their professional development and growth as well as a readiness to help them succeed in the business. In its most extreme cases, this helping ideology motivates owners to encourage stylists to pursue entrepreneurship and to open their own salons. Lana, an owner in her mid-forties who has owned her salon for nearly 15 years, is perhaps the most interesting embodiment of this philosophy in practice. Her salon is large and upscale, with two floors, and is usually bustling with her clientele of mostly middle-class African American women. Lana is opinionated and outspoken and tells her stylists frequently that she expects them to eventually go into business for themselves. When asked the most rewarding part of owning a salon, Lana's reply was similar to that of many other owners: "Having people leave here and become salon owners, and being successful at it. This is really a field where Black women can do well, . . . and it's important that people understand that and respect this industry." Lana also requires that her stylists attend mandatory workshops she sponsors on money management, customer service, and business ethics so that they will have exposure to the knowledge she deems necessary for successful entrepreneurship.

Interestingly, Tanisha and Greta, two other owners interviewed for this study, worked as stylists in Lana's salon before opening their own shops. Lana's relationship with Tanisha in particular exemplifies how the nature of the relationship between owners and stylists is sometimes shaped by a willingness to help other Black women prosper in the industry. Tanisha and Lana first met at church, and after learning that Tanisha styled hair for several of the women in the congregation, Lana decided to take Tanisha under her wing. Tanisha began working full-time for Lana immediately after high school and remained at this salon for three years before moving on to open her own salon. . . .

This helping ideology—owners' desire to help stylists successfully shift to entrepreneurship—stems from a sense of gender/racial solidarity. Yet this gender/racial solidarity also has a dimension of class solidarity. Other research describes gender/racial solidarity among middle- and upper-class Black women as the manifestation of a desire to help their lower-class counterparts (Giddings 1984; McDonald 1997; White 1999). This willingness to help is based on shared experience of racial and gender discrimination but is also characterized by the willingness to reach across class lines to help those less economically privileged. However, in the case of Black female salon owners, their sense of gender/racial solidarity extended to other working-class women in the same class position. Owners did not seek to uplift less fortunate Black women; instead, their sense of gender/racial solidarity compelled them to help other Black women of the same class position. As such, the helping ideology that often underlies the relationship between owner and stylist is very clearly influenced by a sense of race, gender, and class solidarity that facilitates a level of support that Black women in other workplaces rarely experience.

Contradictions, Tensions, and Problems

Predictably, the intersections of race, gender, and class can cause working-class Black women to face certain specific obstacles with entrepreneurship. One such obstacle is the issue of access to start-up capital. Many women interviewed stated unhesitatingly that getting access to start-up funding was the most difficult part of becoming entrepreneurs. . . . Denise, for instance, is 37 and owns a salon that was passed down to her by her mother on retirement. The salon is located outside the city limits in a suburban area that draws a mostly middle-class clientele. . . . Denise owns two other salons in different parts of the city as well, both of which service mostly working-class women. Like most of the other women in this study, Denise relied on unconventional strategies to procure necessary start-up capital. She stated, "The government will give money to[,] seems like everybody else. But when you go to get money for small businesses, *we* [she makes a gesture to indicate the two of us, as Black women] have to go through a lot. So the hard part was trying to get help from the government. That was the hard part. The easy part was having faith in the Lord's will." Like most of the other women interviewed, Denise secured her start-up capital for buying the other salons through meticulous savings as well as loans from family and friends. She interprets her difficulty securing funding as a common pitfall that Black women face at the onset of entrepreneurship. Significantly, she attributes this obstacle to race and gender.

These women's status as working-class Black women also shaped their access to social networks in ways that likely undermined their potential for economic returns as entrepreneurs. . . . In the process of becoming owners, many women relied on those in their social networks for information about available property that they could purchase or rent out to open their salons. Consequently, many women opened salons near the ones where they had previously worked as stylists. This was beneficial in that it helped them remain accessible to former clients, but it also reflected minimal knowledge of real estate in other parts of the city that might have offered other benefits such as better parking, an untapped customer base, or lower property taxes. Maxine states, "I moved right up the street when a place became available. In hindsight, that was not such a good idea. The parking around here is bad. I should not have moved so fast, and if I had to do it again, I wouldn't have rushed. I actually hope to move somewhere else pretty soon." Race and gender play a role here in constraining some owners' business decisions. Social networks that comprise primarily other African

American women can limit the degree and breadth of information and knowledge available to these owners. . . .

Finally, the success these women experience as business owners must be examined in the context of entrepreneurial ventures at large. Although race, gender, and class shape their entry into salon ownership and their relationships with stylists, it is important to note that these women still experience ghettoization within the entrepreneurial sector. Like the majority of Black-owned businesses, these women's entrepreneurial ventures are located in the personal service sector, which tends to generate fewer economic returns than emerging (and male-dominated) fields of minority entrepreneurship such as construction and manufacturing. Furthermore, women's businesses in general tend to be "smaller than men's" and involved in a "narrow range of activity"—characteristics that are applicable to these women's hair salons (Grasmuck and Espinal 2000, 233). For many of these women, owning a hair salon is an economically sound business decision but one that does not exempt them from the pitfalls and challenges that are common to Black female entrepreneurs. Gender, race, and class shape these women's entrepreneurial choices to channel them into a field that, relative to other avenues of entrepreneurship, offers fewer returns and less prestige.

CONCLUSION

The results of this study demonstrate that intersectionality is a key factor in working-class Black women's entrepreneurial activity in two important regards: the transition to entrepreneurship and the nature of relationships with stylists. Through the use of intersectionality as a theoretical lens, it becomes increasingly clear that race, gender, and class have a definitive impact on the market these entrepreneurs try to reach as well as these women's actions, decisions, and experiences as entrepreneurs. The use of an intersectional framework therefore provides greater insight into the social processes of entrepreneurship.

REFERENCES

Boyd, Robert, 2000. Race, labor market disadvantage, and survivalist entrepreneurship: Black women in the urban North during the great depression. *Sociological Forum* 15 (4): 647–70.

Browne, Irene, 1999. Latinas and African American women in the labor market. In *Latinas and African American women at work*, edited by Irene Browne. New York: Russell Sage.

Browne, Irene, and Ivy Kennelly, 1999. Stereotypes and realities: Images of African American women in the labor market. In *Latinas and African American women at work*, edited by Irene Browne. New York: Russell Sage.

Browne, Irene, and Joya Misra, 2003. The intersection of gender and race in the labor market. *Annual Review of Sociology* 29: 487–513.

Budig. Michelle J, 2001. Gender, family status, and job characteristics: Determinants of self-employment participation for a young cohort, 1979–1998. Paper presented at annual meetings of the American Sociological Association, Anaheim, CA, August.

Center for Women's Business Research. 2004. *African American women-owned businesses in the United States, 2004: A fact sheet*, Washington, DC: Center for Women's Business Research.

Chafetz, Janet Salzman, 1997: Feminist theory and sociology: Underutilized contributions for mainstream theory. *Annual Review of Sociology* 23: 97–120.

Collins, Patricia Hill, 1990. *Black feminist theory: Knowledge, consciousness, and the politics of empowerment.* New York: Routledge.

Davies-Netzley, Sally, 2000. *Gendered capital: Entrepreneurial women in American society.* New York: Garland Press.

Giddings, Paula, 1984. *When and where I enter: The impact of Black women on race and sex in America.* New York: Bantam.

Grasmuck, Sherri, and Rosario Espinal, 2000. Market success or female autonomy? Income, ideology, and empowerment among microentrepreneurs in the Dominican Republic. *Gender & Society* 14 (2): 231–55.

Higginbotham, Elizabeth, 1997. Black professional women: Job ceilings and employment sectors. In *Workplace/women's place*, edited by Dana Dunn. Los Angeles: Roxbury.

hooks, bell, 1992. *Black looks: Race and representation.* Boston: South End.

Jewell, K. Sue, 1993. *From mammy to Miss America and beyond.* New York: Routledge.

King, Deborah, 1988. Multiple jeopardy, multiple consciousness. *Signs: Journal of Women in Culture and Society* 14 (1): 42–72.

McDonald, Katrina Bell, 1997. Black activist mothering. *Gender & Society* 11 (6): 773–95.

Reskin, Barbara, and Patricia Roos, 1990. *Job queues, gender queues.* Philadelphia: Temple University Press.

Rollins, Judith, 1985. *Between women: Domestics and their employers.* Philadelphia: Temple University Press.

Rooks, Noliwe, 1996. *Hair raising: Beauty, culture, and African-American women.* New Brunswick, NJ: Rutgers University Press.

Sernau, Scott, 2001, *World's apart: Social inequalities in a new century.* Thousand Oaks, CA: Pine Forge.

Smith, Wade A, 1992. Race, gender, and entrepreneurial orientation. *National Journal of Sociology* 6 (2): 141–55.

St. Jean, Yanick, and Joe Feagin, 1998. *Double burden.* Armonk, NY: M. E. Sharpe.

Weitz, Rose, 2001 Women and their hair: Seeking power through resistance and accommodation. *Gender & Society* 15 (5): 667–86.

White, Deborah Gray, 1999. *Too heavy a load: Black women in defense of themselves, 1894–1994.* New York: W. W. Norton.

Wilson, Midge, and Kathy Russell, 1996. *Divided sisters: Bridging the gap between Black and white women.* New York: Doubleday.

Index

Dear Student,

I hope you enjoyed reading *Race, Class, and Gender: An Anthology*, Seventh Edition. With every book that I publish, my goal is to enhance your learning experience. If you have any suggestions that you feel would improve this book, I would be delighted to hear from you. All comments will be shared with the authors. My email address is Chris.Caldeira@cengage.com, or you can mail this form (no postage required). Thank you.

School and address: _____

Department: _____

Instructor's name: _____

1. What I like most about this book is: _____

_____

_____

2. What I like least about this book is: _____

_____

_____

3. I would like to say to the author of this book: _____

_____

_____

4. In the space below, or in an email to Chris.Caldeira@cengage.com, please write specific suggestions for improving this book and anything else you'd care to share about your experience using this book.

_____

_____

_____

DO NOT STAPLE. PLEASE SEAL WITH TAPE.

FOLD HERE

WADSWORTH
CENGAGE Learning™

NO POSTAGE
NECESSARY
IF MAILED
IN THE
UNITED STATES

BUSINESS REPLY MAIL
FIRST-CLASS MAIL   PERMIT NO. 34   BELMONT CA

POSTAGE WILL BE PAID BY ADDRESSEE

Attn: Chris Caldeira, Sociology

Wadsworth
10 Davis Drive
Belmont CA   94002-9801

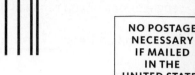

FOLD HERE

OPTIONAL:

Your name: _____ Date: _____

May we quote you, either in promotion for *Race, Class, and Gender: An Anthology,*
Seventh Edition, or in future publishing ventures?

Yes: _____ No: _____

Sincerely yours,

Margaret L. Andersen and Patricia Hill Collins